Optoelectronic Devices

Optoelectronic Devices

Niloy K Dutta
Xiang Zhang

Department of Physics
University of Connecticut, USA

World Scientific

NEW JERSEY • LONDON • SINGAPORE • BEIJING • SHANGHAI • HONG KONG • TAIPEI • CHENNAI • TOKYO

Published by

World Scientific Publishing Co. Pte. Ltd.
5 Toh Tuck Link, Singapore 596224
USA office: 27 Warren Street, Suite 401-402, Hackensack, NJ 07601
UK office: 57 Shelton Street, Covent Garden, London WC2H 9HE

British Library Cataloguing-in-Publication Data
A catalogue record for this book is available from the British Library.

OPTOELECTRONIC DEVICES

ISBN 978-981-3236-69-1

For any available supplementary material, please visit
http://www.worldscientific.com/worldscibooks/10.1142/10894#t=suppl

Desk Editor: Christopher Teo

Typeset by Stallion Press
Email: enquiries@stallionpress.com

Preface

In the past several decades, optoelectronics has totally changed the communications industry and lighting industry and is having a considerable impact on power generation. The optical networking technology represents a revolution inside the optical communications revolution, and it allows the latter to continue its exponential growth. The lighting technology using light-emitting diodes (LEDs) is more efficient, has much longer life than previous schemes, and is replacing incandescent bulbs. Solar power is becoming a significant fraction of total power generation. Optoelectronics is a revolution that has fundamentally transformed the core of the above-mentioned technologies and has impacted the basic science, the enabling technology, and the industry.

The existence and advance of optical fiber communications is based on the invention of low-loss optical fibers, the invention of the laser, particularly the semiconductor junction laser, and other related disciplines such as photodetectors and integrated optics. From the early pioneering ideas, it took more than 25 years to the first commercial deployment of optical communications, the Northeast Corridor system linking Washington and New York in 1983. Another important advance was the deployment of the first Transatlantic submarine optical fiber transmission system in 1988. The optical fiber communication began to seriously impact the way information is transmitted beginning with these milestone achievements.

The market demand for higher capacity transmission was helped by the fact that computers and data centers continued to become more powerful and needed to be interconnected. This is one of the key reasons why the explosive growth of optical fiber transmission technology parallels that of computer processing and other key information technologies. These technologies have combined to meet the explosive global demand for new information services including data, internet, and broadband services — and, most likely, their rapid advance has helped fuel this demand. This demand is continuing its strong growth as internet traffic keeps increasing every year.

The laser emission in various rare earth (Nd, Pr, Er, Ho, Tm, and Yb) doped glasses have been studied since the 1960s. However, the potential for rare earth doped fiber amplifiers and fiber lasers were not recognized until the late 1980s, and it began with a demonstration of very-low-threshold fiber amplifier. This pioneering demonstration conveyed an important message: using well-established low-loss fiber fabrication technology and semiconductor laser fabrication technology, a new class of rare earth doped lasers and amplifiers could be produced for new applications. This message was particularly well received immediately by optical communication technologists, and they went on to develop Er-doped fiber amplifiers for various fiber communication applications.

It was also recognized that an amplifier chain could be used for long ($\sim$1000 to 7000 km) undersea optical communication links and that such a system could be easily upgraded to higher data rates simply by replacing the transmitters and repeaters on the shore — a much easier task than laying a new cable. The first undersea fiber optic link using an amplifier chain went into operation in the mid 1990s.

Optical networking and space communications (satellite to satellite) represent the next advances in optical communications technology. Optical fiber amplifier and semiconductor optical amplifiers are key devices for all-optical networks. The advances in research and many technological innovations have led to superior designs of fiber amplifiers. Today, most optical communication systems use optical fiber amplifiers for signal amplification. High-power short-wavelength amplifiers using Nd or Yb are important for space communication application due to low beam divergence.

Fiber-based Raman lasers and amplifiers represent an important area for lasers and amplifiers in wavelength regions where a rare earth dopant may not be efficient. Using the cascaded Raman process and multiple Stokes shifts, very efficient fiber Raman lasers and amplifiers have been demonstrated. High-power fiber lasers producing kilowatts of power using double-clad fiber geometries is expected to become increasingly important in future industrial applications. Advanced modulation formats are being implemented in data centers. The development of short-pulse fiber laser is expected to become important in many applications such as optical tomography and sensors. Optical fiber Bragg grating-based sensors are used in a wide range of applications including oil and gas exploration.

The term solid-state lighting generally refers to "white" light emission using LEDs. Such sources are quite commonplace today and have been made possible by the invention of LEDs which emit blue light. The light from a blue LED goes through a rare earth-doped glass (phosphor). The phosphors emits in the visible region by absorbing the blue light. The process of visible light generation is about five times more energy-efficient than incandescent light bulbs, has high reliability, and can operate at less than 5 V.

There is a growing need for energy in the world, and since the traditional energy sources based on fossil fuels are limited and will be exhausted in future, solar cells which convert energy from the sun to electric power (using the photovoltaic effect) are considered a promising energy source candidate. Large-scale application of solar cells will also contribute to the diversification of energy sources resulting in more equal distribution of energy sources in the world. Commercial solar cells, based on silicon, are widely used now for individual homes and power plants.

The book is aimed at researchers already engaged in or wishing to enter the field of optoelectronic devices and systems. It is also useful for graduate students, scientists, and engineers interested in optoelectronics. Several appendices have been added where many of the formulas from fundamental principles are derived. An Appendix with a set of problems for each chapter has been added. This allows the book to be used as a textbook. An attempt has been made to make the book to be self-contained, and each chapter has a set of references which can be consulted for further study.

N. K. Dutta
X. Zhang

Contents

Chapter 1

Introduction

1.1 Historical developments

The fiber optic revolution in telecommunication which provided several orders of magnitude improvement in transmission capacity at low cost would not have been possible without the development of reliable optoelectronic devices. The devices include semiconductor lasers, semiconductor photodetectors, optical fibers, optical amplifiers, and photonic integrated circuits. The development of blue light emitting diodes (LEDs) and the subsequent white light generation is revolutionizing the "lighting" industry because of low power consumption and high reliability. Also, optoelectronic devices are used in optical reading and recording (CD players), printers, FAX machines, and in numerous applications in sensors and as high power optical sources. Early optoelectronic devices are the semiconductor laser, the LED, and the photodiode. The laser has a gain medium and an optical feedback mechanism. The latter is generally provided by the cleaved facets of the semiconductor itself or by a grating of suitable periodicity etched close to the gain medium. At threshold, the gain equals loss for a laser, and the gain needed for semiconductor laser action is typically ~5–10 dB. The invention of the semiconductor laser was preceded by that of the LED which emits light by the application of current. The light emission from an LED has a broad spectral width (typically ~30 nm) and that from a semiconductor laser has a narrow spectral width (typically <1 nm).

The photodiode generates current through absorption of incident light and is used in conjunction with light source for many applications. The optical fiber can propagate light with very low loss (~0.1–0.2 dB/km). The optical fiber has a waveguide core of about 10 μm diameter through which the light propagates.

The laser, the photodiode, and the optical fiber constitute the main components of an optical fiber transmission system. The concept of the laser dates back to 1958 [1]. The successful demonstration of a solid-state ruby laser and He–Ne gas laser was reported in 1960 [2, 3]. Laser action in semiconductor was considered by many groups during that period [4–6]. The semiconductor injection laser was demonstrated in 1962 [7–10]. Continuous Wave (CW) operation of semiconductor laser was demonstrated in 1970 [11, 12] using the concept of double heterostructure [13]. Since then the semiconductor injection laser has emerged as an important device in many optoelectronic systems, such as optical recording, high-speed data transmission through fibers, optical sensors, high-speed printing, and guided wave signal processing. An important impact of semiconductor lasers is in the area of fiber optic transmission systems where the information is sent through encoded light beams propagating in glass fibers. These lightwave transmission systems, which have been installed throughout the world, offer a much higher transmission capacity at a lower cost than coaxial copper cable transmission systems.

The advantages of semiconductor lasers over other types of lasers, such as gas lasers, dye lasers, and solid-state lasers, lie in their considerably smaller size and lower cost, and their unique ability to be modulated at gigahertz speeds simply by modulation of the injection current. These properties make the laser diode an ideal device as a source in several optoelectronic systems, especially optical fiber transmission systems. Semiconductor laser properties are discussed in several books [14–17].

Historically, the research on photodiodes followed that of the semiconductor laser. Early work was carried out using Si and Ge material systems [18, 19]. A majority of the follow-on work has been using the InGaAsP/InP material system [20]. This material system is suitable for producing photodetectors with high response in the 1300–1600 nm wavelength range. The fiber optic transmission systems operate in this wavelength range due to the low loss of optical fibers.

The optical fiber used in optical transmission systems is based on silica (SiO_2) material. Low loss silica fibers were first proposed in 1966 [21] and by early ~1970s, low loss fibers had an attenuation of 20 dB/km. Since then, the manufacturing processes have been refined and developed so that transmission fibers have typical losses of <0.15 dB/km in the 1550 nm region of the optical spectrum [21–28]. This book provides a comprehensive account of design, fabrication, and performance of semiconductor optoelectronic devices and optical fiber-based devices.

1.2 Semiconductor materials

The choice of materials for semiconductor light emitters (lasers and LEDs) is principally determined by the requirement that the probability of radiative recombination should be sufficiently high. This also provides that there is enough gain at low current. This is usually satisfied for "direct gap" semiconductors. The various semiconductor material systems along with their range of emission wavelengths are shown in Figure 1.1. The lines represent the range of band gaps that can be obtained by varying the composition (fraction of the constituting elements) of the material. Many of these material systems are ternary (three-element) and quaternary (four-element) crystalline alloys that can be grown lattice-matched over a binary substrate.

Many of these materials were first used to make semiconductor lasers [7–10, 14–17]. The optical gain in lasers and amplifiers occurs at wavelengths close to the band gap. Thus, a suitable set of materials must be chosen to get optical gain at the desired wavelength. Another important criterion in selecting the semiconductor material for a specific heterostructure design is related to lattice matching, i.e. the crystalline materials that form the heterostructure must have nearly equal lattice constants. The binary substrates that are

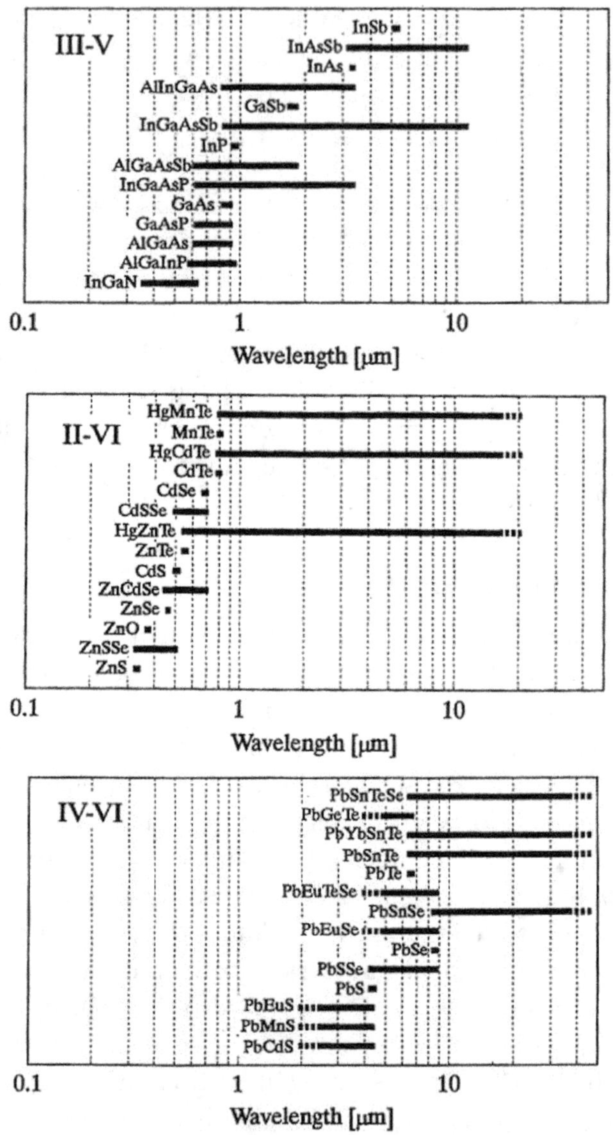

Figure 1.1: Semiconductor materials used in laser fabrication at different regions of the spectrum. The three figures refer to compound semiconductors formed using group III and group V elements (III–V), group II and group VI elements (II–VI) and group IV and group VI elements (IV–VI) of the periodic table [29]. The quantum cascade lasers are fabricated using III–V semiconductors and they operate on intraband transitions in a quantum well.

commonly used are GaAs and InP. They can be grown in a single crystal form from a melt. The ternary or quaternary semiconductors are epitaxially grown over the binary substrate for semiconductor laser or photodetector fabrication. In the epitaxial growth process, the single crystal nature is preserved at the interface of the two materials. This leads to near absence of defect sites at the interface. The development of epitaxial growth techniques has been of major significance for the development of semiconductor photonic devices, such as lasers, amplifiers, and photodetectors.

The commonly used techniques for epitaxial growth are liquid phase epitaxy (LPE) [30], vapor-phase epitaxy (VPE) [31], molecular beam epitaxy (MBE) [32], and metal organic chemical vapor deposition (MOCVD) [33]. Early work on lasers and amplifiers were carried out using the LPE growth technique. The MBE technique is very useful for the growth of very thin semiconductor layers and was first used to fabricate quantum well structures [34]. The MOCVD technique and its variants are generally used for commercial production of lasers, amplifiers, and photodetectors. These growth techniques are described in detail in Chapter 2. In addition to growth, processing techniques of semiconductor materials for laser and photodetector fabrication were developed. The processing techniques include current confinement, low resistance contacts, and etching of the semiconductor material to form specific geometries [35–40].

In the absence of lattice match, the quality of the hetero-interface is poor, i.e. there are lattice defects which serve as non-radiative recombination sites. In certain instances, it is possible to grow epitaxially thin layers of one semiconductor on top of another in the presence of a lattice mismatch of a few percent. Such growth is called strained layer epitaxy, it is discussed in Section 4.4 of Chapter 4. Figure 1.2 shows the relationship between band gap (E_g) and lattice constant (a) of several ternary compound semiconductors. The dots correspond to the binary compounds and the lines represent the ternary compounds.

The ternary semiconductor $Al_xGa_{1-x}As$ can be grown over a GaAs substrate over the entire range of Al fraction x because

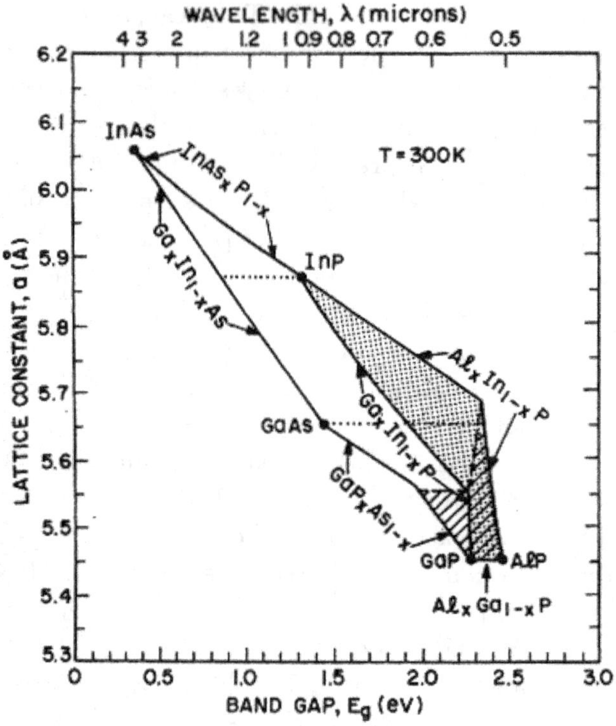

Figure 1.2: Band gap and lattice constants of several ternary and quaternary compounds formed from the binary compounds. For $In_{1-x}Ga_xAs_yP_{1-y}$ quaternary material, the clear region is the region of direct band gap [16].

GaAs and AlAs have nearly the same lattice constant. However, the $Al_xGa_{1-x}As$ material has a direct band gap for compositions with $x < 0.45$. Semiconductors with direct band gap are needed for efficient light emission. Heterostructures using the AlGaAs/GaAs material system were the first to be studied extensively. Semiconductor lasers, amplifiers, bipolar transistors, and field effect transistors have been made using this material system.

In the 1970s, considerable resources were devoted to lasers and photodetector fabrication using the quaternary $In_{1-x}Ga_xAs_y$ P_{1-y}/InP material system. This material system is particularly suitable for lasers in the 1300 and 1550 nm wavelength range, which are in the region of low dispersion and low optical loss for silica optical fibers, respectively. $In_{1-x}Ga_xAs_yP_{1-y}$ can be grown lattice-matched over an InP single crystalline substrate for $In_{1-x}Ga_xAs_yP_{1-y}$ compositions for which $x \sim 0.45$y. Semiconductor lasers and detectors based on this material system are widely used in current fiber optic communication systems. Semiconductor amplifiers are important for all-optical networks, which operate near 1550 nm wavelength range. Thus, much of the developments on optoelectronic devices in the 1980s, and continuing on, have been using the $In_{1-x}Ga_xAs_yP_{1-y}/InP$ material system.

1.3 Semiconductor device operating principles

A typical laser chip is about 250 μm long with cleaved ends (facets). It has a p-cladding layer, an n-cladding layer, and a gain region, all of which are epitaxially grown over a binary substrate.

The schematic diagram of a semiconductor laser chip is shown in Figure 1.3. The laser has a p–n junction which is forward biased during operation. The injected current produces gain in the gain region.

The majority carriers are holes in the p-cladding layer, and they are electrons in the n-cladding layers. The electrons and holes are injected into the gain region, which is made of a lower band gap semiconductor than the cladding layers. The colocated electrons and holes recombine. This results in spontaneous emission of light and optical gain for light propagating in the gain region. It is a fortunate coincidence that semiconductor with lower band gap also has a higher

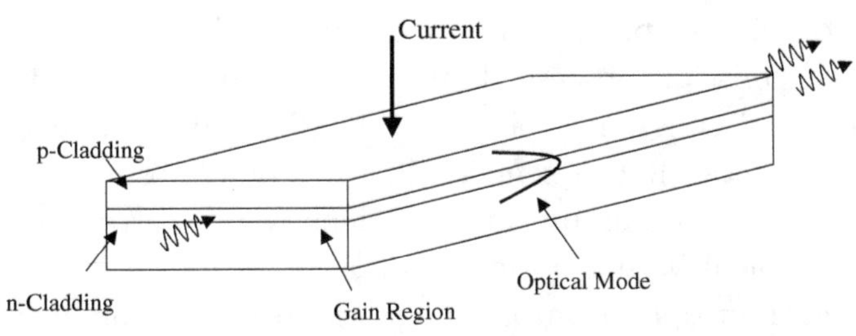

Figure 1.3: Schematic diagram of a semiconductor laser with no waveguide in the lateral (along the plane of p–n junction) direction.

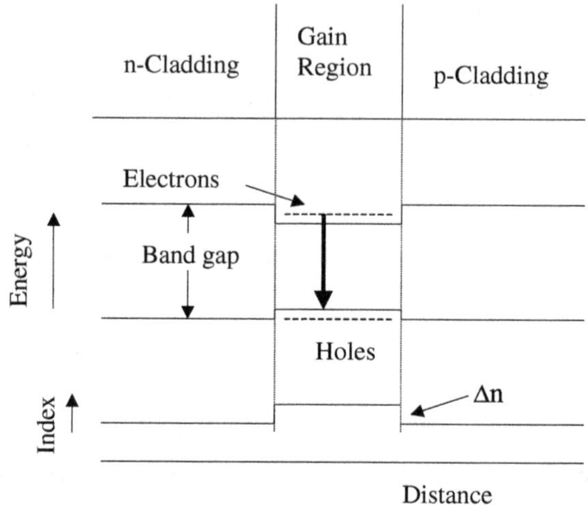

Figure 1.4: Schematic illustration of confinement of the electrons and holes and also simultaneous confinement of optical mode in a double heterostructure.

index than semiconductors with higher band gap. The small index difference produces a waveguide for the propagating light. The light is guided in this waveguide and it experiences amplification (gain). Thus, the double heterostructure material with n-type and p-type high band gap semiconductors around a low band gap semiconductor is instrumental in simultaneous confinement of the charge carriers (electrons and holes) and the optical signal. This is illustrated in Figure 1.4.

The electrons are located in the conduction band and holes are present in the valence band. In the gain region, the electrons and holes recombine to produce photons through both spontaneous and stimulated emission process (Figure 1.5). The optical gain is due to the stimulated emission process. In the absence of current (electron and hole injection), the semiconductor would absorb the incident photons. Certain minimum number density of electrons (and holes) is needed to achieve net optical gain.

Low current operation of a laser requires that the optical signal remain confined in the vicinity of the gain region. In the double heterostructure, this is accomplished by slightly higher index of the

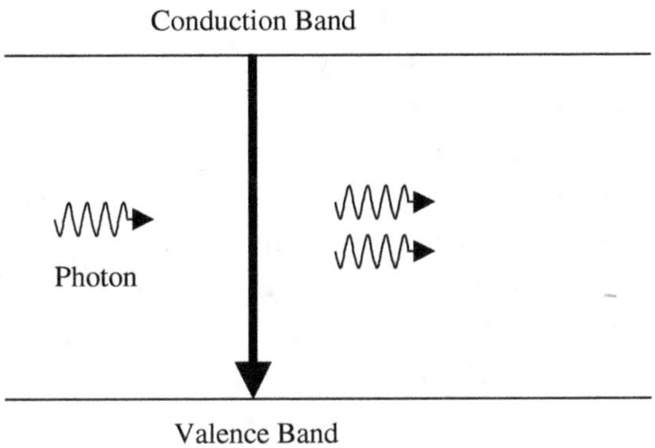

Figure 1.5: Schematic illustration of gain (or amplification). An electron–hole pair recombines to generate a photon. This photon is emitted in the same direction as the incident photon. The process is known as stimulated emission.

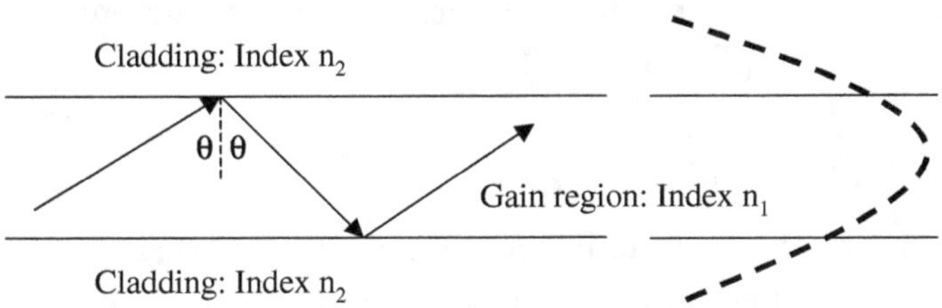

Figure 1.6: Waveguiding in a double heterostructure. The gain region of index n_1 has cladding layers of index n_2 on both sides. The index $n_1 > n_2$. The total internal reflection of a ray incident at an angle θ [with Sin $\theta > (n_2/n_1)$] propagating in the gain region is shown. An alternate view is that a confined optical mode (with intensity profile shown as dashed in the figure) propagates in the waveguide.

gain material compared to that for the cladding materials. These three layers form a waveguide. The optical mode propagation in a three-layer waveguide using the ray approach is shown in Figure 1.6. The rays incident at an angle θ, with $\sin\theta > (n_2/n_1)$, undergo total internal reflection at the interface of the cladding and active region. Thus, rays with these sets of angles of incidence continue to propagate through the amplifier and get amplified. Other rays escape from the gain region. An alternative view is an optical mode with a certain intensity profile determined by the thickness of the gain region and the indices can propagate through the amplifier without any diffraction loss. The intensity profile of a propagating mode is also sketched in Figure 1.6. So, far optical waveguiding normal to the p–n junction has been discussed. Optical waveguiding in the lateral direction (along the plane of the p–n junction) is also important for the fabrication of high-performance lasers, which operate in a single transverse mode. Various schemes for lateral optical guiding and current confinement have been developed. These are described in detail in Chapter 3. Semiconductor optical amplifier is very similar to a laser except it has no reflecting facets [41, 42].

The InGaAs/InP-based photodiode used in many applications also has a double heterostructure grown over InP substrate. The light absorbing layer is the lower band gap InGaAs (typically ~1–2 μm thick) with n-InP and p-InP cladding layers. The absorbed light produces electron–hole pairs which create a photocurrent in a circuit connected to the photodiode [20].

1.4 Optical fibers

Optical fibers are principally made of glass (SiO_2). In optical fibers, the light is confined and guided by a circular waveguide consisting of the core and cladding regions. The core region has a slightly higher index (~0.3%) than the cladding region, and the two regions form a waveguide. Index-raising dopants for SiO_2 glass are germanium, phosphorus, and aluminum, and index-lowering dopants are boron and fluorine. A typical fiber has a Ge-doped SiO_2 core and SiO_2 cladding. For a typical single-mode transmission fiber, the core has a diameter of 10 μm and the cladding has a diameter of 125 μm. The optical mode propagation in a fiber waveguide using the ray approach is shown in Figure 1.7. The rays incident at an angle θ, with Sin $\theta > (n_2/n_1)$, undergo total internal reflection at the interface of the cladding and core region. Thus, rays with these sets of angles of incidence continue to propagate through the fiber. Other rays escape from the core region.

An alternative view is that an optical mode with a certain intensity profile determined by the diameter of the core region and

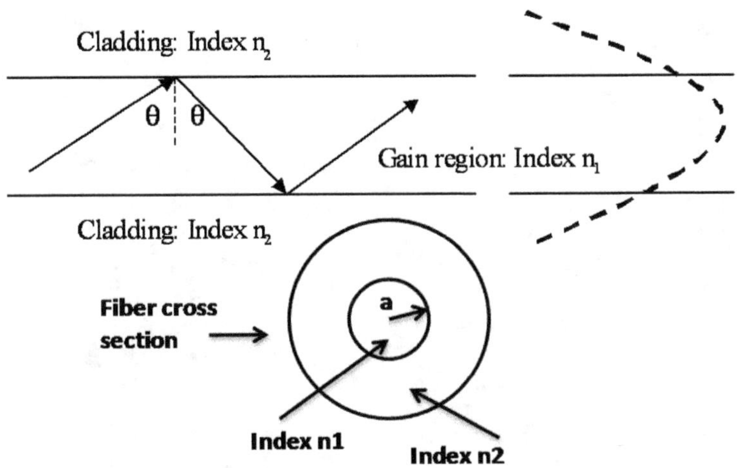

Figure 1.7: Waveguiding in a fiber waveguide. The core region of index n_1 has cladding region of index n_2 around it. The index $n_1 > n_2$. The total internal reflection of a ray incident at an angle θ (with Sin $\theta > (n_2/n_1)$) propagating in the core region is shown. An alternate view is that a confined optical mode with intensity profile (shown as dashed in the figure) propagates in the waveguide.

the indices can propagate through the fiber without any diffraction loss. The intensity profile of a propagating mode is also sketched in Figure 1.7 (see Chapter 9 for details). A fraction of the incident optical signal can, in general, be coupled to the propagating mode. The remaining fraction is lost. Installed optical fibers have a polymer jacket around the cladding. Current optical fibers have optical loss of ~0.15 dB/km near the 1550 nm region of wavelength.

As a result, much of the optoelectronic devices for fiber-based telecommunications operate near this wavelength.

1.5 Operating principles of fiber amplifiers

Rare earth ions doped in crystalline or glassy materials have been studied extensively for optical applications. The doped glasses provide broad emission and absorption spectrum, and optical gain when optically pumped using suitable wavelengths and pump powers. Typical glass hosts are oxide and fluoride glasses. The ground states

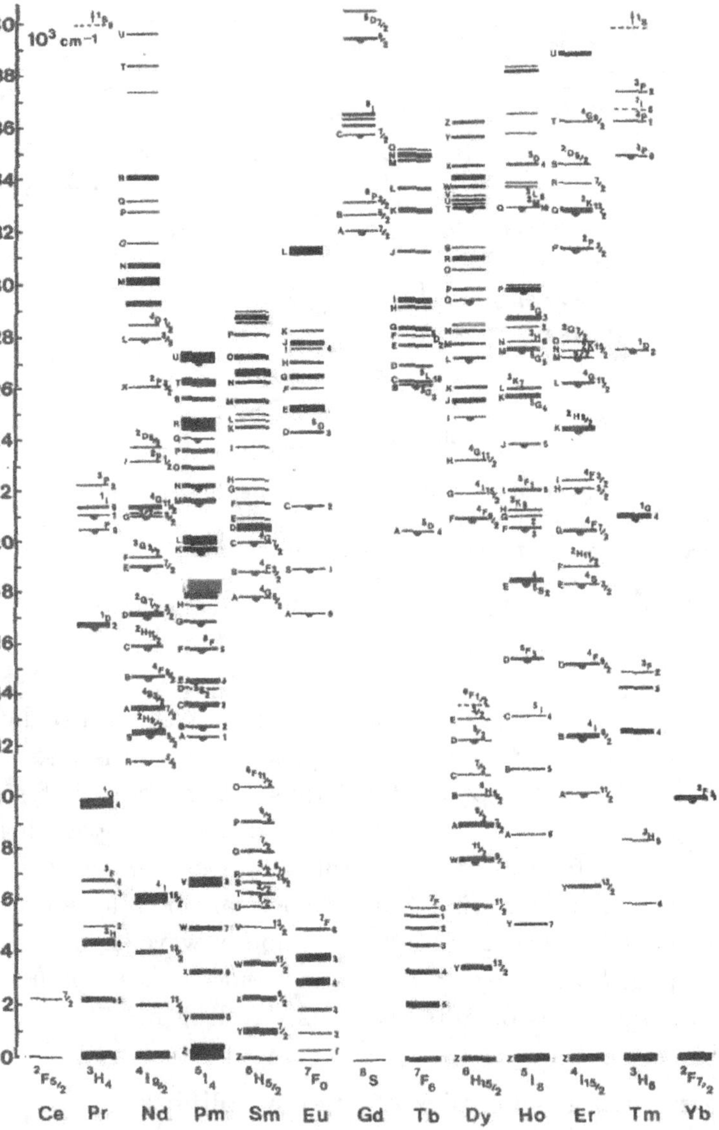

Figure 1.8: Dieke diagram of trivalent rare earth ions [43, 44]. The width of each level represents crystal field splitting [45].

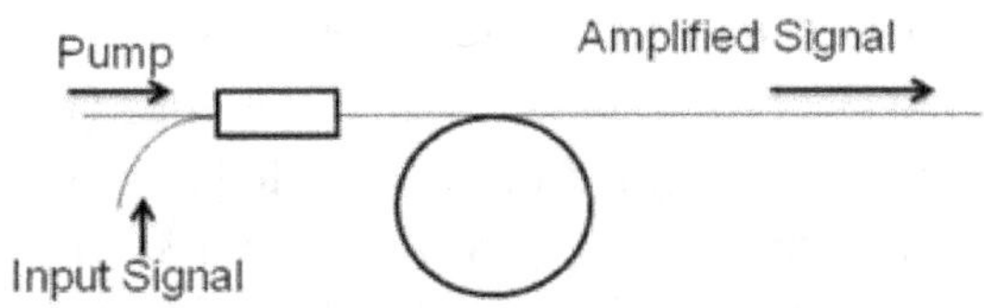

Figure 1.9: Schematic of a rare earth-doped fiber amplifier. Coiled rare earth-doped fiber (circle) is shown. A wavelength division multiplexed (WDM) coupler couples both pump and input signal into the doped fiber.

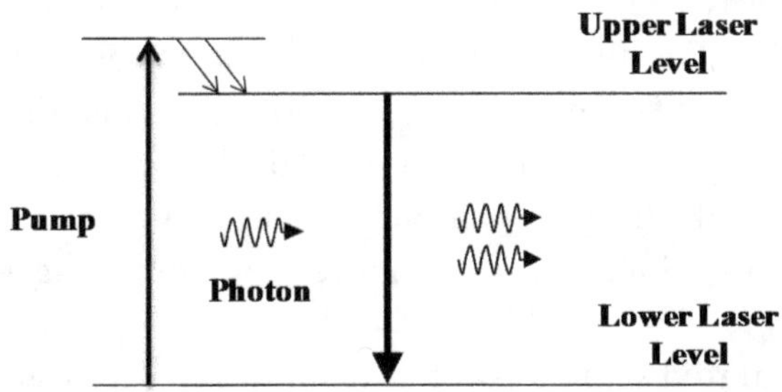

Figure 1.10: Schematic illustration of gain (or amplification). An electron makes a transition to generate a photon. This photon is emitted in the same direction as the incident photon. The process is known as stimulated emission.

of rare earth ions are characterized by an open 4f shell. The free ion energy level structures of rare earth ions are quite complex [43]. The spin–orbit interaction creates a fine structure. The fine structure levels are labeled not only by their J values but also by the L and S values.

The atomic forces split the $4f^N$ orbital to $^{2S+1}L_J$ levels, then the weaker crystal field splits each free ion level into a large number of stark levels. The crystal field splitting is about 100–500 cm^{-1} [43]. The energy level diagram for several trivalent ions, due to Dieke, is shown in Figure 1.8. Figure 1.8 is often called Dieke diagram [43]. Rare earth ions in glass comprise an important class of laser materials. The ordinary glass (SiO_2) has the advantage that it can be drawn into low loss optical fibers providing long-interaction lengths. In a fiber amplifier, the core (as stated in Section 1.4.1) is doped with the rare earth dopant (Er, Nd, Yb, etc.). Er is generally used as a dopant for amplification near 1550 nm.

The schematic of a typical Er-doped fiber amplifier is shown in Figure 1.9. It consists of a pump laser diode and typically a ~10 m long Er-doped fiber. The light from the diode laser and the input light near 1550 nm are both coupled into the fiber using a coupler. The pump light excites the rare earth ions to a high energy level which decays fast to an intermediate energy level (higher energy level of lasing transition) from which it decays to a lower lasing energy level with the emission of a photon. In the presence of another photon of the same wavelength, stimulated emission is observed as shown in Figure 1.10. For fiber laser operation, dichroic mirrors (mirrors with high reflectivity for lasing light and high transmission for pump light) are added at the input and output. The typical reflectivity for the laser light at the pump input end is >90% and at the output end, it is ~50%. Some fiber laser designs use a ring geometry (see Chapter 9).

The stimulated emission (or gain) in a fiber amplifier arises from electron transitions in the rare earth ion in the doped core region, and the light generated is confined and guided by a circular waveguide consisting of the core and cladding regions.

1.6 Applications

Optical fiber transmission systems are one of the major application areas for optoelectronic devices. Terrestrial transmission systems use wavelength division multiplexing (WDM), i.e. multiple wavelengths carrying separate data are transmitted on a single fiber to increase the transmission capacity. The erbium-doped optical amplifier (EDFA) is most commonly used in the optical communication systems because it amplifies signals near 1550 nm which is the region of low loss transmission in optical fibers. Many studies on applications of EDFA in optical fiber systems have been reported [46–49]. The EDFA is well suited for application in single channel fiber systems for data rates exceeding 100 Gb/s and for multichannel WDM transmission systems.

The EDFA is also used as a preamplifier for increasing the receiver sensitivity as a signal amplifier in long-distance transmission

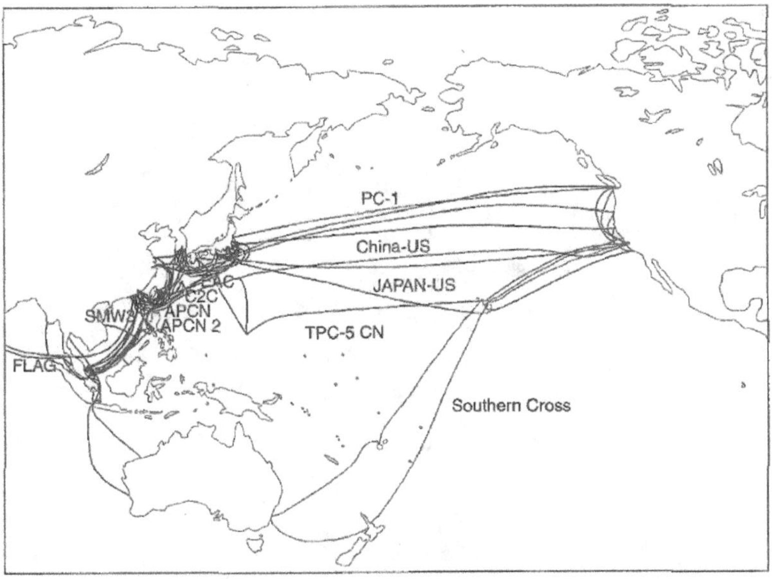

Figure 1.11: Optical amplifier submarine fiber network in the Pacific Ocean [50].

systems, and as a replacement for repeaters in long-distance undersea fiber optic transmission systems. Schematic diagrams of submarine optical fiber network in the Pacific and Atlantic oceans are shown in Figures 1.11 and 1.12, respectively.

White light generation using LEDs is another area of worldwide commercial application. LEDs are poised to capture a majority share of the lighting market. Figure 1.13 shows the development of LEDs over six decades beginning with the first demonstration of visible laser diode in 1962. With the development of high-efficiency blue LEDs in the 1990s and the subsequent use of that LED along with existing ones in other regions of the visible spectrum, or with phosphors, white light creation using current injection through a p–n junction became a reality. Since then, they have been commercialized for many lighting applications including household use.

Optoelectronic devices continue to be developed for new applications and for improved efficiency (e.g. smaller size, reduced cost, and more functionality) in current applications.

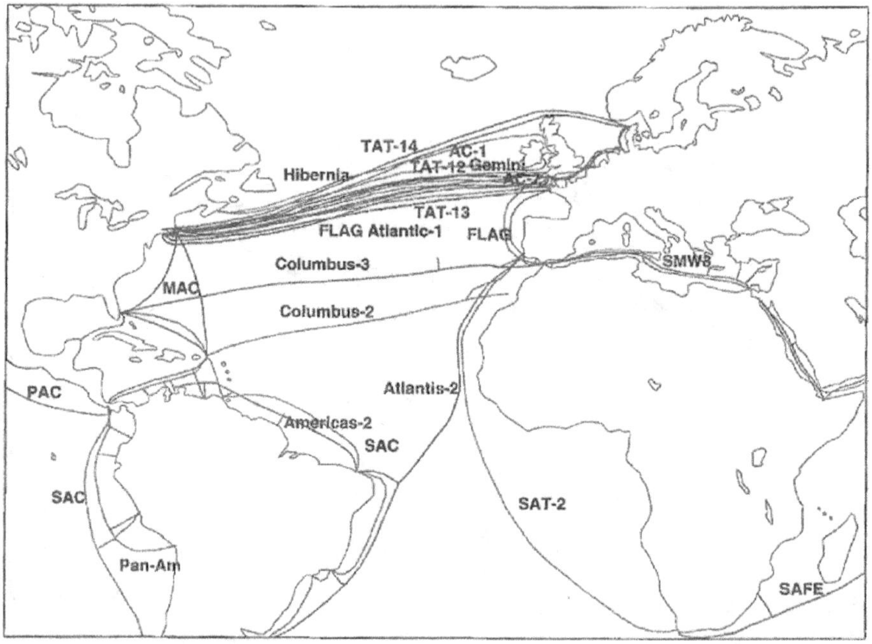

Figure 1.12: Optical amplifier submarine fiber network in the Atlantic Ocean [50].

1.7 Book overview

Optoelectronic devices and optical fibers are important for a wide range of applications including fiber optic communications, free space communications, optical networks, solid-state (LED) lighting, high power laser systems, and sensors. The book describes the advances in research, technology, and commercial trends in the area of optoelectronic and optical fiber-based devices and systems. Several appendices, which describe both the fundamentals and the details, have been added to make the book a classroom text for senior undergraduate or graduate students. These appendices are used to derive the equations applicable to a particular optoelectronic device from a fundamental set of equations (such as Maxwell equations and the Schrodinger equation).

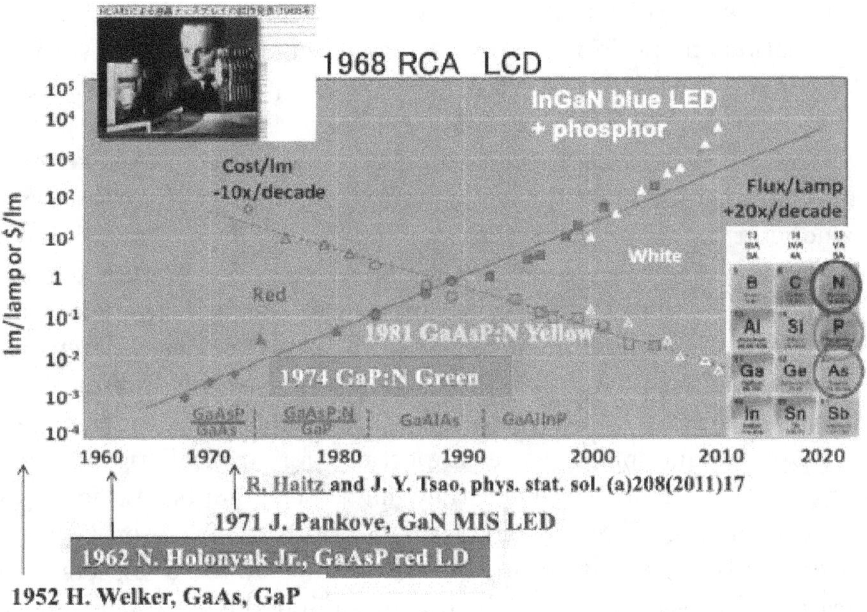

Figure 1.13: Development of LEDs [51].

The fundamental operating principles of semiconductor optical materials used for generation and absorption of light are described in Chapter 2. The principle of operation of light generating devices such as LEDs and semiconductor lasers are described. The epitaxial growth of semiconductor materials is described. Various growth processes and LED fabrication methods are described. Both infrared and visible LED materials and characteristics are described. The growth, design, and performance of blue LEDs based on InGaN/GaN/AlGaN are described. A major commercial application of blue LEDs is in "solid-state lighting" through white light generation using optical pumping of rare earth-doped glass.

Chapter 3 describes the fabrication and performance characteristics of semiconductor lasers. The mechanism for gain, the threshold condition, and optical gain spectrum are described. Index-guided laser structures and their fabrication are described. Single wavelength lasers based on distributed feedback mechanism are described. The fabrication and performance of lasers with quantum well and strained layer gain region are described. The rate equations for pulse generation and small signal modulation for a laser are developed in Chapter 3. Relaxation oscillation and transient effects are discussed.

Chapter 4 builds upon Chapter 3 and further discusses semiconductor lasers. Surface emitting lasers, used for optical interconnections between computer backplanes, and quantum cascade lasers which operate in intersubband transitions in the mid-infrared are discussed. Quantum dot and nano-structure-based lasers are discussed. High power semiconductor lasers used as a pump source for fiber amplifiers, solid- state lasers, and fiber lasers are described.

Optical modulators are important for optical transmission systems, and more recently, they have been important for optical networks and optical logic systems. Both electroabsorption and electro-optic modulators are discussed in Chapter 5. Lasers integrated with electroabsorption modulators on the same chip are an important source for fiber optic communication systems. Mach–Zhender modulators fabricated using InP/InGaAsP semiconductor, GaAs/AlGaAs semiconductor and Ti-$LiNbO_3$ are described. The latter is used in fiber optic transmission systems. Advanced modulation formats (such as phase shift-based formats and multiple intensity levels) for data transmission and the modulator technologies needed are discussed.

Photodetectors are important optoelectronic devices which convert the incident optical signal to an electrical signal which is then processed electronically to retrieve the relevant information for communication or sensor systems. Two principle types of photodiodes are the P–i–N photodiode and the avalanche photodiode (APD). Both devices rely on a reverse biased p–n junction for their operation. The principle of operation, the design, and the performance of these devices are described in Chapter 6. A more common commercial photodetector is a solar cell which converts the light from the sun into electrical power suitable for commercial (or home) use. The performance and design of these devices are described.

Semiconductor optical amplifier (SOA) is a device similar to that of a laser. The fundamental operating principles of SOAs are described in Chapter 7. The optical gain spectrum, the saturation of gain at high powers, and the mechanism for gain are discussed. The rate equations for pulse propagation in SOAs are developed in Chapter 7. Effect of multiple pulses on gain and phase changes, multiwavelength operation is discussed. Various factors for optical noise in amplifiers are described. The commercial use of SOAs is projected to be in the form of functional devices in all-optical networks. Such functional applications include wavelength conversion, transparent switches, optical demultiplexing, optical clock recovery and others. Many of these applications utilize gain saturation phenomenon (which results in cross-gain and cross-phase modulation) or nonlinear four-wave mixing in optical amplifiers. The application of amplifiers for functional performance is discussed in Chapters 7 and 11.

There have been a significant number of developments in the technology of integration of semiconductor lasers, amplifiers and other related devices on the same chip. These chips allow higher levels of functionality than that achieved using single devices. The name photonic integrated circuit (PIC) is generally used when all the integrated components are photonic devices, e.g. lasers, detectors, amplifiers, modulators, and couplers. The design and performance of several photonic integrated circuits are described in Chapter 8.

The fabrication of transmission and rare earth-doped fibers are described in Chapter 9. Single-mode fiber pre-form fabrication methods are described. The drawing of fibers from these pre-forms is described. Incorporation of rare earth dopants by various methods is described. Microstructure fiber fabrication is described. The fabrication of fiber Bragg gratings and its use are described. Mode-locking of fiber lasers is described. Nonlinear processes in single-mode fibers and fiber amplifiers are discussed in Chapter 9. These include four-wave mixing, stimulated Raman effect and stimulated Brillouin effect. Raman lasers based on the stimulated Raman effect in silica fibers are important for new emission wavelengths from fiber lasers.

Fiber Bragg gratings are generally used to provide optical feedback in Raman laser. Cascaded Raman laser designs are described. Supercontinuum generation which includes many of the above nonlinear processes is discussed in Appendix 7. Fiber-based sensors are described.

The fiber amplifier is generally long (few m in length). In order to make compact amplifiers, doped planar waveguide devices ($\sim$10 cm long) have been investigated. The fabrication and performance of silica-based doped planar waveguide amplifiers and lasers are discussed in Chapter 10. The discussion includes high power amplifiers and planar waveguide lasers.

Application of optoelectronic devices in optical transmission systems is described in Chapter 11. These include long-distance transmission, undersea transmission, coherent transmission and analog transmission. Many of these applications incorporate optical fiber amplifiers in commercial systems. Functional applications of SOAs for all-optical networks such as wavelength conversion, optical demultiplexing, optical clock recovery are described. The application of SOAs in optical network applications is discussed in Chapter 11. Optical transmission in data centers and optical transmission using advanced modulation formats are discussed.

Appendices, which describe both the fundamentals and the details, have been added to make the book a classroom text for senior undergraduate or graduate students. These appendices are used to derive the equations applicable to a particular optoelectronic device from fundamentals. The list of appendices are as follows: (1) Waveguide Modes in Planar Waveguide, (2) Waveguide Modes in Optical Fibers, (3) Optical Pulse Propagation Equations, (4) Three-Level and Four-Level Laser Equations, (5) Spontaneous Emission and Absorption for Parabolic Bands, (6) Quantum Cascade Laser (QCL) Gain Calculation, (7) Supercontinuum Generation in Optical Fibers, (8) Supercontinuum Generation in Planar Waveguides, (9) Theory of Harmonic and Rational Harmonic Mode Locking, (10) Reflectivity of a Periodic Structure, (11) Laser Rate Equations, (12) Nonlinear Effects in Fibers, (13) Nonlinear Effects in SOAs, (14) Energy Levels in a Quantum Well, (15) Problems. A set of problems for each chapter has been added in Appendix 15.

1.8 Future challenges

Tremendous advances in Optoelectronic device technologies have occurred over the last several decades. The advances in research and the many technological innovations have led to improved designs of semiconductor lasers, detectors, optical amplifiers, photonic integration, fibers, and optical waveguide-based circuits. Perhaps the most important development over the last decade has been the commercialization of LED-based lighting technology which has high efficiency and considerable long-operating life compared to previous lighting technologies.

The need for higher capacity is continuing to encourage research in wavelength division multiplexed (WDM)-based and optical time division multiplexing (OTDM)-based transmission, which need optical demultiplexer and high power tunable lasers. An important research area would continue to be the development of phase modulation-based systems using Mach-Zehnder interferometers and high-efficiency optical amplifiers.

Improving laser and SOA performance at high temperature is an important area of investigation. The InGaAsN-based system is a promising candidate for making gain regions near 1300 nm with good high-temperature performance. Strained quantum well structures provide ways of making polarization independent amplifiers. Quantum dot-based lasers may operate at a higher speed, and SOAs using quantum dot active regions have fast gain and phase recovery times which makes them suitable for fast optical systems.

Improving Nd- or Yb-doped fiber amplifier performance for space communication application is important. These materials emit at a shorter wavelength, and thus, the diffraction limited spot size is smaller for these amplifiers. Along the same line of thought, the development of shorter wavelength optical fiber amplifiers is important. Although significant developments have occurred in high power optical fiber lasers, the need for higher power for commercial (metal cutting and shaping) applications will continue to drive the research on higher power fiber lasers using Nd or Yb-based systems and possibly others.

For compact fiber amplifiers and lasers, the planar waveguide geometry is an important candidate. Progress in waveguide design and packaging, which includes optimum pump coupling and heat removal, can make these devices competitive with fiber-based amplifiers with considerable reduction in size.

Finally, much of the advances in the optoelectronic device development would not have been possible without the advances in materials and processing technology. The challenges of much of the current optoelectronic device research are intimately linked with the challenges in materials growth which include not only the investigation of new material systems but also improvements in existing technologies, such as quantum dots and microstructure fibers, to make them more reproducible and predictable.

References

1. A. L. Schawlow and C. H. Townes, *Phys. Rev.*, **112** (1958) 1940.
2. T. H. Maiman, *Nature*, **187** (1960) 493.
3. A. Javan, W. R. Bennett, Jr., and D. R. Herriot, *Phys. Rev. Lett.*, **6** (1961) 106.
4. N. G. Basov, O. N. Krokhin, and Yu. M. Popov, *Sov. Phys. JETP*, **13** (1961) 1320.
5. M. G. A. Bernard and G. Duraffourg, *Phys. Status Solidi.*, **1** (1961) 699.
6. W. P. Dumke, *Phys. Rev.*, **127** (1962) 1559.
7. R. N. Hall, G. E. Fenner, J. D. Kingsley, T. J. Soltys, and R. O. Carlson, *Phys. Rev. Lett.*, **9**, (1962) 366.
8. M. I. Nathan, W. P. Dumke, G. Burns, F. H. Dill, Jr., and G. Lasher, *Appl. Phys. Lett.*, **1** (1962) 63.
9. T. M. Quist, R. H. Rediker, R. J. Keyes, W. E. Krag, B. Lax, A. L. McWhorter, and H. J. Ziegler, *Appl. Phys. Lett.*, **1** (1962) 91.
10. N. Holonyak Jr. and S. F. Bevacqua, *Appl. Phys. Lett.*, **1** (1962) 82.
11. Zh. I. Alferov, V. M. Andreev, E. L. Portnoi, and M. K. Trukan, *Sov. Phys. Semiconductors*, **3** (1970) 1107.
12. I. Hayashi, M. B. Panish, P. W. Foy, and S. Sumski, *Appl. Phys. Lett.*, **17** (1970) 109.
13. H. Kroemer, *Proc. IEEE*, **51** (1963) 1782.

14. H. Kressel and J. K. Butler, *Semiconductor Lasers and Heterojunction LEDs*, Academic Press, New York, 1977.
15. H. C. Casey, Jr., and M. B. Panish, *Heterostructure Lasers*, Academic Press, New York, 1978.
16. G. P. Agrawal and N. K. Dutta, *Semiconductor Lasers*, Van Nostrand Reinhold, NY, 1986 (2nd Edn., 1992).
17. L. A. Coldren and S. W. Corzine, *Diode Lasers and Photonic Integrated Circuits*, John Wiley, New York, 1995.
18. H. Melchior, M. B. Fischer and F. R. Arams, *Proc. IEEE*, **58** (1970) 1466.
19. T. Mikawa, S. Kagawa, T. Kaneda, T. Lakuri, H. Ando, and O. Mikami, *IEEE J. Quant. Electron.*, **QE-17** (1981) 210; L. K. Anderson and B. J. McNurry, *Proc. IEEE*, **54** (1966) 1335.
20. K. Taguchi, Chapter 8, in A. K. Dutta, N. K. Dutta, and M. Fujiwara (eds.), *WDM Technologies: Active Optical Components*, Academic Press, New York, 2002.
21. C. K. Kao and G. A. Hockmann, *Proc. IEEE,* **133** (1966) 1151.
22. D. B. Keck, *IEEE Comm. Mag.*, **23** (1985) 17.
23. J. B. MacChesney and P. B. O'Connor, US Patent 4,217,027 (1980).
24. S. R. Nagel, Chapter 4, in S. E. Miller, and I. P. Kaminow (eds.), *Optical Fiber Telecommunications*, 2nd Edn.
25. K. J. Beales and C. R. Day, *Phys. Chem. Glasses*, **21** (1980) 5.
26. K. J. Beales, S. F. Carter, P. W. France, and S. Partington *J. Non-crystalline Solids*, **70** (1985) 253.
27. K. L. Walker, G. M. Hornsy, and F. T. Geyling, *J. Colloid. Int. Sci.,* **69** (1979) 138.
28. K. L. Walker, F. T. Geyling, and S. R. Nagel, *J. Amer Cer. Soc.*, **63** (1980) 552.
29. P. G. Eliseev, Chapter 2 in E. Kapon (ed.), *Semiconductor Lasers II*, Academic Press, New York, 1998.
30. H. Nelson, *RCA Rev.*, **24** (1963) 603.
31. W. F. Finch and E. W. Mehal, *J. Electrochem Soc.*, **111** (1964) 814.
32. A. Y. Cho, *J. Vac. Sci. Technol.*, **8** (1971) 531.
33. R. D. Dupuis, *J. Crystal Growth*, **55**, (1981) 213.
34. R. Dingle, W. Wiegmann, and C. H. Henry. *Phys. Rev. Lett.*, **33** (1974) 827.
35. J. C. Dyment, *Appl. Phys. Lett.*, **10** (1967) 84.
36. J. J. Hsieh, J. A. Rossi, and J. P. Donnelly, *Appl. Phys. Lett.*, **28** (1976) 709.
37. Y. Suematsu, *Proc. IEEE*, **71** (1983) 692.
38. I. P. Kaminow, R. E. Nahory, M. A. Pollack, L. W. Stulz, and 1. C. Dewinter, *Electron. Lett.*, **15** (1979) 763.
39. Y. Suematsu, K. Iga, and K. Kishino, in T. P. Pearsall (ed.), *GalnAsP Alloy Semiconductors*, John Wiley & Sons, New York, 1982.
40. R. J. Nelson and N. K. Dutta, in W. T. Tsang (ed.), *Semiconductor and Semimetals*, Vol. 22, Part C, Academic Press, New York, 1985.
41. T. Saitoh and T. Mukai, *J. Lightwave Technol.*, **6** (1988) 1656.
42. M. O. Mahony, *J. Lightwave Technol.,* **5** (1988) 531.

43. B. Henderson and R. Bertram, *Crystal Field Engineering of Soild-state Laser Materials*, Cambridge University Press, Cambridge, 2000.
44. G. H. Dieke, *Spectra and Energy Levels of Rare Earth Ions in Crystals*, John Wiley, New York, 1968.
45. B. Henderson and G. F. Imbusch, *Optical Spectroscopy of Inorganic Solids*, Oxford University Press, Oxford, 1989.
46. E. Desurvire, *Erbium Doped Fiber Amplifiers*, John Wiley and Sons, USA, 1994.
47. P. C. Becker, N. A. Olsson, and J. R. Simpson, *Erbium Doped Fiber Amplifiers*, Academic Press, New York, 1997.
48. M. J. Digonnet (ed.), *Rare Earth Doped Fiber Lasers and Amplifiers*, Marcel Dekker, Inc., New York, 1993.
49. N. K. Dutta, *Fiber Amplifiers and Fiber Lasers*, World Scientific, Singapore, 2015.
50. S. Shimoda and H. Ishio, *Optical Amplifiers and Their Applications*, John Wiley and Sons, USA, 1994.
51. H. Amano, *Sci. Direct*, **62** (2016) 126.

Chapter 2

Light Emitting Diode (LED) and Materials

2.1 Introduction

When a current is injected into a direct band gap semiconductor with a p–n junction, the electron and holes recombine with the spontaneous emission of light at wavelengths near the band gap of the semiconductor. This is the principle of operation of a light emitting diode, also known as LED, and a semiconductor laser. The latter was invented in 1962 [1–4]. Initially in the 1970s, the LEDs were developed as light sources for telecommunication applications [5]. They were fabricated using GaAs or InP-based semiconductors and emitted near 850 and 1300 nm, respectively. Over the last few years, the most important application of LEDs has been as sources of visible light [6–10]. These visible light sources are replacing the tungsten filament-based or fluorescent light sources for household, automobiles, and commercial applications. LEDs fabricated using GaN-based semiconductors emit in the blue or green. A combination of three LED generated primary colors (blue, green, and red) produces visible light and such a combination is used for regular visible light use. An alternative approach is to have the light from a blue LED impinge on a phosphor, which has a broad emission spectrum in the visible. The term often used to describe these LED-based lighting is solid-state lighting.

The two primary optical sources used in fiber optic communications are the semiconductor laser and the LED [5, 11–16]. The LEDs are used as sources for low data rate (<200 Mb/s) and short distance applications, and lasers are used for high data rate and long distance applications [17–20]. The principle of operation, fabrication, and performance of GaAs-, InP-, and GaN-based LEDs are discussed in this chapter.

2.2 Materials and fabrication

The initial demonstration of LEDs was carried out using a p–n junction produced by diffusion [1]. The lattice matched crystalline growth of one semiconductor over another is called epitaxy. The development of epitaxial growth techniques has been of major significance for the development of semiconductor photonic devices, such as LEDs, lasers, amplifiers, and photodetectors. The commonly used techniques for epitaxial growth are liquid-phase epitaxy (LPE) [21], vapor-phase epitaxy (VPE) [22], molecular beam epitaxy (MBE) [23], and metal organic chemical vapor deposition (MOCVD) [24]. Commercial LEDs were fabricated using LPE growth technique in the 1970s. With the development of MOCVD and MBE growth techniques in the 1980s, the reproducibility of the fabrication process has improved, and it has also led to the development of GaN-based LEDs. The latter is important for LED-based lighting technology.

2.2.1 *Principal material systems*

Extensive work on LED fabrication was carried out in the 1970s using the GaAs/AlGaAs material system. These LEDs had GaAs active region surrounded by p-$Al_{1-x}Ga_xAs$ and n-$Al_{1-x}Ga_xAs$ cladding layers. Much of the recent experimental work on LEDs for communication have been carried out using the $In_{1-x}Ga_xAs_yP_{1-y}$ material system with the optical emission centered around 1300 nm. The LPE growth of AlGaAs/GaAs and InGaAsP/InP heterostructures were studied by several researchers in the 1960s and 1970s [25–47]. Much of this work enabled determination of material parameters (for a review see Refs. [11–13]).

The $Al_xGa_{1-x}As$ material can be grown lattice matched to GaAs for a range of compositions. The composition of the ternary solid is determined by X-ray analysis. The measured lattice constant $a(x, y)$ [46] of $Al_xGa_{1-x}As$ lattice matched to GaAs agrees with Vegard's law which is given by

$$a(x) = x\,a(\text{AlAs}) + (1 - x)a\,(\text{GaAs}) \tag{2.1}$$

where a (AlAs), a (GaAs) are the lattice constants of the binary semiconductors AlAs and GaAs, respectively. For $x < 0.45$, $Al_xGa_{1-x}As$ is a direct band gap semiconductor. The direct band gap of $Al_xGa_{1-x}As$ grown lattice matched over GaAs has been measured using photoluminescence. In this technique, electrons and holes are created by photoexcitation in the conduction and valence bands, respectively, and when they recombine, light at energy corresponding to the band gap is emitted. From the data, the band gap (at 300°K) of $Al_xGa_{1-x}As$ is given by (for $x < 0.45$)

$$E_{\text{g}}(x) = 1.424 + 1.247\,x \tag{2.2}$$

where E_{g} is the band gap in eV. The refractive index of AlAs is 3.178 and that of GaAs is 3.655. The refractive index of $Al_xGa_{1-x}As$ is found to be a linear combination of that for AlAs and GaAs similar to Vegard's Law above, i.e. $n(x) = xn(\text{AlAs}) + (1 - x)\,n\,(\text{GaAs})$.

The InGaAsP material can be grown lattice matched to InP for a range of compositions. The composition of the quaternary solid is determined by X-ray analysis. The measured lattice constant $a(x, y)$ [47] of $In_{1-x}Ga_xAs_yP_{1-y}$ lattice matched to InP agrees with Vegard's law which is given by

$$a(x, y) = xy\,a(\text{GaAs}) + x(1 - y)a(\text{GaP}) + (1 - x)ya(\text{InAs}) + (1 - x)(1 - y)a(\text{InP}) \tag{2.3}$$

where a (GaAs), a (GaP), a (InAs), and a (InP) are the lattice constants of the binary semiconductors GaAs, GaP, InAs, and InP, respectively. Using $a\ (x, y) = a$ (InP) in Eq. (2.3), the following relationship between x and y for $In_{1-x}Ga_xAs_yP_{1-y}$ compositions lattice matched to InP is obtained:

$$x = 0.4526y/(1 - 0.031y) \tag{2.4}$$

or $x \sim 0.45y$. This relationship has been experimentally verified [29].

The direct band gap of $In_{1-x}Ga_xAs_yP_{1-y}$ grown lattice matched over InP has been measured using photoluminescence. From the data, the band gap (at 300°K) of $In_{1-x}Ga_xAs_yP_{1-y}$ lattice matched to InP can be represented by the following expression [47]:

$$E_g(\text{in eV}) = 1.35 - 0.72y + 0.12y^2 \tag{2.5}$$

Knowledge of the refractive index as a function of composition of $In_{1-x}Ga_xAs_yP_{1-y}$ is important for understanding the waveguiding properties of InGaAsP/InP semiconductor heterostructures. The following analytic form is often used for the refractive index ($n(y)$) of $In_{1-x}Ga_xAs_yP_{1-y}$ lattice matched to InP (for which $x \sim 0.45y$) at a wavelength corresponding to the band gap [48, Chapter 4, Ref. 12]

$$n(y) = 3.4 + 0.256y - 0.095y^2 \tag{2.6}$$

Measurements of refractive index of $In_{1-x}Ga_xAs_yP_{1-y}$ lattice matched to InP has also been carried out which agrees with Eq. (2.6). $In_{0.5}(Ga_{1-x}Al_x)_{0.5}P$ active regions on GaAs substrate can be used to make lasers in the visible wavelength range (Section 4.4 of Chapter 4). The band gap E_g and the refractive index (n) of InGaAlP as a function of Al fraction x is given by

$$E_g = 1.90 + 0.60x \quad \text{and} \quad n = 3.65 - 0.37x \tag{2.7}$$

where $x < 0.7$. InGaAlP has indirect band gap for $x > 0.7$ and is therefore not suitable for laser action for $x > 0.7$. AlGaN and InGaN can be grown nearly lattice matched over GaN using the MOCVD growth techniques described in the next section. The $Al_xGa_{1-x}N$ semiconductor has a band gap of 3.42 eV (~365 nm) at $x = 0$ (GaN) and 6.2 eV (~200 nm) at $x = 1$ (AlN). The energy gap E_g as a function of alloy composition of the alloy $(AlN)_x(GaN)_{1-x}$ is given by

$$E_{\mathrm{g}}(x) = E_{\mathrm{g}}(\mathrm{GaN}) + (\mathrm{c} - \mathrm{b})x + bx^2 \tag{2.8}$$

where $c = E_{\mathrm{g}}(\mathrm{AlN}) - E_{\mathrm{g}}$ (GaN) and b is the bowing parameter. For AlGaN, values for b are in the range -1.0 to $+\ 1.3$ eV [13].

2.2.2 *Epitaxial growth methods*

This section describes briefly the various epitaxial growth techniques (LPE, MOCVD, and MBE) used to fabricate the semiconductor material. In LPE, the epitaxial layer is grown by cooling a saturated solution of the composition of the layer to be grown while that solution is in contact with substrate. In VPE, the epitaxial layer is grown by the reaction of gaseous elements or compounds at the surface of a heated substrate. The VPE technique has also been called chemical vapor deposition (CVD) depending on the constituents of the reactants. A variant of the technique is MOCVD [24], which has been very successful for lasers in which metal alkyls are used as the compound source. In MBE, the epitaxial layer growth is achieved by the reaction of atomic or molecular beams of the constituent elements (of the layer to be grown) with a crystalline substrate held at high-temperature in ultrahigh vacuum.

2.2.2.1 *Liquid phase epitaxy (LPE)*

The LPE apparatus used for growing heterostructures has the ability to perform successive growth of several epitaxial layers. A multibin apparatus used for growing double heterostructures is shown in Figure 2.1. It consists of a furnace, a graphite "boat", a slider which holds the substrate (i.e. the single crystalline material on which the epitaxial layers are grown), and several thermocouples, etc. The graphite boat has a number of reservoirs each of which contains a saturated solution of the corresponding epitaxial layer to be grown. The graphite boat is inside the furnace, which is heated to the melting point (~650°C) of the material to be grown. The substrate (seed) is placed in a graphite slider that has a groove to hold the substrate. The slider is attached to a long rod that allows an operator to position

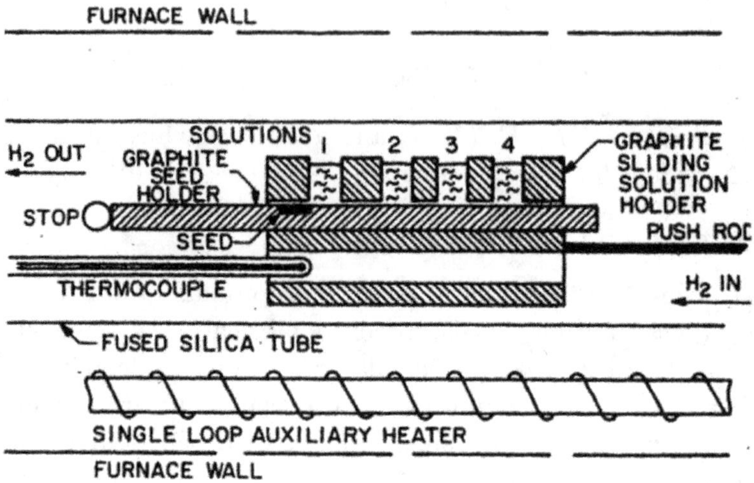

Figure 2.1: Schematic of a multibin LPE apparatus used for growing double heterostructures [49].

the substrate under different reservoirs from outside the furnace. This allows growth of successive layers of desired thickness and composition on the substrate. Hydrogen or helium is generally used as an ambient gas during the growth process.

2.2.2.2 *Vapor-phase epitaxy*

In VPE, the source chemicals from which the layers are grown are gaseous. The technique has been used for the growth of several III–V compound semiconductors [50–58]. Chlorides or hydrides of the elements are often used as sources for the growth of the semiconductor. In the chloride method, $AsCl_3$ or PCl_3 is passed over elemental Ga or In to form chlorides. These metal chlorides then react with AsH_3 and PH_3 near the InP substrate to form epitaxial layers of InGaAsP on InP. The metal chlorides can also be formed by using GaAs or InP sources instead of elemental Ga or In.

Figure 2.2 shows the schematic of a VPE reactor. The composition of the layers to be grown is changed by altering the flow rates of the reactants. P-type doping (Zn-doped) is carried out by flowing H_2 over hot Zn metal, which carries Zn into the reaction chamber

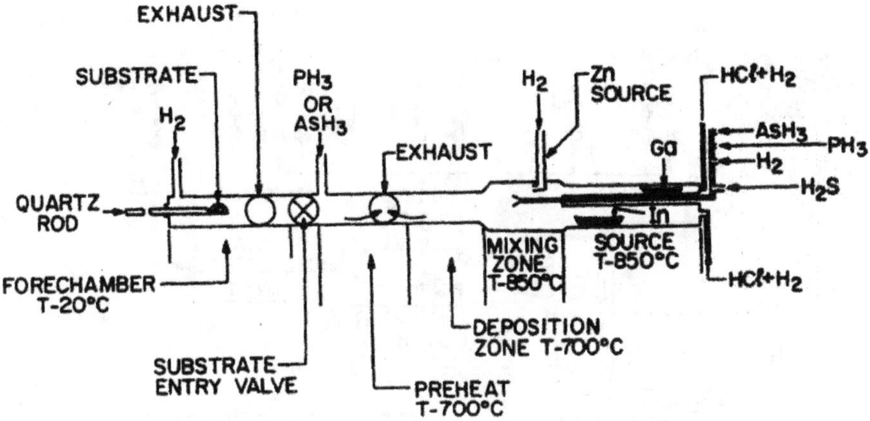

Figure 2.2: Schematic of a VPE reactor [50].

and n-type doping (S-doped) is done by flowing H_2S into the reaction chamber. VPE technique is suitable for fabrication of materials with very low background doping, such as that needed for reducing the dark current in photodetectors.

2.2.2.3 *Metal organic chemical vapor deposition*

MOCVD, also known as metal organic vapor phase epitaxy (MOVPE), uses metal alkyls as sources from which epitaxial layers are formed [59–71]. The low-pressure MOCVD technique where the gas pressure is ~0.1 atm has been used for the growth of AlGaAs and InGaAsP. MOCVD is a very versatile technique; it has been used to grow a wide variety of II–VI and III–V compound semiconductors.

Figure 2.3 shows the schematic of a low-pressure MOCVD system. The quartz reaction chamber contains the InP substrate placed on a radio frequency (RF) heated (~600°C) carbon susceptor. Group III alkyls ($Ga(C_2H_5)_3$ and $In(C_2H_5)_3$) and group V hydrides (AsH_3 and PH_3) are introduced into the reaction chamber. The gas flow near the substrate is laminar with velocities in the range of 1–15 cm/s for a pressure of ~0.1 atm. At the hot substrate surface, the metal alkyls and hydrides decompose, producing elemental In, Ga, As, and P. The elemental species deposit on the surface forming an epitaxial

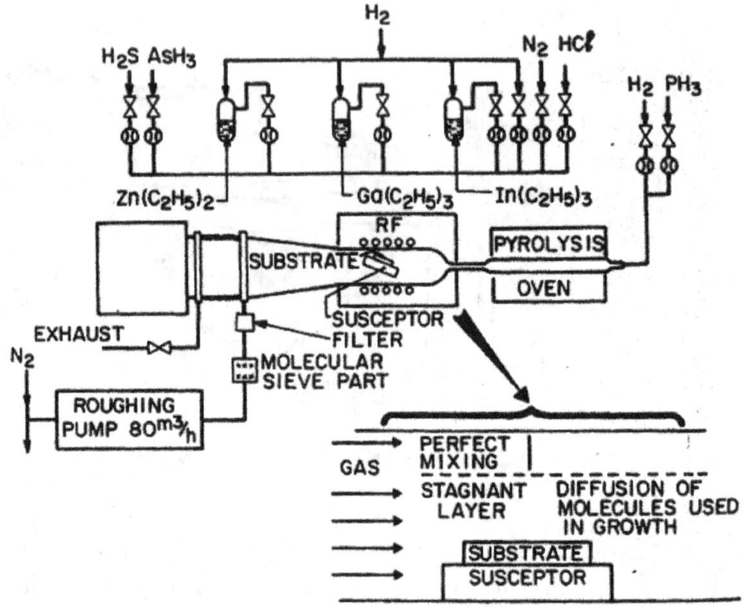

Figure 2.3: Schematic of a MOCVD growth system [67].

layer. The gas flow rates are controlled to produce InGaAsP layers of different compositions. $Zn(C_2H_5)_3$ and H_2S are used as sources for p-type (Zn-doped) doping and n-type (S-doped) doping, respectively.

2.2.2.4 *Molecular beam epitaxy*

In the MBE technique, epitaxial layers are grown by impinging atomic or molecular beams on a heated substrate kept in ultra-high vacuum [72–77]. The constituents of the beam "stick" to the substrate resulting in a lattice matched layer. Since different elements have different sticking coefficients, the beam intensities are separately controlled to take into account the differences.

MBE technique was first used to make AlGaAs heterostructures (Figure 2.4). The substrate is attached to a molybdenum (Mo) block, which is RF heated inside a vacuum chamber at a pressure of 10^{-7}–10^{-10} Torr. The electron-diffraction gun, the Auger analyzer, and the mass spectrometer can be used to study the layer characteristics during growth.

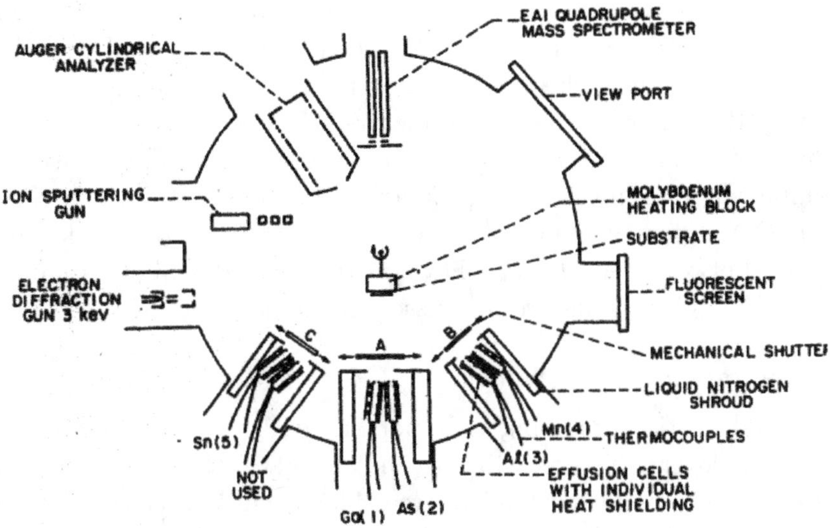

Figure 2.4: Schematic of a MBE growth chamber used for epitaxial growth of AlGaAs [75].

The sources Ga, Al, As, etc. are kept in independently heated effusion cells enclosed in shrouds containing liquid nitrogen. The dopants also have separate effusion cells. The effusion cells are heated to high enough temperature to have adequate beam flux rate on the substrate.

For the growth of GaAs on AlGaAs, the Ga flux is in the range of 10^{12}–10^{14} atoms/(cm^2-s). The flux needed depends on the sticking coefficient. For many group III elements, e.g. Al and Ga, the sticking coefficient is near unity. However, the sticking coefficients differ significantly for group V elements. Growth of high-quality InGaAsP layer on InP by MBE has been demonstrated using gas sources for As_2 and P_2 beams. The As_2 and P_2 beams are generated by decomposing AsH_3 and PH_3 in a heated chamber and then allowing them to leak out to the MBE growth chamber. The decomposition chambers are operated at 900–1200°C and are filled to a pressure of ~0.3–2 atm. Chemical beam epitaxy (CBE) is a variant of MBE [78]. In this technique, all sources are gaseous and are derived from groups III and V alkyls. The growth chamber is similar to that of an MBE system.

The most commonly used technique for growing a variety of semiconductors today is the MOCVD growth technique and the MBE growth technique. These techniques have been used to grow very thin (<10 nm) layers of semiconductors bounded by higher-gap semiconductors. They are called quantum well (QW) layers because they exhibit quantum mechanical properties due to small thickness. These thin QW layers can be grown even if there is slight lattice mismatch between the layers. They are then called strained quantum well (SQW) layers and are discussed in the next section.

2.2.2.5 *Strained layer epitaxy*

Although lattice matching is important for growth of high-quality epitaxial layers, it is possible to make high-quality semiconductor heterostructures with a small degree of lattice mismatch among them [79–85]. This lattice mismatch introduces a strain in the material; hence, these are called strained layer heterostructures. The strain alters the band structure in a favorable way so as to produce higher gain than in the absence of strain. Typical values of tolerable strain ($\Delta a/a$, where a is the lattice constant of the substrate and Δa is the difference in lattice constant between the epitaxial layer and that of the substrate) are less than 1.5%. The larger the amount of strain, the thinner is the layer that can be grown free of dislocations on the substrate. Beyond a certain thickness, known as critical thickness, large numbers of dislocation are generated. These dislocations reduce the luminescence property of the grown material.

InGaAsP layers have been grown on InP with both compressive and tensile strain. Generally, the grown heterostructures have very thin active regions, and they are often called SQW active regions. The light emission wavelength is in 1200–1600 nm range. The SQW layers are grown by the MOCVD growth process. The ternary semiconductor InGaAs with a composition of $In_{0.53}Ga_{0.47}As$ is lattice matched to InP. By varying the flow rate of tri-methyl In (TMIn), InGaAs layers with In fraction of <0.53 (tensile strain) and InGaAs layers with In fraction >0.53 (compressive strain) can be grown [79]. Another well-studied material system for strained layer epitaxial growth is

$In_{1-x}Ga_xAs$ grown over GaAs [86–90]. This system is used to make lasers emitting near 980 nm and for quantum dot (QD) structures.

2.2.2.6 *Growth by mass transport and self-assembly*

For QW structures, the continuous set of energies is related to carrier motion in two dimensions in the plane of the well, the carrier confinement in the other direction result in a discrete set of levels. For quantum wire (QWR) structures, the carrier motion is allowed only in one dimension, i.e. along the length of the wire. The carrier confinement in the other two directions results in a discrete set of levels. For QD structures, the carriers are confined in all three directions. Over the last decade, low-dimensional QW structures have found application in optoelectronic devices, such as lasers, amplifiers, and FETs. Lasers with quantum structures are expected to have improved performance characteristics, such as low threshold operation, low-temperature sensitivity, and low spectral width [91–98].

Many of these improved performances have been realized in QW structures. The advantages of QWRs are enhanced with respect to QWs because of their confinement in two dimensions. In addition, theory predicts that QDs being confined in all three dimensions should have even superior properties for laser applications. The superior property of QDs as it relates to semiconductor optical amplifiers and lasers are high-speed of response since the gain and phase recovery times are short compared to other structures [99].

For the growth of QW structures, thin layers (few nm thick) need to be grown on a planar substrate. This is accomplished by growth techniques, such as MOCVD or MBE discussed earlier. Growth times are adjusted so that the desired thickness is achieved. The typical growth rate in MBE process is $\sim 1\,\mu m/h$ and that for MOCVD process is $\sim 0.1\,\mu m/min$. The growth rate for LPE process is generally too large, $\sim 1\,\mu m/min$, to allow suitable reproducible growth of the thin layers needed for QW structures.

Many efforts have been devoted to the realization of QWR structures [91, 92]. These include self-organized growth, growth on a ridge structure, growth on step structures, and growth in V-grooves

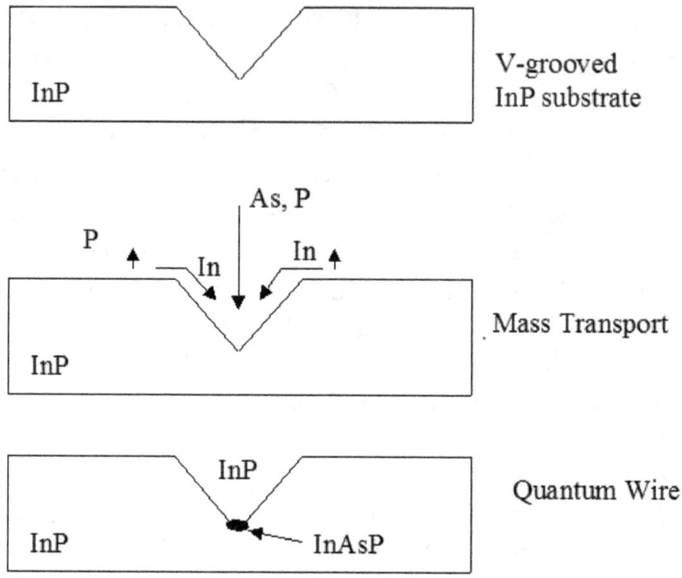

Figure 2.5: Schematic of the mass transport process for the formation of InAsP/InP QWRs.

[100–104]. The QWR lasers using InAsP/InP and InGaAs/InP material system have been fabricated using growth in V-grooves. Fabrication of QWR structure by mass transport is shown in Figure 2.5. V-grooves are etched on an InP substrate by wet chemical etching. The substrate is then heated in an MOCVD reactor to a temperature of ~600°C in the presence of As and P. No group III sources are present. Mass transport is believed to take place as follows: InP on the top edges of the ridge release P to form metallic In, which then migrates to the bottom to recapture the P. If there is As present in the ambient, it captures As at a certain rate. Thus, InAsP is formed at the bottom of the V-groove under certain partial pressure of As and P.

QWR lasers need very good current confinement, i.e. a large fraction of the injected current must flow through the active region and not around it. QWR lasers with semi-insulating Fe-doped InP current confining layer has been reported [105]. The schematic of a device fabricated using the InGaAsP/InP material system is shown

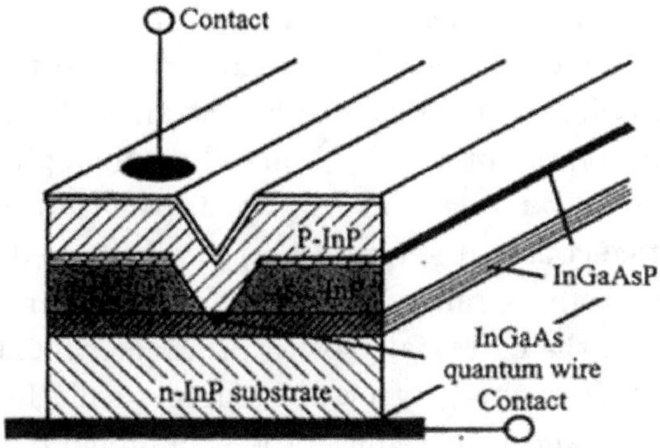

Figure 2.6: Structure of a QWR laser on a V-groove substrate [92, 105].

in Figure 2.6. First, a n-InGaAsP layer followed by a Fe:InP layer is grown on a (100) oriented n-InP substrate by MOCVD growth technique [105]. V-grooves are then etched on the wafer using a SiO_2 mask and wet chemical etching. Then, the undoped InGaAs QWR active region is grown in the tip of the V-groove. Thus is followed by the growth of a thin undoped InGaAsP layer, p-InP layer, and a p-InGaAs layer. The last layer serves as a contact layer for current injection. A modification of the structure, which includes ion implantation to create a high resistance in the p-InP layer (for better current confinement) has been reported [104].

In addition to V-grooves, QWR structures have been grown in ridges or side walls. Growth of InGaAs and InAlAs on the (100) oriented mesa has been reported. The growth experiments showed the exact shape (flat topped or pyramid shape) is important in determining the growth conditions for the growth of a particular material.

Lasers and optical amplifiers using QWR active regions generally have low gain due to difficulty in current confinement at high currents. In addition, the optical mode confinement factor is small. However, for the same reason they are likely to have high saturation powers. Although the advantages of low-dimensional structures (such as QD and QWR) were determined early on [93–95], the experimental challenge has been to fabricate QDs ($\sim$10 nm in dimension) of uniform size in a reproducible way.

There has been considerable work on fabrication of semiconductor nanostructures using the phenomenon of island formation during strained layer epitaxy, a process called Stranski–Krastanow growth mode. A formation of clusters is observed during epitaxial growth of a semiconductor layer (e.g. InGaAs) on top of another (e.g. GaAs) that has a lattice constant several (3–5) percent smaller. For the first few layers, the atoms arrange themselves in a planar layer called the wetting layer. As the epitaxial growth proceeds, the atoms bunch up to form clusters. The cluster growth is energetically favorable since it relaxes the strain and thus reduces the strain energy. Since the QDs appear spontaneously, the process is often called a self-assembly process. Although this process generally produces dots of different sizes, lasers and amplifiers with QD active regions have been fabricated [92]. The self-assembly process is observed in several material systems. For example, GaAs and InAs to form InGaAs QDs, InP and GaP to form InGaP QDs and Ge and Si which form GeSi QDs.

The advantages of self-assembly are: a large number of QDs can be formed in a single step without lithography, and they can be covered epitaxially by the host material without any interface defects. The size of the dots can be controlled within a certain range by

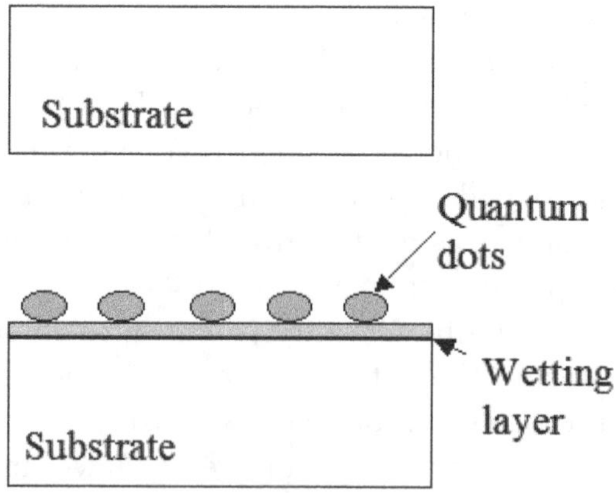

Figure 2.7: Schematic of QD formation. The top figure shows the planar substrate. The lower figure shows QDs on the wetting layer. The dots are covered in a high band gap material for laser or amplifier devices.

controlling the amount of deposited material. The schematic of QD formation is shown in Figure 2.7. The emission wavelength depends on the size of the dot, and thus the emission wavelength can be controlled to a certain extent for the same material system. All techniques of doping and energy band modifications from planar layer growth can, in principle, be adapted to QD growth by self-assembly.

2.3 Operating principles of LEDs

This section describes current injection, carrier confinement, and light emission in LEDs.

2.3.1 *p–n junction*

Under high current injection through a p–n junction a region near the depletion layer can have a high density of electrons and holes. These electrons and holes can recombine radiatively if the interfaces are free of traps. First, LEDs were demonstrated using a p–n homojunction, i.e. the material on both the p-side and n-side of the junction are the same. An important advance was made by the concept of heterojunctions, where the p–n junction is made out of two dissimilar materials [106]. Heterojunctions made CW laser operation and high brightness LEDs possible. The energy-band diagram of a p–n double heterojunction is shown in Figure 2.8. Since a p–n heterojunction is a p–n junction in a single crystal between two dissimilar semiconductors, the fabrication of heterojunctions had to wait until the development of epitaxial growth techniques. The energy-band diagram of a double heterostructure LED at zero bias is shown in Figure 2.8. The device consists of a narrow-gap semiconductor (p-type, n-type, or undoped) sandwiched between higher-gap p-type and n-type semiconductors.

The narrow-gap semiconductor is the light-emitting region (active region) of the laser. For the purpose of illustration, the active region is shown p-type (used in early GaAs lasers) in Figure 2.8, although that may not be the case for LEDs. The band diagram under forward bias is shown in Figure 2.8(b). Electrons and holes are injected at the heterojunction and are confined in the active

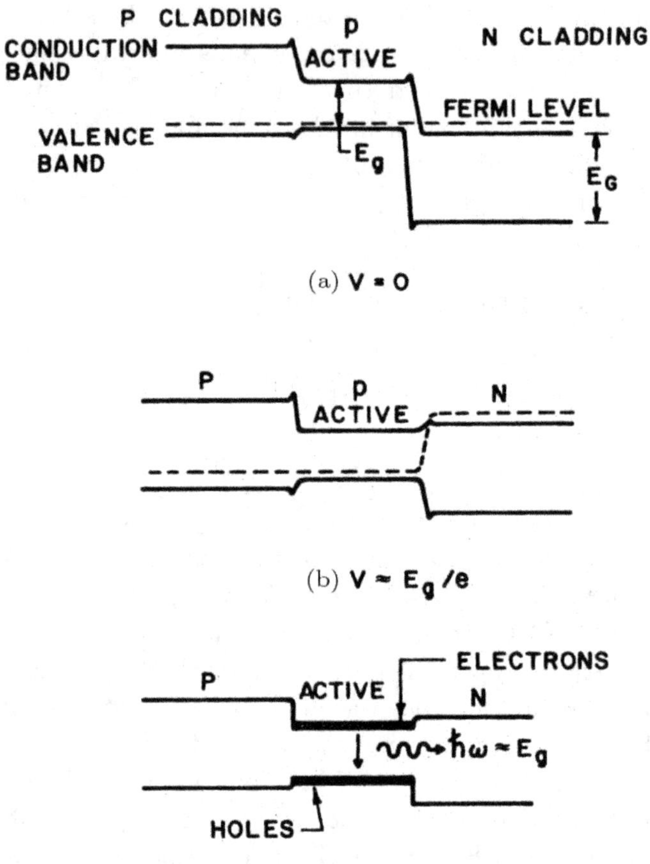

Figure 2.8: Energy band diagram of a double heterostructure (a) zero bias, (b) forward bias $V \sim E_g/e$, and (c) forward bias with electrons and holes. The dashed line represents the Fermi level.

region, where they recombine to produce light emission. For a double heterostructure (device with double heterojunction), the size of the region of optical emission is determined by the thickness of the lower band gap region [21–23] and hence more light emission is observed compared to a p–n homojunction, where the recombination takes place in the depletion layer (~100 Å thick). In addition, the refractive index of the lower-gap active region is higher than that for the n-type and p-type higher-gap confining (also known as cladding) layers. These three layers form a waveguide for the propagating optical mode, which is important for edge-emitting LEDs (also known as superluminescent diode) and lasers. This is discussed in Section 3.2 of Chapter 3. The current-voltage characteristics of a p–n double heterostructure LED is given by

$$I = I_s[\exp(\beta V_d) - 1] \tag{2.9}$$

where V_d is the voltage across the junction of the diode, I_s is the saturation parameter. The quantity $\beta = e/\eta kT$, where e is the charge of an electron, k is the Boltzmann constant and T is the temperature. The quantity η is called the ideality factor. $\eta \sim 2$ for ideal heterojunction LEDs and lasers. The measured voltage, $V = V_d + IR$, where R is the resistance (e.g. sum of layer resistances and contact resistance of the LED). Usually, $\exp(\beta V_d) \gg 1$ and the LED emits light for $V_d \sim E_g$, i.e. for a blue LED emitting at 400 nm, the voltage $V_d \sim 3.1V$. For $R = 2\Omega$ and $I = 100$ mA, this gives $V = 3.3$ V. This is further discussed in Appendix 5.

2.3.2 *Recombination and light emission*

The basis of light emission in semiconductors is the recombination of an electron in the conduction band with a hole from the valence band; the excess energy is emitted as a photon (light quantum). The process is called radiative recombination. The energy vs. wave-vector diagram of the electrons and holes in a cubic (zinc blende type) semiconductor is shown in Figure 2.9 [107]. For direct gap semiconductors, the bottom of the conduction band and the top of the valence band are at the same point in momentum space or $\vec{k}$ space ($\vec{k} = 0$ in Figure 2.9). This allows both energy and momentum conservation in the process of photon emission by electron–hole recombination. For indirect gap semiconductors (e.g. silicon), the momentum conservation can be achieved with the assistance of a phonon (lattice vibration), which significantly decreases the probability of radiative recombination. The valence band in many III–V semiconductors is

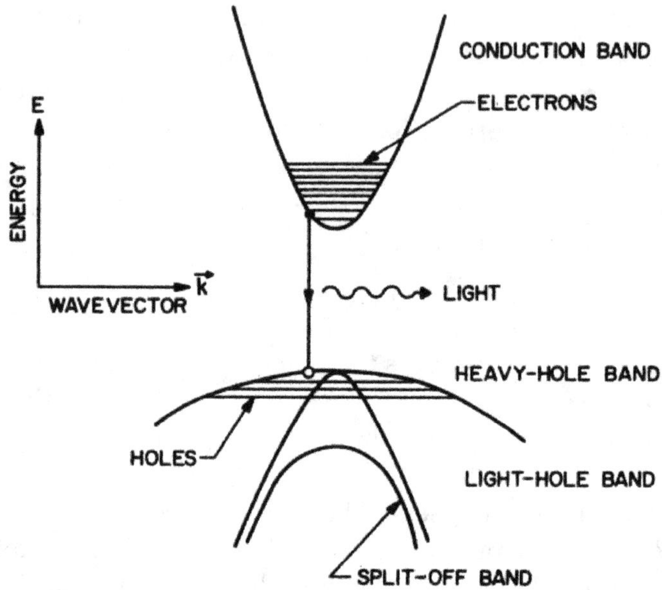

Figure 2.9: Energy vs. wavevector of the four major energy bands for a zinc blende type of direct gap semiconductor [107].

represented by three major subbands. They are the heavy hole band, the light hole band, and the spin-split off band. A heavy hole, as the name implies, has larger effective mass than a light hole. The total spontaneous radiative recombination rate R is given by [108–110, also Appendix 5].

$$R = \int r_{\text{spon}}(E)dE$$

where $r_{\text{spon}}(E)$ is the spontaneous emission rate at a photon energy E. It is given by [109, 110]

$$r_{\text{spon}}(E) = \frac{4\pi n e^2 E}{m_o^2 \varepsilon_0 h^2 c^3} \times \int_{-\infty}^{+\infty} \rho_c(E')\rho_v(E'')|M(E',E'')|^2 f(E')[1 - f(E'')]dE' \tag{2.10}$$

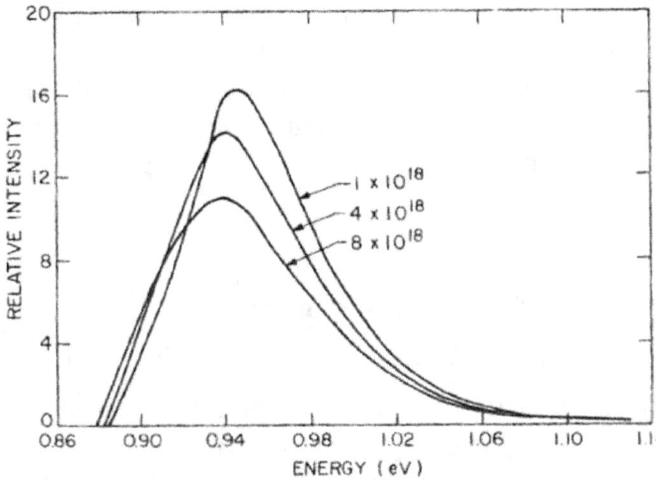

Figure 2.10: The calculated spontaneous emission spectrum for InGaAsP semiconductor with a band gap of 0.96 eV for p-type doping levels of 1×10^{18}, 4×10^{18}, and 8×10^{18} cm^{-3} [110]. These are typical doping levels.

where m_0 is the free-electron mass, h is the Planck's constant, e is the charge of electron, ε_0 is the permittivity of free space, E is the photon energy, n is the refractive index at energy E and ρ_c and ρ_v are the densities of state for unit volume per unit energy in the conduction and valence band, respectively. $f(E')$ is the probability that a state of energy E' is occupied by an electron and M is the effective matrix element between the conduction band state of energy E' and the valence band state of energy E''. The calculated spontaneous emission spectrum using Eq. (2.9) is shown in Figure 2.10. The parameters used are that for an InGaAsP semiconductor with a band gap of 0.96 eV (~1300 nm).

At unity quantum efficiency, the total spontaneous radiative recombination rate R equals the excitation rate (i.e. current injected). The latter is usually expressed as a nominal current density J_n [109, 110]:

$$J_n[A/(\text{cm}^2\mu\text{m})] = eR \tag{2.11}$$

An electron-hole pair can recombine non-radiatively, meaning that the recombination can occur through any process that does not emit a photon. The effect of non-radiative recombination on the performance of LEDs is to decrease the light output for a given current, i.e. the internal quantum efficiency is less than unity. The transition probability and hence the spontaneous recombination rate for parabolic band approximation is calculated in Appendix 5.

2.4 LEDs for infrared wavelengths

2.4.1 *GaAs-based LEDs*

AlGaAs/GaAs-based LEDs were the first to be studied after the demonstration of epitaxial growth of lattice matched AlGaAs on GaAs. The epitaxial growth involves growing n-GaAs buffer layer, followed by N-AlGaAs (Al mole fraction ~0.3), followed by p-GaAs or undoped-GaAs (light emitting region about 0.5 μm thick), followed by p-AlGaAs (Al mole fraction ~0.3), followed by p^+-GaAs layer. The last heavily-doped ($>5 \times 10^{18}$ cm^{-3}) GaAs layer serves as a contact (metal-semiconductor contact) layer. The heavy doping is needed to reduce the contact resistance. Sn is typically used as an n-type dopant and Zn is typically used as a p-type dopant. The substrate, i.e. the crystalline material on which the layers are grown, is typically sulfur-doped (n-type) single crystalline GaAs. The higher-gap P-AlGaAs and N-AlGaAs layers are generally called cladding layers and they serve to confine electrons and holes in the lower-gap active region (GaAs) for recombination and thereby enhanced light emission. A schematic of a GaAs LED is shown in Figure 2.11. A contact area (through which the current is injected) of about 50 μm diameter is made on the p^+-GaAs. For the rest of the area, the current flow is blocked by a SiO_2 layer, which is deposited on the wafer by a CVD technique. Since the GaAs substrate absorbs the emitted light, a well is etched on the substrate side of the wafer for light collection/emission as shown in Figure 2.11 [5, 111, 112]. Masking and etching needed for device fabrication are done by standard photolithographic techniques. The final devices are about 0.5×0.5 mm^2 in size. The LED chips are produced by saw cutting the wafer (i.e.

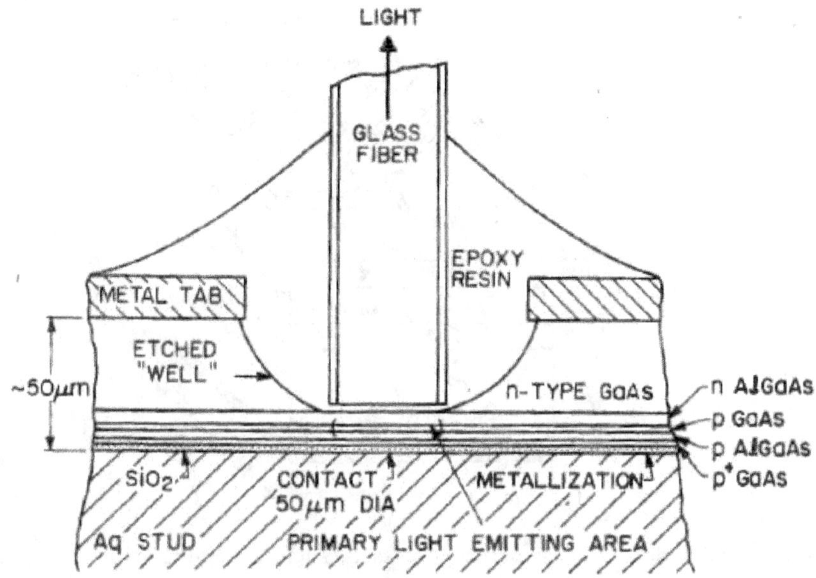

Figure 2.11: Schematic structure of a GaAs/AlGaAs surface-emitting LED [111].

substrate with all the grown layers and metal contacts). A multi-mode fiber (typically with 50 μm diameter core and 125 μm diameter cladding) can be used to collect the light as shown in Figure 2.11. The LED emits a few mW at $\sim$100 mA near 850 nm and has a broad spectrum about 50–70 nm wide.

For higher power operation and somewhat higher modulation bandwidth (discussed later), an edge-emitting LED has been fabricated [113, 114]. A schematic of such a device is shown in Figure 2.12. For this device, the layer growth, contact metallization, and device processing procedure is very similar to that described earlier. The wafer is separated into rectangular chips (devices), the side edges are saw cut, the light emitting edges may be cleaved and then antireflection coated so as not to produce laser action. Edge-emitting LED structure is very similar to that of a laser except that laser action does not take place. Early surface- and edge-emitting LEDs were fabricated used LPE growth technique. Current devices are

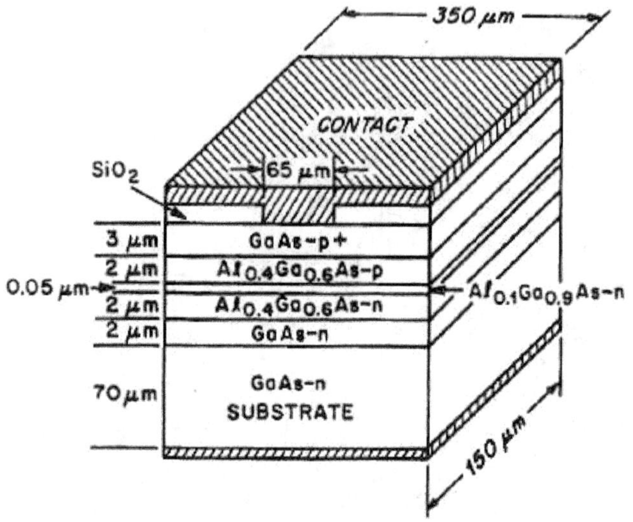

Figure 2.12: Schematic structure of a GaAs/AlGaAs edge-emitting LED [113].

generally fabricated by MOCVD growth technique. The modulation bandwidth of edge-emitting LEDs is generally higher than that of the surface-emitting LEDs. Also, LEDs driven at high current density have a higher modulation bandwidth due to shorter carrier lifetime [115]. The measured modulation bandwidth as a function of current density is shown in Figure 2.13. This is further discussed in Appendix 5.

2.4.2 *InP-based LEDs*

Optical fiber has lower optical loss at 1300 nm compared to that at 850 nm. This led to the development of InGaAsP/InP-based LEDs in the 1970s, which was soon followed by lasers near 1300 nm. The quaternary semiconductor $\mathrm{In}_{1-x}\mathrm{Ga}_x\mathrm{As}_y\mathrm{P}_{1-y}$ has the same lattice constant as binary semiconductor InP for $x \sim 0.45y$. Thus, epitaxial growth over InP is feasible. For $y = 0.6$, InGaAsP has a band gap of 0.96 eV (i.e. it emits near 1300 nm). The schematic of the layer structures of an InGaAsP/InP LED is shown in Figure 2.14. The layers are grown over an n-InP substrate. The first layer is n-InP, followed by undoped or lightly doped $\mathrm{In}_{1-x}\mathrm{Ga}_x\mathrm{As}_y\mathrm{P}_{1-y}$ (with $y = 0.6$,

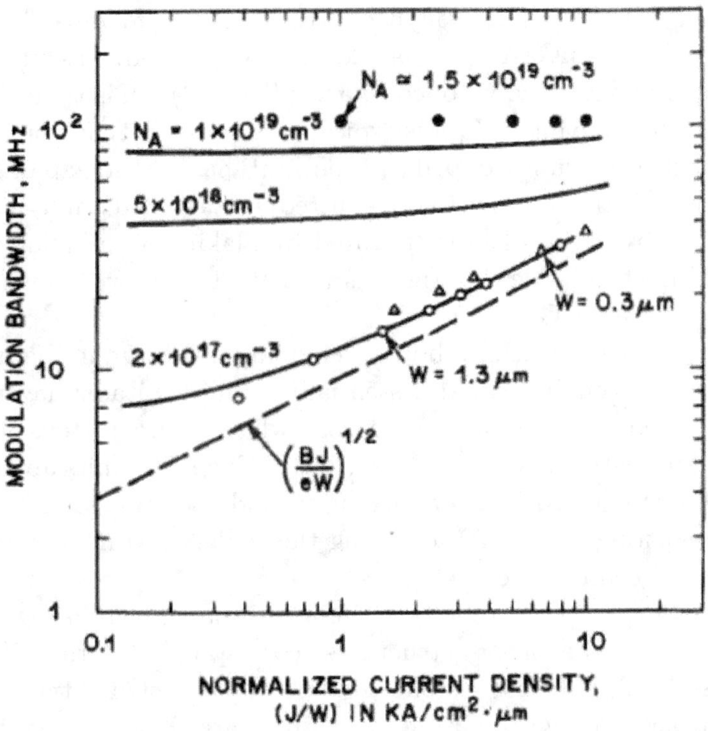

Figure 2.13: Modulation bandwidth as a function of normalized current density for GaAs/AlGaAs LEDs [115].

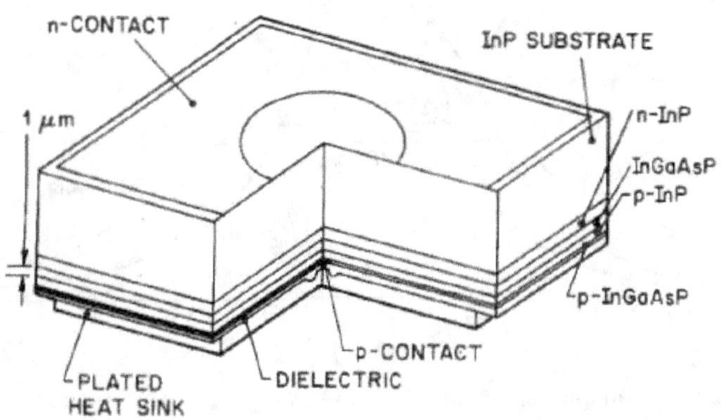

Figure 2.14: Schematic structure of an InGaAsP/InP surface-emitting LED [116].

$x \sim 0.45y$) active region (light emitting region), followed by p-InP, followed by p^+-InGaAs. The last layer serves as a low resistance contact layer and is heavily doped ($>5 \times 10^{18}$ cm^{-3}). Sn is typically used as an n-type dopant and Zn is typically used as a p-type dopant. The InP substrate is sulfur-doped (n-type). InP has a band gap of 1.35 eV and thus it is transparent to the emitted light at 1300 nm (~0.96 eV). The light from the LED is captured by making an annular contact on the substrate side, i.e. there is no need to remove the substrate as is the case for GaAs LEDs.

Both top emitting and bottom emitting (as shown in Figure 2.14) LEDs have been fabricated. Essentially, similar epitaxial layer structure is used in both cases with somewhat different steps for final device fabrication after materials growth. The LED emits about a few mW of power at 100 mA of current around 1300 nm. Edge-emitting LEDs emitting near 1300 nm using the InGaAsP/InP material system have been reported.

The device structure is very similar to that shown in Figure 2.12 except that the layer structures correspond to that for the InGaAsP/InP material system. The spectral emission of typical edge- and surface-emitting InGaAsP/InP LEDs are shown in Figure 2.15.

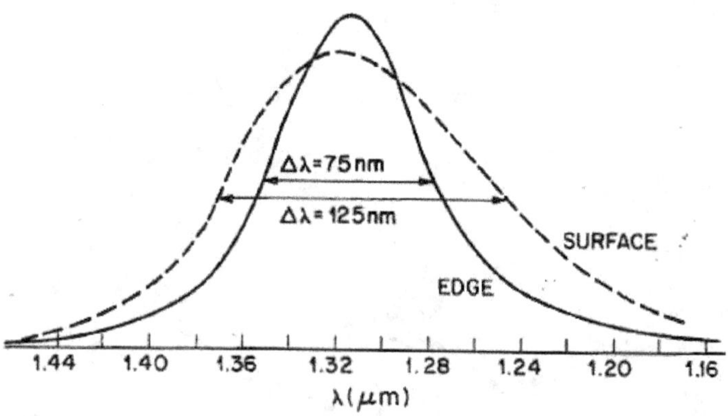

Figure 2.15: Output spectrum for InGaAsP/InP edge- (solid) and surface- (dashed) emitting LEDs. Note the edge-emitting LED has a narrower emission spectrum [115]. This is due to small amount of amplification as the light propagates.

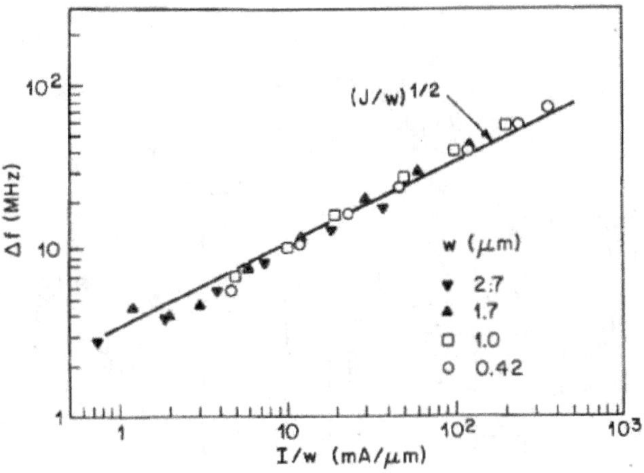

Figure 2.16: Modulation bandwidth as a function of normalized current density for InGaAsP/InP LEDs [117].

Modulation characteristics of InGaAsP LEDs have been reported [116, 117]. The data are shown in Figure 2.16. This is further discussed in Appendix 5.

2.5 LEDs for visible wavelengths

LEDs for visible wavelengths have been fabricated using different material systems. Using AlGaAs active layers with high Al fraction (<0.45), light emission near the red to orange range can be obtained. The materials GaP and InGaAlP have been used for yellow light emission. Also, II–VI-based semiconductors, such as ZnSe, ZnS, and ZnS_xSe_{1-x} have been used for blue and green light emission. A major advancement in white light LEDs took place with the demonstration of blue light emission from GaN, AlGaN, and InGaN material systems. Commercial LED used for white light generation is fabricated using this material system.

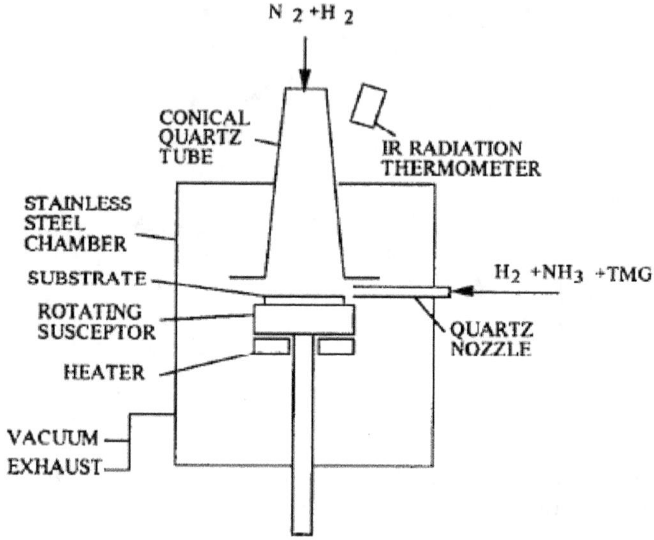

Figure 2.17: Schematic of an MOCVD reactor used for GaN growth [119].

2.5.1 *GaN-based LEDs*

The MOCVD reactor type used for original GaN growth is shown in Figure 2.17 [119]. The growth of GaN takes place on a sapphire substrate heated to ~1000°C. The active gaseous components are trimethylgallium (TMGa) and NH_3 in a hydrogen ambient atmosphere. A flow of N_2 and H_2 mixture is also used on top of the substrate. For the growth of AlGaN and InGaN needed for double heterostructure fabrication, trimethylindium (TMIn), trimethylaluminum (TMAl), and NH_3 are used as sources. For p-type doping of InGaN, Zn is used as a dopant (using diethylzinc (DEZn) source) and for p-type doping of AlGaN, Mg is used as a dopant [using bis-cyclopentadienyl magnesium (Cp2Mg) source]. The reactor used for the fabrication of double heterostructure LED and subsequently laser is similar to that shown in Figure 2.17. The basic chemical reaction for deposition is: $Ga(CH_3)_3 + NH_3 = GaN + 3CH_4$.

The schematics of two types of commercial AlGaN/InGaN LED are shown in Figure 2.18. The light emitting region is Zn-doped

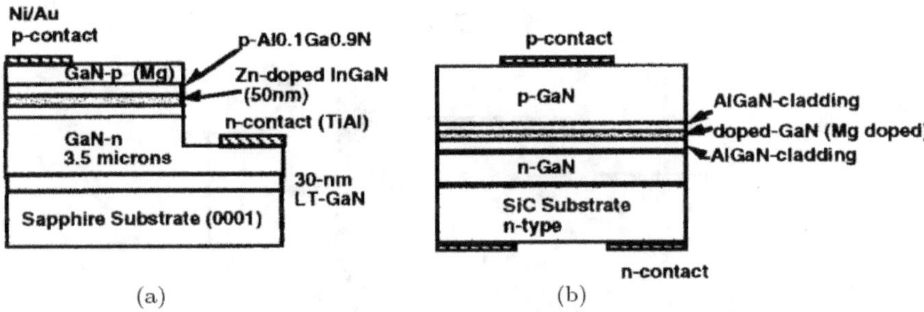

Figure 2.18: Two types of commercially available blue LEDs. (a) Laterally contacted devices with Zn-doped InGaN active region on sapphire substrates, and (b) vertically contacted devices on conducting SiC substrates [120].

InGaN. Since sapphire is insulating, the current contact was made by etching down to the GaN layer. About 2.5 mW of power was obtained for 40 mA of injected current [6]. The peak of the luminescence spectrum is at 450 nm and the full width at half maximum is 70 nm [6]. GaN/AlGaN/InGaN semiconductors (grown on GaN) are the leading materials for fabricating high-efficiency blue light emitters. MOCVD is the leading growth technology for fabricating these high-quality heterostructure-based devices.

2.6 Solid state lighting

The initial development of optoelectronic devices, in the 1970s and 1980s, such as optical fibers, semiconductor lasers, and photodetectors were carried out for optical transmission systems. Since then, the applications of optoelectronic devices have emerged in many different areas. The principal among them is the development of light emitting diodes for general lighting applications which has been made possible by the development of GaN-based LEDs in the visible wavelength range in the 1990s. This section covers the applications of LED devices along with rare earth-doped glass (Chapter 9) in solid state lighting.

The term solid-state lighting generally refers to "white" light emission using light emitting diodes. Such sources are quite a commonplace today, and have been made possible by the invention of

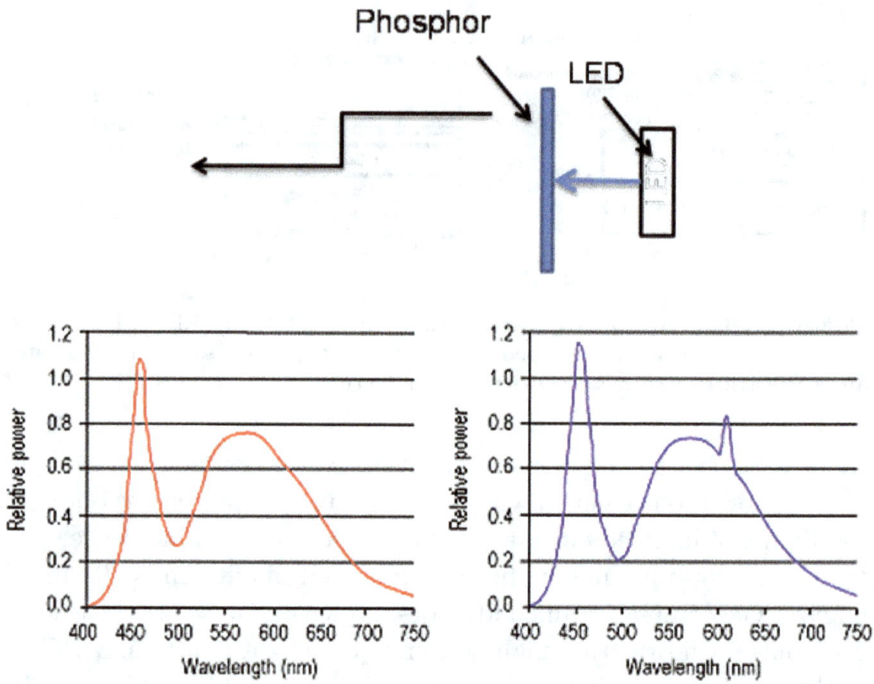

Figure 2.19: (Top) Schematic of blue LED (~450 nm) and Ce-doped garnet as a phosphor. (Bottom) Spectrum of the emitted light from two types of phosphor. The remaining blue LED light transmitted through phosphor is also included in the spectrum.

LEDs which emit blue light. An $In_xGa_{1-x}N/GaN$ heterostructure emits blue light. By changing the value of x (from ~0.02 to 0.2) emission wavelengths from ~390 to 440 nm can be obtained. Longer wavelength emission is obtained using higher x values. In addition, the wavelength can be varied by changing the QW thickness of the active region $In_xGa_{1-x}N$ for an $In_xGa_{1-x}N/GaN$ multiquantum well structure.

A configuration that is used for white light generation is to use a Ce-doped garnet as a phosphor with blue light from the LED incident on it. The schematic of such a configuration along with the spectrum of light emission is shown in Figure 2.19.

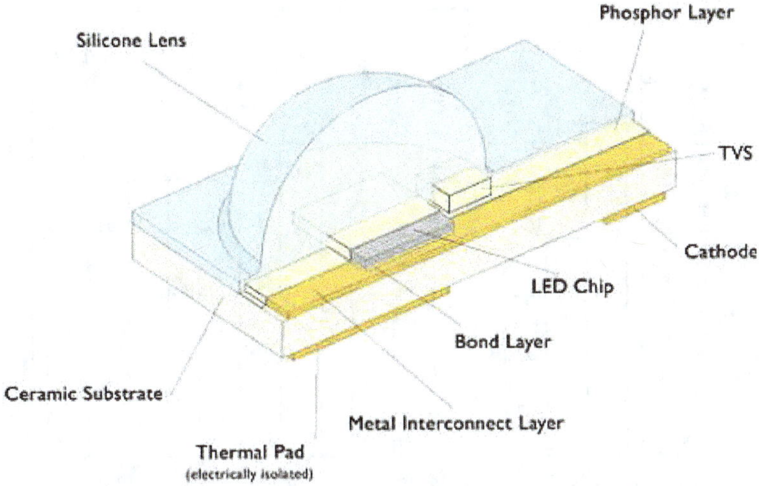

Figure 2.20: Schematic of a blue LED chip with phosphor next to it (Phillips Corp.). The lens helps collimate the light.

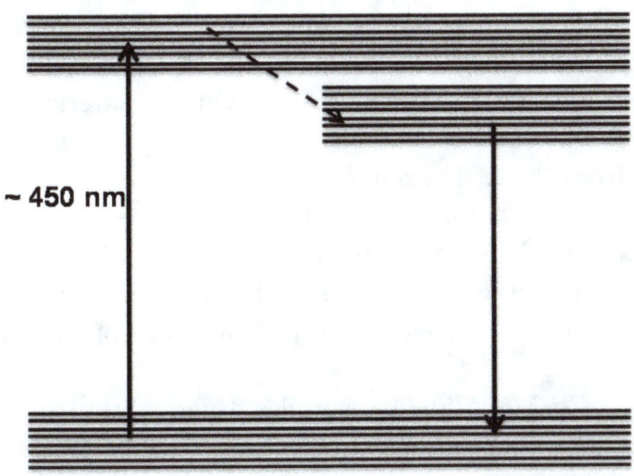

Figure 2.21: Schematic of energy levels of Ce in garnet. The atomic energy levels are broadened by crystal field.

The schematic of blue LED with phosphor for white light emission is shown in Figure 2.20. The LED is bonded to a ceramic substrate. Metal interconnect layers along with a silicone lens to collimate the white light emitted by phosphor is also shown.

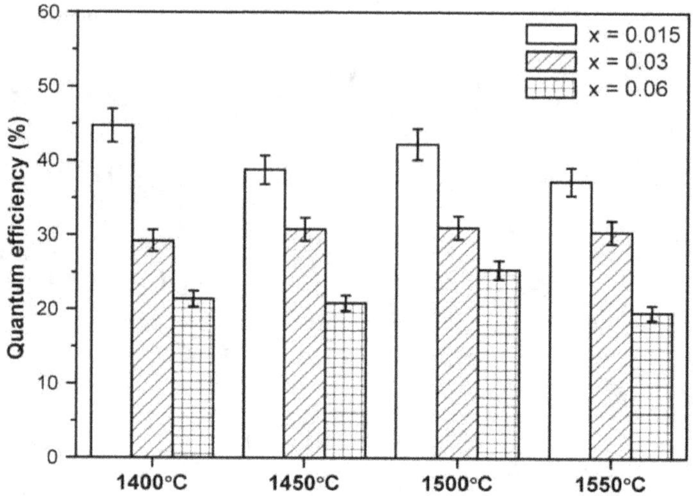

Figure 2.22: Quantum efficiency for different annealing temperature and Ce concentration are shown. Quantum efficiency of ∼40% is obtained.

The phosphor is very close to the LED. Modified versions of the schematic shown in Figure 2.20 have been commercialized for various applications. The energy levels of Ce in garnet material are shown in Figure 2.21. The energy levels are broadened by the crystal field. The blue light from the LED excites the Ce atoms to a high energy state from which they relax non-radiatively to an intermediate state. The transitions from this state result in a broad emission spectrum characteristic of the white light. One of the reasons for using Ce-doped garnet as a phosphor is high quantum efficiency of the luminescence process.

For $Y_{3-x}Ce_xMg_2AlSi_2O_{12}$ samples, the quantum efficiencies as a function of annealing temperature and Ce^{3+} concentration are shown in Figure 2.22. Quantum efficiency close to ∼40% is feasible. The emission spectrum depends on the exact composition of the garnet material and the annealing conditions.

The data are shown in Figure 2.23. The glass matrix is $Y_3Mg_2AlSi_2O_{12}$. The excitation wavelength is at 450 nm. The top figure is emission spectrum for Ce-doped $Y_{2.985}Ce_{0.015}Mg_2AlSi_2O_{12}$ annealed at different temperatures. The bottom figure is emission

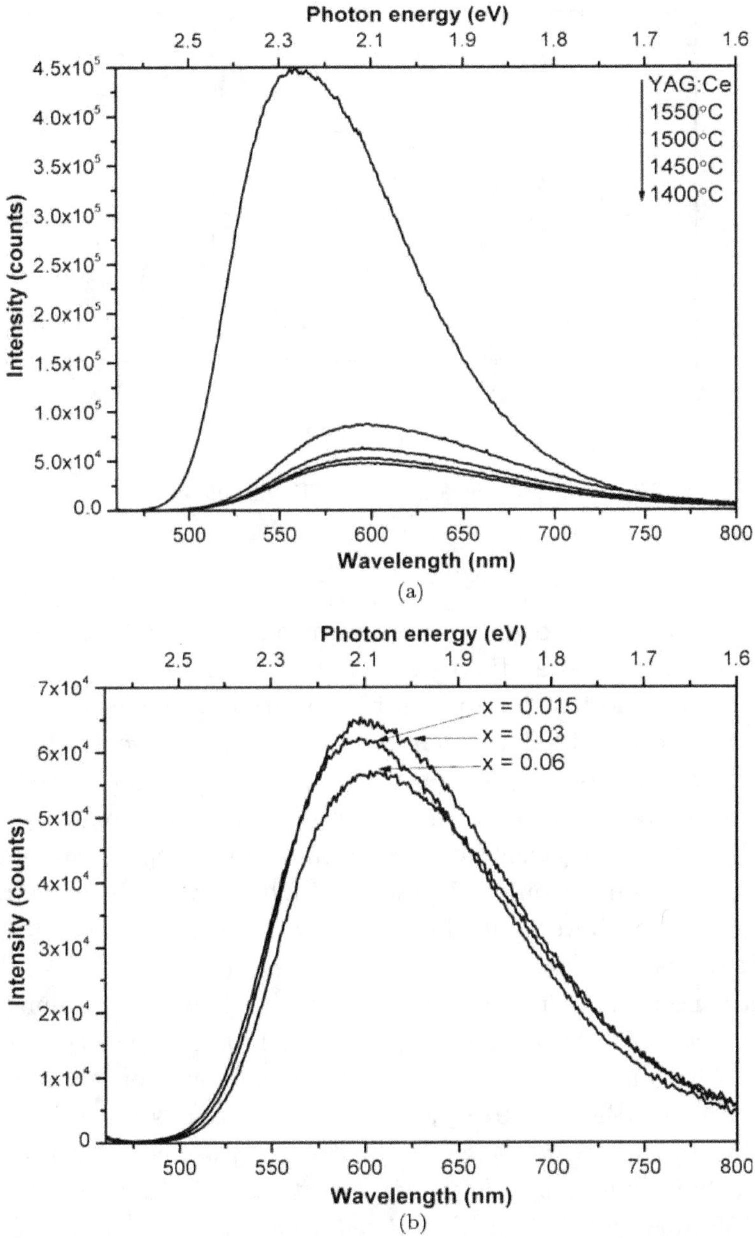

Figure 2.23: Emission spectrum for Ce-doped garnet (a) for different annealing temperatures as noted and (b) for different Ce concentrations annealed at 1500°C.

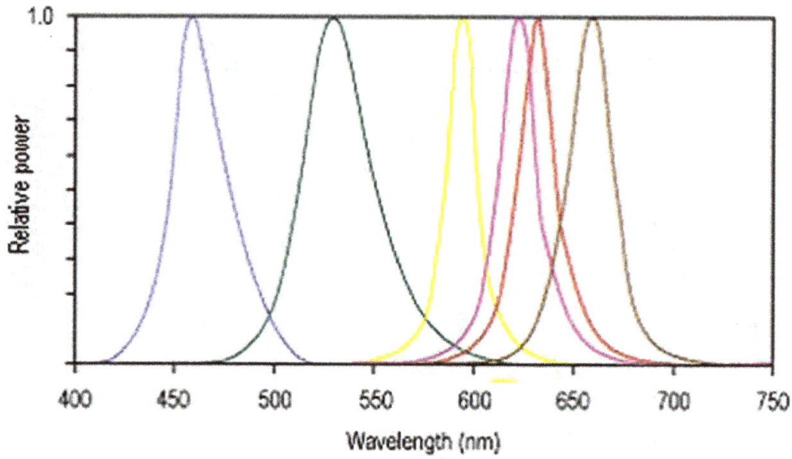

Figure 2.24: Spectrum of LEDs of different color.

spectrum as a function of Ce concentration in $Y_{3-x}Ce_xMg_2AlSi_2O_{12}$ phosphors annealed at 1500°C.

Another mechanism for light generation is by combining the output of several LEDs, which emit at different wavelengths. Spectral power distributions of several types of LEDs (fabricated using different materials) are shown in Figure 2.24. By locating red, green, and blue LEDs adjacent to one another, and properly mixing the amount of their output, the resulting light is white in appearance. Although the above could be made more efficient than phosphor-based LEDs, it is likely to be more expensive.

It is useful to compare the various white light sources along with the typical units used. Historical ($\sim$ from 1700 s) intensity of visible light is in candela. The 16th General Conference on Weights and Measures (CGPM) in 1979, defined candela (cd) as:

The candela is the luminous intensity, in a given direction, of a source that emits monochromatic radiation of frequency 540×10^{12} Hz (555 nm) *and that has an intensity in that direction of* 1/683 W per steradian. The old name of candela is foot-candle. Lumen is the generally used intensity measure of visible (white) light sources.

1 Lumen = 1 cd per sr, A full sphere has a solid angle of $4 \cdot \pi$ steradians
1 cd emitted in all directions = 12.56 Lumens
1 W per steradian = 683 Lumens
1 Lumen ~1/683 W ~1.5 mW/steradian

A typical commercial incandescent bulb rated at 40 W (of electrical power) produces 450 Lumens, i.e. ~11 Lumens/W. A typical commercial LED-based white light needs ~4–5 W of electrical power to produce 450 Lumens, i.e. ~110–90 Lumens/W. The laboratory record value for LED-based lighting is ~300 Lumens/W. The projected lifespan of a typical LED is 50,000 h compared to 1200 h for incandescent light bulbs.

Another measure of efficiency is defined as the luminous efficiency; it is the fraction of the total emitted radiation that falls within the wavelength response of a typical human eye. Luminous efficiency is given by

$$\frac{\int_0^\infty y_\lambda J_\lambda d\lambda}{\int_0^\infty J_\lambda d\lambda} \tag{2.12}$$

where y_λ is the spectral response of the eye and J_λ is the spectral response of the source. Spectral response of the human eye and that of a blackbody at different temperatures are shown in Figure 2.25. For an ideal monochromatic source at 555 nm, the luminous efficiency is 100%.

The luminous efficiency of a 40 W light bulb is ~2%. The luminous efficiency of sun at 5800°K is ~13% and that of a white LED is ~38%. The narrow spectral distribution of emitted radiation makes the white light LED more efficient for visible light use than an incandescent source.

LED is the first semiconductor infrared light source demonstrated in 1962. Since then, it has been used in many applications, such as indicator lights for appliances including telephones, and low data rate communications. The demonstration and development of GaN-based blue LEDs in the late 1990s have revolutionized the lighting industry. Blue LED along with Ce-doped garnet as a phosphor produces visible

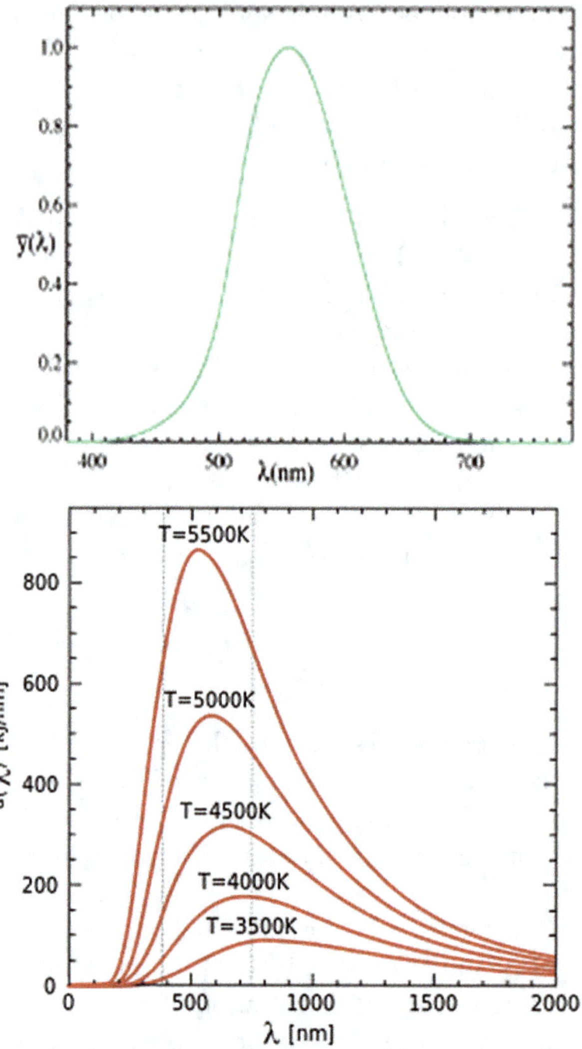

Figure 2.25: Spectral response of human eye (y_λ — top) and spectral distribution of emission, (J_λ — bottom) of a blackbody at different temperatures. The sun is at $T = 5800°$K.

white light. It is nearly seven times more efficient than incandescent light bulbs and has an operating life that is more than ten times larger than light bulbs. In addition, it needs only ~3 V for operation.

LED-based lighting is changing the way we light our homes, workplaces, factories, and is likely to fulfill all our lighting needs in the future.

References

1. N. Holonyak Jr. and S. F. Bevacqua, *Appl. Phys. Lett.*, **1** (1962) 82.
2. R. N. Hall, G. E. Fenner, J. D. Kingsley, T. J. Soltys, and R. O. Carlson, *Phys. Rev. Lett.*, **9** (1962) 366.
3. M. I. Nathan, W. P. Dumke, G. Burns, F. H. Dill, Jr., and G. Lasher, *Appl. Phys. Lett.*, **1** (1962) 63.
4. T. M. Quist, R. H. Rediker, R. J. Keyes, W. E. Krag, B. Lax, A. L. McWhorter, and H. J. Ziegler, *Appl. Phys. Lett.*, **1** (1962) 91.
5. T. P. Lee, C. A. Burrus, and R. H. Saul Chapter 4, in A. G. Chenoweyth and S. E. Miller (eds.), *Optical Fiber Communications II.*
6. S. Nakamura, T. Mukai, and M. Senoh, *Appl. Phys. Lett.*, **64** (1994) 1687.
7. H. Amano, *Progr. Cryst. Growth. Char. Mater.*, **62** (2016) 126.
8. S. Nakamura, M. Senoh, and T. Mukai, *Jpn. J. Appl. Phys.*, **32** (1993) L8.
9. H. Amano, M. Kito, K. Hiramatsu, and I. Akasaki, *Jpn. J. Appl. Phys.*, **28** (1989) L2112.
10. I. Akasaki, H. Amano, Y. Koide, K. Hiramatsu, and N. Sawaki, *J. Cryst. Growth.*, **98** (1989) 209.
11. H. Kressel and J. K. Butler, *Semiconductor Lasers and Heterojunction LEDs*, Academic Press, New York, 1977.
12. G. P. Agrawal and N. K. Dutta, *Semiconductor Lasers*, 2nd Edn., Van Nostrand Reinhold, NY, 19861992.
13. H. C. Casey, Jr. and M. B. Panish *Heterostructure Lasers*, Academic Press", New York, 1978; G. Steude, D. M. Hofmann, B. K. Meyer, H. Amano, I Akasaki, *Physica Status Solidi (b)*, **205** (1998) R7.
14. L. A. Coldren and S. W. Corzine, *Diode Lasers and Photonic Integrated Circuits*, John Wiley, NY, 1995.
15. G. H. B. Thompson, *Physics of Semiconductor Laser Devices*, John Wiley & Sons, New York, 1980.
16. G. P. Agrawal and N. K. Dutta, *Semiconductor Lasers*, Section 2.5, Van Nostrand Reinhold, NY, 1992.
17. A. K. Dutta, N. K. Dutta, and M. Fujiwara, *WDM Technologies — Optical Networks*, Academic Press, NY, 2002.
18. R. Ramaswami and K. N. Sivarajan, *Optical Networks*, Academic Press, 2002.
19. J. D. Gibson, *Communications*, Handbook CRC Press, 1997.
20. I. P. Kaminow and T. Li, *Optical Fiber Telecommunications*, Academic Press, San Diego, CA, 2002.
21. H. Nelson, *RCA Rev.*, **24** (1963) 603.
22. W. F. Finch and E W. Mehal, *J. Electrochem Soc.*, **111** (1964) 814.

23. A. Y. Cho, *J. Vac. Sci. Technol.*, **8** (1971) 531.
24. R. D. Dupuis, *J. Crys. Growth*, **55** (1981) 213.
25. J. M. Woodall, H. Rupprecht, and W. Reuter, *J. Electrochem. Soc.*, **116** (1969) 899.
26. H. Nelson, U.S. Patent 3,565,702 (1971).
27. I. Hayashi, M. B. Panish, P. W. Foy, and S. Sumski, *Appl. Phys. Lett.*, **17** (1970) 109.
28. M. Feng, L. W. Cook, M. M. Tashima, and G. E. Stillman, *J. Electron. Mater.*, **9** (1980) 241.
29. I. Ladany and F. Z. Hawrylo, *J. Cryst. Growth*, **54** (1981) 69.
30. J. J. Hsieh, *Appl. Phys. Lett.*, **28** (1976) 283.
31. R. J. Nelson, *Appl. Phys. Lett.*, **35** (1979) 654.
32. J. J. Hsieh, *J. Cryst. Growth*, **27** (1974) 49.
33. D. L. Rode, *J. Cryst. Growth*, **20** (1973) 13.
34. W. A. Tiller, *J. Cryst. Growth*, **2** (1968) 69.
35. M. B. Small and I. Crossley, *J. Cryst. Growth*, **27** (1974) 35.
36. R. L. Moon, *J. Cryst. Growth*, **27** (1974) 62.
37. P. Besomi, J. Degami, N. K. Dutta, W. R. Wagner, and R. J. Nelson, *J. Appl. Phys.*, **56** (1984) 2879.
38. A. Doi, N. Chin One, A. Aiki, and R. Ito, *Appl. Phys. Lett.*, **34** (1979) 393.
39. M. A. DiGuiseppe, H. Temkin, and W. A. Bonner, *J. Cryst. Growth*, **58** (1982) 279.
40. A. R. Clawson, W. Y. Lum, and G. W. McWilliams, *J. Cryst. Growth*, **46** (1979) 300.
41. G. A. Antypas, *Appl. Phys. Lett.*, **37** (1980) 64.
42. P. Besomi, R. B. Wilson, W. R. Wagner, and R. J. Nelson, *J. Appl. Phys.*, **54** (1983) 535.
43. G. A. Antypas and R. L. Moon, *J. Electrochem. Soc.*, **120** (1973) 1574.
44. G. A. Antypas and J. Edgecumbe, *J. Cryst. Growth*, **34** (1976) 132.
45. K. Nakajima, T. Tanahashi, K. Akita, and T. Yamaoka, *J. Appl. Phys.*, **50** (1979) 4975.
46. K. Nakajima, T. Kusunoki, K. Akita, and T. Kotani. *J. Electrochem. Soc.*, **125** (1978) 123.
47. R. E. Nahory, M. A. Pollack, W.D. Johnston Jr., and R. L. Barns, *Appl. Phys. Lett.*, **33** (1978) 659.
48. P. Chandra, L. A. Coldren, and K. E. Strege, *Electron Lett.*, **17** (1981) 6.
49. I. Hayashi, M. B. Panish, P. W. Foy, and S. Sumski, *Appl Phys. Lett.*, **17** (1970) 109.
50. G. H. Olsen, Chapter 1, in T. P. Pearsall (ed.), *GaInAsP Alloy Semiconductors*, John Wiley & Sons, New York, 1982.
51. J. J. Tietjen and 1. A. Amick, *J. Electrochem. Soc.*, **113** (1966) 724.
52. G. H. Olsen and T. J. Zamerowski, *IEEE J. Quant. Electron.*, **QE-17** (1981) 128.
53. H. Enda, *Jpn. J. Appl. Phys. Part 1*, **18** (1979) 2167.

54. R. E. Enstrom, C J. Nuese, J. R. Appert, and N. J. DiGuiseppe, *Electrochem. Soc. Ext. Abstr.*, **78–1** (1978) 504.
55. S. S. Hyder, R. R. Saxena, and C. B. Cooper, *Appl. Phys. Lett.*, **34** (1979) 584.
56. H. A. Seki, Konkitu, and M. Matsumura, *J. Cryst. Growth*, **54** (1981) 615.
57. W. D. Johnston Jr. and K. E. Strege, *38th Annual IEEE Device Research ConI Abstr.*, Vol. IVB-3, Cornell University, Ithaca, NY, 1980.
58. T. Mizutani, M. Yoshida, A. Usai, H. Watanabe, T. Yuasa, and I. Hayashi, *Jpn. J. Appl. Phys. Part 2*, **19** (1980) L113.
59. H. M. Manasevit and W. I. Simpson, *J. Electrochem. Sac.*, **116** (1969) 1725.
60. Y. Seki, K. Tanno, K. Iida, and E. Ichiki, *J. Electrochem. Soc.*, **122** (1975) 1108.
61. R. D. Dupuis, P. D. Dapkus, R. D. Yingling, and L. A. Moudy. *Appl. Phys. Lett.*, **31** (1978) 201.
62. R. D. Dupuis, *J. Cryst. Growth*, **55** (1981) 213.
63. P. D. Dapkus, *Annu. Rev. Mater. Sci.*, **12** (1982) 243.
64. J. P. Hirtz, M. Razeghi, M. Bonnet, and J. P. Duchemin, Chapter 3 in T. P. Pearsall (ed.), *GalnAsP Alloy Semiconductors*, John Wiley & Sons, New York, 1982; S. Hersee and J. P. Duchemin. Annu. *Rev. Mater. Sci.*, **12** (1982) 65.
65. J. P. Hirtz, J. P. Larivain, J. P. Duchemin, T. P. Pearsall, and M. Bonnet, *Electron. Lett.*, **16** (1980) 415.
66. J. P. Hirtz, J. P. Larivain, D. Leguen, M. Razeghi, and J. P. Duchemin, in *Gallium Arsenide and Related Compounds 1980*, Conference Series Vol. 56, Institute of Physics. Bristol, 1981, pp. 29–35.
67. J. P. Hirtz, M. Razeghi, J. P. Larivain, S. Hersee, and J. P. Duchemin, *Electron. Lett.*, **17** (1981) 113.
68. M. Razeghi, S. Hersee, J. P. Hirtz, R. Blondeau, B. de Cremoux, and I. P. Duchemin, *Electron. Lett.*, **19** (1983) 336.
69. T. Fukui and Y. Horikoshi, *Jpn. J. Appl. Phys. Part 1*, **19** (1980) 395.
70. C. B. Cooper, R. R. Saxena, and M. J. Ludowise, *Electron. Lett.*, **16** (1980) 892.
71. J. Yoshino, T. Lawnot, and H. Kukimoto, *J. Cryst. Growth*, **55** (1981) 74.
72. J. R. Arthur, *J. Appl. Phys.*, **39** 4032 (1968).
73. J. R. Arthur and J. J. LePore, *J. Vac. Sci. Technol.*, **6** (1969) 545.
74. A. Y. Cho, *J. Vac. Sci. Technol.*, **8** S31 (1971).
75. A. Y. Cho, *J. Appl. Phys.*, **41** (1970) 2780.
76. A. Y. Cho, *J. Appl. Phys.*, **42** (1971) 2074.
77. A. Y. Cho and J. R. Arthur, in J. O. McCladin and G. Somorjai (eds.), *Progress in Solid State Chemistry*, Vol. 10, Pergamon Press, New York, 1975, p. 157.
78. W. T. Tsang, *Appl. Phys. Lett.*, **45** (1984) 1234.
79. T. Tanbun-Ek, R. A. Logan, H. Temkin, S. N. G. Chu, N. A. Olsson, A. M. Sergent, and K. W. Wecht, *IEEE J. Quant. Electron.*, **QE-26** (1990) 1323.
80. G. C. Osbourne, *J. Vac. Sci. Technol.*, **B1** (1983) 379.

81. W. T. Massalink, T. Henderson, J. Klem, R. Fischer, P. Pearah, H. Morkoc, M. Hafich, P. D. Wang, and G. Y. Robinson, *Appl. Phys. Lett.*, **45** (1984) 1309.
82. S. E. Fischer, D. Fekete, G. B. Feak, and J. A. Ballantyne, *Appl. Phys. Lett.*, **50** (1984) 714.
83. J. W. Matthews and A. E. Blakeslee, *J. Cryst. Growth*, **27** (1974) 118.
84. T. Tanbun-Ek, R. A. Logan, H. Temkin, S. N. G. Chu, N. A. Olsson, A. M. Sergent, and K. W. Wecht, *IEEE J. Quant. Electron.*, **QE-26** (1990) 1323.
85. P. R. Berger, K. Chang, P. Bhattacharya, J. Singh, and K. K. Bajaj, *Appl. Phys. Lett.*, **53** (1988) 684.
86. S. Uehara, Chapter 8, in S. Shimada and H. Ishio (eds.), *Optical Amplifiers and Their Applications*, John Wiley, New York, 1994.
87. G. W. Yang, J. Y. Xu, Z. T. Xu, L. H. Chen, and Q. M.Wang, *J. Appl. Phys.*, **83** (1998) 8.
88. G. Lin, S. Yen, C. Lee, and D. Liu, *IEEE Photon. Technol. Lett.*, **8** (1996) 1588.
89. M. C. Wu, Y. K. Chen, M. Hong, J. P. Mannaerts, M. A. Chin, and A. M. Sergent, *Appl. Phys. Lett.*, **59** (1991) 1046.
90. K. Shigihara, K. Kawasaki, Y. Yoshida, S.i Yamamura, T. Yagi, and E. Omura, *IEEE J. Quant. Elect.*, **38** (2002) 1081.
91. E. Kapon, Chapter 10, in P. Zory (ed.), *Quantum Well Lasers* Academic Press, NY, 1993.
92. V. M. Ustinov, A. E. Zhukov, A. Yu. Egorov, and N. A. Maleev, *Quantum Dot Lasers*, Oxford University Press, Oxford, 2003.
93. Y. Arakawa, H. Sasaki, *Appl. Phys. Lett.*, **40** (1982) 939.
94. M. Asada, Y. Miyamoto, and Y. Suematsu, *IEEE J. Quant. Elect.*, **QE-22** (1986) 1915.
95. A. Yariv, *Appl. Phys. Lett.*, **53** (1988) 1033.
96. Y. Miyamoto, Y. Miyake, M. Asada, and Y. Suematsu, *IEEE J. Quant. Elect.*, **QE-25** (1989) 2001.
97. M. Asada, Y. Miyamoto, and Y. Suematsu, *Jpn. J. Appl. Phys.*, **24** L95 (1985).
98. M. Cao, Y. Miyake, S. Tamura, H. Hirayama, S. Arai, Y. Suematsu, and Y. Miyamoto, *Trans. IEICE*, **E73** (1990) 63.
99. M. Sugawara, T. Akiyama, N. Hatori, Y. Nakata, H. Ebe, and H. Ishikawa, *Meas. Sci. Technol.*, **13** (2002) 1683; M. Sugawara, H. Ebe, N. Hatori, M. Ishida, Y. Arakawa, T. Akiyama, K. Otsubo, and Y. Nakata, *Phys. Rev. B.*, **69** (2004) Art. No. 235332.
100. M. Ishikawa, W. Pan, Y. Kaneko, H. Yaguchi, K. Onabe, R. Ito, and Y. Shiraki, *Jpn. J. Appl. Phys.*, **37** (1998) 1556.
101. T. Toda and Y. Nakano, *Proc. 11th Int. Conf. InP and Related Materials*, Davos, Switzerland, May 16–20, (1999), p. 17.
102. S. Tiwari and G. D. Petiti, *Appl. Phys. Lett.*, **64** (1994) 3536.
103. E. Kapon, D. M. Hwang, and R. Bhat, *Phys. Rev. Lett.*, **63** (1989) 430.

104. E. Kapon, S. Simhony, R. Bhat, and D. M. Hwang, *Appl. Phys. Lett.*, **55** (1989) 2715.
105. D. Piester, P. Bonsch, T. Schrimpf, H. Wehmann and A. Schlachetzki, *IEEE J. Select. Top. Quant. Electron.*, **6** (2000) 511.
106. H. Kroemer, *Proc. IEEE*, **51** (1963) 1782.
107. E. O. Kane, *J. Phys. Chem. Solids*, **1** (1957) 249.
108. M. G. A. Bernard and G. Duraffourg, *Phys. Stat. Solidi.*, **1** (1961) 699.
109. H. C. Casey, Jr. and M. B. Panish, *Heterostructure Lasers*, Part A, Chapter 3, Academic, New York, 1978.
110. G. P. Agrawal and N. K. Dutta, Chapter 3, in Van Nostrand (ed.), *Semiconductor Lasers*, 2nd Edn., Reinhold, NY, 1992.
111. C. A. Burrus and B. J. Miller, *Opt. Comm.*, **4** (1971) 307.
112. C. A. Burrus and R. W. Dawson, *Appl. Phys. Lett.*, **17** (1970) 97.
113. M. Ettenberg, H. Kressel, and J. P. Wittke, *IEEE J. Quant. Electron.*, **12** (1976) 360.
114. T. P. Lee, C. A. Burrus, and B. I. Miller, *IEEE, J. Quant. Electron.*, **9** (1978) 820.
115. T. P. Lee and A. G. Dentai, *IEEE J. Quant. Electron.*, **14** (1978) 150.
116. A. G. Dentai, T. P. Lee, and C. A. Burrus, *Electron. Lett.*, **13** (1977) 484.
117. O. Wada, S. Yamakoshi, M. Abe, Y. Yishitoni, and T. Sakurai, *IEEE J. Quant. Electron.*, **QE-17** (1981) 174.
118. O. Wada, H. Hamaguchi, Y. Nishitani, and T. Sakurai, *IEEE Electron Dev. Lett.*, **EDL-3** (1982) 129.
119. S. Nakamura, Y. Harada, and M. Seno, *Appl. Phys. Lett.*, **58** (1991) 2021.
120. S. Denbaars, *Proc. IEEE*, **85** (1997) 1740.

Chapter 3

Semiconductor Laser — I

3.1 Introduction

Significant advances in research results, development, and application of optical sources have occurred over the last three decades. The two primary optical sources used in telecommunications are the semiconductor laser and the light-emitting diode (LED). The LEDs are used as sources for low-data rate (<200 Mb/s) and short-distance applications, and lasers are used for high-data rate and long-distance applications. The fiber optic revolution in telecommunication which provided several orders of magnitude improvement in transmission capacity at low cost would not have been possible without the development of reliable semiconductor lasers. Semiconductor injection lasers continue to be the laser of choice for various system applications primarily because of their small size, simplicity of operation, and reliable performance. For most transmission system applications, the laser output is encoded with data by modulating the current. However, for many high-data rate applications, which require long-distance transmission, external modulators are used to encode the data.

Chapters 3 and 4 describe the fabrication, performance characteristics, current state of the art, and research directions for semiconductor lasers. The focus of this chapter is laser sources needed for fiber optic transmission systems. These devices are fabricated using the InP material system. For early work and thorough discussion of semiconductor lasers, see Refs. [1–4].

The semiconductor injection laser was invented in 1962 [5–7]. With the development of epitaxial growth techniques and the subsequent fabrication of double heterojunction, laser technology advanced rapidly in the 1970s and 1980s [1–4]. The demonstration of CW operation of the semiconductor laser in the early 1970s [8] was followed by an increase in development activity in several industrial laboratories. This intense development activity in the 1970s was aimed at improving the performance characteristics and reliability of lasers fabricated using the AlGaAs material system [1]. These lasers emit near 800 nm and were deployed in early optical fiber transmission systems (in the late 1970s and early 1980s).

The optical fiber has zero dispersion near 1300 nm wavelength and has lowest loss near 1550 nm wavelength. Thus, semiconductor lasers emitting near 1300 and 1550 nm are of interest for fiber optic transmission applications. Lasers emitting at these wavelengths are fabricated using the InGaAsP/InP materials system and were first fabricated in 1976 [9]. Much of the fiber optic transmission systems around the world that are in use or are currently being deployed utilize lasers emitting near 1300 or 1550 nm.

3.1.1 *Semiconductor laser fundamentals*

A typical laser (shown in Figure 3.1) has a medium which provides optical gain and a set of mirrors on each side of the gain medium which form a Fabry–Perot resonator and provide optical feedback. The two together form the optical cavity of the laser. The light generated by spontaneous emission in the gain medium is amplified by the gain medium (the atoms emit by stimulated emission normal to the mirrors) and it bounces between the mirrors to form the laser light. A fraction of this light is transmitted by the mirrors and is observed in the form of laser emission. Let the optical cavity be in the z-direction. Then in the plane wave approximation, the (assuming linear polarization) electric field of the laser emission along $+z$

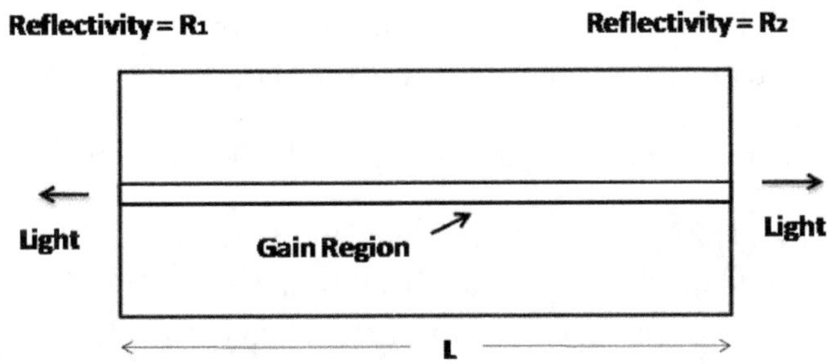

Figure 3.1: Schematic of a typical laser. R_1 and R_2 are reflectivity of the mirrors which provide optical feedback. The laser emission is normal to the optical cavity along the z-direction.

is given by

$$E(z,t) = E_0 \exp[i(\omega t - nk_0 z)] = E_0 \exp[i(\omega t - n_R k_0 z)] \exp(n_I k_0 z) \tag{3.1}$$

where ω is the frequency ($\omega = 2\pi\nu$) of the light, $k_0 = 2\pi/\lambda_0$, where λ_0 is the wavelength in free space, and $n = n_R + in_I$, where n_R and n_I are the real and imaginary part of the refractive index of the gain medium, respectively.

The change in intensity, $I(z) = |E(z)|^2$ as a function of travel distance (δ z) in the gain medium is

$$I(z + \delta z) = I(z) \exp(2n_I k_0 \delta z) \tag{3.2}$$

The quantity $2n_I k_0 = g_n$, where g_n is the net gain as the wave travels in the gain medium. The net gain is the difference between the gain created by optical transitions (g) and optical losses (absorption due to other materials in the gain medium) and scattering of laser light in the gain medium. This combined optical loss is noted as α. A fraction of the optical mode (known as confinement factor, Γ) of the laser is confined in the active region of the laser (see Section 3.2). Thus, the optical gain of the laser mode is Γg, where g is the material gain. Thus, net gain $g_n = \Gamma g - \alpha$. In steady state, round-trip gain must equal round-trip loss. The gain when the laser begins laser action is known as gain at threshold (g_{th}). The threshold gain (g_{th}) is given by

$$R_1 R_2 \exp[2(\Gamma g_{\text{th}} - \alpha)L] = 1 \tag{3.3}$$

where R_1 and R_2 are reflectivity of the mirrors and L is the cavity length. Equation (3.3) may be rewritten as

$$\Gamma g_{\rm th} = \alpha + (1/2L)\ln(1/(R_1R_2)) = \alpha + \alpha_m \tag{3.4}$$

where $\alpha_m = (1/2L)\ln(1/(R_1R_2))$ is also known as "mirror losses". It essentially reflects the optical loss due to laser output. In addition, in steady state, after one round trip in the cavity, the electric field must be identical to the initial electric field, i.e. $E(z) = E(z+2L)$, where L is the length of the optical cavity. This leads to another condition for laser action, i.e.

$$2n_Rk_0L = 2m\pi \tag{3.5}$$

where m is an integer. The above equation says that the round-trip change in phase must be an integer multiple of 2π which is the condition for Fabry–Perot resonance.

Thus, the Fabry–Perot modes of the optical cavity are the laser wavelengths, λ_0 ($k_0 = 2\pi/\lambda_0$). The laser operates at specific discrete wavelengths determined by Eq. (3.5). They are called the longitudinal modes of the laser. From Eq. (3.5), the longitudinal mode spacing (for large m) is given by

$$\Delta\lambda = \lambda^2/(2n_RL) \tag{3.6}$$

For a typical gain medium, the optical gain occurs over a small range of wavelengths. The gain as a function of wavelength is called gain spectrum. A typical laser operates at several longitudinal modes (discrete wavelengths separated by $\Delta\lambda$) centered on the peak of the gain spectrum. This is shown schematically in Figure 3.2. For a semiconductor laser gain medium, the gain increases with increasing current (I). It can be approximated as

$$g = b(I/I_0 - 1) \tag{3.7}$$

where $I_0 \sim 1$ kA/cm^2 and $b \sim 100$ cm^{-1}. As current is increased, the gain in a semiconductor laser increases until it reaches a certain current (called threshold current) when the laser begins to operate

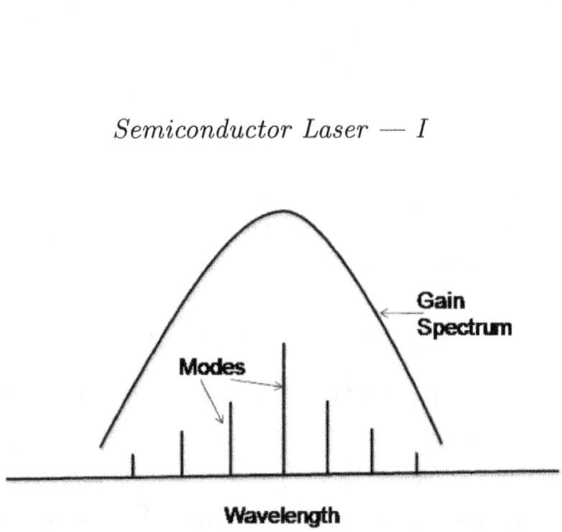

Figure 3.2: Schematic of the gain spectrum and a typical laser spectrum. The longitudinal modes near the maximum gain generally have higher intensity. The mode spacing is $\Delta\lambda$ (Eq. (3.6)).

as a laser. Also, n_R changes with current. The z dependence of the electric field in a laser is given by (using Eq. (3.1))

$$E(z) = E_0 \exp(-i\beta z) \quad \text{with } \beta = n_R k_0 + i(\Gamma g - \alpha)/2 \tag{3.8}$$

For currents above threshold current (I_{th}), the total laser output (from two mirrors) at a current I, is approximately given by

$$P_{\text{out}} = (h\nu/e)(I - I_{\text{th}})\eta_d \quad \text{with } \eta_d = \alpha_m(\alpha + \alpha_m) \tag{3.9}$$

The quantity η_d is called the differential quantum efficiency. For a real laser, the measured differential quantum efficiency (η_d) needs to be multiplied by a quantity called internal efficiency (η_i). The internal efficiency represents the fraction of injected carriers that recombine radiatively. In many cases, $\eta_i \sim 1$. For a typical single-mode semiconductor laser, the dimensions are: (see also Sections 3.2 and 3.3) 0.1 μm active region thickness, 1 μm width of the active region, and 250 μm length (L) of active region. Thus, the area of gain region $= 10^{-9}$ cm^2. For a laser with cleaved facets ($R_1 = R_2 = 35\%$), $\alpha_m = (1/2L)\ln(1/R_1R_2) = 21$ cm^{-1}. Typically, $\alpha = 20$ cm^{-1}, thus, $\eta_d = 0.51$. For $h\nu = 0.96$ eV, i.e. for laser emitting near 1300 nm (1 eV photon energy), the power output per unit current above threshold is ~0.25 mW/mA/facet. The measured slope of power output vs. current above threshold is also often called differential efficiency. For reasons, such as leakage current and heating (discussed later in this chapter), the power output is not linearly proportional to current at high currents. The average intracavity power ($P_{\rm in}$) is related to the intracavity photon density ($N_{\rm ph}$) as

$$P_{\rm in} = h\nu V{\rm N}_{\rm ph}/\tau_p \tag{3.10}$$

where $h\nu$ is the photon energy, V is the volume of the active region, and, τ_p is the photon lifetime in the cavity given by $\tau_p^{-1} = v_g(\alpha+\alpha_m)$, where v_g is the group velocity. Since the photons escape the laser cavity from the output facets at the rate of $v_g\alpha_m$, the power output from each facet for equal reflectivity is given by

$$P_{\rm out} = (1/2)h\nu v_g\alpha_m VN_{\rm ph} \tag{3.11}$$

For many applications, the laser facets are coated with a dielectric material so that most of the power comes out of one facet. For facet reflectivity of R_1 and R_2, the ratio of the output power (P_1 and P_2) from the respective facets is given by (see Appendix 1)

$$\frac{P_1}{P_2} = \frac{(1-R_1)\sqrt{R_2}}{(1-R_2)\sqrt{R_1}} \tag{3.12}$$

The sum of the power outputs $P_1 + P_2 = 2P_{\rm out}$ given by Eq. (3.11).

The materials growth and current injection for lasers are similar to that of LEDs and is described in Chapter 2. The following section describes optical waveguides.

3.2 Optical waveguide

Conceptually, stimulated emission in a semiconductor laser arises from electron–hole radiative recombination in the active region, and the light generated is confined and guided by a dielectric waveguide (Figure 3.3). The threshold current of an injection laser was historically reduced using a double heterostructure (DH) for carrier confinement [10]. The region of optical gain is determined by the thickness of the active region in DH lasers. In addition, the refractive index of the lower-gap active region is higher than that for the n-type and p-type higher-gap confining (also known as cladding) layers. The three layers form an optical waveguide. The energy distribution of the fundamental mode of the waveguide is also sketched in Figure 3.3.

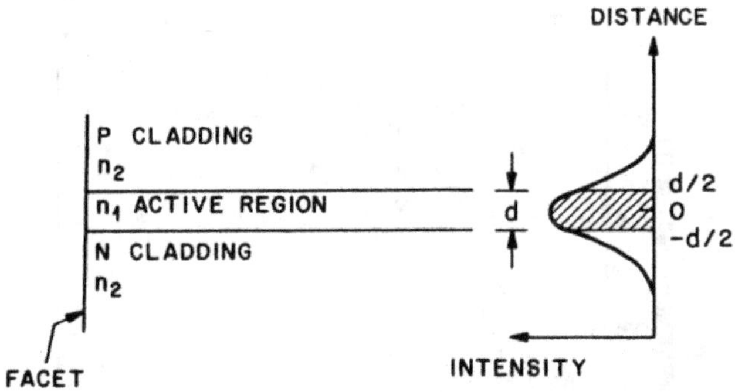

Figure 3.3: The dielectric waveguide of a semiconductor laser. n_2 is the refractive index of the cladding layers and n_1 that of the active region. $n_1 > n_2$. The cladding layers are of a higher band gap material than the active region. The intensity distribution of the fundamental mode is shown. The cross-hatched region represents the fraction of the mode (Γ) within the active region.

A fraction of the optical mode is confined in the active region. Two types of fundamental transverse modes can propagate in the waveguide: the transverse electric (TE) and the transverse magnetic (TM) modes.

The expression for the confinement factor (Γ), the fraction of mode in the active region, has been previously calculated [11, 12]. The calculation of Γ is discussed in Appendix 1. Figure 3.4 shows the calculated Γ as a function of active layer thickness for the TE and TM modes for the $\lambda = 1300$ nm InGaAsP DH with p-InP and n-InP cladding layers. For a three-layer slab waveguide with cladding layer index n_2 and active layer index n_1, the confinement factor of the fundamental mode can be approximated by [12]

$$\Gamma = D^2/(2 + D^2) \tag{3.13}$$

with $D = k_0\ (n_1^2 - n_2^2)^{1/2} d$, where d is the thickness of the active region and $k_0 = 2\pi/\lambda_0$, where λ_0 is the wavelength in free space.

The optical gain of a signal in the fundamental mode traveling through the amplifier is given by

$$g_m = \Gamma_g \tag{3.14}$$

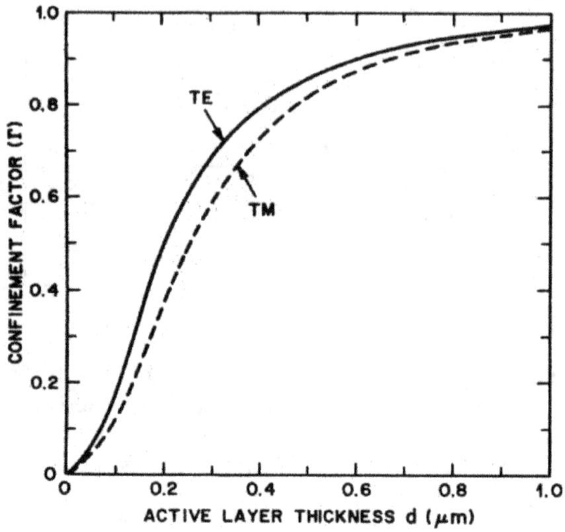

Figure 3.4: Confinement factor of the fundamental TE and TM modes for a waveguide with InGaAsP ($\lambda = 1300$ nm) active layer and InP cladding layers as a function of the thickness of the active region [11]. The refractive indexes of the active and cladding region are 3.51 and 3.22, respectively.

The mode gain (g_m) is proportional to the mode confinement factor Γ. The semiconductor LED, laser, and optical amplifiers used in current applications utilize a DH for carrier injection and carrier confinement.

3.3 Laser designs and emission characteristics

The DH material can be processed to produce lasers in several ways. Perhaps the simplest is the broad area laser (Figure 3.5) which involves putting contacts on the p- and n-side and then cleaving. Such lasers do not have transverse mode confinement or current confinement which leads to high threshold and nonlinearities in light vs. current characteristics.

Several laser designs have been developed to address these problems. Among them are the gain-guided laser, the weakly index-guided laser, and the buried heterostructure (BH) (strongly index guided) laser. A typical version of these laser structures is shown

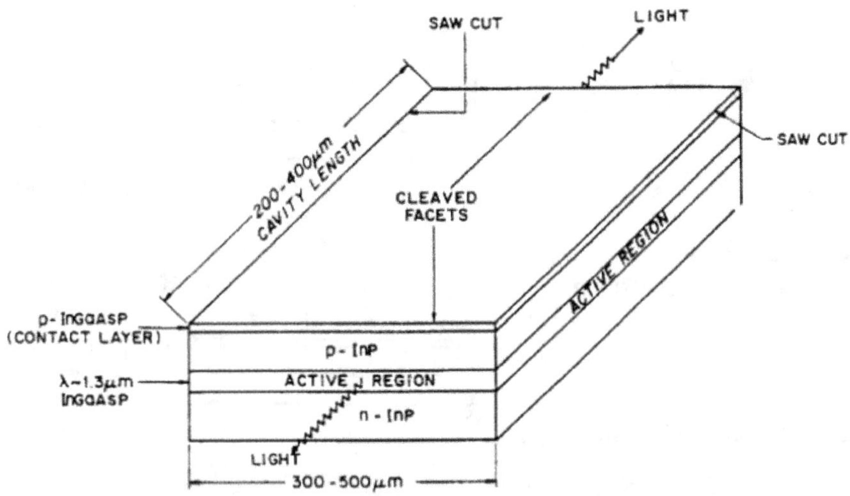

Figure 3.5: Schematic of a broad area laser.

in Figure 3.6. The gain-guided structure uses a dielectric layer for current confinement. The current is injected in the opening in the dielectric (typically 6–12 μm wide), which produces gain in that region and hence, the lasing mode is confined to that region. The weakly index-guided structure has a ridge etched on the wafer, a dielectric layer surrounds the ridge. The current is injected in the region of the ridge, and the optical mode overlaps the dielectric (which has a low index) in the ridge. This results in weak index guiding.

The strongly index guided, BH, design shown in Figure 3.6 has the active region surrounded (buried) by lower index layers. The fabrication process of double channel planar BH (DCBH) laser, involves growing a DH, etching a mesa using a dielectric mask, and then regrowing the layer surrounding the active region using a second epitaxial growth step. The second growth can be a single Fe-doped InP (Fe:InP) semi-insulating layer or a combination of p-InP, n-InP, and Fe:InP layer. Generally, MOCVD growth process is used for the growth of the regrown layer. Researchers have often given different names to the particular BH laser design that they discovered. These are described in detail in Ref. [12]. For the structure of Figure 3.7,

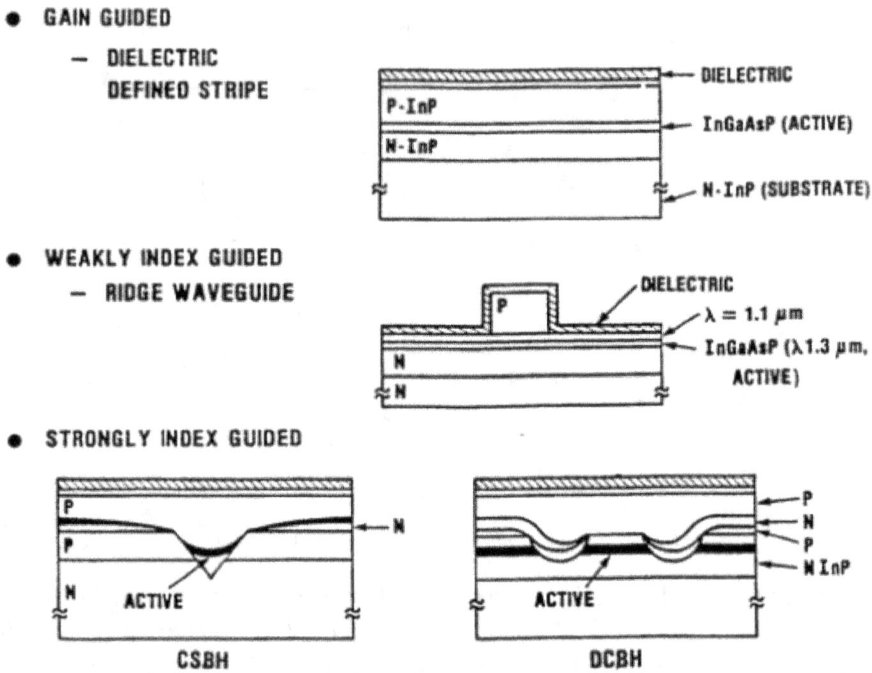

Figure 3.6: Several laser designs using InGaAsP/InP material system.

the Fe-doped InP layer provides both optical confinement to the lasing mode and current confinement to the active region. BH lasers are used in communication system applications because a properly designed strongly index-guided BH design has superior mode stability, higher bandwidth, and superior linearity in light vs. current (L vs. I) characteristics compared to the gain-guided and weakly index-guided designs. Scanning electron micrograph of a capped mesa buried heterostructure (CMBH) laser along with the laser structure is shown in Figure 3.7. The current blocking layers in this structure consist of i-InP (Fe-doped InP), n-InP , i-InP, and n-InP layers.

These sets of blocking layers have the lowest capacitance and are therefore needed for high-speed operation. An optimization of the thickness of these layers is needed for highest speed performance. The laser fabrication involves the following steps. One micron wide mesas are etched on the wafer and the current blocking layers consisting of

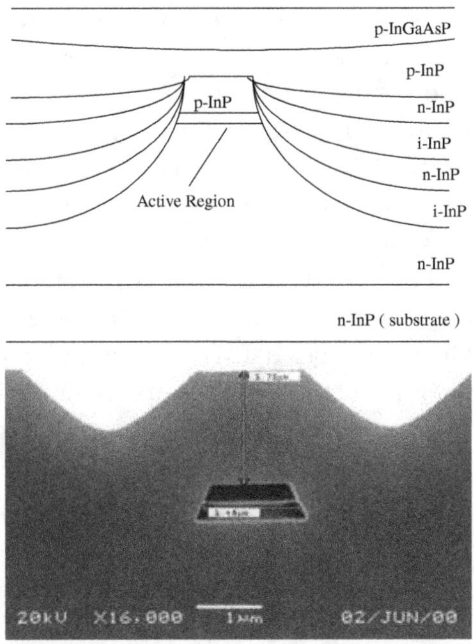

Figure 3.7: Schematic and scanning electron photomicrograph of a buried heterostructure (BH) laser.

i-InP, n-InP, i-InP, and n-InP layers are grown on the wafer with an oxide layer on top of the mesa in place. The oxide layer is then removed and a third growth of p-InP cladding layer and p-InGaAs contact layer is carried out. The wafer is then processed using standard lithography, metallization, and cleaving techniques to produce the lasers.

The light vs. current characteristics at different temperatures of an InGaAsP BH laser emitting at 1.3 μm is shown in Figure 3.8. Typical threshold current of a BH laser at room temperature is in the 5–10 mA range. For gain-guided and weakly index-guided lasers, typical room temperature threshold currents are in the 50–100 and 25–50 mA range, respectively. The external differential quantum efficiency defined as the derivative of the L vs. I characteristics above threshold is $\sim$0.25 mW/mA/facet for a cleaved uncoated laser emitting near 1300 nm. A semiconductor laser with cleaved facets generally emits in a few longitudinal modes of the cavity. Typical spectrum of a laser with cleaved facets is shown in Figure 3.8. The discrete emission wavelengths are separated by the longitudinal cavity mode spacing which is ~10 A for a laser ($\lambda \sim 1300$ nm) with 250 μm cavity length. Lasers can be made to emit in a single frequency using frequency selective feedback, for example, using a grating internal to the laser cavity as described in later sections. For BH devices, the confinement of the injected current to the active region is achieved through reverse biased heterojunctions or forward biased InP homojunctions (which turn on at a voltage higher than that needed for laser action). The difference between the total current I injected into the lasers and the current going through the active region (I_{a}) is the leakage current (I_{l}), i.e.

$$I_{\mathrm{l}} = I - I_a \tag{3.15}$$

The magnitude of I_a and its variation with I depends on the carrier concentration of the current blocking layers, their thickness, and also on the presence of any defects in the current confining junctions or in any of the layers. In addition to leakage current, heating of the active region (with increasing current) results in a sublinearity of the L vs. I characteristics of the laser. A large leakage current would lead to low optical gain at high injected currents and is therefore not desirable. The current blocking layers in the lasers are designed to minimize the leakage current [4].

3.4 Recombination mechanisms and optical gain

Recombination mechanisms in direct-gap semiconductors are described in this section. Recombination mechanisms can, in general, be divided into two categories: radiative and non-radiative. The radiative recombination of the electrons and holes are responsible for stimulated optical emission or gain. Sufficient numbers of electrons and holes must be present in the semiconductor for net optical gain. This is accomplished using a p–n DH discussed in Section 2.3. The I–V characteristics of lasers are discussed in Appendix 5.

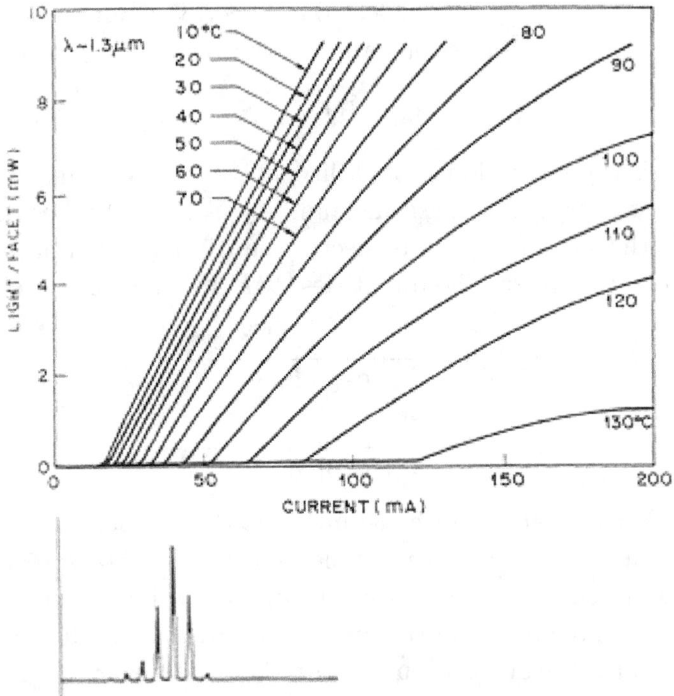

Figure 3.8: Light vs. current (L vs. I) characteristics of a semiconductor laser fabricated using InGaAsP/InP material system emitting near 1.3 μm. The figure also shows the emission spectrum at 3 mW. The various peaks represent different longitudinal modes of the cavity separated by ~10 A. The center wavelength is 1300 nm.

3.4.1 *Radiative recombination*

The basis of light emission in semiconductors is the recombination of an electron in the conduction band with a hole from the valence band; the excess energy is emitted as a photon (light quantum). The process is called radiative recombination as described previously in Chapter 2.

3.4.1.1 *Condition for gain*

Consider the transition shown in Figure 2.9 of Chapter 2 for a photon whose energy $h\nu = E = E_g + E_c + E_v$, where E_g is the band gap, E_c, E_v are energies of the electron and hole, respectively. The photon can be absorbed creating an electron of energy E_c and hole of energy $E_{\rm v}$. The stimulated emission rate is given by

$$R_e = Bf_cf_v\rho(E) \tag{3.16}$$

where B is the transition probability, $\rho(E)$ is the density of states of photons of energy E, and $f_c(E_c)$ and $f_{\rm v}(E_v)$ are the Fermi factors, which is the probability that electron and hole states of energy E_c and E_v are occupied. The quantities $f_c(E_c)$ and $f_v(E_v)$ are given by

$$f_c(E_c) = \frac{1}{1+\exp[(E_c - E_{fc})/kT]} \tag{3.17}$$

$$f_v(E_v) = \frac{1}{1+\exp[(E_v - E_{fv})/kT]} \tag{3.18}$$

where k is the Boltzmann constant, T is the temperature, and E_{fc}, E_{fv}, are the quasi Fermi levels of the electrons and holes, respectively. The stimulated emission process involves the recombination of an electron and hole pair and the absorption process creates an electron and hole pair. The absorption rate for photons of energy E is given by

$$R_a = B(1-f_c)(1-f_v)\rho(E) \tag{3.19}$$

The condition for net stimulated emission or gain is

$$R_e > R_a \tag{3.20}$$

Using Eqs. (3.16) and (3.19) this becomes

$$f_c + f_v > 1 \tag{3.21}$$

Using Eqs. (3.17) and (3.18) for f_c and f_v, the condition for gain at photon energy E becomes

$$E_{fc} + E_{fv} > E - E_g \tag{3.22}$$

This condition was first derived in Ref. [13]. The condition for zero net gain or zero absorption at photon energy is

$$E_{fc} + E_{fv} = E - E_g \tag{3.23}$$

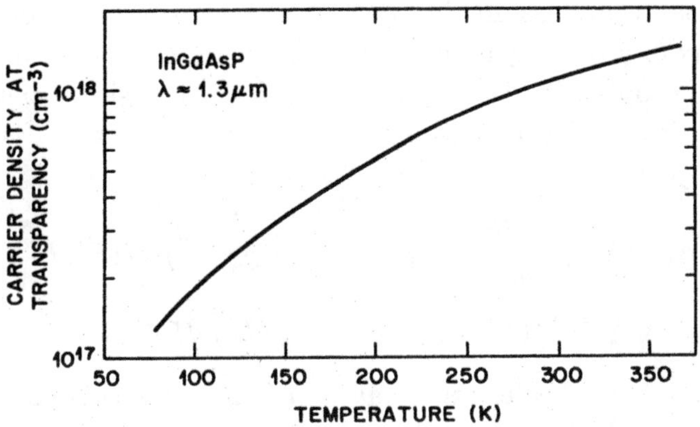

Figure 3.9: The calculated injected carrier density for transparency as a function of temperature for undoped $\lambda = 1300$ nm InGaAsP.

Since the minimum value of E is E_g, the condition for gain at any energy is

$$E_{fc} + E_{fv} = 0. \tag{3.24}$$

This is often known as the condition for transparency. With increasing injected carrier densities (electrons and holes) due to current injection both E_{fc} and E_{fv} increase. This is further discussed in Appendix 5. Figure 3.9 shows the variation of the injected carrier density for transparency n_t as a function of temperature for undoped $\lambda = 1300$ nm ($E_g = 0.96$ eV) InGaAsP. The parameter values used in the calculation are $m_c = 0.061\ m_0$, $m_{\text{hh}} = 0.45\ m_0$, $m_{\text{lh}} = 0.08\ m_0$, where m_0, m_{hh}, and m_{lh} are the free-electrons, heavy-hole, and the light-hole mass, respectively. Figure 3.9 shows that n_t is considerably smaller at low temperatures. The threshold carrier density is ~20–30% higher than n_t.

3.4.1.2 *Gain calculation*

The quantities associated with a radiative recombination are the absorption spectrum, emission spectrum, gain spectrum, and the total radiative emission rate. The calculation of the above for transitions between parabolic bands (i.e. when both conduction and valence bands are assumed to be parabolic in E vs. $\mathbf{k}$) is described in Appendix 5. For the general case of non-parabolic bands and exciton states, the optical absorption or gain for a transition between the valence band and the conduction band at an energy E is given by [4, 14–17]

$$\alpha(E) = \frac{e^2 h}{2\varepsilon_0 m_0^2 cnE} \int_{-\infty}^{+\infty} \rho_c(E')\rho_v(E'')|M(E', E'')|^2 \times [f(E'' = E' - E) - f(E')]dE' \tag{3.25}$$

where m_0 is the free-electron mass, h is the Planck's constant, e is the electron charge, ε_0 is the permittivity of free space, E is the photon energy, n is the refractive index at energy E, and ρ_c and ρ_v are the densities of state for unit volume per unit energy in the conduction and valence band, respectively. $f(E')$ is the probability that a state of energy E' is occupied by an electron and M is the effective matrix clement between the conduction band state of energy E' and the valence band state of energy E''. Several models for the density of states and matrix element for transition between band-tail states exist [14–17]. The latest of such models that take into account the contributions of the parabolic bands and impurities is a Gaussian fit of the Kane form to the Halperin–Lax model of band tails [16]. The matrix element M may be expressed as a product of two terms: $M = M_b M_{\text{env}}$. The quantity M_b arises from the band-edge Bloch functions and M_{env} arises from the envelope wave functions. For III–V semiconductors using the Kane model [14],

$$|M_b|^2 = \frac{m_0^2 E_g}{12 m_c} \frac{E_g + \Delta}{E_g + \frac{2}{3}\Delta} \tag{3.26}$$

where m_c is the conduction-band effective mass, m_0 is the free-electron mass, E_g is the energy gap, and Δ is the spin–orbit coupling. The envelop matrix element for the band-tail states have been calculated by Stern [15].

The calculated spectral dependence of absorption or gain at various injected carrier densities is shown in Figure 3.10 for InGaAsP ($\lambda \sim 1300$ nm) with acceptor and donor concentrations of 2×10^{17}

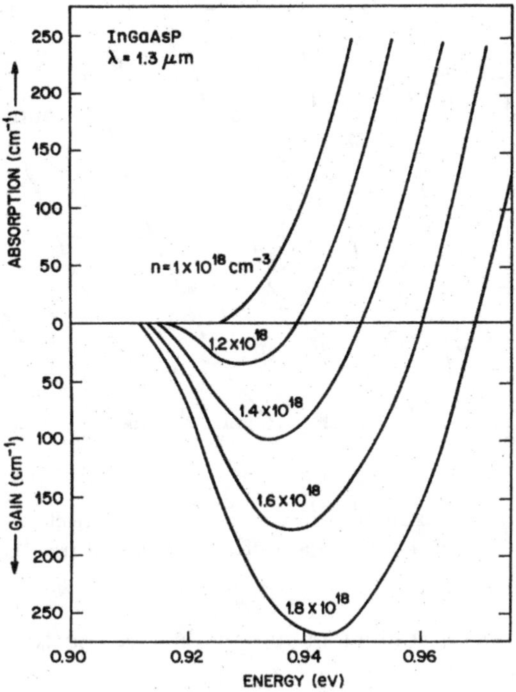

Figure 3.10: Calculated gain or absorption as function of photon energy for $\lambda = 1300$ nm ($E_g = 0.96$ eV) InGaAsP at various injected carrier densities [17].

cm^{-3}. The material parameters used in the calculation are $m_c = 0.061m_0$, $m_{\text{hh}} = 0.45m_0$, $m_{\text{lh}} = 0.08m_0$, $\Delta = 0.26$ eV, $E_g = 0.96$ eV, and $\varepsilon = 11.5$. Figure 3.10 shows that the gain peak shifts to higher energies with increasing injection. This is observed in semiconductor amplifiers.

Figure 3.11 shows the maximum gain as a function of injected carrier density at different temperatures. Note that considerably lower injected carrier density is needed at a lower temperature to achieve the same optical gain. This is the origin of the lower threshold current at low temperature. Similar results are obtained for different materials. It is often convenient to use a linear relationship between the gain g and the injected carrier density n of the form

$$g = a(n - n_0) \tag{3.27}$$

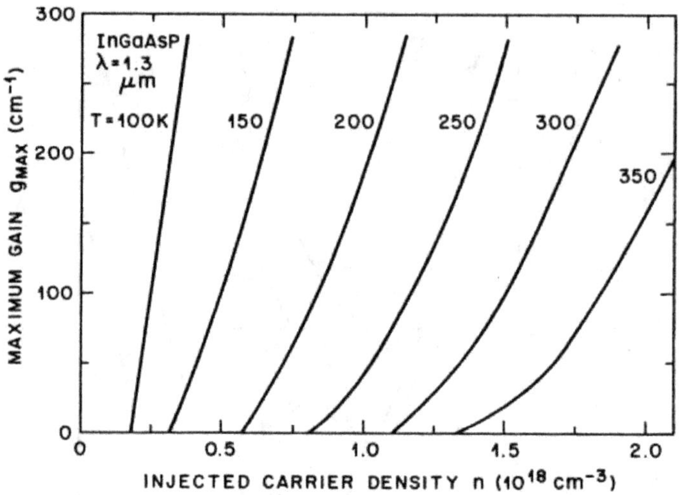

Figure 3.11: The maximum gain as a function of injected carrier density for undoped InGaAsP ($\lambda = 1300$ nm) at different temperatures [17].

where a is the gain coefficient and n_0 is the injected carrier density at transparency.

3.4.1.3 *Spontaneous emission rate*

At unity quantum efficiency, the total spontaneous radiative recombination rate R equals the excitation rate. The latter is usually expressed in terms of the nominal current density J_n [4, 14–17]:

$$J_n[A/(\text{cm}^2\mu\text{m})] = eR \tag{3.28}$$

with $R = \int r_{\text{spon}}(E)dE$, where e is the charge of an electron and the thickness of the active region is assumed to be 1 μm. $r_{\text{spon}}(E)$ is the spontaneous emission rate at a photon energy E. It is given by [Eq. 2.3.2, 4, Appendix 5]

$$r_{\text{spon}}(E) = \frac{4\pi n e^2 E}{m_o^2 \varepsilon_0 h^2 c^3} \int_{-\infty}^{+\infty} \rho_c(E')\rho_v(E'')|M(E', E'')|^2 f(E') \times [1 - f(E'')]dE' \tag{3.29}$$

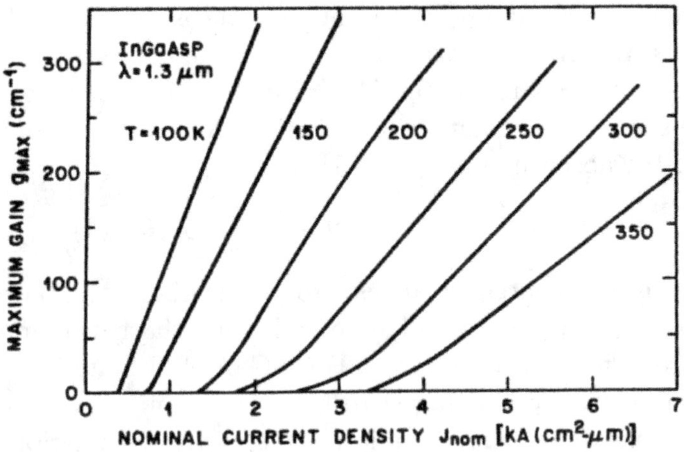

Figure 3.12: The maximum gain as a function of nominal current density for InGaAsP ($\lambda = 1300$ nm) at different temperatures [17].

The calculated maximum optical gain g as a function of the nominal current density at various temperatures is shown in Figure 3.12 for InGaAsP ($\lambda \sim 1300$ nm). The calculation is for an undoped lightly compensated material with 10^{17} cm^{-3} of acceptors and donors, respectively. Note that gain varies linearly with J_n above a certain gain. From the above figure, the relationship between gain and nominal current density for InGaAsP ($\lambda \sim 1300$ nm) at 300°K is approximately given by

$$g = 0.08(J_n - 3100)\text{cm}^{-1} \tag{3.30}$$

where J_n is expressed in amperes per square centimeter per micron. The total spontaneous radiative recombination rate R can be approximated by

$$R = Bnp \tag{3.31}$$

where B is the radiative recombination coefficient and n, p are the electron and hole densities, respectively. For undoped semiconductors, Eq. (3.31) becomes $R = Bn^2$. For GaAs, the measured $B = 1 \times 10^{-10}$ cm$^3/s$. Calculation of the radiative recombination rate shows that B decreases with increasing carrier density [4]. This has been confirmed by carrier lifetime measurements [4]. In DH materials, the injected carriers can recombine by both radiative and non-radiative recombination. The injected current density J is simply the sum of the radiative R and non-radiative R_{nr} recombination rates in the absence of carrier leakage:

$$J = e(R + R_{\text{nr}})d = J_r + J_{\text{nr}} \quad (J_r = J_n d) \tag{3.32}$$

where e is the charge of an electron, d is the active layer thickness, and J_r, J_{nr} are the radiative and non-radiative components of the current density, respectively. For a GaAs DH material where the quantum efficiency is believed to be close to unity, i.e. most of the injected carriers recombine radiatively, $J \sim J_r$. The relation between the optical gain and the nominal current density for GaAs material system has been calculated by Stern [15]. It is given by

$$g = 0.045(J_n - 4200)\text{cm}^{-1} \tag{3.33}$$

where J_n is expressed in amperes per square centimeter per micron.

It is useful to make the connection with the operation of a GaAs laser for which the non-radiative recombination is believed to be very small. For a laser, the optical gain provided by current injection must equal optical loss. The latter includes mirror losses. This is expressed as

$$\Gamma g_{\text{th}} = \alpha_a \Gamma + (1 - \Gamma)\alpha_c + \frac{1}{L}\ln\frac{1}{R} \tag{3.34}$$

where g_{th} is the threshold gain in the active region; α_a, α_c are the absorption losses in the active and cladding regions, respectively; L is the cavity length; and R is the mirror facet reflectivity. Typically, $L \sim 250$ μm and $R \sim 0.3$. For a 0.2-μm-thick active layer, Γ(TE) is $\sim$0.6, and using $\alpha_a = \alpha_c = 20$ cm^{-1}, the calculated g_{th} $\sim$114 cm^{-1}.

The threshold current of a 250-μm-long GaAs–AlGaAs DH laser with $2\mu\text{m} \times 0.2\mu\text{m}$ active region can now be calculated using Eq. (3.33). The calculated $J_{\text{th}} = 1.35$ kA/cm^2 (for a gain $g_{\text{th}} \sim 114$ cm^{-1}, i.e. in this case, gain $\sim$84 cm^{-1}/(kA/cm^2)), and the threshold current is 6.7 mA. This compares well with the experimentally observed threshold currents in the range 5–10 mA for GaAs–A1GaAs DH lasers with good current confinement.

Figure 3.12 shows that the current density needed to achieve a certain optical gain increases with temperature. The same figure also illustrates that the threshold current density of a laser is expected to increase with increasing temperature. For InGaAsP lasers, a significant amount of non-radiative recombination is believed to be present at high carrier densities needed for gain. The non-radiative recombination can increase the current needed for gain (as shown in Eq. (3.32)) and also its temperature dependence.

The temperature dependence of threshold current (I_{th} (T)) of a laser is usually expressed by a relation:

$$I_{\mathrm{th}}(T) = I_{\mathrm{th0}} \exp(T/T_0) \tag{3.35}$$

where I_{th0} is a constant. T_0 is a parameter know as characteristic temperature. T_0 is low if the temperature dependence of threshold current is high. T_0 has a value of ~200°K for GaAs (850 nm) lasers and T_0 ~60°K for InGaAsP 1300 and 1550 nm lasers. The high-temperature dependence of threshold current (low T_0) of long wavelength InGaAsP lasers is generally attributed to non-radiative Auger recombination discussed in Section 3.4.2. The band gap of a semiconductor decreases with increasing temperature, the rate is about 0.35 nm/C.

3.4.2 *Non-radiative recombination*

An electron–hole pair can recombine non-radiatively, meaning that the recombination can occur through any process that does not emit a photon. The effect of non-radiative recombination on the performance of lasers is to increase the current needed for a certain optical gain. If τ_{nr} is the carrier lifetime associated with the non-radiative process, the increase in current density is given approximately by

$$J_{\mathrm{nr}} = \frac{en\,d}{\tau_{\mathrm{nr}}} \tag{3.36}$$

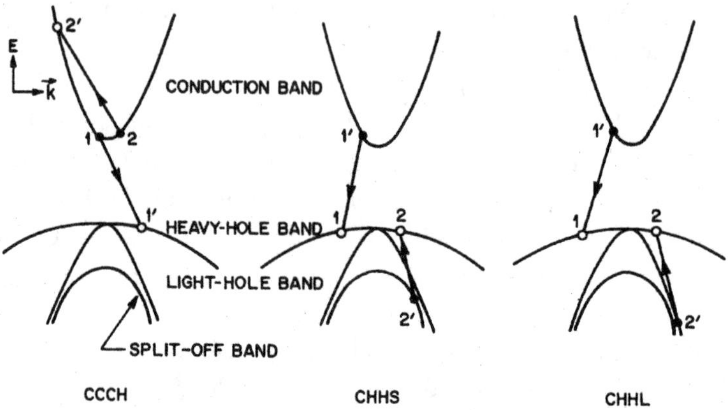

Figure 3.13: Band-to-band Auger recombination processes in a direct gap semiconductor [21].

where n is the carrier density, d is the active layer thickness, and e is the electron charge. The non-radiative recombination processes described in this section are the Auger effect, surface recombination, and recombination at defects or traps.

3.4.2.1 *Auger effect*

Since the pioneering work by Beattie and Landsberg [18, also 19–23] it is generally accepted that Auger recombination can be a major non-radiative recombination mechanism in narrow-gap semiconductors. The band-to-band Auger processes in direct-gap semiconductors are shown in Figure 3.13. The three processes are labeled CCCH, CHHS, and CHHL, where C stands for the conduction band and H, L, S stand for heavy-hole, light-hole, and split-off valence band hole, respectively. The CCCH mechanism involves three electrons and a heavy hole and is dominant in n-type material. The process was first considered by Beattie and Landsberg [18].

The CHHS process involves one electron, two heavy holes, and a split-off band hole. The CHHL process is similar to the CHHS process except that it involves a light hole. The CHHS and CHHL mechanisms are dominant in p-type material. Under the high injection conditions present in lasers and amplifiers, all of these mechanisms must be considered.

Band-to-band Auger processes are characterized by a strong temperature dependence and band gap dependence, the Auger rate decreasing rapidly either for low temperature or for high band-gap materials. These dependencies arise from the energy and momentum conservation that the four free particle states involved (1, 2, 1', 2' in Fig. 3.13) must satisfy. The momentum and energy conservation laws give rise to a threshold energy E_T for each of the processes. For the CCCH process, if we assume $E_1 \sim E_2 = 0$, only holes with energies greater than $\sim (E_T - E_g) = \delta E_g$ can participate (δ is a constant that depends on effective masses). The number of such holes varies approximately as $\exp(-\delta E_g/kT)$ for non-degenerate statistics and thus the Auger rate varies approximately as $\exp(-\delta E_g/kT)$ in the non-degenerate case.

The Auger rate calculations [21] can be applied to any direct-gap semiconductor for which the Kane model of the band structure is valid. The various band-to-band Auger processes are characterized by a strong dependence on both the band gap (E_g) and the temperature T. In the non-degenerate approximation, the dependence of the Auger rate R_A on E_g and T may be approximately written as

$$R_A \propto \exp(-\Delta E/kT) \tag{3.37}$$

where for the CCCH process

$$\Delta E = \frac{m_c}{2m_{c0} + m_v - m_c} E_g$$

for the CHHS process

$$\Delta E = \frac{m_s}{2m_v + m_{c0} - m_s} (E_g - \Delta)$$

and for the CHHL process

$$\Delta E = \frac{m_l}{2m_v + m_{c0} - m_l} E_g$$

where m_{c0}, m_v are the conduction-band and the valence-band (heavy-hole) mass at the band edge and m_c, m_s, and m_l are the effective masses of an electron, split-off band hole, and the light hole at the energy $E_T = E_g + \Delta E$ where ΔE is given by the above equations for the respective processes and E_g is the band gap. Thus,

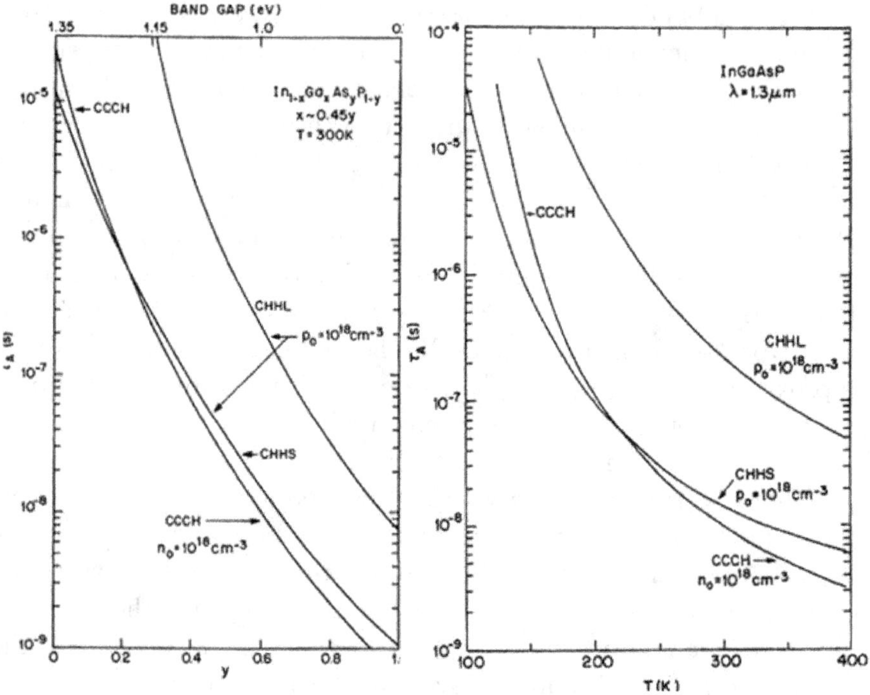

(a) (b)

Figure 3.14: (a) The calculated Auger recombination lifetime for the three processes CCCH, CHHS, and CHHL at a carrier density of 10^{18} cm^{-3} as a function of y (the arsenic composition) for InGaAsP lattice matched to InP. (b) The calculated Auger lifetime at a carrier density of 10^{18} cm^{-3} as a function of temperature for the three Auger processes for InGaAsP which emits near 1300 nm [21].

the non-parabolicity of the band structure plays a significant role in determining the Auger recombination rate. The calculated Auger recombination lifetime for the three processes CCCH, CHHS, and CHHL at a carrier density of 10^{18}cm^{-3} as a function of y (the arsenic composition) for InGaAsP lattice matched to InP is shown in Figure 3.14. The band gap of InGaAsP decreases with increasing y as [24]

$$E_g(y) = 1.35 - 0.75y + 0.12y^2 \tag{3.38}$$

for compositions that are lattice matched to InP ($x \sim 0.45y$). The Auger lifetime decreases, i.e. Auger recombination rate increases with decreasing band gap. First, consider n-type materials where the CCCH process is dominant. In the non-degenerate approximation with parabolic bands ($m_c = m_{c0}$), Eq. (3.37) can be further simplified to yield

$$R_A \propto \exp\left|-\frac{\mu}{1+\mu}\frac{E_g}{kT}\right| \tag{3.39}$$

where $\mu = m_{c0}/m_v$. Since, in general, the conduction-band effective mass decreases with decreasing band gap, it follows that the lower band-gap materials will have a larger band-to-band Auger rate. A similar result holds for p-type materials for the CHHL Auger process. The calculated Auger lifetime at a carrier density of 10^{18}cm^{-3} as a function of temperature for the three Auger processes for InGaAsP which emits near 1300 nm is shown in Figure 3.14. The Auger lifetime decreases, i.e. the Auger recombination rate increases with increasing temperature.

In the absence of momentum conservation, there is no threshold energy E_T [18, 19–23]. Thus, the strong temperature dependence does not appear if the individual particle states are not states of definite momentum, for example, if they are trap states, or if momentum conservation is satisfied through phonon assistance, i.e. for phonon-assisted Auger process. The phonon-assisted process dominates at high band gap, and the band-to-band processes dominate for low band gap semiconductors. The active region of an InGaAsP laser is nominally undoped. Under high injection the Auger rate R_a varies approximately as

$$R_a = Cn^3 \tag{3.40}$$

where n is the injected carrier density. Calculations of the Auger coefficient (C) using the Kane band model yield a value of 10^{-28} cm^6/s [4, 21] for $\lambda \approx 1300$ nm InGaAsP. A value of $\sim 2.5 \times 10^{-29}$ cm^6/s for $\lambda \approx 1300$ nm InGaAsP using a different model for the band structure has been calculated [23]. The experimental values of the Auger coefficient for this material are in the range: 2×10^{-29} – 7×10^{-29}cm^6/s [4, 21]. For GaAs, $C \approx 10^{-31}$ cm^6/s. From Eq. (3.40), the current lost to Auger recombination at threshold is given by

$$J_A = edCn^3 \tag{3.41}$$

where d is the active layer thickness and n is the carrier density at threshold. The effect of Auger recombination on shorter wavelength lasers (e.g. GaAs lasers which emit near 850 nm) is small compared to that for longer wavelength lasers (e.g. InGaAsP lasers which emit near 1300 nm) because the Auger coefficient is smaller for high band gap materials. In addition, Auger recombination produces hot carriers. Carrier temperatures as high as 400°K have been reported for 300°K lattice temperature for 1nGaAsP/InP DH under high injection [24, 25].

3.5 Rate equations and dynamic characteristics

For many applications, the laser output is modulated by modulating the current. This section discusses the dynamic characteristics of semiconductor lasers. Assuming the laser is oscillating in a single mode, the rate equations for power output (P) and the total carrier density (N) are given by (Appendix 11)

$$\dot{P} = (G - \gamma)P + R_{\rm sp} \tag{3.42}$$

$$\dot{N} = I/e - \gamma_e N - GP \tag{3.43}$$

where G is the optical gain which depends on N (total number of carriers), γ is the total optical loss, $R_{\rm sp}$ is the spontaneous emission rate, I is the injected current, γ_e is the inverse of the lifetime (τ_e^{-1}) of the carriers. The quantity γ_e is a function of n where $n = N/V$ is the carrier density per unit volume. Equation (3.43) may be written as

$$\frac{dn}{dt} = \frac{I}{eV} - \gamma_e(n)n - GP/V \tag{3.44}$$

with $\gamma_e\ (n) = A_{\rm nr} + Bn + Cn^2$, where $A_{\rm nr}$ is the non-radiative recombination (constant lifetime) rate and the other two terms are due to radiative and Auger recombination. Often a constant lifetime recombination model is used, in such cases $\gamma_e = \tau^{-1}$, where τ is the average lifetime which is independent of n. Below threshold ($P = 0$), the following equation for carrier density (n) is obtained

$$\frac{dn}{dt} = \frac{I}{eV} - n/\tau \tag{3.45}$$

When the current is turned on, it takes a certain amount of time (τ_d-called delay time) for the carrier density to reach threshold (n_{th}) and hence for laser action to begin. Using $\gamma_e = \tau^{-1}$, the following simple expression for τ_d is obtained by integrating Eq. (3.45) from $n = 0$ to $n = n_{\text{th}}$

$$\tau_d = \tau \ln\left(\frac{I}{I - I_{\text{th}}}\right) \tag{3.46}$$

where $I_{\text{th}} = \text{eV n}_{\text{th}}/\tau$ is the threshold current. If the laser already has a CW bias current of I_0 ($= \text{eV}n_0/\tau$), the above equation is modified to

$$\tau_d = \tau \ln\left(\frac{I - I_0}{I - I_{\text{th}}}\right) \tag{3.47}$$

A semiconductor laser exhibits damped periodic oscillation before settling to a steady state value. The steady state solutions (P_{s}, N_{s}) of Eqs. (3.42) and (3.43) are:

$$\begin{aligned} P_s &= R_{\text{sp}}/(\gamma - G) \\ N_s &= (I/e - GP_s)/\gamma_e \end{aligned} \tag{3.48}$$

The rate Eqs. (3.42) and (3.43) are now linearized using $P = P_s + p_1$, and $N = N_s + n_1$ and using $G(N) = G\ (N_s) + G_{\text{n}}\ (N - N_s)$, where $G_n = dG/dN$

$$\begin{aligned} \dot{p}_1 &= G_N P_s n_1 - \Gamma_P p_1 \\ \dot{n}_1 &= \Gamma_N n_1 - G_s p_1 \end{aligned} \tag{3.49}$$

where

$$\begin{aligned} \Gamma_N &= \gamma_e + G_N P_s \\ \Gamma_P &= R_{\text{sp}}/P_s, \quad G_s = G(N_s) \end{aligned}$$

Using $p_1(t) = p_0 \exp(-ht)$ and $n_1(t) = n_0 \exp(-ht)$ and using $h = \Gamma_R + i\Omega_R$, the following values are obtained for Γ_R and Ω_R

$$\Gamma_R = \frac{1}{2}(\Gamma_N + \Gamma_P)$$
$$\Omega_R = (GG_N P)^{1/2} \tag{3.50}$$

The results show that the output of the laser undergoes damped oscillations before a steady state is reached. Since the power output P is proportional to $(I - I_{\text{th}})$, the relaxation (or damped) oscillation frequency Ω_R varies as $(I - I_{\text{th}})^{1/2}$. In terms of the previously described gain coefficient a (i.e. g = a $(n - n_{\text{o}})$ in Sect. 3.4), the quantity G_n is given by $G_n = \Gamma v_g a/V$, where Γ is the confinement factor of the mode, v_g is the group velocity, and V is the volume of active region.

3.5.1 *Small signal modulation*

The small signal modulation response of a laser is obtained using Eqs. (3.42) and (3.43) and the following phase change equation (Appendix 11)

$$\frac{d\phi}{dt} = -\frac{\bar{\mu}}{\mu_g}(\omega - \Omega) + \frac{1}{2}\alpha(G - \gamma) \tag{3.51}$$

with $\alpha = \frac{dn_R/dN}{dn_I/dN}$
where Ω is the specific longitudinal mode oscillation frequency, ω is the laser frequency, $\bar{\mu}$ is the mode index, μ_g is the group index, and α is the linewidth enhancement factor (also known as α-factor) defined above. It is the ratio of the change in real part of the index (n_R) to change in imaginary part (n_I) of the index. Assume a small modulation current $(I_m(t))$ over a CW bias current (I) applied to a laser, i.e. $I(t) = I + I_m\ (t)$. Assume $I > I_{\text{th}}$, then Eqs. (3.42) and (3.43) for $I_m \ll I$ and $\omega = \Omega$ become

$$\delta\dot{P} = -\Gamma_P \delta P + (G_N P)\delta N$$
$$\delta\dot{N} = -\Gamma_N \delta N - G\delta P + I_m(t)/e$$
$$\delta\dot{\phi} = \frac{1}{2}\alpha G_N \delta N \tag{3.52}$$

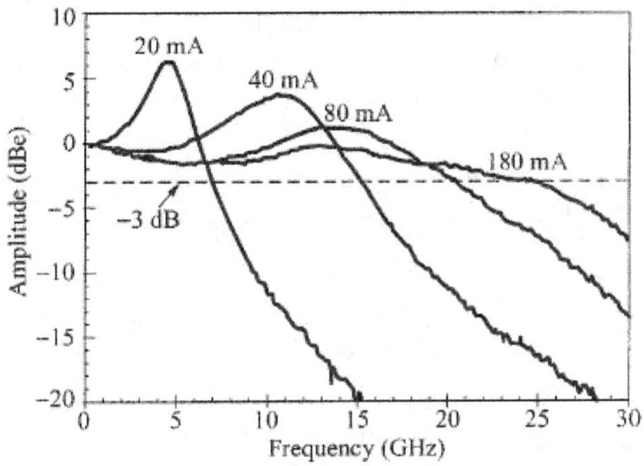

Figure 3.15: The measured modulation response of a BH laser emitting near 1550 nm. Note that there is a peak in response in agreement with the model described earlier [26].

Now consider, $I_m(t) = I_p \exp(-i\omega_m t)$, i.e. $I_m(t)$ is sinusoidal with frequency ω_m. Then the above equations can be solved and the following expression for the real valued modulated power is obtained

$$\delta P(t) = \delta P_0 \sin(\omega_m t + \phi) \tag{3.53}$$

with

$$\delta P_0 = \frac{G_N P I_p/e}{[(\omega_m^2 - \Omega_R^2 - \Gamma_R^2)^2 + 4\omega_m^2 \Gamma_R^2]^{1/2}}$$

$$\frac{\delta P_0(\omega_m = \Omega_R)}{\delta P_0(\omega_m = 0)} \simeq \frac{\Omega_R}{2\Gamma_R}$$

Thus, the modulation response (δP_0 as a function of ω_m) is relatively flat for $\omega_m \ll \Omega_R$, peaks near $\omega_m \sim \Omega_R$, and drops off sharply for $\omega_m \gg \Omega_R$, as $\delta P_0 \sim \omega_m^{-2}$. The small signal modulation response of an InGaAsP/InP BH laser with semi-insulating (Fe:InP) current blocking layers is shown in Figure 3.15 [26]. The laser emits near 1550 nm.

The time dependent phase change ($\delta\phi(t)$) results in a variation in frequency $\delta\nu(t)$ around the frequency of laser oscillation ($\nu = \omega/2\pi$)

$$\delta\nu(t) = \delta\dot{\phi}(t)/2\pi \tag{3.54}$$

The time dependent frequency shift is called frequency chirping. An expression for frequency chirp can be obtained using the Fourier relation

$$\delta\dot{\phi}(t) = \frac{1}{2\pi}\int_{-\infty}^{\infty} i\omega\phi(\omega)\exp(i\omega t)d\omega \tag{3.55}$$

For the sinusoidal modulation, discussed earlier, an expression for $\phi(\omega)$ can obtained using Eq. (3.52). The variation of frequency $\delta\nu(t)$ is then given by

$$\delta\nu(t) = \delta\nu_0 \sin(\omega_m t + \phi)$$

The expression for $\delta\nu_0$ is

$$\delta\nu_0 = \frac{\alpha I_P G_N}{4\pi e}\left(\frac{\omega_m^2 + \Gamma_P^2}{(\omega_m^2 - \Omega_R^2 - \Gamma_R^2)^2 + 4\omega_m^2\Gamma_R^2}\right)^{1/2} \tag{3.56}$$

Using $\omega_m \ll \Omega_R$, $\Gamma_R \ll \Omega_R$, and $\Omega_R^2 = GG_nP$, $\delta\nu_0$ has the following form

$$\delta\nu_0 = \frac{\alpha Ip}{4\pi eGP}(\omega_m^2 + \Gamma_P^2)^{1/2} \tag{3.57}$$

Thus, $\delta\nu_0$ is proportional to the amplitude of modulation current and the α-factor.

3.6 Multiquantum well lasers

Beginning with the 1980s, lasers with very thin active regions, quantum well (QW) lasers, were being developed in many research laboratories [27–36]. QW lasers have active regions of ~100 A thick which restricts the motion of the carriers (electrons and holes) in a direction normal to the well. This results in a set of discrete energy levels and the density of states is modified to a "two-dimensional-like" density of states. This modification of the density of states results in several improvements in laser characteristics such as lower threshold current, higher efficiency, higher modulation bandwidth, and lower CW and dynamic spectral width. All of these improvements were first predicted theoretically and then demonstrated experimentally [37–46].

A regular DH laser consists of an active layer sandwiched between higher-gap cladding layers. The active layer thickness is typically 0.1–0.5 μm. DH lasers have been fabricated with an active layer thickness ~100 Å or less. The carrier (electron or hole) motion normal to the active layer in these structures is restricted. This may be viewed as carrier confinement in a one-dimensional potential well, and hence these lasers are called QW lasers [36–44]. When the thickness L_z of a narrow-gap semiconductor layer confined between two wide-gap semiconductors becomes comparable to the de Broglie wavelength ($\lambda \sim h/p \sim L_z$), quantum-mechanical effects are expected to occur. The energy levels of the carriers confined in the narrow-gap semiconductor can be determined by separating the Hamiltonian into a component normal to the layer (z component) and into the usual (unconfined) Bloch function components (x, y) in the plane of the layer. The resulting energy eigenvalues are:

$$E(n, k_x, k_y) = E_n + \frac{\hbar^2}{2m_n^*}(k_x^2 + k_y^2) \tag{3.58}$$

where E_n is the n_{th} confined-particle energy level for carrier motion normal to the well. m_n^* is the effective mass of the nth level, $\hbar$ is Planck's constant (h) divided by 2π, and k_x, k_y are the usual Bloch function wave-vectors in the x- and y-directions. Figure 3.16 shows schematically the energy levels E_n of the electrons and holes confined in a QW. The confined particle energy levels E_n are denoted by E_{1c}, E_{2c}, E_{3c} for electrons; E_{1hh} , E_{2hh} for heavy holes; and E_{1lh}, E_{2lh}, for light holes. The calculation of these quantities is a standard problem in quantum mechanics for a given potential barrier (ΔE_c, ΔE_v). For an infinite potential well, the following simple result is obtained:

$$E_n = \frac{h^2 n^2}{8L_z^2 m_n^*} \tag{3.59}$$

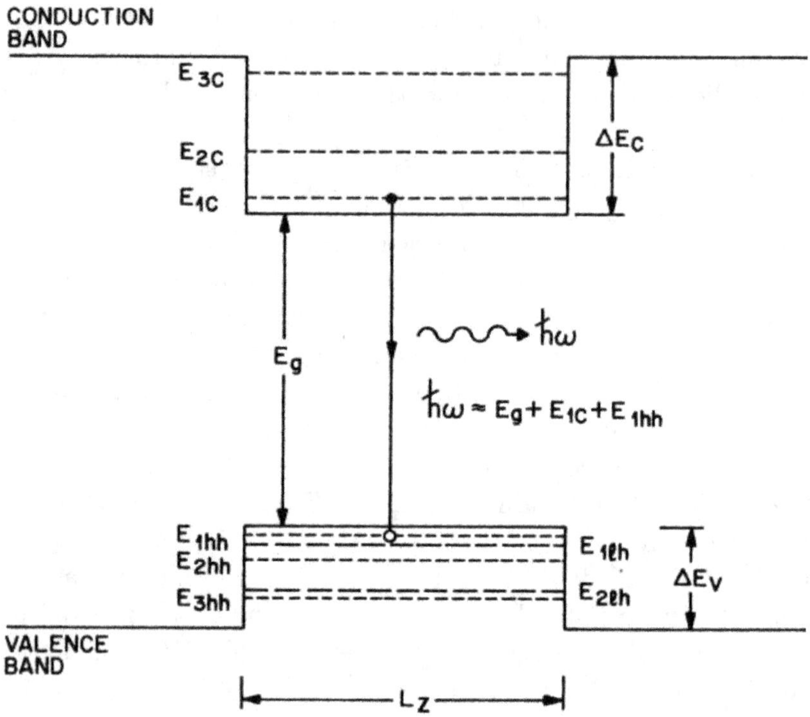

Figure 3.16: Energy levels in a quantum well structure [27, 28].

The energy levels for a well of finite depth are discussed in Appendix 14. The separation between the lowest conduction band level and the highest valence band level is given by

$$E_q = E_g + E_{1c} + E_{1\text{hh}} \approx E_g + \frac{h^2}{8L_z^2}\left(\frac{1}{m_c} + \frac{1}{m_h}\right) \tag{3.60}$$

It follows that in a QW structure the energy of the emitted photons can be varied simply by varying the well width L_z. This has been observed experimentally for InGaAs active region sandwiched between InP layers [47].

The discrete energy levels result in a modification of the density of states in a QW to a "two-dimensional-like" density of states. This modification of the density of states results in several improvements

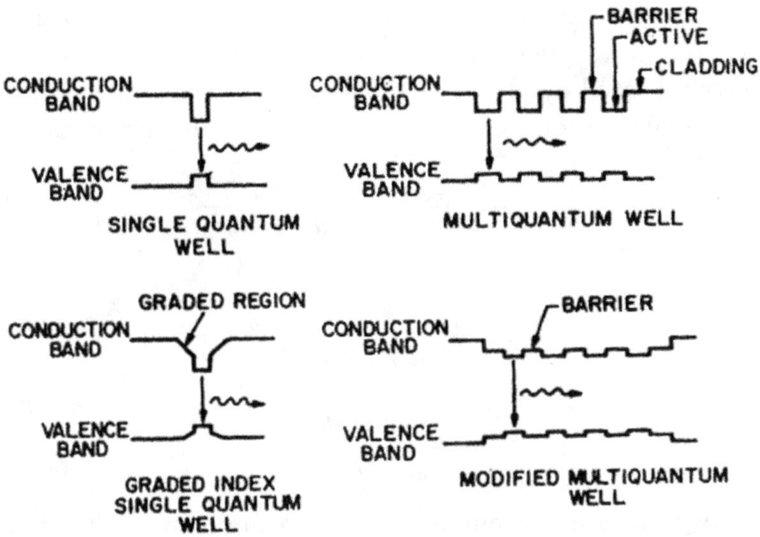

Figure 3.17: Various types of quantum well amplifier active region designs [4].

in laser characteristics such as lower threshold current, higher efficiency, higher modulation bandwidth and lower CW and dynamic spectral width [48–56]. Various types of QW laser and also amplifier active region designs have been reported. The conduction and valence band diagram of these designs along with their names are shown in Figure 3.17.

3.6.1 *Multiquantum well designs*

QW lasers have been fabricated with both single quantum well (SQW) active region and with multiple quantum well (MQW) active region. The mode confinement factors of SQW and MQW lasers are now discussed. For a SQW laser of active layer thickness d, the confinement factor is

$$\Gamma \cong 2\pi^2(n_a^2 - n_c^2)d^2/\lambda_0^2 \tag{3.61}$$

where n_a, n_c are indices of the cladding and active region, λ_0 is the free space wavelength, and d is the thickness of the SQW. For an MQW active region, the confinement factor can be calculated by solving the wave equation in a multi-layer waveguide consisting of the active, the barrier, and the cladding layers. The procedure is quite complicated due to the large number of layers involved.

The following formula gives reasonably accurate results:

$$\Gamma(\mathrm{MQW}) = \gamma \frac{N_a d_a}{N_a d_a + N_b d_b} \tag{3.62}$$

with

$$\gamma = 2\pi^2 (N_a d_a + N_b d_b) \frac{\overline{n}^2 - n_c^2}{\lambda_0^2} \tag{3.63}$$

$$\overline{n} = \frac{N_a d_a n_a + N_b d_b n_b}{N_a d_a + N_b d_b} \tag{3.64}$$

where N_a, N_b are the number of active and barrier layers in the MQW structure and d_a, d_b are the thickness of the active and barrier layers, n_a, n_b, n_c are the refractive indices of the active, barrier, and cladding layers, respectively. The quantity $\overline{n}$ can be interpreted as the effective index of the MQW region (consisting of all the active layers and barrier layers) and γ is the confinement factor in this layer of index $\overline{n}$. The confinement factor Γ of the active region is obtained by multiplying γ by the ratio of the active region thickness to the total thickness of the MQW region (weighted by index).

3.6.2 *Optical gain and auger recombination*

In a QW structure, the kinetic energy of the electrons and holes for motion normal to the well (z-direction) is quantized into discrete energy levels. Each of these discrete levels forms a set of two-dimensional subbands. This modifies the density of states from the regular three-dimensional case. Using the principle of box quantization for kinetic energies in the x- and y-direction, the number of electrons per unit area in the x–y plane for the ith subband within an energy interval dE is given by

$$D_i(E)dE = 2 \frac{d^2\mathbf{k}}{(2\pi)^2} \tag{3.65}$$

where the factor 2 arises from two spin states and $\mathbf{k} = (k_x, k_y)$ is the momentum vector. Using the parabolic approximation, namely $E = \frac{\hbar^2 k^2}{2m_{ci}}$ and the relation $d^2\mathbf{k} = 2\pi k dk$

$$D_i = \frac{m_{ci}}{\pi\hbar^2} \tag{3.66}$$

where m_{ci} is the effective mass of the electron in the ith subband of the conduction band. Thus, the density of states per unit volume g_{ci} is given by

$$g_{ci} = \frac{D_i}{L_z} = \frac{m_{ci}}{\pi\hbar^2 L_z} \tag{3.67}$$

A similar equation holds for the valence band. The density of states for regular three-dimensional case can be similarly obtained (Appendix 5). It is given by $\rho_c(E)$ where,

$$\rho_c(E) = 4\pi \left(\frac{2m_c}{\hbar^2}\right)^{3/2} E^{1/2} \tag{3.68}$$

The main difference between the optical gain of a QW structure and regular DH structure arises from the difference between the density of states in the two cases. For a QW structure, the density of states in a subband does not depend on energy of the electron and that for a regular DH (thick active region) it varies with energy E of the electron as $E^{1/2}$. This is sketched in Figure 3.18. This modification in the density of states can significantly alter the total recombination rate in a QW structure compared to that for regular DH with thick active regions.

The expression for optical absorption or gain is given by Eq. (3.25). For gain, the simplistic view is that the number of electrons in the conduction band at certain energy must be sufficiently high. The number of electrons at a given energy is the product of the density of states and Fermi factor. The latter represents the occupancy probability of an electron of certain energy. The Fermi factor at room temperature approximately has the dependence on energy E of the electron as exp $(-E/kT)$. The density of states for both QW and regular DH, the Fermi factors, and the product of the two

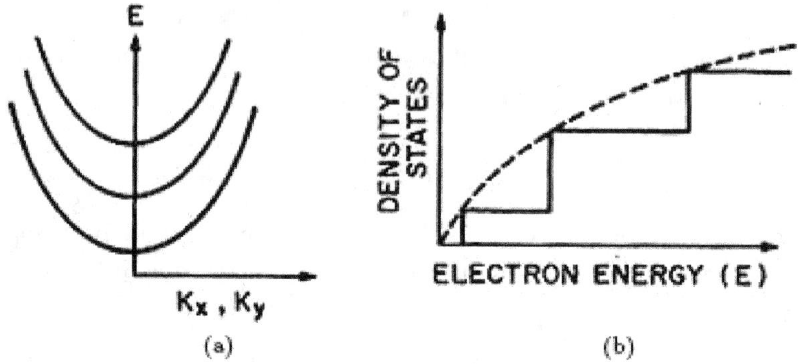

Figure 3.18: (a) Schematic representation of two-dimensional parabolic subbands. (b) The density of states in QW (solid line) and that for thick double heterostructure (dashed line).

are sketched in Figure 3.19. For a regular DH, the density of states is small for energies where the Fermi factor is large, whereas for QW the density of states is large where the Fermi factor is large. This results in high gain at the bottom of the band for a QW whereas for a DH the conduction band must be filled with electrons up to a certain energy level for sufficient optical gain. The QW structure, as a result has high gain at low currents compared to a regular DH. The calculated gain using a simple model is shown in Figure 3.20.

In the simplified gain model [49], the spectral broadening effects and effects resulting from the anisotropy of QW have been ignored. The gain spectrum and the spontaneous emission spectrum exhibit unrealistic sharp edges as a result of the step-like density of states in a QW. The spectral broadening results mostly from intraband carrier relaxation due to scattering. These include carrier–carrier, carrier-phonon, and carrier-impurity scattering. The energy broadening due to intraband scattering δE is much larger than that due to radiative transitions. This is due to vastly different lifetimes. The radiative recombination lifetime $\tau_r \sim 1-2$ ns and the intraband scattering time is $\tau_{\text{in}} \sim 0.1-1$ ps. The quantity δE is related to the lifetime by the uncertainty principle

$$\delta E = \hbar/\tau_{\text{in}} \tag{3.69}$$

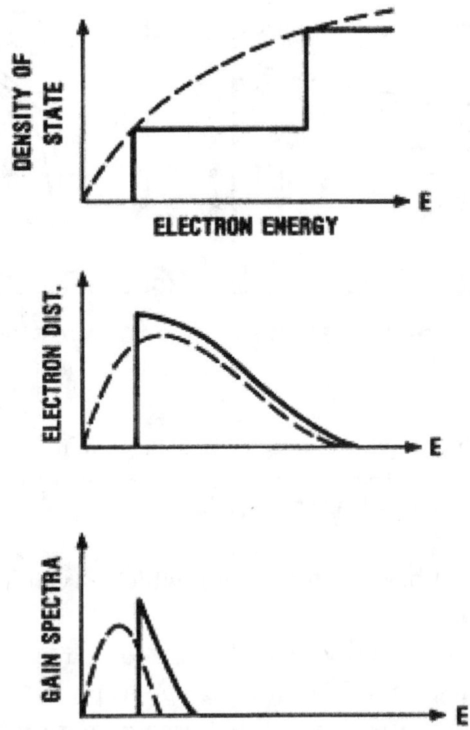

Figure 3.19: The density of states, the electron distribution (product of density of states and Fermi Factor), and gain spectrum are plotted for both QW (solid lines) and regular DH (dashed lines) structure.

Without broadening, the number of electrons per unit energy and per unit volume ($n(E)$) with energies between E and $E + dE$ is

$$n(E) = \rho(E) f_c(E) \tag{3.70}$$

where $\rho(E)$ is the density of states and $f_c(E)$ is the Fermi factor. With broadening the above equation changes to

$$n(E) = \rho(E) f_c(E) L(E - E_1) \tag{3.71}$$

where

$$L(E - E_1) = \frac{\delta E}{2\pi} \frac{1}{(E - E_1)^2 + \left(\frac{\delta E}{2}\right)^2} \tag{3.72}$$

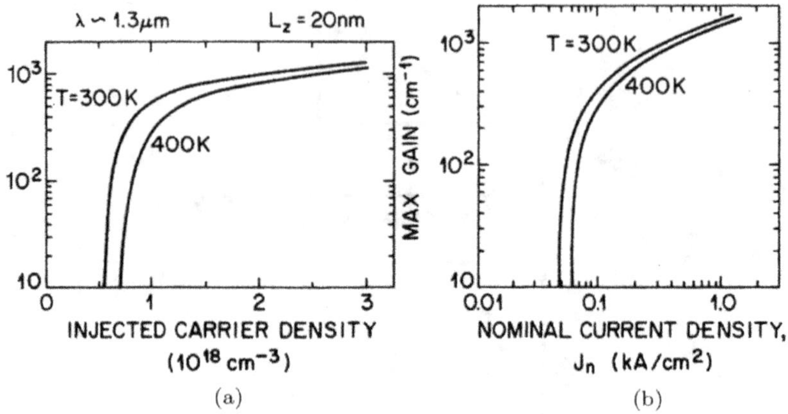

(a) (b)

Figure 3.20: The calculated maximum gain as a function of injected carrier density and nominal current density at two temperatures for a 1.3 μm InGaAsP/InP quantum well structure [49].

is the normalized Lorentzian function which is valid for $\delta E \ll E$ and $\delta E = \hbar/\tau_{\text{in}}$.

For the broad energy distribution of carriers, the unbroadened gain function ($g_u\ (E)$) and the spontaneous emission function ($R_{\text{sp},u}(E)$) is integrated with the Lorentzian broadening factor to obtain the gain spectrum and the spontaneous emission spectrum.

$$g(E) = \int_{-\infty}^{\infty} g_u(E')L(E - E')dE' \tag{3.73}$$

$$R_{\text{sp}}(E) = \int_{-\infty}^{\infty} R_{\text{sp},u}L(E - E')dE' \tag{3.74}$$

The calculated gain is shown in Figure 3.21. The planar symmetry of the electronic wavefunction in a QW structure results in a polarization dependence of the optical transitions. A modification of the transition matrix elements takes place resulting in a difference for the dipole transition matrix elements for the electron–light hole (e-lh) and electron–heavy hole (e-hh) transitions. The e-lh transition is shifted to higher energy from that of the e-hh transition due to the lower effective mass of the light holes. Light polarized in the plane of the QW (TE mode in a waveguide) experiences higher gain than

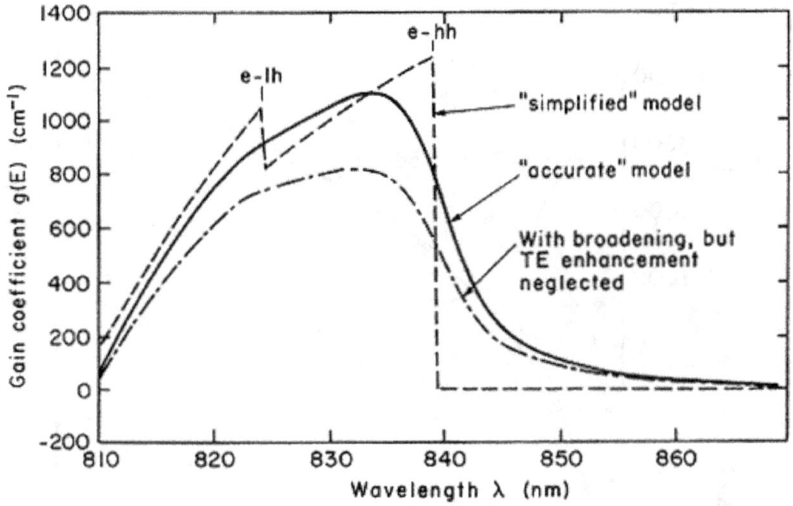

Figure 3.21: Quantum well gain spectrum at $N = 3.04 \times 10^{18}$ cm^{-3} injected carrier density. The intraband scattering time is 0.1 ps. The material is GaAs QW between AlGaAs barrier layers [53].

that for the light polarized perpendicular to the plane of the well (TM mode).

A TE mode gain spectrum calculated using the more complete model is compared with the spectrum calculated using the simple model in Figure 3.21 under the same injection. The peak of the gain spectrum is reduced by broadening. However, the enhancement of the TE mode transition matrix element compensates for the decrease in gain due to broadening. The maximum TE gain as a function of injected carrier density is plotted in Figure 3.22 for the simplified model and for the more accurate model.

Similar calculations have been reported for the InGaAsP material system. The radiative recombination rate in a QW structure taking into account transition between all subbands (in the Γ valley) as well as the filling of the L valleys has been calculated [54]. Figure 3.23 shows the dependence of gain on well thickness for different injected carrier densities. The modulation in the maximum gain is due to the recombination of electrons in the various subbands in the conduction band and heavy and light holes in the valence band. The steep

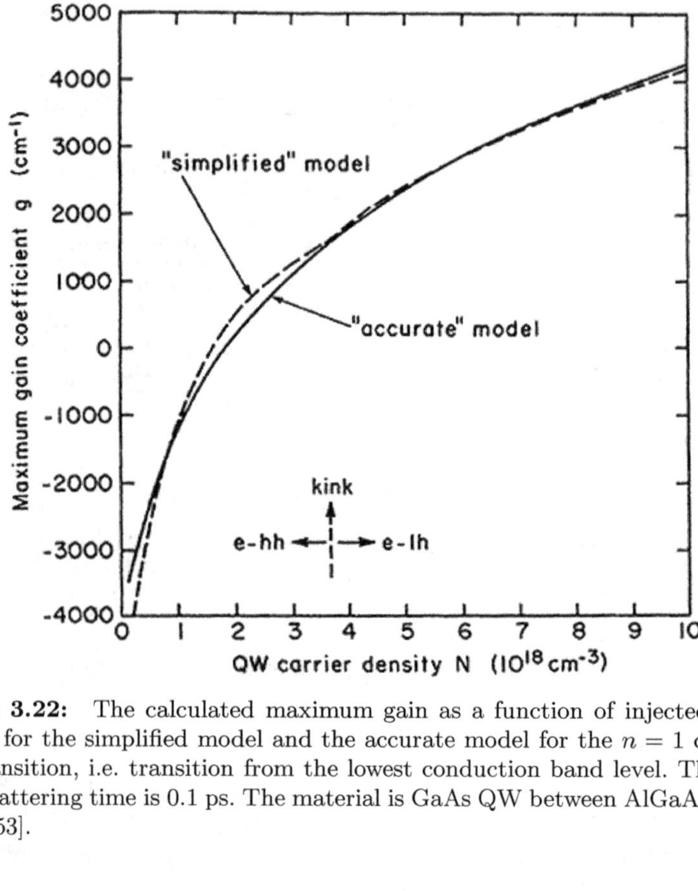

Figure 3.22: The calculated maximum gain as a function of injected carrier density for the simplified model and the accurate model for the $n = 1$ quantum well transition, i.e. transition from the lowest conduction band level. The intraband scattering time is 0.1 ps. The material is GaAs QW between AlGaAs barrier layers [53].

decrease in gain for small well thickness is due to the distribution of carriers in the L valley which reduces the number of carriers in the Γ valley.

Non-radiative Auger recombination of an electron–hole pair, as the name implies, is characterized by the absence of an emitted photon in the recombination process. As in the case of bulk semiconductors, there are several types of Auger recombination processes. The three major types are band-to-band Auger processes, phonon-assisted Auger processes, and trap-assisted Auger processes. The band-to-band Auger processes in direct gap QW semiconductors are shown in Figure 3.24. The three processes are labeled CCCH, CHHS,

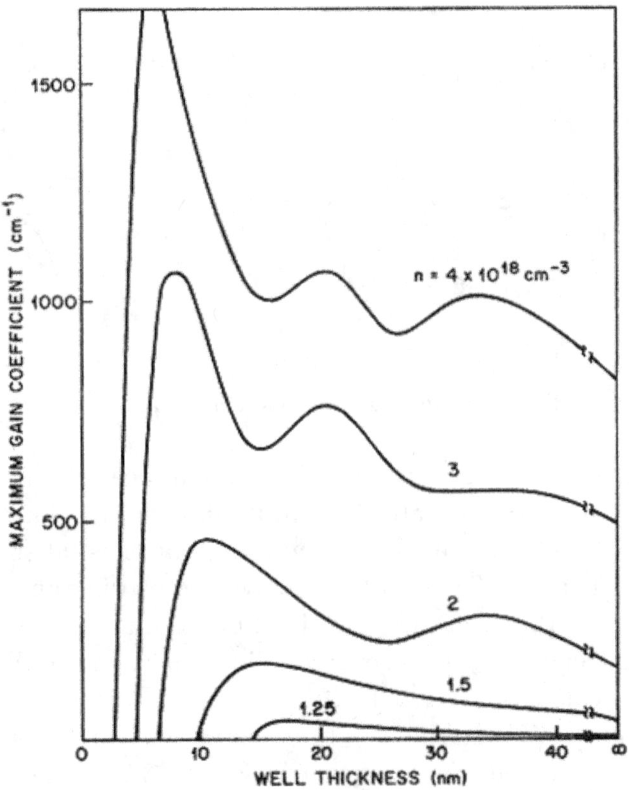

Figure 3.23: Maximum gain for an undoped InGaAs/InP ($\lambda \sim 1070$ nm) quantum well as a function of well thickness for several carrier densities [54].

and CHHL, where C stands for the conduction band and H, L, S stand for heavy- hole, light-hole, and split-off valence band hole, respectively. The CCCH mechanism involves three electrons and a heavy hole and is dominant in n-type material. The CHHS process is dominant in p-type material. Under high injection conditions present in lasers and amplifiers, all three processes must be considered.

Similar to the case for regular DH (3-dimensional density of states) the Auger rate in QW structures is also expected to increase with increasing temperature and for low band gap materials. However, the Auger rate itself is different. In the absence of momentum conservation, there is no threshold energy (Section 3.3.1). Thus, the

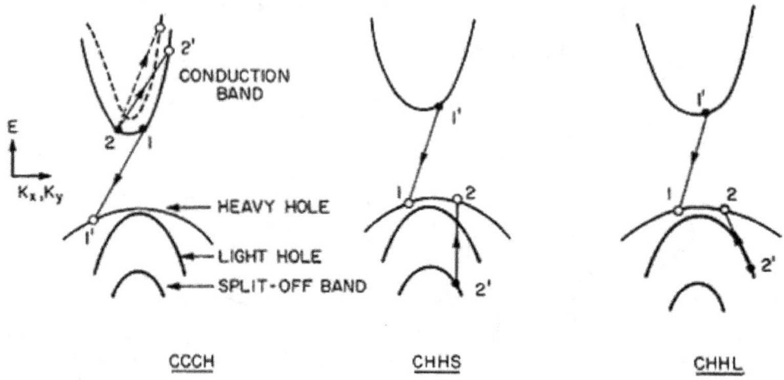

Figure 3.24: Band-to-band Auger processes in a quantum well semiconductor.

strong temperature dependence does not appear if the individual particle states are not states of definite momentum, for example, if they are trap states, or if momentum conservation is satisfied through phonon assistance. The energy vs. wave vector diagram of phonons illustrate that they can carry considerable amount of momentum for a very small energy. The Auger recombination rate for QW structures have been calculated [54–57]. The calculated Auger lifetime τ_A for an InGaAsP QW material as a function of well thickness for several injected carrier densities is shown in Figure 3.25. The undulations in the vicinity of well thickness of ~10 nm are due to the filling of the higher order subbands with increasing n. The value of the Auger coefficient, C, can be estimated from Figure 3.25 using the equation

$$\mathrm{R}_A = \mathrm{n}/\tau_A \tag{3.75}$$

For a well thickness of 30 nm, $n = 2 \times 10^{18}$ cm^{-3} (typical values), $C \sim 2.5 \times 10^{-29}$ cm^{-6} s^{-1} from Figure 3.25. It should be noted that due to various uncertainties in the matrix elements and effective masses only order of magnitude estimates of the Auger rate are feasible. Phonon-assisted Auger recombination rates for QW structures have also been investigated [4, 55]. The Auger processes can be represented by transitions described previously for regular DHs. As expected, the phonon-assisted Auger rates dominate at low temperatures and band-to-band Auger rates are larger at high

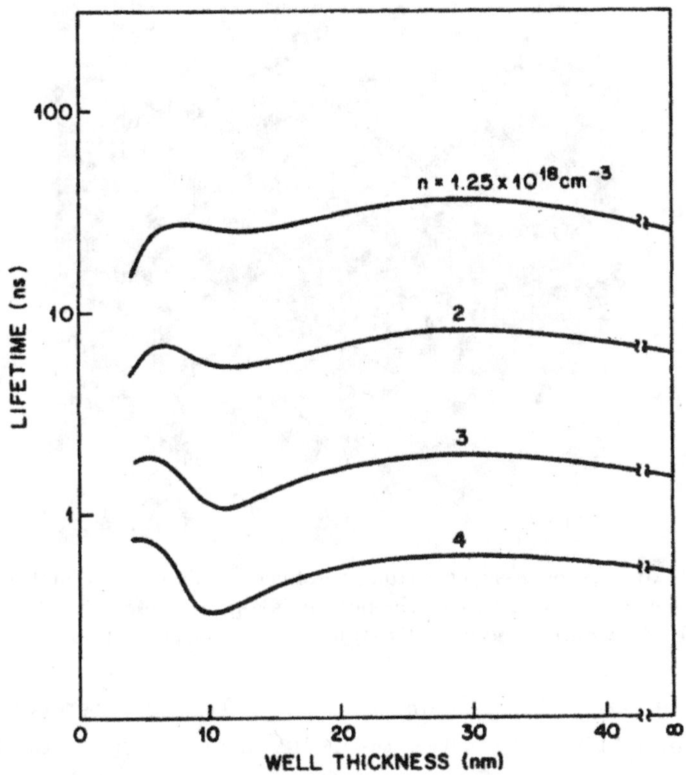

Figure 3.25: Non-radiative Auger lifetime as a function of well thickness for several carrier density values for λ ~1070 nm InGaAsP [54].

temperatures. Trap-assisted Auger processes are also feasible for QW structures.

3.6.3 *InGaAsP multiquantum well lasers*

The development of InGaAsP QW lasers was made possible by the development of MOCVD and gas source MBE growth techniques. The transmission electron micrograph (TEM) of a multiple QW laser structure is shown in Figure 3.26. Shown are four InGaAs QWs grown over n-InP substrate. The well thickness is 70 A and they are separated by barrier layers of InGaAsP ($\lambda \sim$ 1100 nm). MQW lasers with threshold current densities of 600 A/cm^2 have

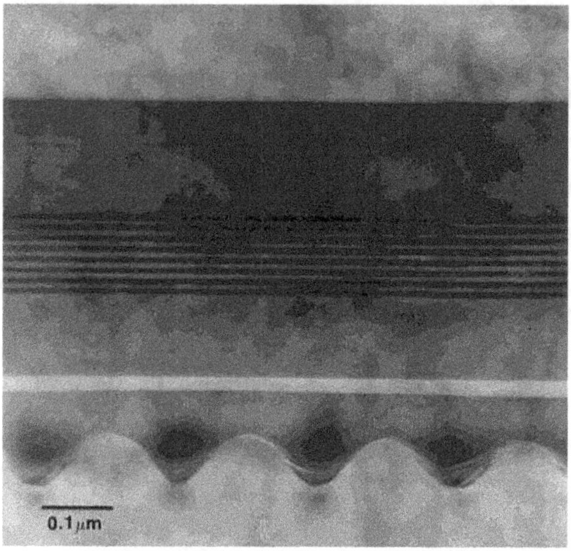

Figure 3.26: Transmission electron photomicrograph of multiquantum well layers. The corrugated structure at the bottom is a grating etched on the substrate needed for distributed feedback (DFB) laser described in Section 3.7.

been fabricated. The schematic of an MQW BH laser is shown in Figure 3.27. The composition of the InGaAsP material from the barrier layers to the cladding layer (InP) is gradually varied in this structure over a thickness of $\sim$0.1 μm. This produces a graded variation in index (GRIN structure) which results in a higher optical confinement of the fundamental mode than that for an abrupt interface design.

Larger mode confinement factor of the GRIN design results in lower threshold current. The laser has an MQW active region and it utilizes Fe- doped semi-insulating (SI) InP layers for current confinement and optical confinement. The laser emits near 1500 nm, has a threshold current of 8 mA at 20°C, and emits >20 mW. The MQW lasers have lower threshold currents than regular DH lasers. Also, the two-dimensional like density of states of the QW lasers makes the transparency current density of these lasers significantly lower than that for regular DH lasers. This allows the fabrication of very low threshold lasers using high reflectivity coatings. The optical gain

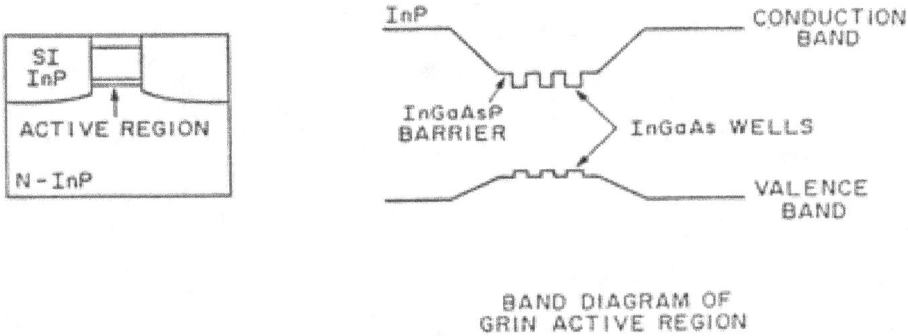

Figure 3.27: Schematic of a multiquantum well buried heterostructure laser. The graded index (GRIN) region is shown on the left. SI InP: Semi-insulating Fe-doped InP.

(g) of a laser at a current density J is given by

$$g = a(J - J_0), \tag{3.76}$$

where a is the gain constant and J_0 is the transparency current density. Although a logarithmic dependence of gain on current density [53] is often used in order to account for gain saturation, a linear dependence is used here for simplicity. The cavity loss α is given by

$$\alpha = \alpha_c + (1/L)\ln(1/R_1R_2), \tag{3.77}$$

where α_c is the free carrier loss, L is the length of the optical cavity, and R_1, R_2 are the reflectivity of the two facets. At threshold, gain equals loss, hence it follows from (3.74) and (3.75) that the threshold current density (J_{th}) is given by

$$J_{\text{th}} = \alpha_c/a + (1/La)\ln(1/R_1R_2) + J_0 \tag{3.78}$$

Thus, for a laser with high reflectivity facet coatings ($R_1, R_2 \sim 1$) and with low loss ($\alpha_c \sim 0$), $J_{\text{th}} \sim J_0$. For a QW laser, $J_0 \sim 50$ A/cm^2 and for a DH laser, $J_0 \sim 700$ A/cm^2 , hence, it is possible to get much lower threshold current using QW as the active region. The light vs. current characteristics of a QW laser with high reflectivity coatings on both facets is shown in Figure 3.28 [58].

The threshold current at room temperature is ~1.1 mA. The laser is 170 μm long and has 90% and 70% reflective coating at the facets.

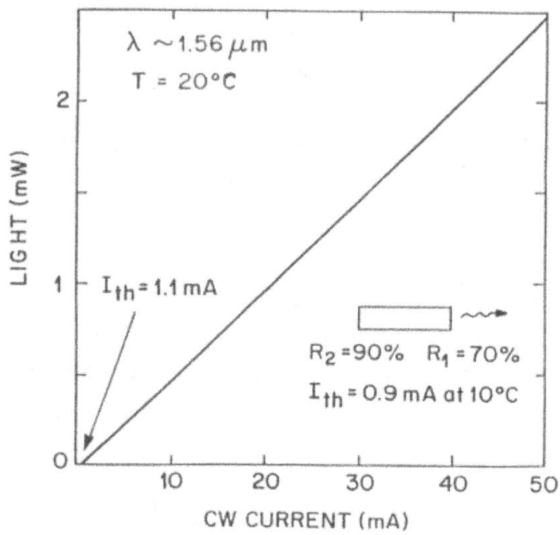

Figure 3.28: Light vs. current characteristics of a multiquantum well laser with high reflectivity coatings [58].

This laser has a compressively strained MQW active region. For lattice matched MQW active region, a threshold current of 2 mA has been reported [34]. Such low threshold lasers are important for array applications. QW lasers have been fabricated which have higher modulation bandwidth than regular DH lasers. The current confinement and optical confinement in this laser are carried out using MOCVD grown Fe-doped InP layers similar to that shown in Figure 3.27. The laser structure is then further modified by using a small contact pad and etching channels around the active region mesa. These modifications are designed to reduce the capacitance of the laser structure for high-speed operation. A 3-dB bandwidth of 25 GHz is obtained [59].

3.6.4 *Strained quantum well lasers*

QW lasers have also been fabricated using an active layer whose lattice constant differs slightly from that of the substrate and cladding layers. Such lasers are known as strained quantum well (SQW) lasers. Over the last two decades, SQW lasers have been

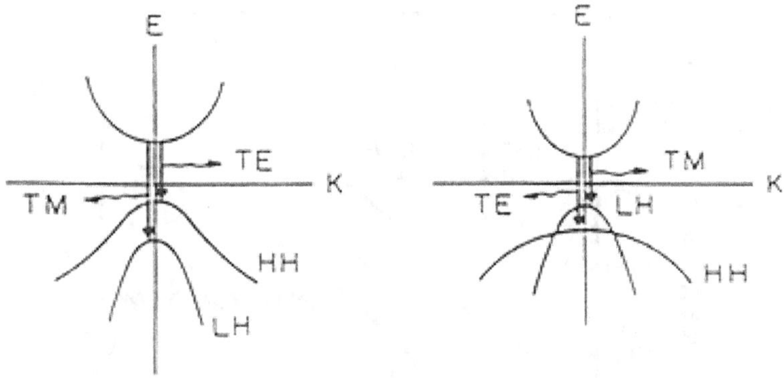

Figure 3.29: Schematic of the band structure of tensile and compressively strained III–V semiconductors.

extensively investigated all over the world [60–68]. They show many desirable properties such as (i) a very low threshold current density and (ii) a lower linewidth than regular MQW lasers both under CW operation and under modulation. The origin of the improved device performance lies in the band-structure changes induced by the mismatch-induced strain [68, 69].

Figure 3.29 shows the band structure of a semiconductor under tensile and compressive strains. Strain splits the heavy-hole and the light-hole valence bands at the Γ point of the Brillouin zone where the band gap is minimum in direct band gap semiconductors. The calculated optical gain as a function of carrier density and radiative current density is shown in Figure 3.30 for strained and unstrained QW material. Two material systems have been widely used for SQW lasers: (i) InGaAs grown over InP by the MOCVD growth technique [59–61] and (ii) InGaAs grown over GaAs by the MOCVD or the MBE growth technique [70–77]. The former material system is of importance for low-chirp semiconductor lasers for lightwave system applications, while the latter material system has been used to fabricate high-power lasers emitting near 980 nm, a wavelength of interest for pumping Er-doped fiber amplifiers.

The alloy $In_{0.53}Ga_{0.47}As$ has the same lattice constant as InP. Semiconductor lasers with an $In_{0.53}Ga_{0.47}As$ active region have been grown on InP by the MOCVD growth technique. Excellent material

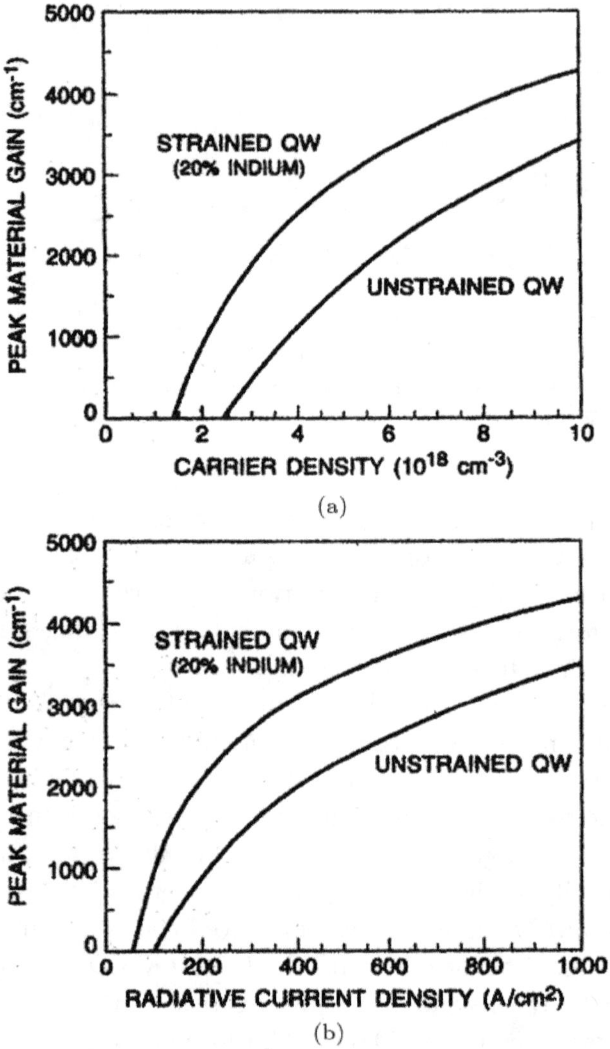

Figure 3.30: Optical gain calculated as a function of injected carrier density (top) and radiative current density (bottom) for strained and unstrained quantum well lasers [67].

quality is also obtained for $In_{1-x}Ga_xAs$ alloys grown over InP by MOCVD for non-lattice-matched compositions. In this case, the laser structure generally consists of one or many $In_{1-x}Ga_xAs$ QW layers with InGaAsP barrier layers whose composition is lattice matched

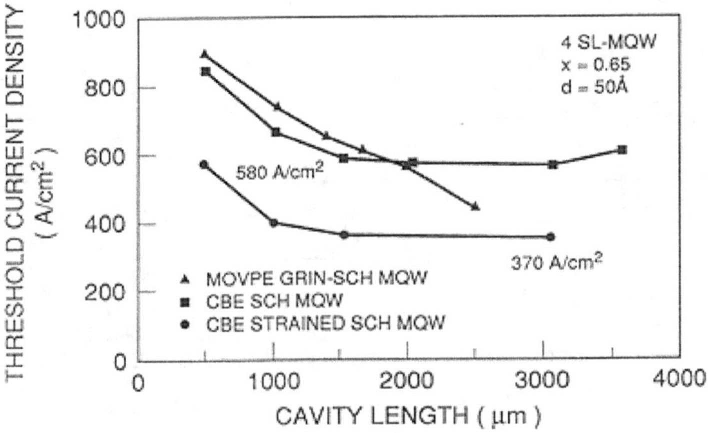

Figure 3.31: Threshold current density as a function of cavity length for strained and regular quantum well laser [62].

to that of InP. For $x < 0.53$ the active layer in these lasers is under tensile stress, while for $x > 0.53$ the active layer is under compressive stress. Superlattice structures of InGaAs/InGaAsP with tensile and compressive stress have been grown by MOCVD growth technique over an n-type InP substrate. Figure 3.31 shows the broad-area threshold current density as a function of cavity length for strained MQW lasers with four $In_{0.65}Ga_{0.35}As$ [39] QWs with InGaAsP ($\lambda \sim 1250$ nm) barrier layers. The active region in this laser is under 0.8% compressive strain. Also, shown for comparison is the threshold current density as a function of cavity length of lattice-matched lasers with $In_{0.53}Ga_{0.47}As$ QWs. The entire laser structure, apart from the QW composition, is identical for the two cases. The threshold current density is lower for the compressively strained MQW structure than that for the lattice-matched MQW structure.

BH lasers have been fabricated using compressive and tensile strained MQW lasers. Lasers with compressive strain have a lower threshold current than that for lasers with tensile strain. This can be explained by splitting of the light-hole and heavy-hole bands under stress [47, 48]. However, more recent studies have shown that it is

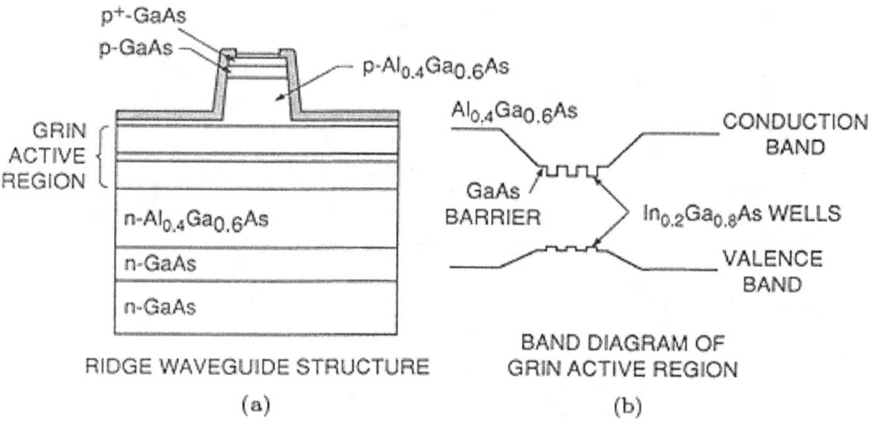

Figure 3.32: Schematic of an InGaAs/GaAs strained quantum well laser structure. The figure on the left is the active region design and the figure on the right is a schematic of a ridge waveguide design.

possible to design tensile strained lasers with lower threshold current [37, 42].

SQW lasers fabricated using $In_{1-x}Ga_xAs$ layers grown over a GaAs substrate have been extensively studied [41, 43, 49–54]. The lattice constant of InAs is 6.06 A and that of GaAs is 5.654 A. The $In_{1-x}Ga_xAs$ alloy has a lattice constant, between these two values, and to a first approximation, it can be assumed to vary linearly with x. Thus, an increase in the In mole fraction x increases the lattice mismatch relative to the GaAs substrate and therefore produces larger compressive strain on the active region. A typical laser structure grown over the n-type GaAs substrate is shown in Figure 3.32 for this material system.

It consists of an MQW active region with one to four $In_{1-x}Ga_xAs$ wells separated by GaAs barrier layers. The entire MQW structure is sandwiched between n- and p-type $Al_{0.3}Ga_{0.7}As$ cladding layers, and the P-cladding layer is followed by a p-type GaAs contact layer. Variations of the above structure with different cladding layers or large optical cavity designs have been reported. Emission wavelength depends on the In composition, x. As x increases, the emission wavelength increases and for x larger than a certain value (typically ~0.25), the strain is too large to yield high-quality material. For x ~0.2, the emission wavelength is near 980 nm, a wavelength region of interest for pumping fiber amplifiers (Chapter 9). Threshold current density as low as 47 A/cm^2 has been reported for $In_{0.2}Ga_{0.8}As/GaAs$ strained MQW lasers [73]. High-power lasers have been fabricated using $In_{0.2}Ga_{0.8}As/GaAs$ MQW active region. Single-mode output powers of greater than 500 mW have been demonstrated using a ridge-waveguide-type laser structure.

Frequency chirp of strained and unstrained QW lasers has been investigated. SQW lasers (InGaAs/GaAs) exhibit the lowest chirp (or dynamic linewidth) under modulation. The high efficiency, high power, and low chirp of strained and unstrained QW lasers make these devices attractive candidates for fiber optic transmission applications.

3.7 Single frequency and tunable lasers

Semiconductor lasers fabricated using the InGaAsP material system is widely used as sources in many lightwave transmission systems. One measure of the transmission capacity of a system is the data rate. Thus, the drive towards higher capacity pushes the systems to higher data rates where the chromatic dispersion of the fiber plays an important role in limiting the distance between regenerators. Sources emitting in a single wavelength help reduce the effects of chromatic dispersion and are therefore used in most systems operating at high data rates (>1.5 Gb/s).

The single wavelength laser source used in most commercial transmission systems is the DFB laser (see for example Ref. [4]) where a diffraction grating etched on the substrate close to the active region provides frequency selective feedback which makes the laser emit in a single wavelength. If Λ is the period of the grating and n is the effective index of the optical mode, the grating reflects light at the free space wavelength (λ) of

$$\lambda = 2n\Lambda \tag{3.79}$$

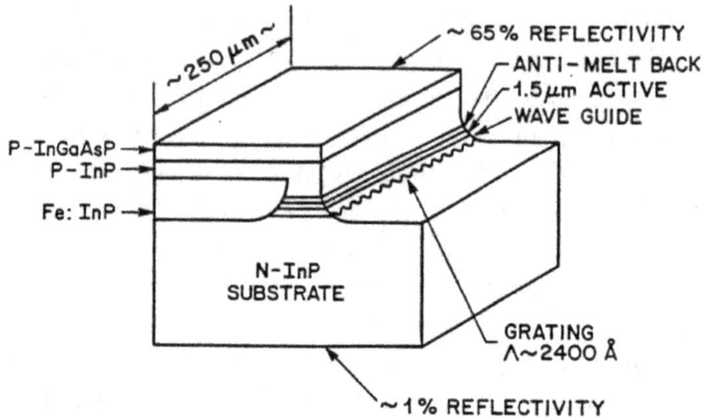

Figure 3.33: Schematic of a distributed feedback laser (DFB laser).

For $\lambda = 1550$ nm and $n \sim 3.3$, this corresponds to $\Lambda \sim 240$ nm. The schematic of a DFB laser structure emitting near 1550 nm is shown in Figure 3.33. The fabrication of the device involves the creation of a grating with a periodicity of 240 nm on a (100) oriented n-InP substrate using optical holography and wet chemical etching (discussed in Chapter 9). Rest of the fabrication follows the steps described in previous sections. The final laser chips have antireflection coating (<1%) at one facet and high reflection coating (~65%) at the other facet. The asymmetric facet coatings help single-mode operation by removing the degeneracy between the two modes in the stop band [4].

The CW light vs. current characteristics of a laser is shown in Figure 3.34. Also shown is the measured spectrum at different output powers. The threshold current of these lasers is in the 15–20 mA range.

For high fiber coupling efficiency, it is important that the laser emits in the fundamental transverse mode. The laser of Figure 3.34 operates in the fundamental transverse mode in the entire operating power range from threshold to 60 mW. The full width at half maximum of the beam divergences normal and parallel to the junction plane are 40° and 30°, respectively.

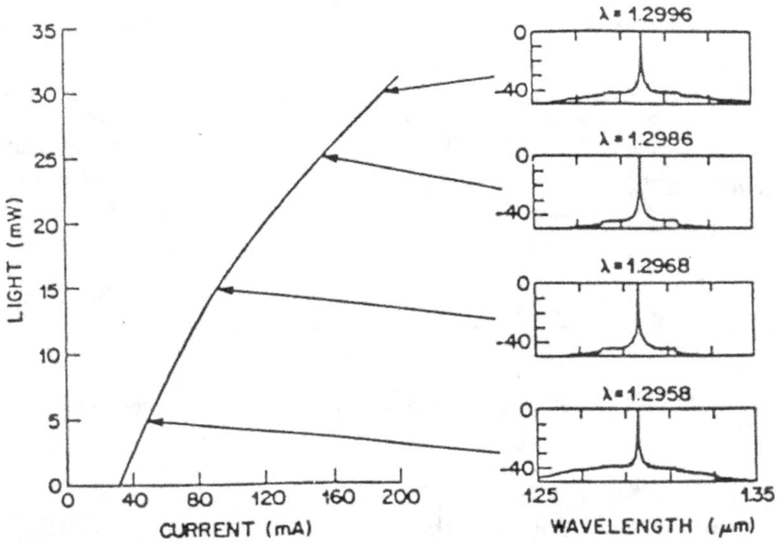

Figure 3.34: Measured Light vs. current characteristics along with optical spectrum at different currents.

3.7.1 *Tunable lasers*

Tunable semiconductor lasers are needed for many applications. Examples of application in lightwave transmission systems are (i) wavelength-division multiplexing where signals at many distinct wavelengths are simultaneously modulated and transmitted through a fiber and (ii) coherent transmission systems where the wavelength of the transmitted signal must match that of the local oscillator. Equation (3.79) shows the grating reflects at wavelength λ which is proportional to n, the effective index. Thus, the laser wavelength can be tuned or varied by varying the effective index n. The index depends on the carrier density (i.e. the current) in the active/waveguide region and temperature of the active region. Thus the index, and hence the laser emission wavelength, can be tuned (or varied) by varying the current through the active region or a waveguide region in a suitably designed laser. Several types of tunable laser structures have been reported in the literature [78–83]. Two principal schemes are (i) multisection DFB laser and (ii) multisection

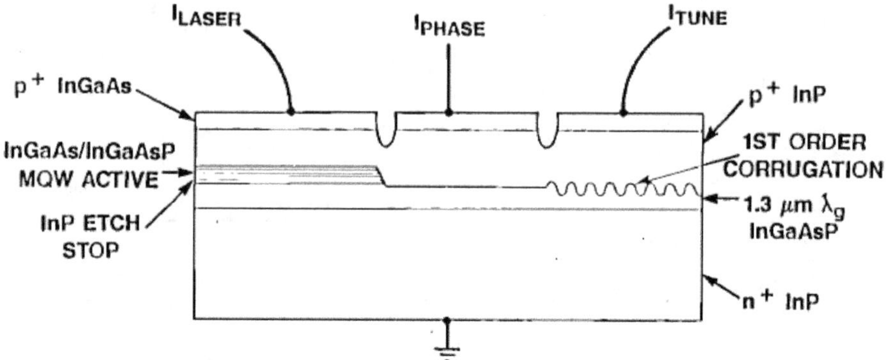

Figure 3.35: Schematic of a three-section distributed Bragg reflector (DBR laser [78]).

distributed Bragg reflector (DBR) laser. The multisection DBR lasers generally exhibit higher tunability than the multisection DFB lasers. The design of a multisection DBR laser is shown in Figure 3.35 schematically [78].

The three sections of this device are (i) the active region section that provides the gain, (ii) the grating section that provides the tunability, and (iii) the phase-tuning section that is needed to access all wavelengths continuously. The current through each of these sections can be varied independently. The tuning mechanism can be understood by noting that the emission wavelength λ of a DBR laser is given by $\lambda = 2n\Lambda$, where Λ is the grating period and n is the effective refractive index of the optical mode in the grating section. The latter can be changed simply by varying the current in the grating section. For a laser to operate, the round trip phase must also be a multiple of 2π. The phase-tuning section allows the adjustment of this phase. In addition, a change in active region current also changes the index of the waveguide. The wavelength tunability of an experimental three-section DBR laser is shown in Figure 3.36 [78]. Measured wavelengths are plotted as a function of phase-section current for different currents in the tuning section. A tuning range in excess of 6 nm can be obtained by controlling currents in the grating and phase-tuning sections.

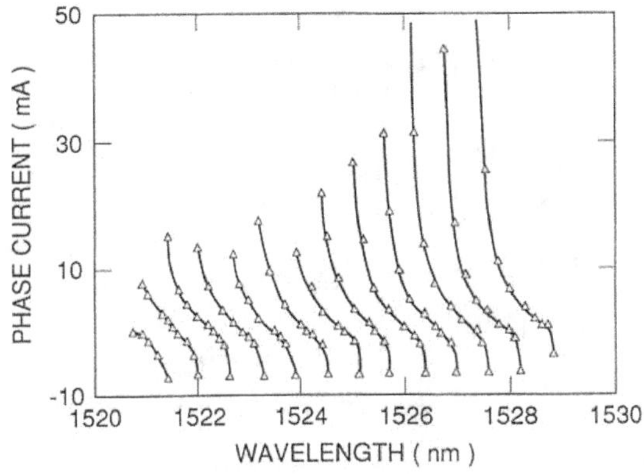

Figure 3.36: Tuning characteristics of the 3-section DBR laser shown in Figure 3.35 [78, 80].

An important characteristic of lasers for applications requiring a high degree of coherence is the spectral width (linewidth) under CW operation. The CW linewidth depends on the rate of spontaneous emission in the laser cavity. Typical CW spectral width of a DFB laser is a few MHz.

Multiple tunable reflectors have been used as wavelength selective elements, and have been incorporated into a single optical cavity. A wide discontinuous tuning can be obtained in a DBR like structure by using two DBRs located at each end of the cavity. If both gratings have a comb reflection spectrum with different spacings of the reflection peaks, lasing is possible at wavelengths where spectrum simultaneously exhibits one reflection peak [84–90]. Comb-reflection spectrum can be realized by a spatial modulation of the Bragg grating. It can be a spatial amplitude modulation (AM, [84, 85]) or spatial frequency modulation (FM, [86, 87]). The former is often known as SG (sampled grating) DBR laser [88], and the latter as SSG (super structure grating) DBR laser [88]. Experimental wavelength tuning characteristics *f* an SSG DBR laser are shown in Figure 3.37. A tuning range of 95 nm is shown.

The wavelength of a DFB laser varies with temperature due to change in index with temperature, and the rate is about 0.1 nm/C. Also, a CW current increases the temperature and causes a shift

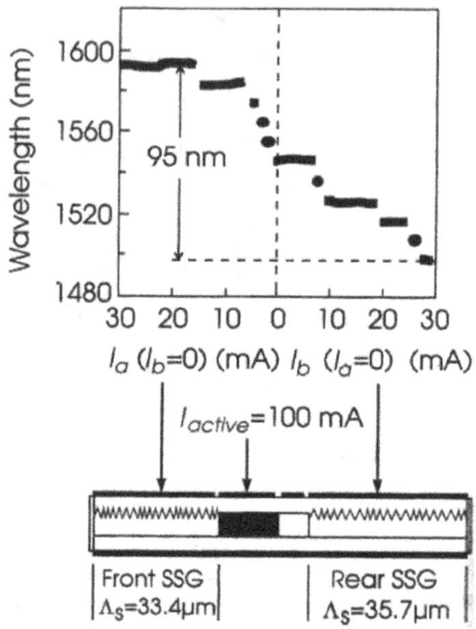

Figure 3.37: Tuning characteristics of an SSG DBR laser. The schematic of the device is shown in the lower figure [88, 89].

in emission wavelength; the rate is about 0.01 nm/C for a typical bonded laser.

3.8 Laser reliability

The performance characteristics of injection lasers used in lightwave systems can degrade during their operation. The dominant mechanism responsible for the degradation is determined by any or all of the several fabrication processes including epitaxial growth, wafer quality, device processing, and bonding [90–102]. In addition, the degradation rate of devices processed from a given wafer depends on the operating conditions, viz., the operating temperature and the injection current. Although many of the degradation mechanisms are not fully understood, extensive amounts of empirical observations exist in the literature which have allowed the fabrication of InGaAsP laser diodes with extrapolated median lifetimes in excess of 25 years at an operating temperature of 20°C [90] (the reliability requirements of a submarine lightwave transmission system are discussed in a special issue Ref. [91]). It is important to have a reasonably accurate estimate of the operating lifetime of lasers (typically >10 years) before they are used in practical systems. This reliability assurance is usually carried out by operating the devices under a high stress (e.g. high temperature) which enhances the degradation rate so that a measurable value can be obtained in an operating time of a few hundred hours. The degradation rate under normal operating conditions can then be obtained from the measured high-temperature degradation rate using the concept of activation energy [102], i.e $R(T) = R_0 \exp(E_a/kT)$, where $R(T)$ is the degradation rate at temperature T, k is the Boltzmann constant, R_0 is a constant, and E_a is the activation energy. The lifetime varies inversely with the degradation rate.

Some lasers exhibit an initial rapid degradation after which the operating characteristics of the lasers are very stable. Given a population of lasers, it is possible to quickly identify the "stable" lasers by a high-stress test (also known as the purge test) [91, 101]. The stress test implies that operating the laser under a set of high-stress conditions (e.g. high current, high temperature, high power) would cause the weak lasers to fail and stabilize the possible winners.

The expected operating lifetime of a semiconductor laser is generally determined by accelerated aging at high temperatures and by using an activation energy. The lifetime (t) at a temperature T is experimentally found to vary as $\exp(-E_a/kT)$, where E_a is the activation energy and k is the Boltzmann constant [102]. The operating current of good lasers increases at a rate of less than 1%/khr of aging time at 60°C operating temperature. Assuming a 50% change in operating current as the useful lifetime of the device and activation energy of 0.7 eV, this aging rate corresponds to a light-emitting lifetime of greater than 100 years at 20°C. The light output vs. current characteristics of a laser changes after stress aging. Aging data for InGaAsP lasers, emitting at 1300 nm, used in the first submarine fiber optic cable is shown in Figure 3.33 [90].

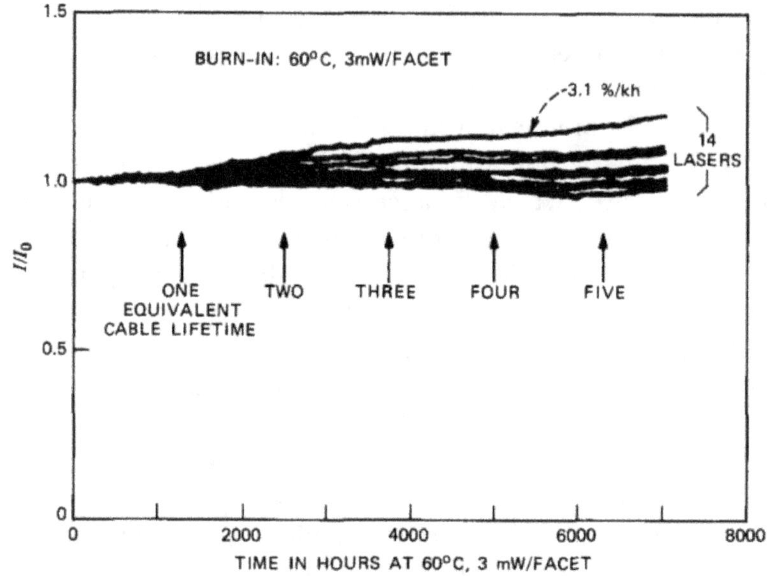

Figure 3.38: Aging data of lasers used in the first transatlantic fiber optic cable in 1988 [90]. An equivalent cable lifetime is 25 years at 10°C.

A parameter that determines the performance of the DFB laser is the side mode suppression ratio (SMSR), i.e. the ratio of the intensity of the dominant lasing mode to that of the next most intense mode. The SMSR of good DFB lasers does not change significantly after aging, which confirms the spectral stability of the emission. For some applications such as coherent transmission systems, the absolute wavelength stability of the laser is important. DFB lasers have been found to have this required wavelength stability.

Semiconductor lasers continue to be the laser of choice for applications in optical communication systems, sensor systems, and many consumer applications such as printers, displays, bar code scanners, etc. They are far more reliable than other types of lasers such as gas lasers and dye lasers. Semiconductor lasers are also used as high-power pump sources for solid-state lasers (slab-based or fiber-based) which are important for many applications. These are discussed in Chapters 4 and 9. With the development of materials and processing technologies, semiconductor lasers at new range of wavelengths and with new designs will continue to emerge as important light sources for new applications.

References

1. H. Kressel and J. K. Butler, *Semiconductor Lasers and Heterojunction LEDs*, Academic Press, NY, 1977.
2. H. C. Casey, Jr. and M.B. Panish, *Heterostructure Lasers*, Academic Press, NY, 1978.
3. G. H. B. Thompson, *Physics of Semiconductor Lasers*, John Wiley and Sons, NY, 1980.
4. G. P. Agrawal and N. K. Dutta, *Long Wavelength Semiconductor Lasers*, van Nostrand Reinhold Co. 1986 (2nd Edn., 1993.
5. N. Holonyuk, Jr. and S. F. Bevacqua, *Appl. Phys. Lett.*, **1** (1962) 82.
6. M. I. Nathan, W. P. Dumke, G. Burns, F. H. Dill, Jr., and G. Lasher, *Appl. Phys. Lett.*, **1** (1962) 63.
7. T. M. Quist, R. H. Rediker, R. J. Keyes, W. E. Krag, B. Lax, A. L. McWhorter, and H. J. Ziegler, *Appl. Phys. Lett.*, **1** (1962) 91 ; R. N. Hall, G. E. Fenner, J. D. Kingsley, T. J. Soltys and R. O. Carlson, *Phys. Rev. Lett.*, **9** (1962).
8. I. Hayashi, M. B. Panish, P. W. Foy, and S. Sumsky, *Appl. Phys. Lett.*, **47** (1970) 109.
9. J. J. Hsieh, *Appl. Phys. Lett.*, **28** (1976) 283.
10. H. Kroemer, *Proc. IEEE*, **51** (1963) 1782.
11. G. P. Agrawal and N. K. Dutta, *Semiconductor Lasers*, Section 2.5, Van Nostrand Reinhold, NY, 1992.
12. D. Botez, *IEEE J. Quant. Electron.*, **QE-17** (1981) 178.
13. M. G. A. Bernard and G. Duraffourg, *Phys. Status Solid*, **1** (1961) 699.
14. E. O. Kane, *J. Phys. Chem. Solids*, 1: 249 (1957)
15. F. Stern, *J. Appl. Phys.* **47**, 5382 (1976); *IEEE J. Quant. Electron.*, **QE-9** (1973) 290.
16. C. J. Hwang, *Phys. Rev. B*, **2** (1970) 4117.
17. N. K. Dutta, *J. Appl. Phys.*, **51** (1980) 6095; **52** (1981) 55.
18. A. R. Beattie and P. T. Landsberg, *Proc. R. Soc. London*, **249** 16 (1959); *Ser. A*, **258** (1960) 486.
19. Y. Horikoshi and Y. Furukawa, *Jap. J. Appl. Phys.*, **18** (1979) 809.
20. G. H. B. Thomson and G. D. Henshall, *Electron. Lett.*, **16** (1980) 42.
21. N. K. Dutta and R. J. Nelson, (a) *Appl. Phys. Lett.* 38: **407** (1981); (b) *J. Appl. Phy.*, **53** (1982) 74 and references therein.
22. A. Sugimura, *IEEE J. Quantum Electron.*, **QE-17** (1981) 627.
23. A. Haug, *Appl. Phys. Lett.*, **42** (1983) 512.
24. R. E. Nahory, M. A. Pollack, W. D. Johnston, Jr., and R. L. Barns, *Appl. Phys. Lett.*, **33** (1978) 659.
25. See Chapter 3, Ref. 4.

26. P. Morton, R. A. Logan, T. Tanbun-Ek, *Electronics Lett.*, **28** (1992) 2156.
27. N. Holonyak, Jr., R. M. Kolbas, R. D. Dupuis, and P. D. Dapkus, *IEEE J. Quant. Electron.*, **QE-16** (1980) 170.
28. N. Holonyak, Jr., R. M. Kolbas, W. D. Laidig, B. A. Vojak, K. Hess, R. D. Dupuis, and P. D. Dapkus, *J. Appl. Phys.*, **5** (1980) 1328.
29. W. T. Tsang, *Appl. Phys. Lett.*, **39** (1981) 786.
30. W. T. Tsang, *IEEE J. Quant. Electron.* **QE-20** (1986) 1119.
31. S. D. Hersee, B. DeCremoux, and J. P. Duchemin, *Appl. Phys. Lett.*, **44** (1984).
32. F. Capasso and G. Margaritondo (eds.), *Heterojunction Band Discontinuation: Physics and Applications,* Amsterdam, North Holland, 1987.
33. N. K. Dutta, S. G. Napholtz, R. Yen, R. L. Brown, T. M. Shen, N. A. Olsson, and D. C. Craft, *Appl. Phys. Lett.*, **46** (1985) 19.
34. N. K. Dutta, S. G. Napholtz, R. Yen, T. Wessel, and N. A. Olsson, *Appl. Phys. Lett.*, **46** (1985) 1036.
35. N. K. Dutta, T. Wessel, N. A. Olsson, R. A. Logan, R. Yen, and P. J. Anthony, *Electron. Lett.*, **21** (1985) 571.
36. N. K. Dutta, T. Wessel, N. A. Olsson, R. A. Logan, and R. Yen, *Appl. Phys. Lett.*, **46** (1985) 525.
37. Y. Arakawa and A. Yariv, *IEEE J. Quant. Electron.*, **QE-21** (1985) 1666.
38. Y. Arakawa and A. Yariv, *IEEE J. Quant. Electron.*, **QE-22** (1986) 1887.
39. A. Yariv, C. Lindsey, and V. Sivan, *J. Appl. Phys.*, **58** (1985) 3669.
40. A. Sugimura, *IEEE J. Quantum Electron.*, **QE-20** (1984) 336.
41. A. Sugimura, *Appl. Phys. Lett.*, **43**, (1983) 728.
42. N. K. Dutta and R. J. Nelson, *J. Appl. Phys.*, **53** (1982) 74.
43. L. C. Chiu and A. Yariv, *IEEE J. Quant. Electron.*, **QE-18** (1982) 1406.
44. N. K. Dutta, *J. Appl. Phys.*, **54** (1983) 1236.
45. A. Sugimura, *IEEE J. Quant. Electron.*, **QE-19** (1983) 923.
46. C. Smith, R. A. Abram, and M. G. Burt, *J. Phys., C* **16** (1983) L171.
47. H. Temkin, N. K. Dutta, T. Tanbun-Ek, R. A. Logan, and A. M. Sergent, *Appl. Phys. Lett.*, **57** (1990) 1610.
48. K. Hess, B. A. Vojak, N. Holonyak, Jr., R. Chin, and P. D. Dapkus, *Solid State Electron.*, **23** (1980) 585.
49. N. K. Dutta, (a) *Electron. Lett.*, **18**, 451 (1982); (b) *J. Appl. Phys.*, **53** (1982) 7211.
50. Y. Arakawa and A. Yariv, *IEEE J. Quant. Electron.*, **QE-21** (1985) 1666.
51. H. Temkin, N. K. Dutta, T. Tanbun-Ek, R. A. Logan, and A. M. Sergent, *Appl. Phys. Lett.*, **57** (1990) 1610.
52. M. Asada and Y. Suematsu, *IEEE, J. Quant. Electron.*, **QE-21** (1985) 434; M. Asada, *IEEE J. Quant. Electron.*, **QE-25** (1989) 2019.
53. R. W. H. Engelmann, C. L. Shieh, and C. Shu, Chapter 3, in P. Zory (ed.), *Quantum Well Lasers*, Academic Press, NY, 1993.
54. A. Sugimura, *IEEE J. Quant. Electron.*, **QE-19** (1983) 932.
55. N. K. Dutta, *J. Appl. Phys.*, **54** (1983) 1236.
56. L. C. Chiu and A. Yariv, *IEEE J. Quant. Electron.*, **QE-18** (1982) 1406.

57. C. Kazmierski, A. Ougazzaden, M. Blez, D. Robien, J. Landreau, B. Sermage, J. C. Bouley, and A. Mirca, *IEEE J. Quant. Electron.*, **27** (1991) 1794.
58. H. Temkin, T. Tanbun-Ek, and R. A. Logan, *Appl. Phys. Lett.*, **56** (1990) 1210.
59. P. Morton, R. A. Logan, T. Tanbun-Ek, P. F. Sciortino, Jr., A. M. Sergent, R. K. Montgomery, and B. T. Lee, *Electron. Lett.*, **29** (1993) 1429.
60. P. J. A. Thijs, L. F. Tiemeijer, P. I. Kuindersma, J. J. M. Binsma, and T. van Dongen, *IEEE J. Quantum Electron.*, **27** (1991) 1426.
61. P. J. A. Thijs, L. F. Tiemeijer, J. J. M. Binsma, and T. van Dongen IEEE, *J. Quant. Electron.*, **QE-30** (1994) 477.
62. W. T. Tsang, L. Yang, M. C. Wu, Y. K. Chen, and A. M. Sergent, *Electron. Lett.*, (1990) 2033.
63. W. D. Laidig, Y. F. Lin, and P. J. Caldwell, *J. Appl. Phys.*, **57** (1985) 33.
41. S. E. Fischer, D. Fekete, G. B. Feak, and J. M. Ballantyne, *Appl. Phys. Lett.*, **50** (1987) 714.
64. N. Yokouchi, N. Yamanaka, N. Iwai, Y. Nakahira, and A. Kasukawa, *IEEE J. Quant. Electron.*, **QE-32** (1996) 2148.
65. K. J. Beernik, P. K. York, and J. J. Coleman, *Appl. Phys. Lett.*, **25** (1989) 2582.
66. J. P. Loehr and J. Singh, *IEEE J. Quant. Electron.*, **27** (1991) 708.
67. S. W. Corzine, R. Yan, and L. A. Coldren, "Optical Gain in III-V Bulk and Quantum Well Semiconductors" in*Quantum Well Lasers*, P. Zory (ed.), Academic Press, NY, USA.
68. H. Temkin, T. Tanbun-Ek, R. A. Logan, D. A. Coblentz, and A. M. Sergent, *IEEE Photonic Technol. Lett.*, **3** (1991) 100.
69. A. R. Adams, *Electron. Lett.*, **22** (1986) 249.
70. M. C. Wu, Y. K. Chen, M. Hong, J. P. Mannaerts, M. A. Chin, and A. M. Sergent, *Appl. Phys. Lett.*, **59** (1991) 1046.
71. N. K. Dutta, J. Lopata, P. R. Berger, D. L. Sivco, and A. Y. Cho, *Electron. Lett.*, **27** (1991) 680.
72. J. M. Kuo, M. C. Wu, Y. K. Chen, and M. A. Chin, *Appl. Phys. Lett.*, **59** (1991) 2781.
73. N. Chand, E. E. Becker, J. P. van der Ziel, S. N. G. Chu, and N. K. Dutta, *Appl. Phys. Lett.*, **58** (1991) 1704.
74. H. K. Choi and C. A. Wang, *Appl. Phys. Lett.*, **57** (1990) 321.
75. N. K. Dutta, J. D. Wynn, J. Lopata, D. L. Sivco, and A. Y. Cho, *Electron. Lett.*, **26** (1990) 1816.
76. C. E. Zah, R. Bhat, B. N. Pathak, F. Favire, W. Lin, D. M. Hwang, T. P. Lee, Z. Wang, D. Darby, D. Flanders, and J. J. Hsieh, *IEEE, J. Quant. Electron.*, **QE-30** (1994) 511.
77. A. Kusukawa, T. Namegaya, T. Fukushima, N. Iwai, and T. Kikuta, *IEEE, J. Quant. Electron.*, **QE-27** (1993) 1528.
78. T. L. Koch, U. Koren, R. P. Gnall, C. A. Burrus, and B. I. Miller, *Electron. Lett.*, **24** (1988) 1431.

79. Y. Suematsu, S. Arai, and K. Kishino, *J. Lightwave. Tech.*, **LT-1** (1983) 161.
80. T. L. Koch and U. Koren, *IEEE J. Quant. Electron.*, **QE-27** (1991) 641.
81. N. K. Dutta, A. B. Piccirilli, T. Cella, and R. L. Brown, *Appl. Phys. Lett.*, **48** (1986) 1501.
82. K. Y. Liou, N. K. Dutta, and C. A. Burrus, *Appl. Phys. Lett.*, **50** (1987) 489.
83. T. Tanbun-Ek, R. A. Logan, S. N. G. Chu, and A. M. Sergent, *Appl. Phys. Lett.*, **57** (1990) 2184.
84. V. Jayaraman, D. A. Cohen, and L. A. Coldren, *Appl. Phys. Lett.*, **60** (1992) 2321.
85. V. Jayaraman, A. Mathur, L. A. Coldren, and P. D. Dapkus, *IEEE Photon. Tech. Lett.*, **5** (1993) 489.
86. Y. Tohmori, Y. Yoshikuni, F. Kano, H. Ishii, T. Tamamura, and Y. Kondo, *Electron. Lett.*, **29** (1993) 352.
87. Y. Yoshikuni *et al.*, *Proc. Optical Fiber Comm. Conf.*, 1993.
88. M. C. Amann, Chapter 3, Single Mode and Tunable Laser Diodes in E. Kapon (ed.), *Semiconductor Laser II*, Academic Press, NY, USA, 1999.
89. H. Ishii, Y. Tohmori, M. Yamamoto, T. Tamamura, and Y. Yoshikuni, *Electron. Lett.*, **30** (1994) 1141.
90. F. R. Nash, W. J. Sundburg, R. L. Hartman, J. R. Pawlik, D. A. Ackerman, N. K. Dutta, and R. W. Dixon, *AT&T Tech. J.*, **64** (1985) 809.
91. N. K. Datta, *AT&T Tech. J.*, **64** (1985) 3.
92. B. C. DeLoach, Jr., B. W. Hakki, R. L. Hartman, and L. A. D'Asaro, *Proc. IEEE.*, **61** (1973) 1042.
93. P. M. Petroff and R. L. Hartman, *Appl. Phys. Lett.*, **2** (1973) 469.
94. P. M. Petroff, W. D. Johnston, Jr., and R. L. Hartman, *Appl. Phys. Lett.*, **25** (1974) 226.
95. J. Matsui, R. Ishida, and Y. Nannichi, *Jpn. J. Appl. Phys.*, **14** (1975) 1555.
96. P. M. Petroff and D. V. Lang, *Appl. Phys. Lett.*, **31** (1977) 60.
97. O. Ueda, I. Umebu, S. Yamakoshi, and T. Kotani, *J. Appl. Phys.*, **53** (1982) 2991.
98. O. Ueda, S. Yamakoshi, S. Komiya, K. Akita, and T. Yamaoka, *Appl. Phys. Lett.*, **36** (1980) 300.
99. S. Yamakoshi, M. Abe, O. Wada, S. Komiya, and T. Sakurai, *IEEE J. Quant. Electron.*, **QE-17** (1981) 167.
100. K. Mizuishi, M. Sawai, S. Todoroki, S. Tsuji, M. Hirao, and M. Nakamura, *IEEE J. Quant. Electron.*, **QE-19** (1983) 1294.
101. E. I. Gordon, F. R. Nash, and R. L. Hartman, *IEEE Electron. Device Lett.*, **ELD-4** (1983) 465.
102. R. L. Hartman and R. W. Dixon, *Appl. Phys. Lett.*, **26** (1975) 239.

Chapter 4

Semiconductor Laser — II

4.1 Introduction

As mentioned in the previous chapter, rapid advances in research results, development, and application of semiconductor lasers have occurred over the last few decades. This chapter describes the fabrication and performance characteristics of vertical cavity surface-emitting lasers, mid-infrared quantum cascade lasers, quantum dot (QD) and nanostructure-based lasers, high power lasers, and mode-locked semiconductor lasers.

4.2 Vertical cavity surface-emitting lasers

Semiconductor lasers described in the previous chapters have cleaved facets that form the optical cavity. The facets are perpendicular to the surface of the wafer and light is emitted parallel to the surface of the wafer. For many applications requiring a two-dimensional laser array or monolithic integration of lasers with electronic components (e.g. optical interconnects), it is desirable to have the laser output normal to the surface of the wafer. Such lasers are generally known as surface-emitting lasers (SEL). A class of SELs also have optical cavity normal to the surface of the wafer [1–8]. These devices are known as vertical-cavity surface-emitting lasers (VCSEL) in order to distinguish them from other surface emitters. These lasers are discussed in this section.

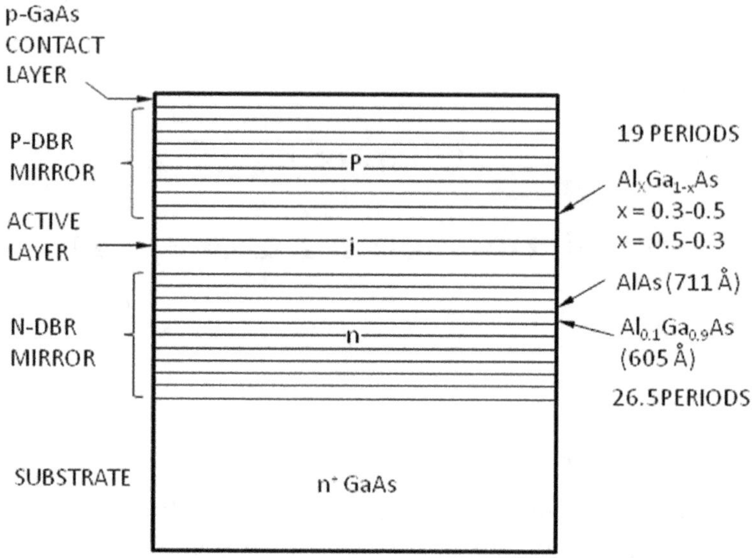

Figure 4.1: A schematic of semiconductor layers for a typical vertical cavity surface-emitting laser. For the DBR mirrors, AlAs has the low index and $Al_{0.1}Ga_{0.9}As$ has the high index.

4.2.1 *Design and Bragg mirror reflectivity*

A generic SEL structure utilizing multiple semiconductor layers to form a Bragg reflector is shown in Figure 4.1. The active region is sandwiched between n- and p-type cladding layers, which are sandwiched between the two n- and p-type multiple semiconductor layers each quarter wavelength thick. These layers form distributed Bragg reflectors for the laser cavity. This structure (Figure 4.1) is shown using the AlGaAs/GaAs material system, which has been very successful in the fabrication of SELs. The Bragg mirrors consist of alternating layers of low-index and high-index materials.

The thickness of each layer is one quarter of the wavelength of light in the medium. Such periodic quarter wave thick layers can have very high reflectivity. For normal incidence, the reflectivity is given by [9, Appendix 10]:

$$R = \frac{(1 - n_4/n_1(n_2/n_3)^{2N})^2}{(1 + n_4/n_1(n_2/n_3)^{2N})^2} \tag{4.1}$$

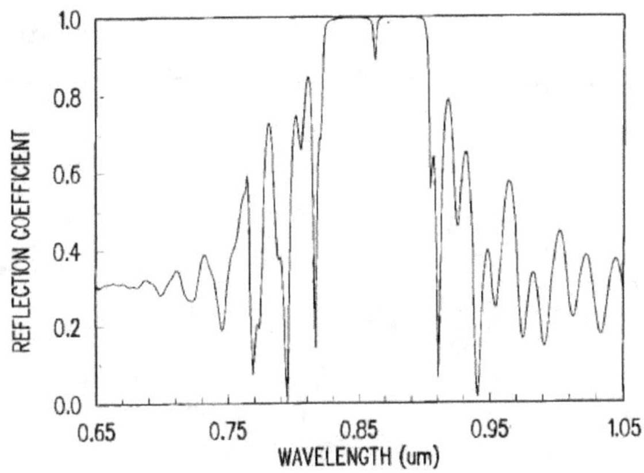

Figure 4.2: Measured reflectivity of a typical Bragg reflector stack of layer used for SEL fabrication.

where n_2, n_3 are the refractive indices of the alternating layer pairs, n_4, n_1 are the refractive indices of the medium on the transmitted and incident sides of the DBR mirror, and N is the number of pairs. As N increases, R increases. Also, for a given N, R is larger if the ratio of n_2/n_3 is smaller. For AlAs/$Al_{0.1}Ga_{0.9}As$ set of quarter wave layers, typically 20 pairs are needed for a reflectivity of ∼99.5%. Various types of AlGaAs/GaAs SELs have been reported [10–18].

For an SEL to have a threshold current density comparable to that of an edge-emitting laser, the threshold gains must be comparable for the two devices. The threshold gain of an edge-emitting laser is ∼100 cm^{-1}. For an SEL with an active-layer thickness of 0.5 μm, this value corresponds to a single-pass gain of ∼0.5%. Thus, for the SEL device to operate with a threshold current density comparable to that of an edge emitter, the reflectivity of mirrors must be >99%. The reflectivity spectrum of an SEL structure is shown in Figure 4.2. The reflectivity is >99% over a 10 nm band. The drop in reflectivity in the middle of the band is due to the Fabry–Perot mode formed by the two Bragg mirrors.

The number of pairs needed to fabricate a high-reflectivity mirror depends on the refractive index of layers in the pair. For large index

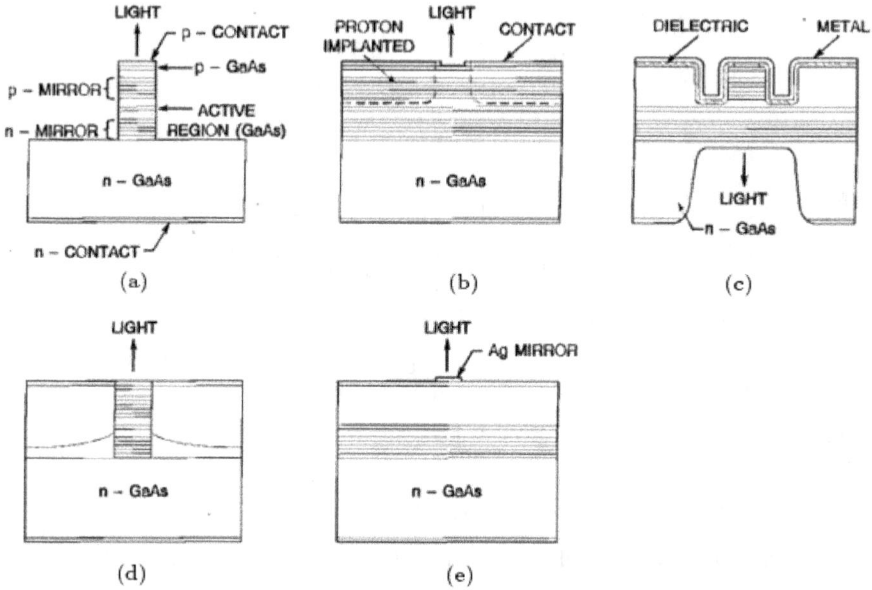

Figure 4.3: Five typical SEL structures for low-threshold laser fabrication.

differences fewer pairs are needed. For example, in the case of CaF_2 and ZnS pairs for which the index difference is 0.9, only six pairs are needed for a reflectivity of 99%. By contrast, for an InP/InGaAsP ($\lambda \sim 1300$ nm) layer pair for which the index difference is 0.25, more than 40 pairs are needed to achieve a reflectivity of 99%. The active region (light-emitting region) and the semiconductor layers needed for Bragg reflector are fabricated in a single-growth step using the molecular beam epitaxy (MBE) or the metal organic chemical vapor deposition (MOCVD) growth techniques. These techniques are described in Chapter 2.

4.2.2 *GaAs-based VCSEL*

The five principal structures (Figure 4.3) used in GaAs/AlGaAs-based SEL fabrication are (a) etched mesa structure, (b) ion-implanted structure, (c) dielectric isolated structure, (d) buried heterostructure, and (e) metallic reflector structure. Variations of these structures and that shown in Figure 4.4 (oxide aperture) are

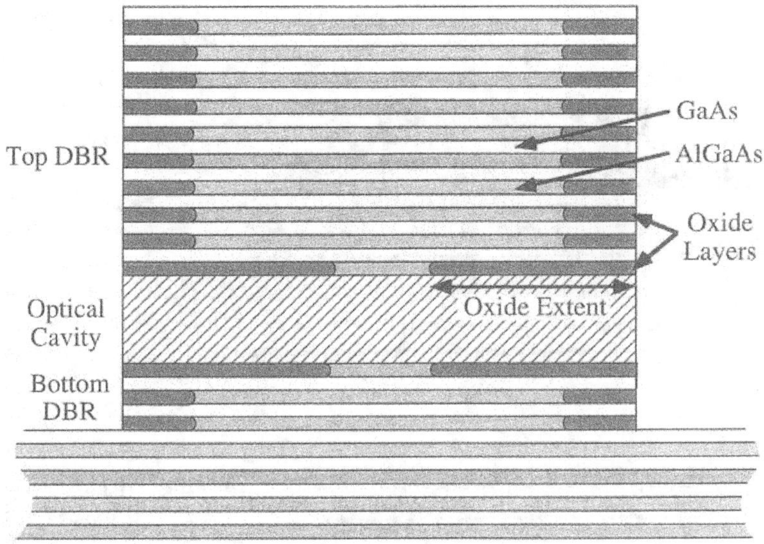

Figure 4.4: Schematic of an oxide aperture SEL structure.

currently used in commercial applications. The threshold current of ~0.3 mA have been reported for InGaAs/GaAs SEL devices. An SEL design has been demonstrated whose output can be focused to a single spot [16]. The laser has a large area (~100 μm dia) and it has a Fresnel zone like structure etched on the top mirror. The laser emits 500 mW in a single mode. An important SEL structure for the AlGaAs/GaAs material system is the oxide aperture device [17] (Figure 4.4). AlAs has the property that it oxidizes rapidly in the presence of oxygen to aluminum oxide which forms an insulating layer. Thus by introducing a thin AlAs layer in the device structure it is possible to confine the current to a very small area. This allows the fabrication of lasers with very low threshold (<0.2 mA) and high bandwidth (~14 GHz).

4.2.3 *InGaAsP-based VCSEL*

A key to the fabrication of low-threshold SELs is the ability to fabricate high-reflectivity mirrors. In the late 1970s, Soda *et al.* [1] reported on an SEL fabricated using the InP material system

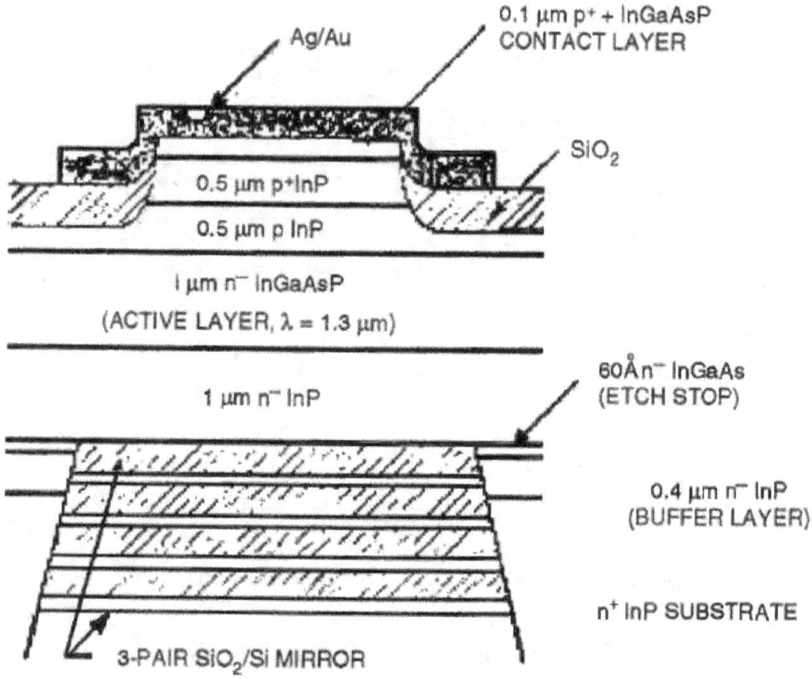

Figure 4.5: Schematic of an InGaAsP/InP SEL structure [1].

(Figure 4.5). The surfaces of the wafer form the Fabry–Perot cavity of the laser. Fabrication of the device involves the growth of a double heterostructure on an n-InP substrate. A circular contact is made on the p-side using a SiO_2 mask. The substrate side is polished making sure that it is parallel to the epitaxial layer, and ring electrodes (using an alloy of Au–Sn) are deposited on the n-side. The laser had a threshold current density of $\sim$1 kA/cm^2 at 77°K and operated at output powers of several mWs.

InGaAsP/InP SELs have been investigated by many researchers over the last few years [1, 19–24]. Many of the schemes utilize alternating layers of InP and InGaAsP to produce Bragg mirrors [21]. The refractive index difference between InGaAsP ($\lambda \sim 1300$ nm) and InP layers is smaller than that in GaAs SELs, hence, InGaAsP/InP SELs utilize more pairs (typically 40–50) to produce a high-reflectivity

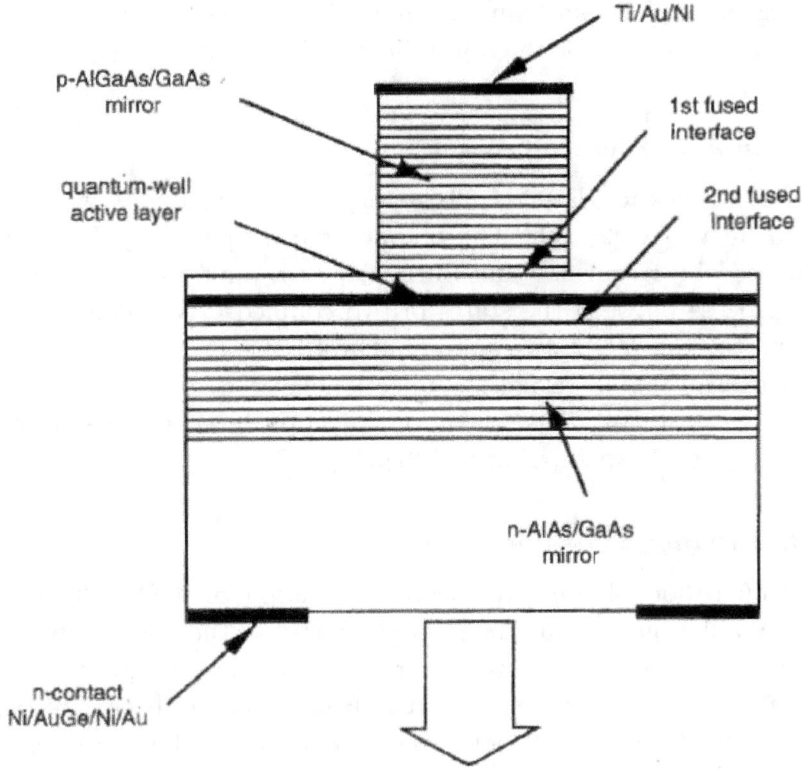

Figure 4.6: Schematic of an SEL emitting near 1500 nm fabricated using wafer bonding [24].

(>99%) mirror. Such mirror stacks have been grown by both chemical beam epitaxy [(CBE) — a variant of MBE)] and MOCVD growth techniques and have been used to fabricate InGaAsP SELs. Room temperature pulsed operation of InGaAsP/InP SELs using these mirror stacks, emitting near 1.5 μm, have been reported [23].

An alternative approach is using the technique of wafer fusion [24, 25]. In this technique, the Bragg mirrors are formed using GaAs/AlGaAs system grown by MBE and the active region of InGaAsP bounded by thin InP layers is formed by MOCVD. The post type structure of a 1500 nm wavelength SEL formed using this technique is shown in Figure 4.6 [24]. The optical cavity is formed by wafer fusion of the InGaAsP quantum well (QW) between the

Bragg mirrors. Room temperature CW threshold current of 2.3 mA has been reported for the 8 μm diameter post device.

4.3 Quantum cascade laser

Quantum cascade laser (QCL) operates on transitions within one band in a QW [26–30]. The operating principle of QCL was first proposed by Kazarinov and Suris [26] as a photon-assisted tunneling process between the states of two coupled QWs. The first realization of the QCL was quite similar to the Kazarinov and Suris idea except that a relaxation/injection layer was inserted between two nearby stages. This relaxation/injection layer was based on the digital-graded gap superlattice [27].

4.3.1 *Principle of operation*

The operation of unipolar electron injection in a QCL to obtain laser oscillations is quite different compared to the normal operation of p–n diode semiconductor lasers. The photon energy in conventional semiconductor lasers results from direct radiative recombination of electrons and holes and is determined by the difference in the conduction- and valence-band energy levels, which is simply the semiconductor band gap. Therefore, the emission wavelength in p–n junction lasers is fixed by the direct semiconductor band gap.

The basic principle of the QCL is illustrated in Figure 4.7, where a tilted conduction band structure (due to applied voltage) as well as the corresponding calculated electron wave-functions are shown. The active region is designed so that there are at least three quasi-bound levels. Laser transition takes place between the levels 3 and 2. To ensure population inversion, the lifetime of the upper state (level 3) must be longer than the lifetime of the lower state (level 2). Through careful design of the superlattice injector region, miniband could be introduced at the energy position of level 3 and therefore increase the lifetime of level 3 by reducing tunneling probability into the continuum. To reduce the lifetime of level 2, the energy separation

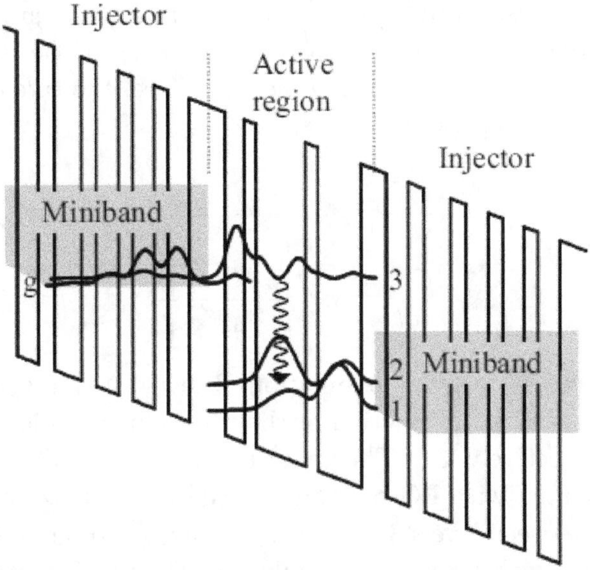

Figure 4.7: Conduction-band profile of one active region with the two adjacent injector regions under an applied electric field of 55 kV cm^{-1}. The moduli squared of the relevant wavefunctions are labeled 1, 2, 3, and g. The wavy arrow denotes the laser transition. The shaded region in the injectors represents a band-like manifold of states. This "miniband" extracts electrons from levels 1 and 2 of a previous active region and injects them into state 3 of the following downstream active region. An energy range with a low density of states above the miniband prevents electrons from tunneling out of level 3. Electrons traverse the structure from left to right. The QW material is InGaAs, the barrier material AlInAs; both are grown lattice matched to InP. The band offset is assumed as 520 meV. The length of one period of active region and injector is 43.4 nm; 32 periods are grown [31].

between levels 2 and 1 is carefully chosen to be close to the LO phonon energy which is 34 meV for AlInAs ternary.

MBE growth technology was originally used in synthesizing QCL structures to meet the precise layer thickness and interface abruptness that determines the QW electron states and, hence, the emission wavelength. Figure 4.7 shows a schematic diagram of a QCL active region grown by MBE. The material system is InGaAs/InAlAs grown over n-InP. These devices have been fabricated into wide-stripe lasers and have operated with peak powers of ~200 mW at 200°K.

In order to estimate the gain of the QCL, a rate equation method is used to determine the population at each level. Below threshold, the stimulated emission is negligible; therefore, the rate equation for levels 3 and 2 are given by:

$$\frac{dN_3}{dt} = \eta_3 I - \frac{N_3}{\tau_{32}} - \frac{N_3}{\tau_{31}} \tag{4.2}$$

$$\frac{dN_2}{dt} = \eta_2 I + \frac{N_3}{\tau_{32}} - \frac{N_2}{\tau_2} \tag{4.3}$$

where N_2 and N_3 are the population of levels 2 and 3, respectively. η_2 and η_3 describe the fraction of the total current injected to levels 2 and 3, respectively [31]. The quantities τ_{32} and τ_{31} describe the lifetimes associated with LO phonon induced transitions between levels 3 to 2 and between levels 3 to 1. τ_2 is the lifetime of level 2 dominated by resonant (with level 1) depopulation. Define τ_3 by the following equation:

$$\frac{1}{\tau_3} = \frac{1}{\tau_{32}} + \frac{1}{\tau_{31}} \tag{4.4}$$

then in steady state from Eqs. (4.2) and (4.3)

$$\begin{aligned} N_3 &= \tau_3 \eta_3 I \\ N_2 &= \tau_2 \left(\eta_2 I + \frac{N_3}{\tau_{32}} \right) \end{aligned} \tag{4.5}$$

and

$$N_3 - N_2 = \tau_3 \eta_3 I \left[1 - \frac{\tau_2}{\tau_{32}} - \frac{\tau_2\,\eta_2}{\tau_3\,\eta_3} \right] \tag{4.6}$$

In the limit of no current injection to level 2, i.e. $\eta_2 \to 0$,

$$N_3 - N_2 = \tau_3 \eta_3 I \left(1 - \frac{\tau_2}{\tau_{32}} \right) \tag{4.7}$$

Then, $\tau_2 < \tau_{32}$ is the condition for population ($N_3 > N_2$) or laser action. Then gain $g(\nu)$ for laser is given by

$$\begin{aligned} g(\nu) &= \sigma(\nu)(N_3 - N_2) \\ &= \sigma(\nu)\tau_3\eta_3 I\left(1 - \frac{\tau_2}{\tau_{32}}\right) \end{aligned} \tag{4.8}$$

where $\sigma(\nu)$ is the stimulated emission cross-section, the gain calculation is further discussed in Appendix 6. The estimated gain for QCL from measurements [27] is ~8.7 cm^{-1}/(kA cm^{-2}). The gain is estimated using the standard laser equation (gain = loss), i.e. (Sections 3.1 and 3.4 of Chapter 3)

$$\Gamma g_{th} = \alpha_a\Gamma + (1-\Gamma)\alpha_c + \frac{1}{2L}\ln\frac{1}{R_1R_2} \tag{4.9}$$

where g_{th} is the threshold gain in the active region; Γ is the confinement, α_a, α_c are the absorption losses in the active and cladding regions, respectively; L is the cavity length; and R_1, R_2 are the mirror facet reflectivity. The quantities α_{w} and α_{m} are known as waveguide losses and mirror losses, and are defined by: $\alpha_{\mathrm{w}} = \alpha_a\Gamma + (1-\Gamma)\alpha_c$ and $\alpha_{\mathrm{m}} = (1/2L)\ln(1/R_1R_2)$. Using the estimated $\Gamma = 0.42$, $\alpha_{\mathrm{m}} = 5.3$ cm^{-1}, and $\alpha_{\mathrm{w}} \sim 19$ cm^{-1} and the measured current density, the gain g_{th} is estimated. Higher gain values (~30 cm^{-1}/(kA cm^{-2}) have been reported [30]. QCL gain calculation is further discussed in Appendix 6.

The emitted photon energy in the QCL is determined by the energy difference between the quasi-confined energy states in the QWs. These QW states can be tuned over a very broad range by "band gap engineering" through appropriate selection of the thickness and band gap of the constituent semiconductors that form the superlattice. The QCL typically contains multiple periods (typically 30) of the QW injection and active emission regions in order to obtain useful laser gains and low-threshold current densities. The operating wavelengths of the QCL have been demonstrated over the range of 3–17 μm. The ability to specifically design the operating wavelength of the QCL for spectroscopic detection of gas molecules is the key reason why the QCL is the laser of choice for gas sensor applications.

4.3.2 *QCL designs and results*

The QCL active regions with multiple layers are fabricated by MBE or MOCVD growth technique. Early QCL lasers had either large stripes or large ridge waveguide structures, which resulted in pulsed operation. A buried heterostructure laser design using both MBE and MOCVD growth was utilized to demonstrate CW room temperature operation [32]. For this device, the lattice matched InGaAs/InAlAs active regions with multiple layers are grown by MBE on an InP substrate. The active region has 35 periods, each consisting of a partially n-doped injector region and the undoped four QW light-emitting regions. The active region (laser core) is surrounded by a waveguide.

The bottom waveguide layers are InGaAs layer followed by the InP substrate and the top waveguide layers are also a grown InGaAs

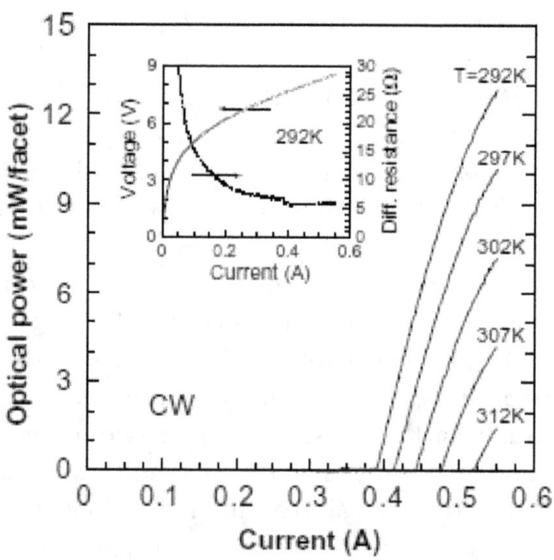

Figure 4.8: CW Light vs. current characteristics of a buried heterostructure QCL at different temperatures. The laser emits near 1094 cm^{-1} [32].

layer and an InP top cladding which was grown by MOCVD. 12 μm wide mesas were then made on the wafer followed by semi-insulating Fe:InP regrowth by MOCVD for lateral current confinement. The wafer is then cleaved to produce 750 μm long lasers. The laser facets were coated with films of ZnSe/PbTe which produces a reflectivity of 70% per pair. The lasers are bonded to a diamond heat sink for CW measurements. The data are shown in Figure 4.8. Room temperature CW operation has also been demonstrated using a ridge waveguide geometry using long cavity length which reduces the thermal impedance (Figure 4.9). The laser is bonded epi-side down on a diamond heat sink [29]. The threshold current density is 1.1 kA/cm^2. Using a ridge width of 200 μm, 34 W of pulsed output power has been reported [29]. The laser uses double-channel geometry to form the ridge as shown in the scanning electron micrograph of Figure 4.9(b).

4.4 Lasers for visible wavelengths

Lasers in the visible wavelengths (in the 760–800 nm range) have been fabricated using AlGaAs/GaAs system with high Al concentration in the active region. For shorter wavelengths to ~630 nm, $In_{0.5}(Ga_{1-x}Al_x)_{0.5}P$ active regions on GaAs substrate can be used [33, 34]. The band gap E_g and the refractive index (n) of InGaAlP as a function of Al fraction x is given by

$$E_g = 1.90 + 0.60x \quad \text{and} \quad n = 3.65 - 0.37x \tag{4.10}$$

where $x < 0.7$ since InGaAlP has indirect band gap for $x > 0.7$. InGaP/GaAs lasers were first demonstrated in 1985. Since then, lasers with InGaAlP active region have been fabricated. The laser wavelength decreases with increasing Al mole fraction x. The data is shown in Figure 4.10 Some of these lasers are fabricated on misoriented GaAs substrate. The misorientation results in slightly shorter wavelength (by ~20–30 nm) than that for (100) oriented substrate. Also, shown in Figure 4.10 is the light vs. current characteristics of an InGaP (x = 0 in InGaAlP composition stated earlier) laser

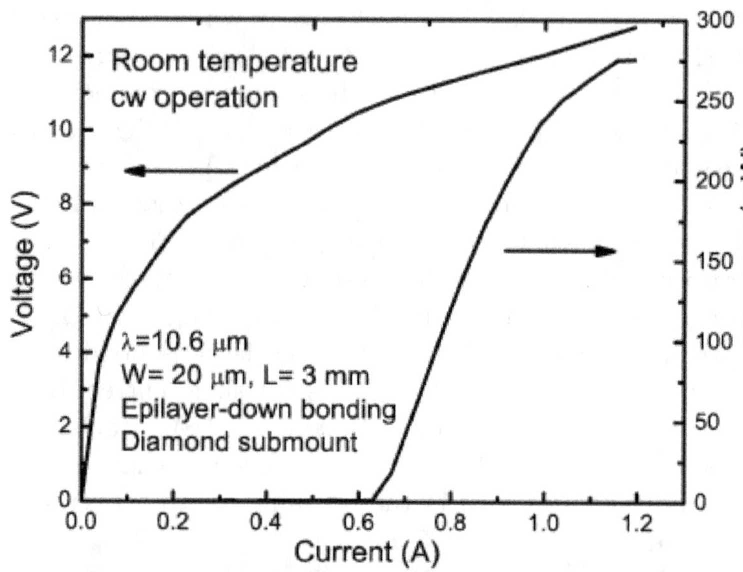

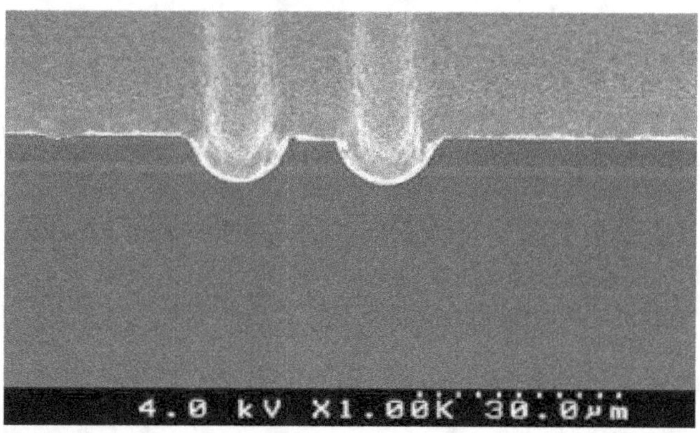

Figure 4.9: (a) L-I-V (light–current–voltage) characteristics of a QCL emitting near 10.6 μm. The ridge width and cavity length are 20 and 3mm, respectively. (b) Scanning electron photomicrograph of double channel geometry [29].

fabricated on a 7° misoriented GaAs substrate. The laser wavelength is 655 nm.

There has been considerable amount of research and demonstration of blue–green lasers using II–VI semiconductors. These semiconductors are ZnS, ZnSe, and ZnSSe, they have been grown

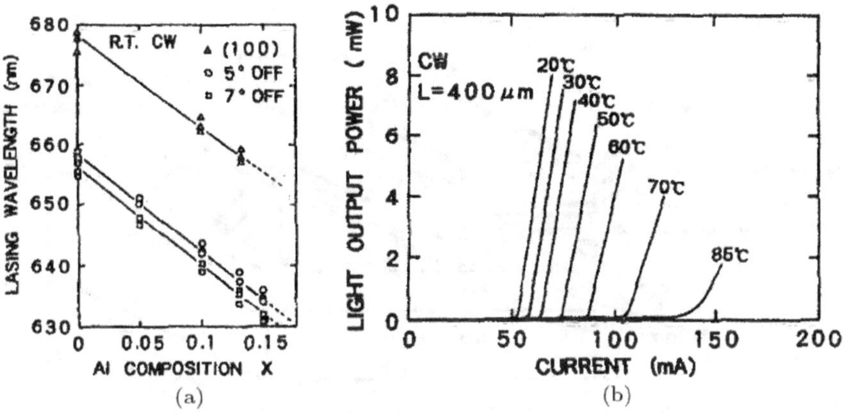

Figure 4.10: (a) Measured wavelength as a function of Al fraction x for several lasers grown over misoriented GaAs substrates. (b) L–I characteristics of an InGaP/GaAs laser fabricated on a 7° misoriented substrate [33].

over GaAs substrates. However, with the demonstration of blue lasers using the nitride systems (GaN/InGaN/AlGaN) [35–38], most of the current development is focused on this system. This section describes the nitride-based lasers.

Stimulated emission from InGaN SQW was observed under pulsed operation in 1996 [35]. Laser emission under pulsed operation using InGaN multiquantum-well (MQW) structure fabricated from III–V nitride materials was demonstrated by Nakamura *et al.* [36]. The laser design is shown in Figure 4.11. The semiconductor layers are grown over a sapphire substrate.

The active region of the lasers has 26 periods of 30 A wide $In_{0.2}Ga_{0.8}N$ wells with 50 A wide $In_{0.05}Ga_{0.95}N$ barriers. Since sapphire is insulating, the n-contact is formed on the top by etching down past the active region. The laser produced 215 mW at 2.3 A of pulsed current. The laser emission wavelength is 417 nm.

Blue laser diodes have been commercialized. The light vs. current characteristics and spectrum of a typical commercial laser is shown in Figure 4.12. Surface-emitting blue–green laser diodes have also been fabricated [40].

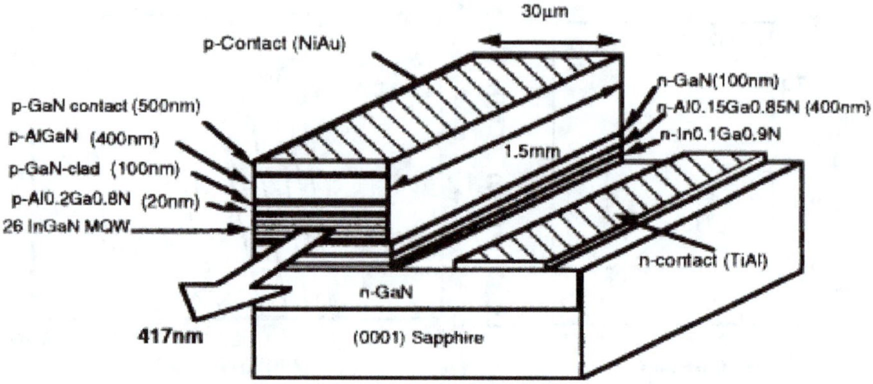

Figure 4.11: Schematic of a blue light-emitting GaN/InGaN/AlGaN laser diode [36].

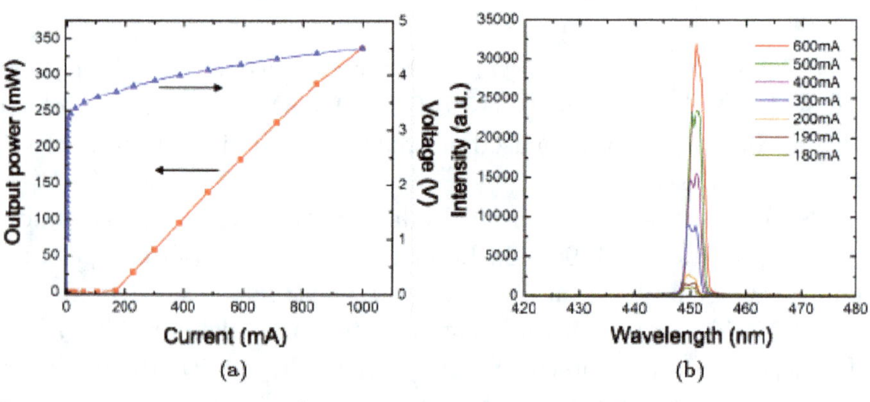

Figure 4.12: $L - I - V$ (light–current–voltage) characteristics of a blue laser (b) emission spectrum of the laser at different currents [39].

4.5 Quantum dot and nanostructure lasers

Extensive research has been carried out on semiconductor lasers and semiconductor optical amplifiers (SOA) with QD active region over the last 20 years. Like lasers with MQW active regions, the peak of the gain spectrum in QD lasers depend on the size of the dots in addition to the material composition of the dots, and the material composition of the region in which the dots are embedded. In order to have sufficient optical gain, multiple layers of QDs are often used in the active (gain) region. The QD material has been used to fabricate both lasers and SOAs (see Chapter 7).

4.5.1 *Gain in quantum wire (QWR) and quantum dot (QD) structures*

The electron motion is restricted further in quantum wire (QWR) and QD structures compared to that for a QW [41–47]. For QWR structures, the electron motion is restricted in two directions and in QD structures, the electron motion is restricted in all directions. Using the similar procedure, as for QWs (Section 3.6.2 of Chapter 3), the density of states of QWR and QD structures are given by [41]

$$\rho_{\mathrm{QWR}} = \sum_n \sum_m \frac{1}{2\pi}\left(\frac{2m}{\hbar^2}\right)^{1/2} n_{\mathrm{QWR}}(E - E_{n,m})^{1/2} \tag{4.11}$$

$$\rho_{\mathrm{QD}} = \sum_n \sum_m \sum_l 2n_{\mathrm{QD}}\delta(E - E_{n,m,l}) \tag{4.12}$$

where n_{QWR} (cm^{-1}) and n_{QD}(cm^{-2}) are the density of QWRs and QDs, respectively, and the summation is over the discrete energy levels. The density of states for regular bulk materials, QW, QWR, and QD material are sketched in Figure 4.13. For QWR, the density of states has sharp maximum at the position of each of the discrete energy states, and for QD the density of states is large at only the discrete energy states. As discussed previously for QW structures, the density of states for QWR and QD structures are not sharp in practice, they get broadened by the interband carrier relaxation time for QWR structures and by the variation in dot sizes and interaction with surrounding material for QD structures. The broadening for QWR structures is given by a term similar to Eq. (3.29) of Chapter 3, i.e.

$$L(E - E_{n,m,l}) = \frac{\hbar}{2\pi\tau_{\mathrm{in}}} \frac{1}{(E - E_{n,m,l})^2 + \left(\frac{\hbar}{2\tau_{\mathrm{in}}}\right)^2} \tag{4.13}$$

where τ_{in} is the intraband relaxation time. The gain spectrum for QWR structures has been calculated. The calculated gain spectrum is

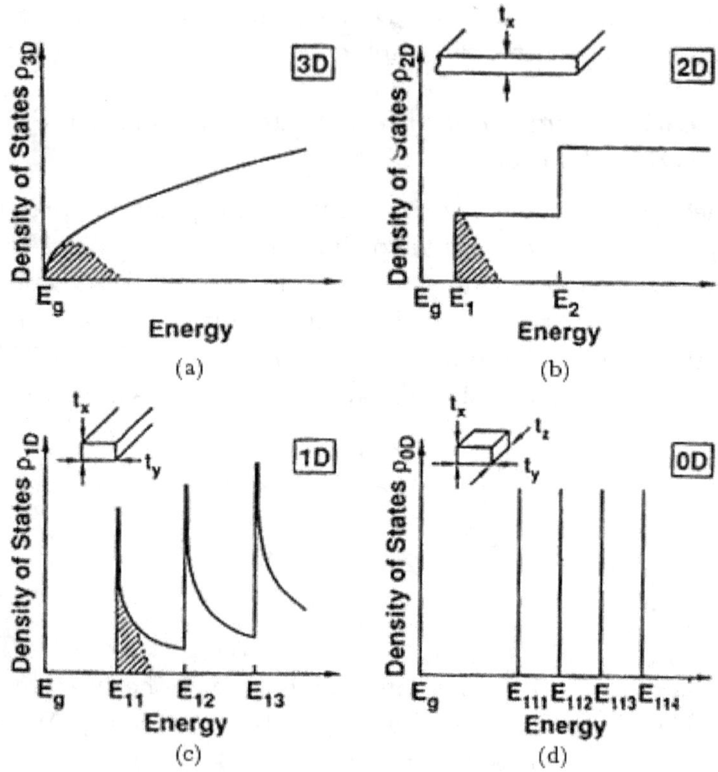

Figure 4.13: Schematic description of the density of states for regular bulk semiconductor, QW, QWR, and QD structures. The dimensionalities of each of these types are: Bulk (3D), QW (2D), QWR (1D), QD (0D) [41].

shown in Figure 4.14. The dimension of the wire is 12×20 nm. In the QWR structure, the discrete energy levels are labeled by (n, m), where the integers n, m correspond to discrete energy levels corresponding to energy quantization in the two orthogonal directions. The lowest energy states are labeled (0,0), (0,1), and (1,0) for the next two levels and so on. The quantity τ_{in}, the intraband relaxation time, is assumed to be 0.1 ps. Size fluctuations in QWR structure (i.e. fluctuation in QWR cross-section) can significantly reduce the peak gain.

In addition, anisotropy of the dipole moment may play a role in determining the observed gain in QWR. Gain anisotropy of $\sim$2 has

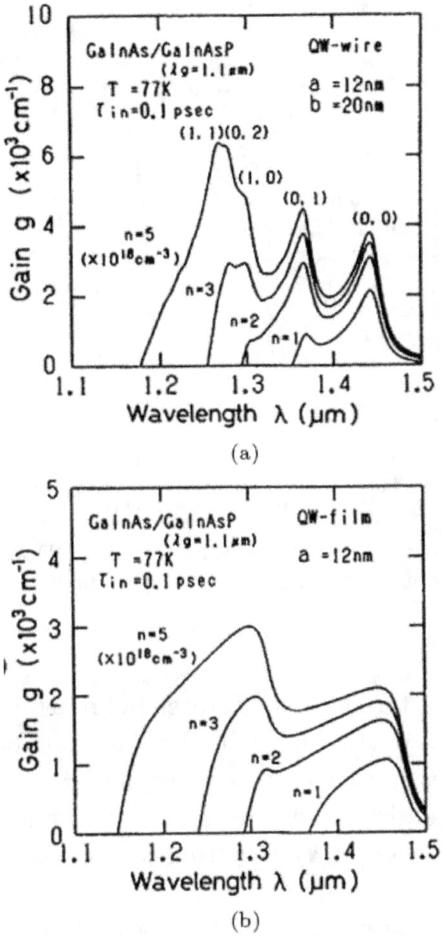

Figure 4.14: Calculated gain spectrum at 77°K for InGaAsP/InGaAs quantum wires (QWRs) for various values of carrier density n. For comparison, the calculated gain spectrum of a InGaAsP/InGaAs quantum well is also shown. The same value of intraband relaxation time (0.1 ps) is used in both calculations [41, 47].

been predicted for QWR of equal width and height, i.e. gain for light polarized along the wire is larger by a factor of ∼2 [46].

QDs fabricated by self-assembly (described in Chapter 2) generally show a fluctuation in size which results in an energy distribution of the discrete energy states in an array of QDs. This, in effect, results

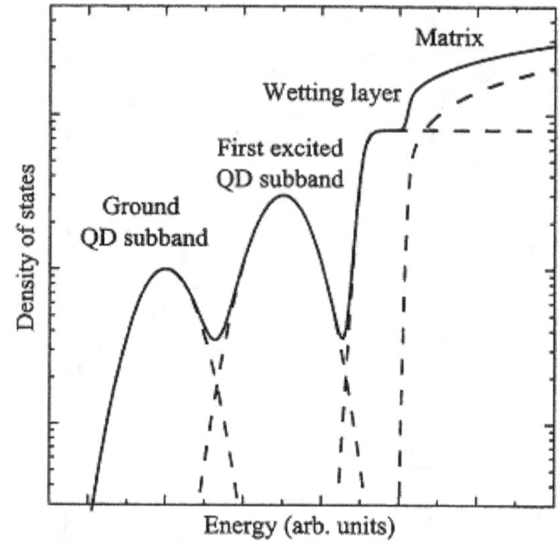

Figure 4.15: Density of states of a self-assembled QD array. The dashed lines represent the states of the ground and excited subbands, the wetting layer, and the matrix [48].

in a broadening of the density of states in an array of QDs. The QDs are formed from a wetting layer by the self-assembly process and the entire assembly is immersed in a cladding layer matrix. Additional broadening of the discrete energy states takes place due to its interaction with the states in the wetting layer and in the surrounding matrix.

The density of states in a QD array is characterized by a set of subbands that result from the inhomogeneously broadened quantum levels of the ground and excited states of each QD. These subbands can overlap depending on the size distribution of the dots.

The density of states of a self-assembled QD array is schematically shown in Figure 4.15. Both the ground state and the first subbands are broadened and at high energy they merge with the bands in the wetting layer. For an ideal QD array, the energy width of the density of states is given by homogeneous broadening (Δ) which is ~0.1 meV. This homogeneous broadening is due to the interaction of the dot with the surrounding material. The saturated gain (G_s) of a QD array is proportional to

$$G_s \sim g n_{QD}/\Delta \tag{4.14}$$

where n_{QD} is the density of QDs and g is the degeneracy of the level.

The current density needed for zero net gain (transparency) is proportional to n_{QD}. Thus for high saturated gain, the current density needed for transparency is also high. In addition to the ground state level, one or more excited state levels can be thermally populated. Since the excited states have a higher degeneracy, the saturated gain of the excited states is higher, i.e. with increasing current the gain of the QD array would increase and its spectrum will shift to higher energies. A schematic representation of the gain for an ideal and self-assembled QD array is shown in Figure 4.16. For a self-assembled array (non-ideal), the broadening factor (Δ) is higher and hence the transparency current density is higher and the saturated gain is lower. The experimental results for InGaAs QD array in an AlGaAs matrix are shown in Figure 4.17. The ground state lasing is observed at low-current density when the lasers had low optical loss and excited state lasing is observed at high-current densities when the lasers have high optical loss. The optical loss for a laser can be varied using facet coating.

The dependence of optical gain on modal G can be empirically fitted [49] to the following expression:

$$G = G_s\{1 - \exp(-\gamma(J/J_{tr} - 1))\}$$

which simplifies for low-current density to

$$G = \gamma G_s(J/J_{tr} - 1) \tag{4.15}$$

where G_s, J_{tr}, and J are the saturated gain, the transparency current density, and the injected current density, respectively. The quantity γ is a parameter. The values of G_s, J_{tr}, and γ for the data shown in Figure 4.17 are 8.2 cm^{-1}, 50 A/cm^2, and 0.9, respectively, for the ground state and 25.8 cm^{-1}, 145 A/ cm^2, and 0.8, respectively,

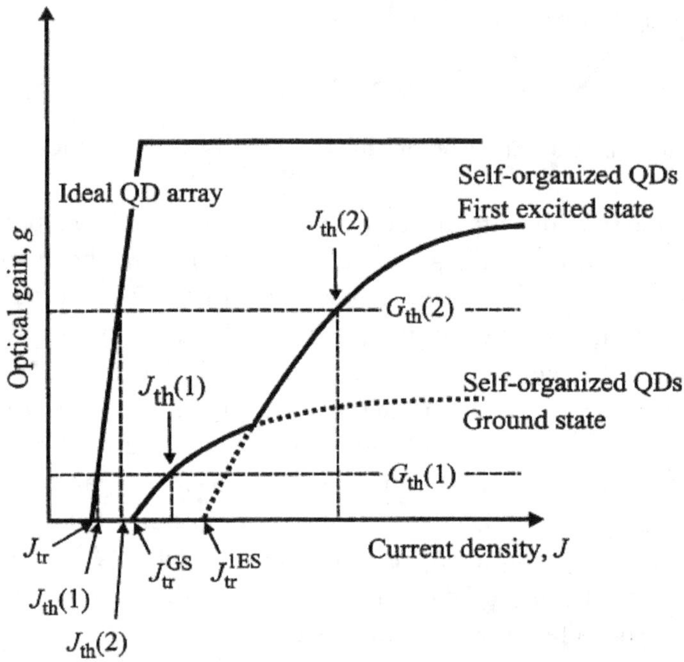

Figure 4.16: Schematic representation of gain as a function of current density for an ideal QD array and for a non-ideal (self-assembled) QD array. The two J_{th} (1) shows the current needed for the same optical gain for the ideal and non-ideal case for the ground state transition. The J_{th} (2) shows the current needed for a certain optical gain for the non-ideal case for the first excited state transition [48].

for the first excited state. Theory of optical gain in QD lasers and SOAs have been developed [50–52]. The carriers are injected into the wetting layers from which they make a fast transfer to the QD. QD SOA has a fast gain and phase recovery, which makes them suitable for high-speed system applications. The primary reason for the faster response of QD SOA compared to that for an SOA with regular active region is the presence of the wetting layer. The wetting layer serves as a carrier reservoir layer. Carriers depleted by the injected optical pulse in the QD ground state are replaced by fast carrier transfer from the wetting layer. The fabrication, design, and performance of QD SOA are discussed in Chapter 7.

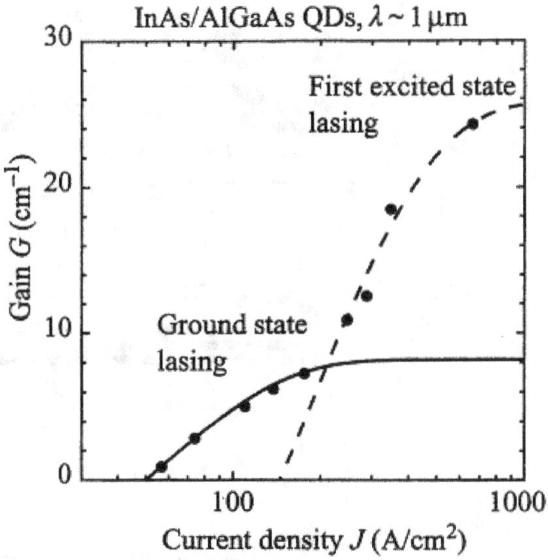

Figure 4.17: Experimental relation between modal gain G and current density J [48].

4.5.2 *Quantum dot laser fabrication and results*

The QD material, as the name implies, consists of dots (sometime irregular) of a semiconductor (e.g. InAs) embedded in another semiconductor (e.g. InGaAs). The growth process utilized for dot formation is generally the Stranski–Krastanov growth (S–K growth) process. In S–K growth process, a "layer-plus-island growth" process occurs. The S–K growth is a two-step process: initially, complete films of InAs (adsorbate), up to several monolayers thick, are grown in a layer-by-layer fashion on a lattice-matched (or nearly lattice-matched) crystalline substrate. Beyond a critical layer thickness, which depends on strain and the chemical potential of the deposited film, growth continues through the nucleation and coalescence of InAs (adsorbate) "islands". This growth mechanism was first noted by Ivan Stranski and Lyubomir Krastanov in 1938. In recent years, MBE technique has been used to grow QD material.

The QD active regions (i.e. QDs and the embedding layers) are often surrounded by a set of separate confinement layers which have

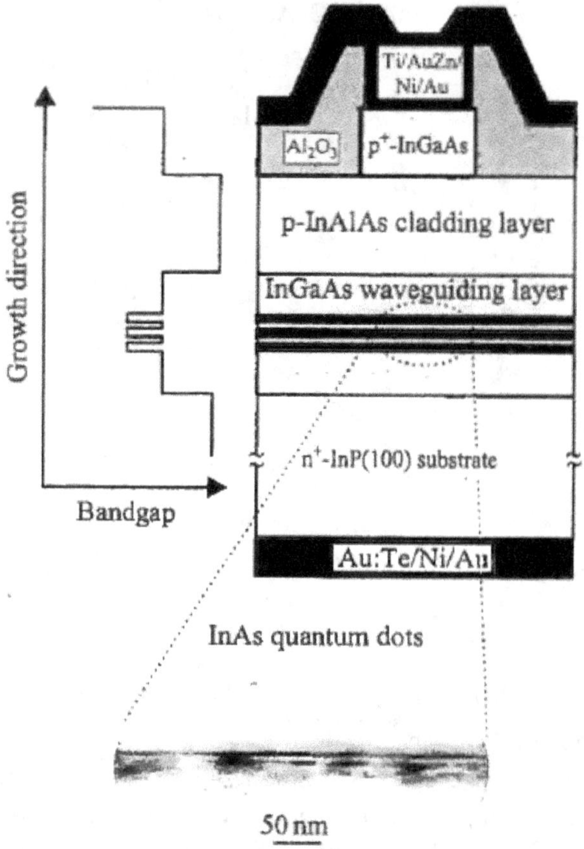

Figure 4.18: Schematic of an InAs QD laser and the transmission electron micrograph of the InAs dot region [53].

n-type and p-type cladding layers on either side similar to that for a regular laser (Chapter 3). The lateral confinement in this heterostructure could be either strongly index guided (buried heterostructure) or weakly index guided (e.g. ridge waveguide type), i.e. the lateral index and current confinement follows the same methods as described earlier (Chapter 3).

The structure of a QD laser along with the scanning electron photomicrographs of the QD active region is shown in Figure 4.18 [53]. The QD layers consist of InAs QDs embedded in InGaAs waveguiding layer. The cladding layers are InAlAs. The InGaAs and InAlAs

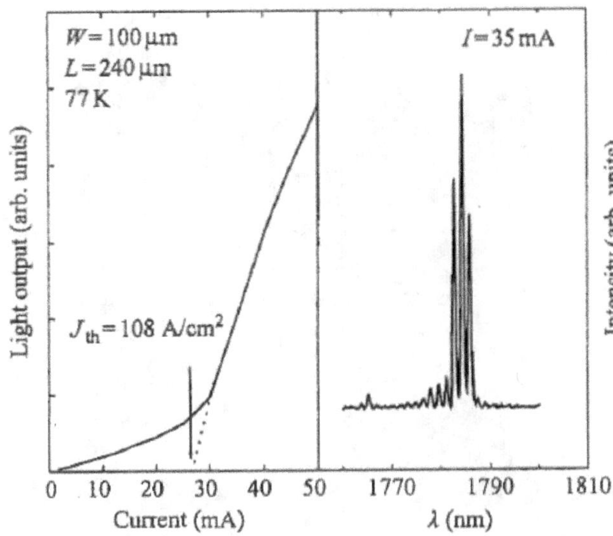

Figure 4.19: Light vs. current characteristics at 77°K. The W and L values are noted in the figure. Also, shown is the laser spectrum. The emission wavelength is near 1.79 μm [53].

layers are lattice matched to the InP substrate. The InAs dots are embedded in the InGaAs matrix. The lateral dimension (W) is provided by etching the InGaAs contact layer. The lasers were cleaved to a cavity length (L) of 240 μm. The light vs. current characteristics and the spectrum at 77°K are shown in Figure 4.19. In addition to QW and QD, QWR lasers have been reported [41, 54, 55].

4.5.3 *Nanostructure lasers*

Various types of photonic crystal based nanostructure lasers have been reported [56–59]. The two-dimensional photonic crystals are formed on a regular multiple QW like layer structure by etching air holes in a regular pattern using reactive ion beam etching. Example of a photonic crystal ring laser is shown in Figure 4.20. The hexagonal ring waveguide is formed by reactive ion beam etching and standard photolithographic techniques [56].

The ring laser can be operated under optical pumping using a 980 nm pump laser at a pump power of ~3 mW. A low-threshold

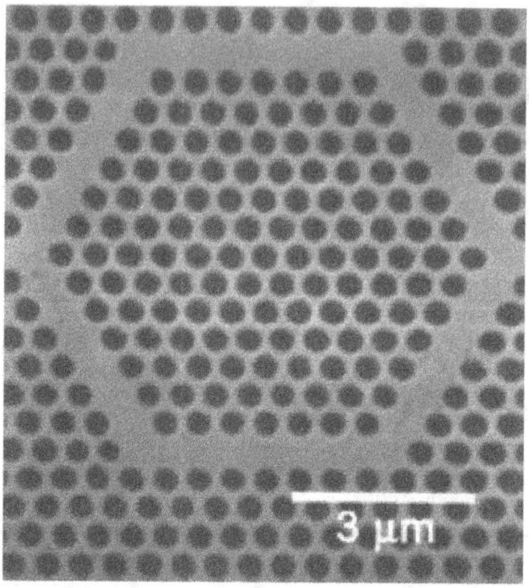

Figure 4.20: Scanning electron photomicrograph of a photonic crystal ring laser fabricated using InGaAsP/InP material system [56].

current-pumped photonic crystal laser has been reported using the InGaAsP/InP material system built on p-InP substrate. In this case, a post (p-InP) connects the photonic crystal material with air holes to the substrate. A submicron wire contact is made on the n-side. The schematic of the device along with the layer structure is shown in Figure 4.21 [57]. Laser action is observed with a threshold current of 260 μA.

4.6 High power semiconductor lasers

Typical semiconductor lasers used in communication or consumer system applications emit a few mW of power. High power (>100 mW) semiconductor lasers are important as pump lasers for rare-doped fiber laser systems, printing, and industrial (metal cutting) applications. Semiconductor lasers have been fabricated that emit several Watts of output power and many such lasers have been combined either on a single chip or on a single output beam to produce 100s of W of output power. Generally, the output of the semiconductor laser chip in such cases does not have a single coherent beam, but these

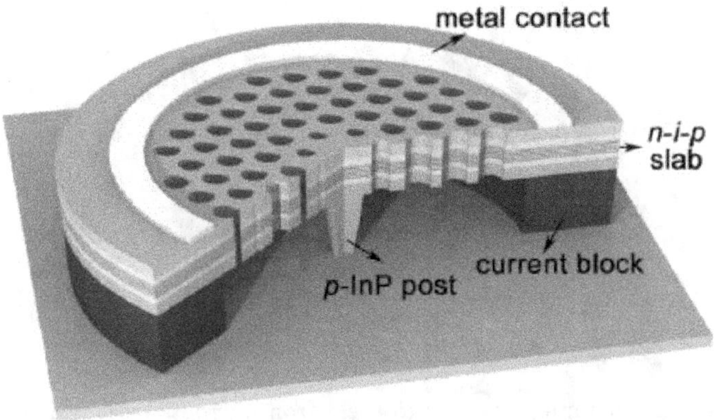

(a)

Material	Thickness	Doping (cm^{-3})	Remark
n-InP	0.1 μm	~ 1.1×10^{18}	Capping layer, removed
n-1.24Q InGaAsP	600 Å	~ 2.7×10^{19}	Delta doping
i-1.24Q InGaAsP	200 Å	Undoped	
1.30Q (-0.5%) InGaAsP	115 Å	Barrier	6 pairs, λc = 1.55 μm
1.65Q (+0.8%) InGaAsP	70 Å	Well	
1.30Q (-0.5%) InGaAsP	115 Å	Barrier	
i-1.24Q InGaAsP	200 Å	Undoped	
p-1.24Q InGaAsP	600 Å	~ 3×10^{18}	
p-InP	1.0 μm	~ 1.5×10^{18}	Sacrificial layer (post)
p-InGaAs	0.1 μm	~ 5×10^{18}	Etch-stop layer
p-Buffer InP	0.3 μm	~ 1×10^{18}	
p-InP substrate			

(b)

Figure 4.21: (a) Schematic of the nanostructure photonic crystal device. (b) The epitaxial layer structure used for fabrication [57].

lasers have been fabricated at wavelengths suitable as pump sources for rare-earth-doped fiber lasers (Chapter 9) and in these cases a single coherent beam with 100s of W output power can be obtained from a fiber laser. This section describes high-power lasers which have been used as pump lasers for rare-earth-doped fibers to fabricate fiber lasers and fiber amplifiers. The typical laser wavelengths are 1480 nm (for a pump source for Er-doped fiber), 980 nm (for a pump source for Er-doped fiber), 960 nm (for a pump source for Yb-doped fiber), and 808 nm (for a pump source for Nd-doped fiber).

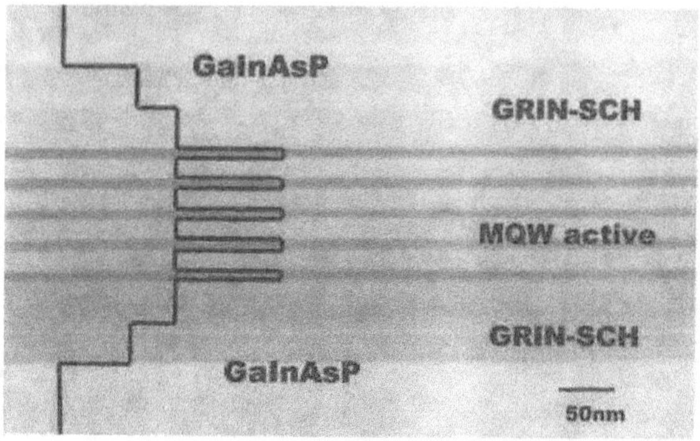

Figure 4.22: Scanning transmission electron micrograph (TEM) of the active region of a 1480 nm pump laser [60].

4.6.1 *1480 nm laser*

A high differential quantum efficiency, low-threshold current, and low leakage current (Chapter 3) are important for high-power operation [60–62]. These quantities are determined by the active region design. In addition, the laser must have low thermal impedance (low resistance contacts) and bonded to a good heat sink so that its temperature does not increase significantly with the application of high current.

The design used for a high power 1480 nm pump has an MQW active region with a graded index region (GRIN) for optical confinement. The scanning transmission electron micrograph (TEM) of an MQW active region is shown in Figure 4.22. The parameters that need to be optimized are doping of layers, number of wells, GRIN design, and the optical cavity length. There are many trade-offs. For example, long cavity length reduces the thermal impedance but increases the threshold current and decreases quantum efficiency due to higher cavity optical loss.

Similarly, higher doping reduces resistance but increases optical loss. Increasing the number of active layer wells increases the threshold current beyond a certain number of wells. An experimental investigation has been carried out using the active region band diagram shown in Figure 4.23. The active layer is 40 A thick InGaAs QWs

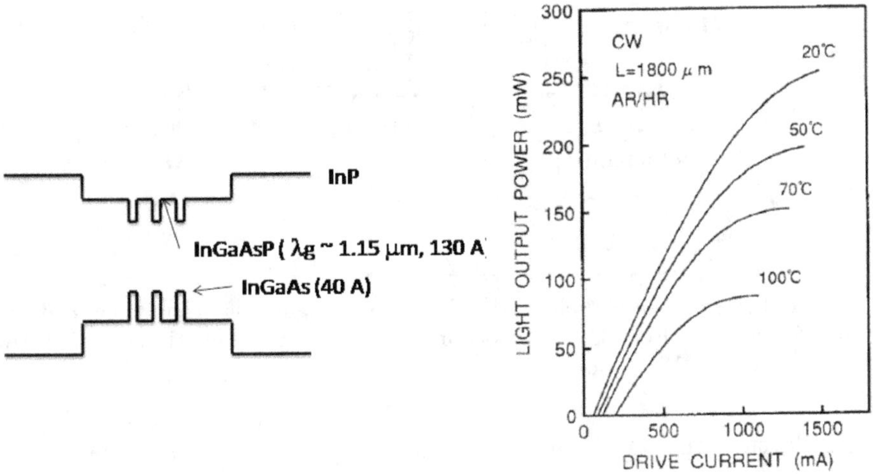

(a) (b)

Figure 4.23: (a) Band diagram of the MQW active region. (b) Light vs. current characteristics of a 1480 nm pump laser at various temperatures of the design shown in Figure 4.22 [61].

with 130 A thick InGaAsP ($\lambda \sim 1.15\ \mu$m, $E_g = 1.08$ eV) barriers. The width of the entire active region is 2000 A. The wafers are grown by MOCVD and strongly index-guided lasers are fabricated using the DCBH structure (Figure 3.6 of Chapter 3). Lasers with 2, 3, 5, and 7 QWs are fabricated. Lasers with 5 QWs were found to have best high power performance. The light vs. current characteristics of an 1800 μm long laser with five QWs at various temperatures is shown in Figure 4.23. The laser has antireflection coating (AR $\sim$4%) at one facet and high reflectivity ($\sim$90%) at the other facet. The laser emits more than 250 mW near room temperature. $\sim$500 mW of single-mode output power has been reported from similar designs [60]. Typical far-field divergence of such a laser is 25° and 20° normal and parallel to the junction plane, respectively [60].

4.6.2 *980 nm laser*

For $\sim$980 nm laser operation, thin layers of $In_{0.2}Ga_{0.8}As$ are grown on GaAs using either MOCVD or MBE growth technique [63–67].

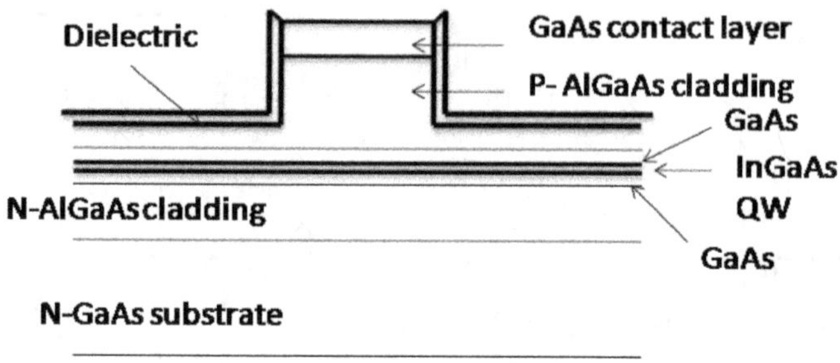

Figure 4.24: A simple schematic of a ridge waveguide laser emitting near 980 nm. The device of Ref. [63] has graded index layers surrounding the active InGaAs quantum well (QW) region.

The lattice mismatch introduces a strain between the grown layer ($In_{0.2}Ga_{0.8}As$) and the substrate (GaAs). As a result, such lasers are often called strained quantum well lasers. Typical design of a single mode 980 nm laser consists of one or two QWs of $In_{0.2}Ga_{0.8}As$ surrounded by GaAs barrier layers. Exact well thickness needed for 980 nm laser is often determined experimentally. It is generally in the 7–10 nm range. For good transverse mode confinement, a GRIN design is often used. The wafer is generally fabricated into a ridge waveguide type laser with a ridge width of $\sim$4 μm for single transverse mode operation.

Specific designs of the active region for high power kink free operation or circular mode (i.e. nearly equal beam divergences along and normal to junction plane) have been reported.

The weak index-guided ridge waveguide design allows a large area of the emitted light at the facet, thus, for a given output power the laser power density is lower at the facet compared to that of a strongly index guided buried heterostructure design. Lower optical power density at the facet reduces the probability of catastrophic mirror damage (COMD) at the facet. Various schemes have been developed to reduce/eliminate the COMD. The schematic of a ridge waveguide laser is shown in Figure 4.24.

This particular active region design provides high kink free output power. The output has an AR ($\sim$4%) coating and the rear facet has a high-reflection (HR) ($\sim$93%) coating of dielectric materials.

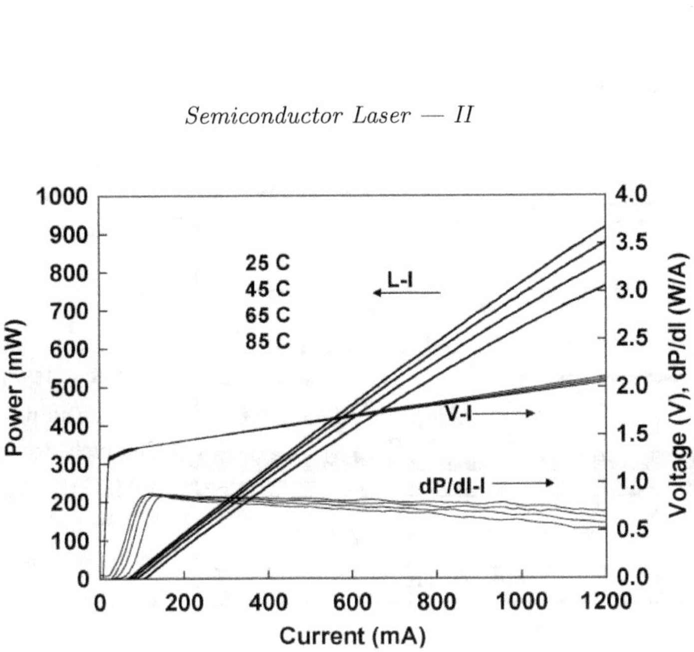

Figure 4.25: Light vs. current characteristics of a 980 nm laser [64].

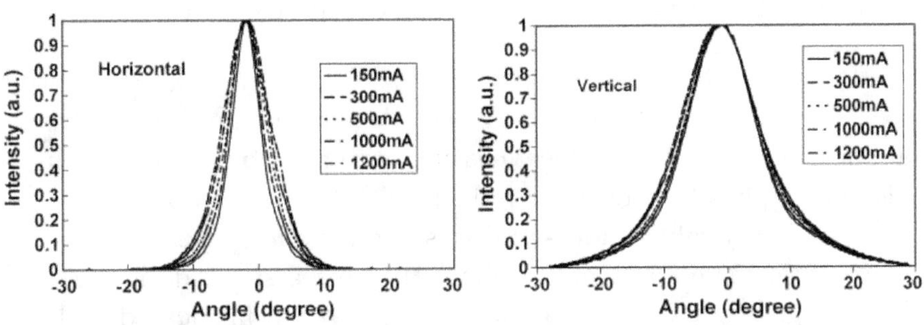

Figure 4.26: The far-field patterns along (horizontal) and normal (vertical) to the junction plane are shown. Single lobed pattern illustrates single transverse mode operation [64]. The full-width at half maximum of beam divergences along the horizontal and vertical planes are ~12° and 18°, respectively.

The light vs. current characteristics of a 980 nm laser at different temperatures is shown in Figure 4.25. Also shown is the voltage–current (V–I) and slope ($\mathrm{d}P/\mathrm{d}I - I$) vs. current characteristics. The slope efficiency near room temperature is ~0.9W/A. The far-field patterns are shown in Figure 4.26.

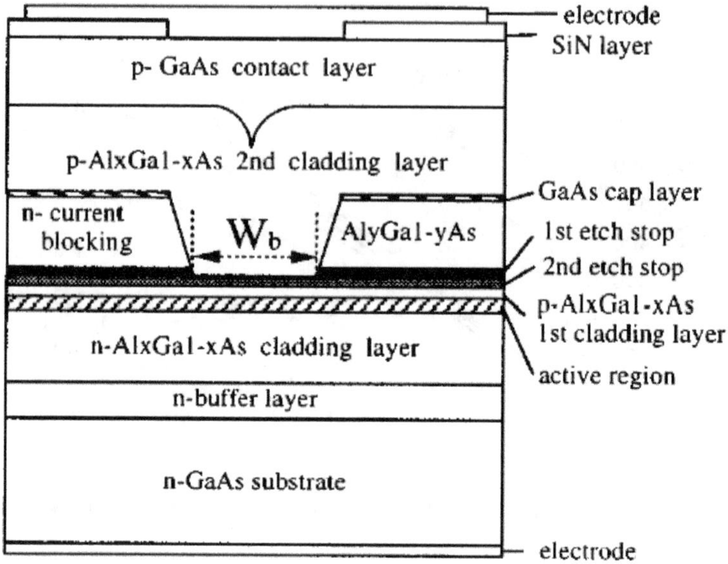

Figure 4.27: Schematic of a weakly index guided buried stripe laser [67].

Reliability of the 980 nm pump laser has been extensively studied. The lasers are found to be reliable for communication system applications. Specialized facet coating in vacuum or impurity induced disordering has been used to reduce COMD at high powers.

Another weakly index-guided structure that operates to high power in a single transverse mode is the buried stripe design [67–72]. The design is shown in Figure 4.27. The device is fabricated with two $In_{0.2}Ga_{0.8}As$ QWs. Both the current blocking layer (n-type shown) and the Si_3N_4 dielectric layer are effective in confining the current to the active region. The buried stripe width (W_b) is 2.2 μm. A single mode output power of >500 mW has been reported using the above design.

The laser had AR coating at the emitting facet and HR coating at the rear facet.

Single mode lasers are typically packaged with a fiber-coupled output (Figure 4.28). The package has a thermoelectric cooler, which keeps the laser at a certain temperature. The temperature control is important for fixed wavelength operation and for power stability.

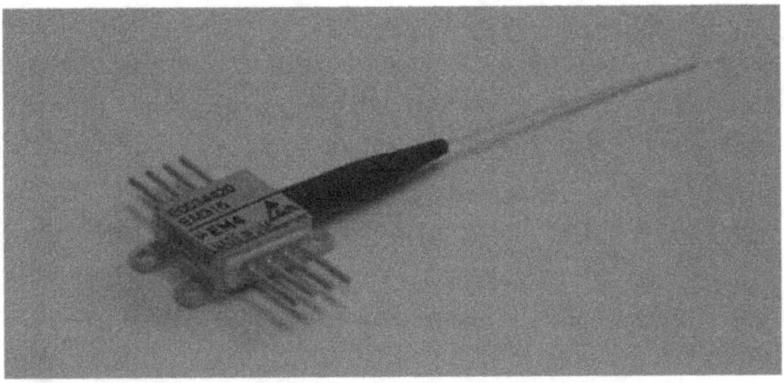

Figure 4.28: Single mode pump laser package. The output is coupled to a single mode fiber. Regular low power lasers used for communication systems are also packaged in a similar fashion.

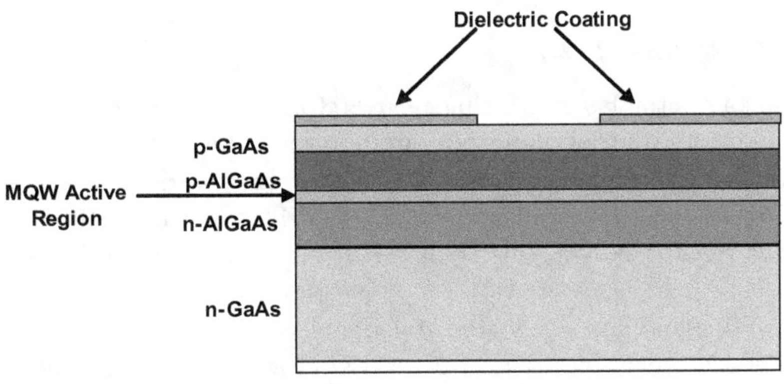

Figure 4.29: Schematic of a 100 μm stripe laser. The current is injected through an opening in the dielectric.

Lasers emitting near 960–970 nm are important as pump lasers for multimode large area Yb-doped fiber amplifiers and also Er–Yb-doped double clad fibers. These lasers are generally fabricated as 100 μm wide stripe devices. A schematic of the device's cross-section is shown in Figure 4.29. The active region is InGaAs with In composition ($\sim$0.2) chosen to produce emission in the 960–970 nm wavelength range. The output facet is AR coated and the rear facet is HR coated.

The laser cavity length is 2 mm. The long cavity length is important for reducing thermal impedance. The lasers typically emit

~9–10 W at 10 A. The lasers are similar to broad area lasers with an opening in the SiO_2 dielectric (~100 μm wide). The current is injected in the active region through this opening. This makes the mode area larger and hence lowers the power density at the facet which prevents high power induced facet degradation. For higher power operation, multiple openings (typically 50–100 μm wide) are present in the dielectric stripe. These openings are typically 50–100 μm apart. Lasers with 10 openings (10 stripes) emit in 10 spots along the active region. The combined output power can be >100 W. These devices are also called laser arrays. These laser arrays have been stacked in two dimensions, and are known to produce nearly 1 kW in quasi-CW operation. These types of laser arrays have been used as a pump for Yb-doped fiber lasers.

4.6.3 *808 nm laser*

Semiconductor laser emitting near 808 nm is important as a pump source for Nd-doped rare earth material. Both single mode and multimode broad stripe 808 nm lasers are fabricated using the AlGaAs/GaAs material system [73–79]. The Al composition in the active region $Al_xGa_{1-x}As$ is chosen so that the emission is near 808 nm. The band gap of $Al_xGa_{1-x}As$ varies with the Al composition (x). The band gap E_g is given by $E_g(x) = 1.424 + 1.247x$ (eV). The layers are grown by MBE or MOCVD growth technique. The layers are N-AlGaAs cladding layer, $Al_xGa_{1-x}As$ active layer, P-AlGaAs cladding layer, and P-GaAs contact layer. The grown wafer is fabricated into ridge waveguide lasers. The reliability of these types of lasers is generally limited by catastrophic facet degradation which occurs at high powers. A special facet coating process has been reported to lessen this problem. It involves the cleaving of the lasers in vacuum (to prevent facet oxidation) and applying a dielectric coating after cleaving. It is the same cleaving and coating process which lessens the degradation of lasers emitting near 980 nm.

For the single mode device described below, the active region is a single QW and is grown by MBE growth technique. The light vs. current characteristics at various temperatures are shown in

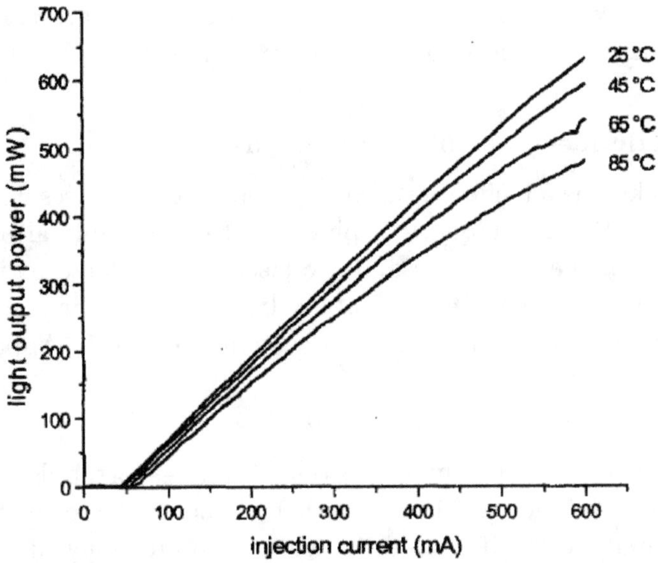

Figure 4.30: Light vs. current characteristics of a laser emitting near 808 nm [77].

Figure 4.30. The laser has an AR coating at the emitting facet and HR coating at the rear facet. The device emits >800 mW at 25°C @ ~900 mA. The slope efficiency is ~1.1 W/A up to ~600 mA at 25°C. At higher current or higher temperature, the slope efficiency decreases due to increase in junction temperature.

The single transverse mode nature of the emission is determined through a measurement of the far-field characteristics. The far-field pattern shows a full-width at half maximum normal to the junction plane and along the junction plane of 27° and 7°, respectively.

Multimode devices with 100 μm stripe width have been reported. They produce several *W* (~12 *W* at 15 A [79]) of output power and generally fail after sometime due to COMD. A laser device with proper facet coating (e.g. with ZnSe in this case) operates to peak power without sudden degradation and rolls over due to heating [80]. Semiconductor lasers operating at high power have been fabricated at many wavelengths. They offer reliable turn-key pump sources for rare-earth-doped amplifiers and lasers. Advances in cooling schemes have resulted in >100 W of output power from rare earth-doped lasers using semiconductor laser pump sources (Chapter 9).

4.7 Mode-locked semiconductor lasers

Mode locking is a technique used to generate short pulses from a laser [80–84]. In this technique, the phases of the longitudinal modes are locked when the gain or loss of the laser is modulated at inverse of the round trip time of the optical cavity. This results in a short pulse operation as described in the next section and also in Appendix 9.

4.7.1 *Principle of mode locking*

Mode locking is a technique to obtain ultra-short pulses of light, which involves the "locking" together of the phases of many cavity longitudinal modes. The electric field of the mth longitudinal mode may be written as

$$E(z,t) = \hat{\varepsilon}_m E_m(z)\sin(\omega_m t + \phi_m) = \hat{\varepsilon}_m E_m \sin k_m z \sin(\omega_m t + \phi_m) \tag{4.7.1}$$

where

$$k_m = m\frac{\pi}{L} \tag{4.16}$$

and

$$\omega_m = k_m c = m\frac{\pi c}{L} \tag{4.17}$$

$\mathrm{m} = 1, 2, 3, \ldots$. For simplicity, it is assumed that the mode fields all have the same magnitude and polarization. Furthermore, consider the case when the phases of all modes are equal, i.e. consider the simple case $\phi_\mathrm{m} = 0$. Then the total electric field in the cavity is

$$E(z,t) = \sum_m E_m(t) = E_0 \sum_m \sin(k_m z)\sin\omega_m t \tag{4.18}$$

where the summation is over all oscillating modes. The integer m typically is a very large number. Consider, $m \to M + n$ and write Eqs. (4.16) and (4.17) as

$$k_m = \frac{(M+n)\pi}{L} \tag{4.19}$$

$$\omega_m = \frac{(M+n)\pi c}{L} \tag{4.20}$$

where M is a very large positive integer ($M \approx 10^6$) and n runs from — $(N-1)/2$ to $(N-1)/2$, corresponding to a total of N modes (few hundreds) centered at the frequency $M\pi c/L$, which is near the peak of the gain spectrum. Then, Eq. (4.18) becomes

$$\begin{aligned} E(z,t) &= E_0 \sum_{-(N-1)/2}^{(N-1)/2} \sin\frac{(M+n)\pi z}{L} \sin\frac{(M+n)\pi ct}{L} \\ &= \frac{1}{2}E_0 \sum_n \left[\cos\frac{(M+n)\pi(z-ct)}{L} - \cos\frac{(M+n)\pi(z+ct)}{L}\right] \end{aligned} \tag{4.21}$$

for the total electric field in the laser cavity. Using the general identity

$$\sum_{-(N-1)/2}^{(N-1)/2} e^{inx} = \frac{\sin(Nx/2)}{\sin(x/2)} \tag{4.22}$$

the sum

$$\begin{aligned} &\sum_{-(N-1)/2}^{(N-1)/2} \cos\frac{(M+n)\pi(z-ct)}{L} \\ &\quad = \mathrm{Re} \sum_{-(N-1)/2}^{(N-1)/2} e^{i(M+N)\pi(z-ct)/L} \end{aligned}$$

$$= \mathrm{Re}\left\{e^{iM\pi(z-ct)/L}\frac{\sin\left[\pi N(z-ct)/2L\right]}{\sin\left[\pi(z-ct)/2L\right]}\right\}$$

$$= \left[\cos\frac{M\pi(z-ct)}{L}\right]\frac{\sin\left[\pi N(z-ct)/2L\right]}{\sin\left[\pi(z-ct)/2L\right]} \tag{4.23}$$

From Eq. (4.23) then,

$$E(z,t) = \frac{\mathrm{E}_0}{2}\left\{\cos k_0(z-ct)\frac{\sin[\pi N(z-ct)/2L]}{\sin[\pi(z-ct)/2L]} - \cos k_0(z+ct)\frac{\sin[\pi N(z+ct)/2L]}{\sin[\pi(z+ct)/2L]}\right\} \tag{4.24}$$

where $k_0 = \pi M/L$.

The function

$$A_N^{(\pm)}(z,t) = \frac{\sin[\pi N(z \pm ct)/2L]}{\sin[\pi(z \pm ct)/2L]} \tag{4.25}$$

appearing in Eq. (4.24) has maxima occurring at

$$z \pm ct = m(2L), \quad m = 0, \pm 1, \pm 2, \ldots. \tag{4.26}$$

Consider now a fixed value of z inside the cavity, for instance, then there are pulses of peak amplitude $NE_0/2$ appearing at time intervals of $2L/c$, each pulse having a duration T/N, as indicated in Figure 4.31. Consider again the scalar electric field in the exponential

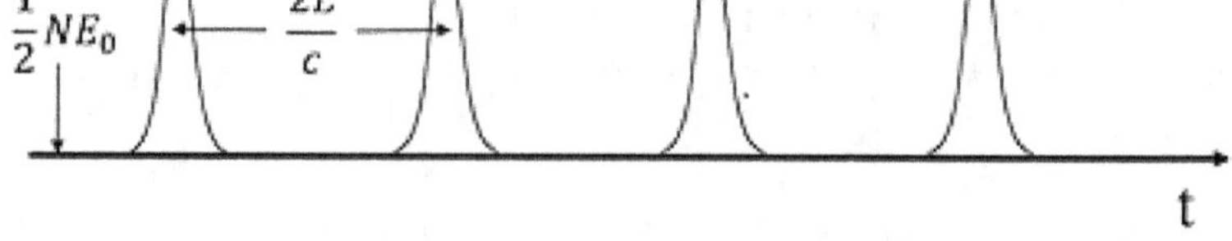

Figure 4.31: A mode-locked pulse train as a function of time, observed at a fixed position z.

form,

$$E(z,t) = E_m e^{i(\omega_m t+\phi_m)} e^{ik_m z} \tag{4.27}$$

suppose that the amplitude E_m is not constant, but rather is modulated periodically in time according to the formula

$$E_m = E_0(1 - \varepsilon \cos \Omega t) \tag{4.28}$$

where Ω is the modulation frequency and E_0 and ε are constants.

It creates an amplitude-modulated field:

$$E_m(z,t) = E_0(1 - \varepsilon \cos \Omega t) e^{i(\omega_m t+\phi_m)} e^{ik_m z} \tag{4.29}$$

Since

$$\cos \Omega t e^{i\omega_m t} = \frac{1}{2} e^{i(\omega_m+\Omega)t} + \frac{1}{2} e^{i(\omega_m-\Omega)t} \tag{4.30}$$

the electric field may be written as a sum of harmonically varying parts:

$$E_m(z,t) = -\frac{1}{2} E_0 [e^{i(\omega_m+\Omega)t} - 2e^{i\omega_m t} + e^{i(\omega_m-\Omega)t}] e^{i\phi_m} e^{ik_m z} \tag{4.31}$$

Equation (4.31) shows the amplitude modulation of the field of frequency ω_m has generated sidebands of frequency $\omega_m \pm \Omega$. If Ω equals $\pi c/L$, i.e. longitudinal mode spacing, the sidebands produced reinforce the adjacent modes and makes them oscillate in the same phase. This creates a situation that the highest gain is obtained if all the modes within the gain spectrum oscillate in the same phase; this is equivalent to a pulsed operation as shown in Figure 4.31. This is the principle of amplitude-modulated mode locking. Often a saturable absorber is used (in the optical cavity) in addition to current modulation for mode locking.

4.7.2 *Mode-locked laser designs and performance*

Many demonstrations of mode-locked semiconductor lasers have been carried out in an external cavity configuration shown in Figure 4.32. An AlGaAs laser chip is used as the gain element. It has an AR coating on the side with the optical cavity. A saturable absorber is

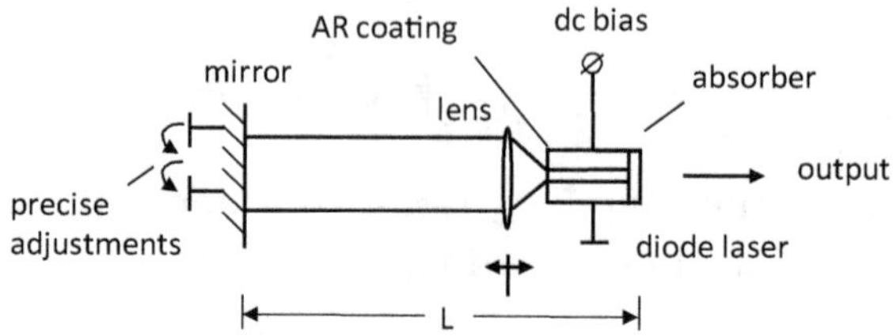

Figure 4.32: Schematic of an external cavity mode-locked laser diode.

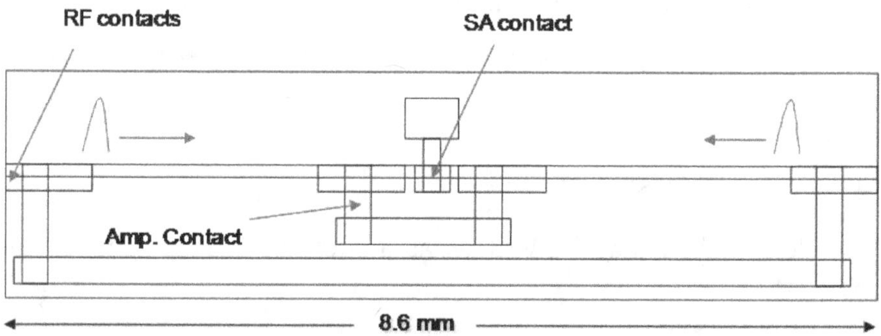

Figure 4.33: Schematic of a colliding pulse mode-locked laser. SA: Saturable absorber [84].

created on one facet by oxygen ion bombardment. The implantation depth is about 10 μm. Round trip time in the cavity is 1.2 ns. The laser has a DC bias and the current is modulated at a frequency corresponding to the inverse round trip time. Precise adjustment of the laser cavity is needed for stable mode-locked operation. Pulses with 700 fs pulse width at ~1 GHz repetition rate are obtained.

Mode-locked semiconductor lasers at transmission data rates (~10 Gb/s and higher) are important for a wide range of fiber transmission systems. These include soliton generation and clock recovery in very high-speed transmission systems. Compact, single chip, mode-locked lasers are desirable for practical systems since they are likely to be more stable than the external cavity designs.

The schematic of a compact mode-locked laser on a chip is shown in Figure 4.33 [83]. The laser is fabricated using the InGaAsP/InP material system. A saturable absorber is placed in the middle of the optical cavity. The mode-locked pulses from opposite directions arrive at the saturable absorber at the same time, which saturates the absorber faster and hence a shorter pulse could be obtained. This principle is called colliding pulse mode locking (CPM). The round trip time in CPM configuration is $t = nL/c$, where L is the length of the linear cavity, n is the index, and c is the speed of light in vacuum. The laser has four amplifier sections. Two sections are in the middle and two sections are at each end of the optical cavity. These sections are joined by a waveguide layer through which light propagates and are pumped with enough current to produce gain. The saturable absorber is in the middle. All sections have both dc and rf contacts. The modulation is applied to the saturable absorber section. The waveguide below the active region exists over the entire length of the chip. The light generated in the active region travels through this waveguide and is reflected back. The optical cavity is formed by the cleaved facets. The four active gain sections are ~350 μm long each, and the saturable absorber section is ~150 μm long. The total length of the chip is ~8.6 mm. The estimated round trip time is 0.1 ns which corresponds to a repetition rate of about 10 GHz for the mode-locked pulses. The laser emits near 1560 nm. Pulse widths of ~1 ps is obtained at ~10 GHz repetition rate.

Mode-locked lasers that operate at a higher speed have been reported using a colliding pulse scheme and a short cavity. The laser schematic is shown in Figure 4.34 [81]. It has two gain sections

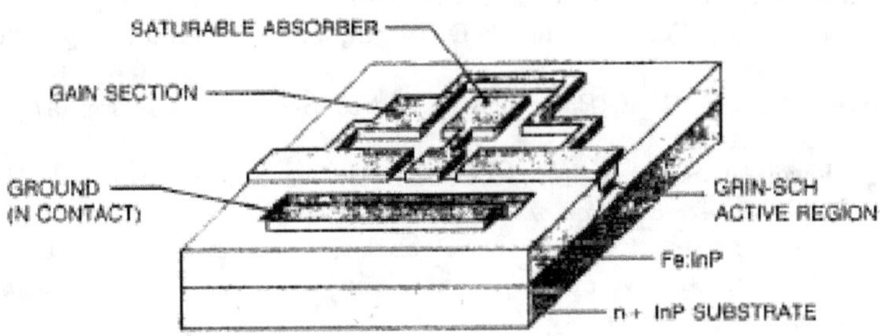

Figure 4.34: Schematic of a colliding pulse mode-locked laser fabricated using InGaAsP material system. The active region has a graded index separate confinement heterostructure (GRIN SCH) [81].

around a saturable absorber region along the optical cavity. The laser is fabricated using the InGaAsP/InP material system and it emits near 1520 nm. The Fe:InP semi-insulating layer is needed to reduce the capacitance. Both laser contacts are on the top surface. Shortest lasers were 534 μm in linear length, this corresponds to a round trip time of 2.87 ps. Transform-limited pulses with 640 fs pulse width at 350 GHz were obtained.

Advances in materials technologies have led to novel semiconductor lasers, such as nanostructure lasers, compact mode-locked lasers, and lasers based on intersubband transitions in QWs. These lasers not only demonstrate advances in semiconductor laser research but also new directions in applications. These advances are expected to continue.

References

1. H. Soda, K. Iga, C. Kitahara, and Y. Suematsu, *Jpn. J. Appl. Phys.*, **18** (1979) 2329.
2. K. Iga, F. Koyama, and S. Kinoshita, *IEEE J. Quant. Electron.*, **24** (1988) 1845.
3. J. L. Jewell, J. P. Harbison, A. Scherer, Y. H. Lee, and L. T. Florez, *IEEE, J. Quant. Electron.*, **27** (1991) 1332.
4. C. J. Chang-Hasnain, M. W. Maeda, N. G. Stoffel, J. P. Harbison, and L. T. Florez, *Electron. Lett.*, **26** (1990) 940.
5. R. S. Geels, S. W. Corzine, and L. A. Coldren, *IEEE J. Quantum Electron.*, **27** (1991) 1359.
6. R. S. Geels and L. A. Coldren, *Appl. Phys. Lett.*, **5** (1990) 1605.
7. K. Tai, G. Hasnain, J. D. Wynn, R. J. Fischer, Y. H. Wang, B. Weir, J. Gamelin, and A. Y. Cho, *Electron. Lett.,* **26** (1990) 1628.
8. K. Tai, L. Yang, Y. H. Wang, J. D. Wynn, and A. Y. Cho, *Appl. Phy. Lett.*, **56** (1990) 2496.
9. M. Born and E. Wolf, *Principles of Optics*, Section 1.6.5,' Pergamon Press, NY, (1977), p. 69.
10. J. L. Jewell, A. Scherer, S. L. McCall, Y. H. Lee, S. J. Walker, J. P. Harbison, and L. T. Florez, *Electron. Lett.*, **25** (1989) 1123.
11. Y. H. Lee, B. Tell, K. F. Brown-Goebeler, J. L. Jewell, R. E. Leibenguth, M. T. Asom, G. Livescu, L. Luther, and V. D. Mattera, *Electron. Lett.*, **26** (1990) 1308.
12. K. Tai, R. J. Fischer, C. W. Seabury, N. A. Olsson, D. T. C. Huo, Y. Ota, and A. Y. Cho, *Appl. Phys. Lett.*, **55** (1989) 2473.
13. A. Ibaraki, K. Kawashima, K. Furusawa, T. Ishikawa, T. Yamayachi, and T. Niina, *Jpn. J. Appl. Phys.*, **28** (1989) L667.

14. E. F. Schubert, L. W. Tu, R. F. Kopf, G. J. Zydzik, and D. G. Deppe, *Appl. Phys. Lett.*, **57** (1990) 117.
15. R. S. Geels, S. W. Corzine, J. W. Scott, D. B. Young, and L. A. Coldren, *IEEE Photo. Tech. Lett.,* **2** (1990) 234.
16. D. Vakhshoori, J. D. Wynn, and R. E. Liebenguth, *Appl. Phys. Lett.*, **65** (1994) 144.
17. Chapters 1 and 2, in D. Deppe and K. Choquette (eds.) *Vertical Cavity Surface Emitting Lasers*, Gordon Breach, NY, 2000.
18. N. K. Dutta, L. W. Tu, G. J. Zydzik, G. Hasnain, Y. H. Wang, and A. Y. Cho, *Electron. Lett.*, **27** (1991) 208.
19. K. Iga, F. Koyama, and S. Kinoshita, *IEEE J. Quant. Electron.*, **QE-24** (1988) 1845.
20. K. Tai, F. S. Choa, W. T. Tsang, S. N. G. Chu, J. DWynn, and A. M. Sergent, *Electron. Lett.*, **27** (1991) 1514.
21. L. Yang, M. C. Wu, K. Tai, T. Tanbun-Ek, and R. A. Logan, *Appl. Phys. Lett.* **56** (1990) 889.
22. T. Baba, Y. Yogo, K. Suzuki, F. Koyama, and K. Iga, *Electron. Lett.*, **29** (1993) 913.
23. Y. Imajo, A. Kasukawa, S. Kashiwa, and H. Okamoto, *Jpn. J. Appl. Phys. Lett.*, **29**, (1990) 1130.
24. D. I. Babic, K. Streubel, R. Mirin, N. M. Margalit, J. E. Bowers, E. L. Hu, D. E. Mars, L. Yang, and K. Carey, *IEEE Photon. Tech. Lett.*, **7** (1995) 1225.
25. Z. L. Liau and D. E. Mull, *Appl. Phys. Lett.,* **56** (1990) 737.
26. R. Kazarinov and R. A. Suris, *Sov. Phys. Semicond.*, **5** (1971) 707.
27. J. Faist, F. Capasso, D. L. Sivco, C. Sirtori, A. L. Hutchinson, and A. Y. Cho, *Science*, **264**, (1994) 553; C. Sirtori *et al. Appl Phys. Lett.*, **73** (1998) 3486.
28. J. S.Yu, A. Evans, S. Slivken, S. R. Darvish, and M. Razeghi, *Appl. Phys. Lett*, **88** (2006) 118.
29. M. Razeghi, *IEEE J. Selec Areas Quant Electron*, **15** (2009) 941.
30. C. Gmachl, F. Capasso, D. L.Sivco, and A. Y. Cho, *Rep. Prog. Phys.*, **64** (2001) 1553.
31. J. Faist *et al.*, *IEEE J. Quant. Electron.,* **34** (1998) 336; J. Faist *et al.*, *Appl. Phys. Lett.,* **68**, (1996) 3680.
32. M. Beck, D. Hofstetter, T. Aellen, J. Faist, U. Oesterle, M. Ilegems, E. Gini, and *H. Melchior Sci.*, **291** (2002) 301.
33. H. Hamada, M. Shono, S. Honda, R. Hiroyama, K. Yodoshi, and T. Yamaguchi, *IEEE J. Quant. Electron.*, **QE-27** (1991) 1483.
34. K. Itaya, M. Ishikawa, and Y. Uematsu, *Electron Lett.*, **26** (1990) 839.
35. I. Akasaki and H. Amano, *Electron. Lett.*, **32** (1990) 1105.
36. S. Nakamura, M. Senoh, S. Nagahama, N. Iwasa, T. Yamada, T. Matsushita, H. Kiyoku, and Y. Sugimoto, *Jpn. J. Appl. Phys. Lett.*, **35** (1996) L74.
37. K. Itaya, M. Onomura, J. Nishio, L. Sugiura, S. Saito, M. Suzuki, J. Rennie, S. Nuone, M. Yamamoto, H. Fujimoto, Y. Kokubun, Y. Ohba, G. Hatakoshi, and M. Ishikawa, *Jpn. J. Appl. Phys.,* **35** (1996) L1315.
38. S. Denbaars, *Proc. IEEE.*, **85** (1997) 1740.

39. C. Lee, C. Zhang, M. Cantore, R. Farrell, S. Ho-Oh, T. l Margalith, J. Speck, S. Nakamura, J. E. Bowers, and S. P. DenBaars, *Opt. Exp.,* **23** (2015) 16232.
40. G. Weng, *et al.*, *Opt. Exp.*, **24** (2016) 15546.
41. E. Kapon, Chapter 10, in P. Zory (ed.), *Quantum Well Lasers*, Academic Press, New York, 1993.
42. Y. Arakawa and H. Sasaki, *Appl. Phys. Lett.***,** **40** (1982) 939.
43. M. Asada, Y. Miyamoto, and Y. Suematsu, *IEEE J. Quant. Electron.*, **QE-22** (1986) 1915.
44. A. Yariv, *Appl. Phys. Lett.*, **53** (1988) 1033.
45. Y. Miyamoto, Y. Miyake, M. Asada, and Y. Suematsu, *IEEE J. Quant. Elect.*, **QE-25** (1989) 2001.
46. M. Asada, Y. Miyamoto, and Y. Suematsu, Jpn. *J. Appl. Phys.,* **24** (1985) L95.
47. M. Cao, Y. Miyake, S. Tamura, H. Hirayama, S. Arai, Y. Suematsu, and Y. Miyamoto, *Trans. IEICE,* **E73** (1990) 63.
48. V. M. Ustinov, A. E. Zhukov, A. Yu. Egorov, and N. A. Maleev, *Quantum Dot Lasers*, Oxford University Press, UK, 2003.
49. A. E. Zhukov, A. R. Kovsh, A. Yu. Egorov, N. A. Maleev, V. M. Ustinov, B. V. Volovik, and M. V. Maximov, *Semiconductors*, **33** (1999) 153.
50. M.Sugawara, T.Akiyama, N.Hatori, Y.Nakata, H. Ebe, and H. Ishikawa, *Meas. Sci. Technol.,* **13** (2002) 1683.
51. M.Sugawara, H. Ebe, N. Hatori, M. Ishida, Y. Arakawa, T. Akiyama, K. Otsubo, and Y. Nakata, *Phys. Rev. B*, **69**, Art No. 235332 (2004).
52. H. Sun, Q. Wang, H. Dong, and N. K. Dutta, *Opt. Exp.*, **13**, (2005) 1892.
53. V. M. Ustinov, A. E. Zhukov, A. Egorov, A. Yu, A. R. Kovsh, S. V. Zaitsev, Zh. I. Alferov, and D. Bimberg, *Electron. Lett.*, **34** (1998) 670.
54. 54. Y. Miamoto, *Jpn. J. Appl. Phys.*, **26** (1985) **L**225.
55. M. Cao *et al.*, *Electron. Lett.*, **24** (1988) 824.
56. S. Kim, H. Ryu, H. Park, G. Kim, Y. Choi, and Y. Lee, *Appl. Phys. Lett.*, **81** (2002) 2499.
57. H. Park, S. Kim, M. Seo, Y. Ju, S. Kim, and Y. Lee, *IEEE J Quant Electron.*, **QE-41** (2005) 1131.
58. K. Inoshita and T. Baba, *IEEE J. Sel. Topics Quant. Electron.*, **9** (2003) 1347.
59. H. G. Park, S. K. Kim, S. H. Kwon, G. H. Kim, S. H. Kim, H. Y. Ryu, S. B. Kim, and Y. H. Lee, *IEEE Photon. Technol. Lett.*, **15** (2003) 1327.
60. A. Kasukawa, Chapter 3, in A. K. Dutta, N. K. Dutta, and M. Fujiwara (eds.), *WDM Technologies: Active Optical Components* Academic Press, NY, 2002.
61. H. Asano, S. Takano, M. Kawaradani, M. Kitamura, and I. Mito, *IEEE Photon. Tech. Letts.*, **3** (1991) 415.
62. T. Tanbun-Ek, R. A. Logan, **N.** A. Olsson, H. Temkin, A. M. Sergent, and K. W. Wecht, *Appl. Phys. Lett.*, **57** (1990) 224.
63. K. Shigihara, K. Kawasaki, Y. Yoshida, S. Yamamura, T. Yagi, and E. Omura, *IEEE J. Quant. Electron.*, **38** (2002) 1081.

64. G. Yang, G. M. Smith, M. K. Davis, A. Kussmaul, D. A. S. Loeber, M. H. Hu, H.-K. Nguyen, C. E. Zah and R Bhat, *IEEE Photon. Tech. Lett.*, **16**, (2004) 981.
65. C. J. Chang-Hasnain, R. Bhat, H. Leblanc, and M. A. Koza, *Electron. Lett.*, **29** (1993) 1.
66. G. Hunziker and C. Harder, *Appl. Opt.*, **34** (1995) 6118.
67. H. Horie,N. Arai, Y. Mitsuishi, N. Komuro, H. Kaneda, H. Gotoh, M. Usami, and Y. Matsushima, *IEEE Photon. Tech. Lett.*, **12**, (2000) 1304.
68. A. Okubo, Y. Yamada, Y. Oeda, K. Igarashi, T. Fujimoto, Y. Yamada, and K. Muro, *Proc. Tech. Dig. Optical Fiber Commun. Conf.*, San Diego, CA, 1999, Paper TuC3.
69. T. Hiramoto, M. Sagawa, T. Kikawa, and S. Tsuji, *IEEE J. Sel. Top. Quant. Electron.*, **5**, (1999) 817.
70. H. Horie, S. Nagao, K. Shimoyama, and T. Fujimori, *Jpn. J. Appl. Phys.*, **38** (1999) 5888.
71. M. Usami, N. Edagawa, Y. Matsushima, H. Horie, I. Sakamoto, T. Fujimori, and H. Gotoh, *Proc. Optical Fiber Commun.*, San Diego, CA, 1999, Paper PD-39.
72. H. Horie, H. Ohta, Y. Yamamoto, N. Arai, B. Kelly, T. Fujimori, H. Gotoh, M. Usami, and Y. Matsushima, *Proc. SPIE*, **3945** (2000) 280.
73. J. Sebastian, G. Beister, F. Bugge, F. Buhrandt, G. Erbert, H. G. Hänsel, R. Hülsewede, A. Knauer, W. Pittroff, R. Staske, M. Schröder, H. Wenzel, M. Weyers, and G. Tränkle, *IEEE J. Sel. Top. Quant. Electron.*, **7** (2001) 334.
74. J. N. Walpole,*Opt. Quant. Electron.*, **28**, (1996) 623.
75. G. Erbert, J. Fricke, R. Hülsewede, A. Knauer, W. Pittroff, P. Ressel, J. Sebastian, B. Sumpf, H. Wenzel, and G. Tränkle, *Proc. SPIE*, **4995** (2003) 29.
76. F. Dittmar, B. Sumpf, J. Fricke, G. Erbert, and G. Tränkle *IEEE Photon. Tech. Lett.* **18**, (2006) 601.
77. S. Pawlik, J. Muller, N. Lichtenstein, D Jaeggi, and B. Schmidt, *Proc. Opt. Fiber Commun.*, (2000) Paper ThP3.
78. Y. He, H. An, J. Cai, C. Galstad, S. Macomber, and M. Kanskar, *Electron Lett.*, **45** (2009) 163.
79. P. Crump H. Wenzel, G. Erbert P. Ressel, M. Zorn, F. Bugge, S. Einfeldt, R. Staske, U. Zeimer, A. Pietrzak, and G. Tränkle, *IEEE Photon. Tech. Lett.*, **20** (2006) 1378.
80. P. B. Hansen, G. Raybon, U. Koren, B. I. Miller, M. G. Young, M. Newkirk, M. D. Chien, B. Tell, and C. A. Burrus, *Appl. Phys. Lett.*, **62** (1993) 1445.
81. Y. K. Chen, M. C. Wu, T. Tanbun-Ek, R. A. Logan, and M. A. Chin, *Appl. Phys. Lett.*, **58** (1991) 1253; Y. K. Chen, M. C. Wu, T. Tanbun-Ek, R. A. Logan, and M. A. Chin, *IEEE Photon. Tech. Lett.*, **3** (1991) 824.
82. P. A. Morton, V. Mizrahi, and T. Tanbun-Ek, *Electron. Lett.*, **28** (1992) 2156.
83. H. Fan, C. Wu, M. El-Aasser, N. K. Dutta, U. Koren, and A. B. Piccirilli, *IEEE Photon. Tech. Lett.*, **12** (2000) 972.
84. R. Kaiser and B. Huttl, *IEEE J. Sel. Top. Quant. Electron.*, **13** (2007) 127.

Chapter 5

Modulators

5.1 Introduction

All optical-dense wavelength-division multiplexing (DWDM) and all optical time-division multiplexing (OTDM) systems are important ways to utilize the vast bandwidth already available in fiber. One of the key features that makes these systems viable is the availability of inexpensive low-chirp (chirp is spectral width under modulation) optical transmitters (TXs). The maximum capacity and repeater span of an optical fiber transmission system are limited by the bit rate, the optical loss, the dispersion, the nonlinear characteristics of the fiber, and the frequency chirp of the source. For example, the single-channel dispersion-limited transmission distance for an ideal 1550 nm source is about 900 km at 2.4 Gb/s and 60 km at 10 Gb/s for standard non-dispersion-shifted single-mode fiber, which has a dispersion of 17 ps/nm/km at 1550 nm. However, the frequency chirp of directly modulated distributed feedback (DFB) laser typically limits the repeater span to <200 km for 2.4 Gb/s system and <50 km for 10 Gb/s system. Dispersion management techniques, such as dispersion compensating fiber, negative-chirp sources, and mid-span spectral inversion can be used, but they have some optical loss and they are transmission distance dependent and wavelength dependent. Thus, it is desirable to have low-chirp TXs. The frequency of chirping associated with the high-speed direct modulation of laser diodes above a few Gb/s has been an area of investigation for many years.

External modulators with high bit rate and low voltage operation are required to construct low-chirp TXs and can be used to produce zero-chirp and negative-chirp to minimize the system degradation associated with fiber dispersion.

In this chapter, semiconductor modulator and $LiNbO_3$ Mach–Zehnder modulator (MZM) devices are described. The semiconductor modulator can be divided into two categories: semiconductor waveguide modulator and the semiconductor surface-normal modulator. For encoding electrical data onto an optical signal in an intensity-modulated form (On–Off Keying, OOK), the modulator generally has a continuous wave (CW) DFB laser at its input. A modulation voltage representing the data is applied to the modulator, and the modulator optical output has the data encoded onto it. The electroabsorption (EA) modulator works on the principle of change in absorption of a semiconductor material by the application of electric field (or voltage). The transmission as a function of voltage is shown in Figure 5.1(a). The MZM works on the principle of phase change by the application of voltage. The phase change is converted to an intensity change using the Mach–Zhender (MZ) interferometer configuration. The ratio P_{out}/P_{in} of the MZMs as a function of voltage is shown in Figure 5.1(b).

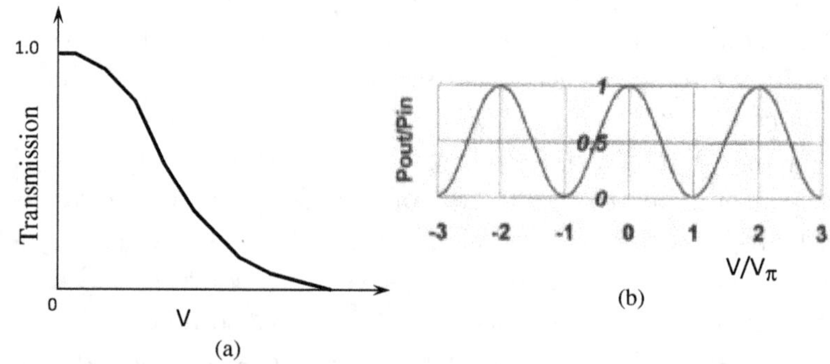

Figure 5.1: (a) Optical transmitted power as a function of voltage for EA modulator. (b) Optical transmitted power as a function of voltage difference (V) between the two arms of the MZM. The ratio of the output to input is plotted as a function of V/V_π. V_π is the voltage needed for π phase shift.

5.2 Semiconductor waveguide modulator

There are two types of semiconductor waveguide modulators. They are the EA modulator and the MZ interferometer-based modulator. As mentioned earlier, the former is based on the change in the absorption near the band gap of a semiconductor (known as EA effect) and the latter is based on phase change of a propagating wave in the presence of an electric field.

5.2.1 *Electroabsorption (EA) modulator*

In 1958, Franz [1] and Keldysh [2] reported the first theoretical studies of the effect of an electric field on the absorption edge of a semiconductor. They predicted that in the presence of an electric field, optical absorption would occur at photon energies less than the band energy gap (Franz–Keldysh effect) due to a decrease in the effective band gap. This effect was experimentally observed by Williams [3, 4], Moss [5], and Boer *et al.* [6]. Based on this effect, high-speed, low-chirp Franz–Keldysh modulators [7, 8] have been developed. The development of crystal growth techniques, such as metal organic chemical vapor deposition (MOCVD) and vapor phase epitaxy (VPE), has made the multiple quantum well (MQW) structure the basis for these modulators. The MQW structure exhibits strong exciton effect, which modifies the fundamental absorption edge of the materials. The exciton effect results from the Coulomb interaction between the electron–hole pair in the crystal, and manifests itself as an increased steepness in the absorption coefficient. When an electric field is applied perpendicular to the MQW structures, the absorption coefficient near the band edge decreases significantly. The EA effect in MQW structures differs from that in the bulk in several ways. Most importantly, the two mechanisms differ depending on whether the field is parallel or perpendicular to the quantum wells, with significantly different behavior appearing for the perpendicular case (the quantum confined Stark effect, QCSE) [9–11]. The QCSE is very large even at room temperature, which makes possible low drive voltage EA modulators. The calculated absorption spectrum of a quantum well in the presence of an electric field is

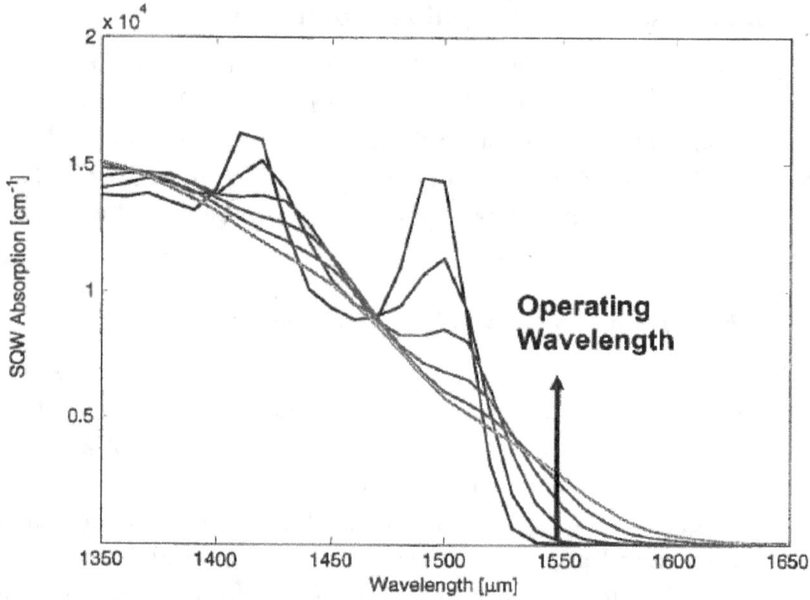

Figure 5.2: Absorption spectrum for various electric fields in a single quantum well (SQW) structure. With increasing field, the absorption increases at 1550 nm [11].

shown in Figure 5.2 [11]. At zero electric field, transitions at exciton peak wavelengths are clearly visible. With increasing field, the exciton peak broadens and the absorption shifts to longer wavelength (smaller energies). Also, shown in Figure 5.2 is the location of an incident laser wavelength, illustrating that with increasing field, the absorption at this wavelength increases.

The EA modulator has a waveguide structure similar to that of a laser. It consists of a P-type cladding, N-type cladding, and an undoped multiquantum well absorbing region. For an EA modulator working near 1550 nm, the cladding layers are P-InP, N-InP and the absorbing region is InGaAsP active quantum wells with higher gap InGaAsP (or InP) barrier layers. Often, there are 3–6 periods of active and barrier layers for high enough absorption (needed for high extinction ratio for optical transmission) with electric field. The composition of active quantum wells and barriers, and device length

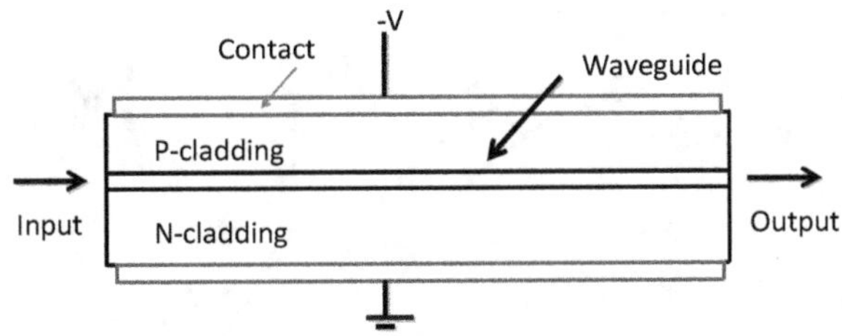

Figure 5.3: Schematic of an EA modulator design. The input and output facets have antireflection (AR) coating.

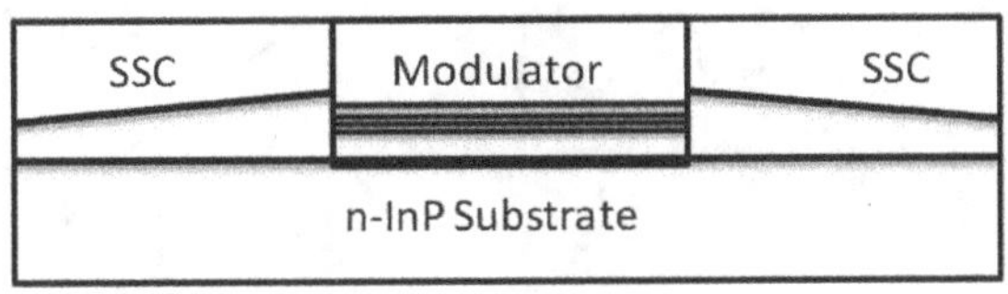

Figure 5.4: EA modulator with spot size converter both at input and output. SSC: spot size converter region.

are chosen to optimize EA device performance. Lateral index guiding is provided by etching and regrowth of semi-insulating (Fe:InP) layers along with other layers. A schematic of an EA modulator is shown in Figure 5.3. A single-mode EA modulator generally has an active region thickness (combined thickness of wells and barriers in the undoped multiquantum wells) of $\sim 0.1\ \mu$m and a lateral waveguide thickness of $\sim 1\ \mu$m.

For some applications, such as for fiber optic communications, the input to the EA modulator is fiber coupled and the output is also coupled to a single-mode fiber. In order to reduce the coupling losses, it is important that the EA modulator waveguide modes match that of the fibers. In order to do so, EA modulators with spot size converters (or an addition to the EA modulator waveguide which matches the EA modulator mode profile to that of a single-mode fiber) at each waveguide end have been fabricated. Tapered waveguides are used to fabricate these spot size converters in the EA modulator wafer (Figure 5.4) [11–15].

The optical loss of an EA modulator depends on the incident wavelength, and it increases with increasing voltage. This is shown in

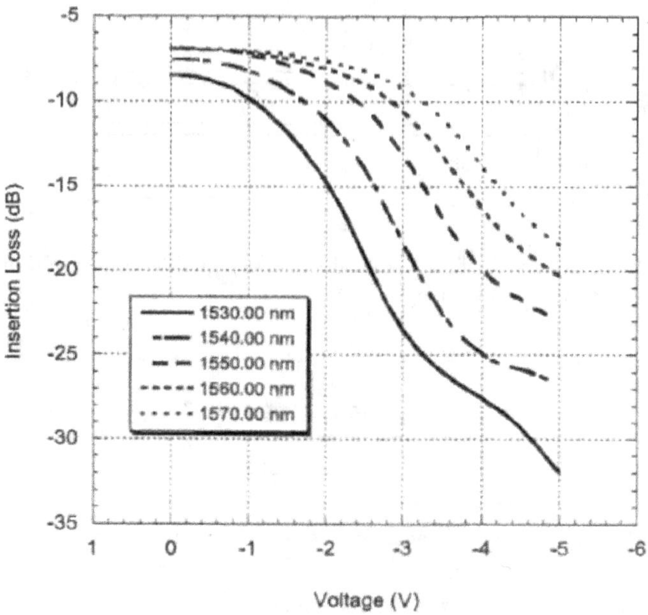

Figure 5.5: Optical loss (shown as dB) as a function of applied voltage to an EA modulator for several wavelengths of incident light. The band gap of the modulator is 1495 nm [11].

Figure 5.5. The figure shows the absorption for incident wavelengths from 1530 to 1570 nm [11]. The band gap of the absorbing material in this case is 1495 nm. As expected, the zero-bias absorption is larger for incident wavelength close to the band gap.

The small signal modulation response of EA modulator has been measured. This is done by applying a small modulation voltage on top of the dc bias and measuring the modulated intensity on the output signal. The relative output modulation intensity as a function of modulation frequency is called the modulation response. A decrease in relative response to half the value (3 dB) is called the bandwidth (or 3 dB bandwidth) of the modulator. A sample of the data for modulation response is shown in Figure 5.6 [15].

High-speed multiple quantum well EA modulators with low drive voltage (<3 V) have been fabricated [16–18]. They have been used as external modulators in optical transmission experiments. In order to

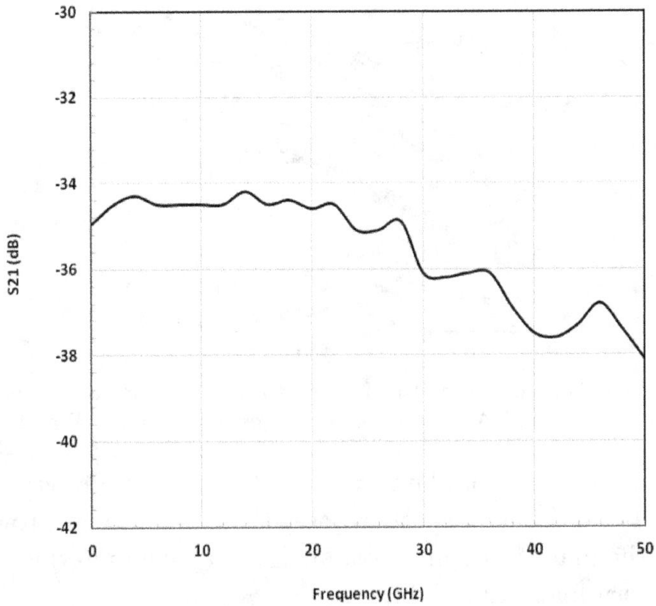

Figure 5.6: Modulation response of an EA modulator. The 3 dB bandwidth is 45 GHz [15].

fully utilize the advantages of an EA modulator, it is often integrated with a single-wavelength DFB laser. These integrated EA-modulated DFB lasers (EML) are important for present-day fiber transmission systems. In the following section, EML fabrication and performance are described.

5.2.1.1 *Integrated EA modulator and laser (EML)*

A typical EML structure is shown in Figure 5.7. Two different growth techniques have been used to fabricate EMLs: etch and regrowth technique and the selective area growth (SAG) technique. In the etch and regrowth process [19], the active layers for one of the devices, typically the laser, are grown first, then selectively etched away, and the active layers of the modulator are grown in the etched area. This technique allows almost complete freedom to separately optimize the laser and modulator active layers, but the abrupt discontinuity in the optical waveguide can lead to low coupling efficiency between

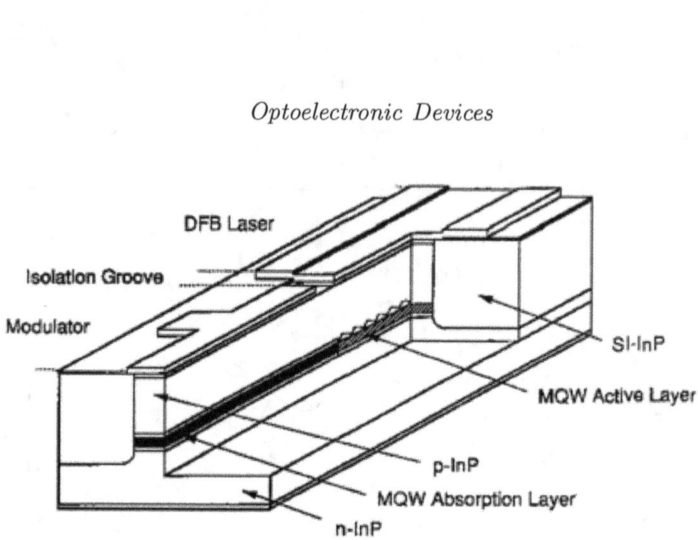

Figure 5.7: The schematic of EML structure. The DFB laser is integrated on the same chip with an EA modulator. Both devices have MQW active region.

the laser and the modulator if the etching and growth rates are not tightly controlled. Furthermore, these additional growth steps increase the manufacturing cost, and the regrown interface may also be a concern for laser reliability.

The SAG process, on the other hand, uses a single growth of the MQW active structure on an oxide-patterned substrate, resulting in a reproducible band gap difference between the laser and the modulator [20]. This comes about because of the enhanced growth rate of the wells and barriers in the vicinity of the oxide pattern, as well as a change in the composition (and hence strain) of the layers caused by the difference in diffusion and sticking coefficients of the Group III precursors. The SAG process gives a continuous laser active layer to modulator transition with high coupling efficiency and improved reliability. Figure 5.8 illustrates the photoluminescence (PL) wavelength shift in the presence of a dielectric mask (similar to that could be used for EML active region growth) during MOCVD growth process. The PL peak wavelengths shown are similar to that needed for EML growth [21].

The fabrication of an EML involves the following steps: First, a grating with a periodicity of $\Lambda(\Lambda = \lambda_{\rm DFB}/2n_{\rm eff}$, $n_{\rm eff}$ is the effective index, typically $\Lambda \sim 2400$ A for 1550 nm laser) is formed on part of the wafer (under the DFB laser region) by optical holography (discussed in Chapter 9) or other techniques (phase

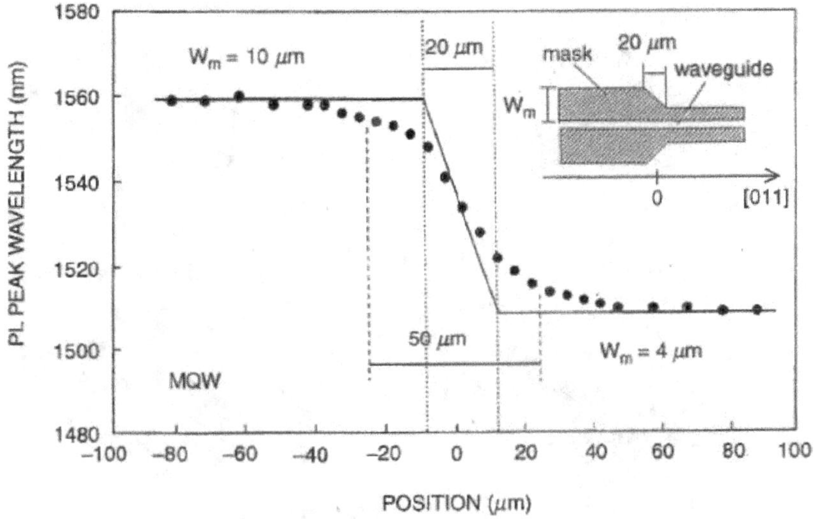

Figure 5.8: The PL peak wavelength shift for an InGaAsP/InP MQW structure along the direction of varying stripe width. The mask shown is a SiO_2 mask on the InP wafer. The mask width (W_m) is varied from 10 to 4 μm. The direction of the active waveguide is [011] in this case [21].

mask or electron-beam etching). Variable-width SiO_2 masks are then formed on the wafer along the active region direction. The mask width W_m is chosen so that in the DFB laser region the gain peak is near 1550 nm, and in the modulator region, the absorber has a band gap of $\Delta\lambda$ (design value) shorter than the gain peak of the laser region. A typical value of $\Delta\lambda$ is ~60–70 nm. Then, the active region waveguide of laser and modulator are simultaneously grown by selective-area MOCVD, which results in no discontinuity between the two waveguides. For the fabrication process shown in Figure 5.9, new SiO_2 stripes (~7 μm wide) are deposited for buried heterostructure formation along the active region direction. Then, p^-InP embedding layer (laterally around the active waveguide) and p^+InGaAs contact layer are grown on the wafer.

A 25 μm wide contact layer is removed where the laser and modulator join to provide electrical isolation (resistance >8 kΩ). Standard metal contacts (Ti/Au) with small bonding pads (for low capacitance) are applied to p-side. The n-side has a broad area

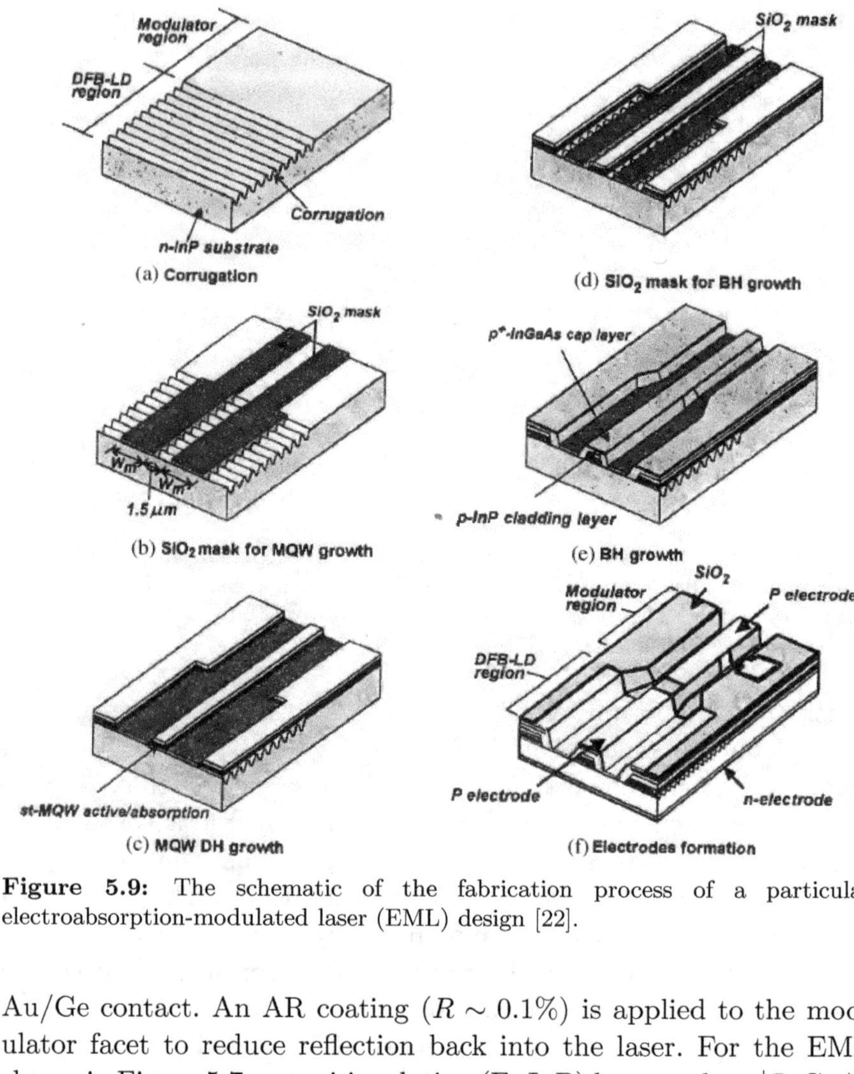

Figure 5.9: The schematic of the fabrication process of a particular electroabsorption-modulated laser (EML) design [22].

Au/Ge contact. An AR coating ($R \sim 0.1\%$) is applied to the modulator facet to reduce reflection back into the laser. For the EML shown in Figure 5.7, a semi-insulating (Fe:InP) layer and a p^+InGaAs contact layer are used during second growth. CW light vs. current characteristics of an EML for several modulator voltages is shown in Figure 5.10.

The light output decreases by ~20 dB with −3 V on the modulator. The modulation speed of an MQW EA modulator is limited by its electrical capacitance. As a result, a short modulator length

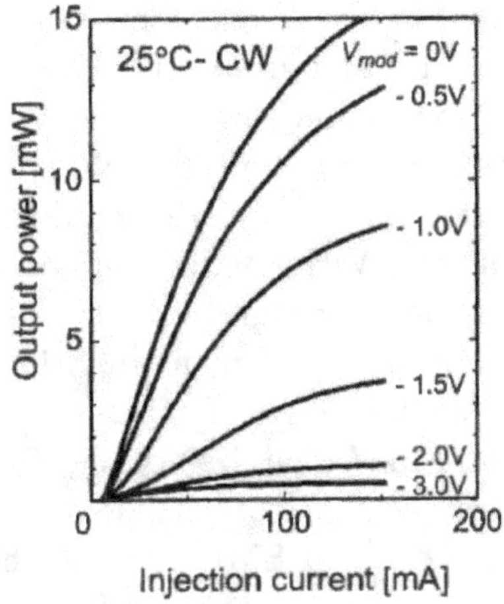

Figure 5.10: Light vs. current characteristics of an EML for various modulator voltages [22].

is preferable for high-speed operation. A short length also results in a lower extinction ratio. A simple and effective method to limit the problem is to increase the number of MQW wells in the EA modulator. At the same time, in an EML, the loss and the extinction characteristics strongly depend on the wavelength difference between the exciton-peak wavelength and the wavelength of the DFB laser, and this wavelength difference is called the detuning wavelength. When the numbers of quantum wells increase, the loss increases for the same detuning wavelength. The small detuning wavelength causes large undesirable loss, and the larger detuning wavelength produces large driving voltage. A high-speed, low-voltage EML is fabricated by optimizing the modulator length, the number of quantum wells, and the band gap [23, 24].

The most important performance parameter for an EML used for optical transmission is the wavelength chirp. The chirp can be categorized as either intrinsic to the EA modulator itself, or due to the parasitic modulation of the DFB laser. The intrinsic chirp of any EA modulator comes from the modification of the absorption band edge by the applied electric field, due to the Franz–Kelydsh effect or the QCSE, and the coupling between the real and imaginary parts of the refractive index, n and k, through the Kramers–Kronig relation. The sign of the linewidth enhancement factor, $\alpha = \mathrm{d}n/\mathrm{d}k$, may be either positive or negative depending on the operating point, and the resulting wavelength chirp will be proportional to $\mathrm{d}n/\mathrm{d}t$ and hence occur only at the leading and trailing edge of the pulses. A negative α can be achieved in MQW EA modulators at large reverse bias or small band gap difference between the laser and the modulator, but there is a penalty paid in terms of modulation insertion loss [25]. The magnitude of the chirp of an EML is much smaller than that of a directly modulated laser, and since the chirp generally occurs at the leading and trailing edges of the pulses, the dispersion-related penalty is further reduced. An EML may also have excess parasitic chirp caused by electric cross-talk or the reflection of modulated light back to the laser [27]. Both of these effects will induce a modulation of the laser carrier density, which will cause a chirp proportional to the much larger α parameter of the laser. In the case of optical reflection, the optical phase as well as the magnitude of the reflected light are important in determining the magnitude and sign of the chirp of the EML. The formation of the electrical isolation and AR coating of the output facet are thus key processing steps in fabricating EMLs. The chirp of EMLs is small enough to allow long-distance transmission. Historically (in the 1990s), the development of EMLs was pursued for enabling long-distance fiber optic transmission systems due to their lower chirp (spectral width under modulation) compared to that for a regular DFB laser. The schematic and bit-error-rate (BER) measurement of a transmission system experiment are shown in Figure 5.11. BER date for 600 km, 800 km, and back-to-back (<0.5 km fiber) transmission is shown in Figure 5.11. The experiment utilized several optical amplifiers for amplifying the signal over the 600 and 800 km transmission distance. Both optical amplifiers and optical transmission are discussed in later chapters.

5.2.2 *Mach–Zehnder modulator*

MZMs have less chirp than EA modulators. When the bit rate is more than 10 Gb/s, it is advantageous to use "chirp-less" modulators, such

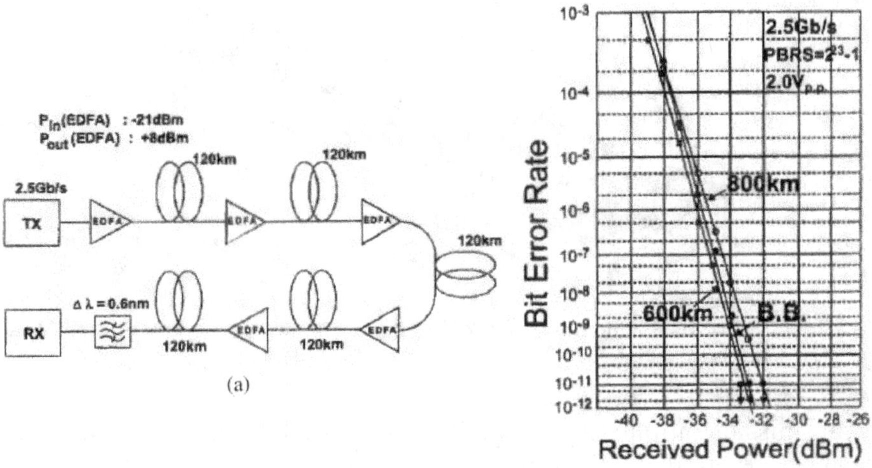

(a) (b)

Figure 5.11: (a) Experimental set up for a 2.5 Gb/s, 600 kn transmission over a regular fiber. (b) Measured BER characteristics. BER of $<10^{-11}$ has been obtained, which demonstrates suitability of EML for long-distance transmission. TX: transmitter with EML as the source. EDFA: Erbium-doped fiber amplifier, RX: photodiode based receiver, B:B back to back i.e. <0.5 km of fiber [26].

as MZ type instead of EA modulators. The MZ interferometer was proposed independently by Mach and Zehnder in 1891, as guided-wave interferometers. These modulators are capable of modulating a laser beam using the electro-optic effect and were first fabricated using lithium niobate ($LiNbO_3$). Recent transmission experiments have demonstrated that MQW p-i-n waveguide MZMs are also well suited for high bit rate, long-distance telecommunication systems [27, 28]. The semiconductor waveguide MZM offers transmission performance comparable to that of the more widely used $LiNbO_3$ MZM and has the potential for integration with a laser source [29–31]. It also has the important feature that the chirp of the transmitted signal is a function of the electro-optics properties of the p-i-n waveguide, the splitting ratio of the two Y-junctions, the differential phase shift between the two arms of the unbiased interferometer, and the format of the modulating voltage applied to each arm of the modulator. This provides considerable flexibility in optimizing device performance for

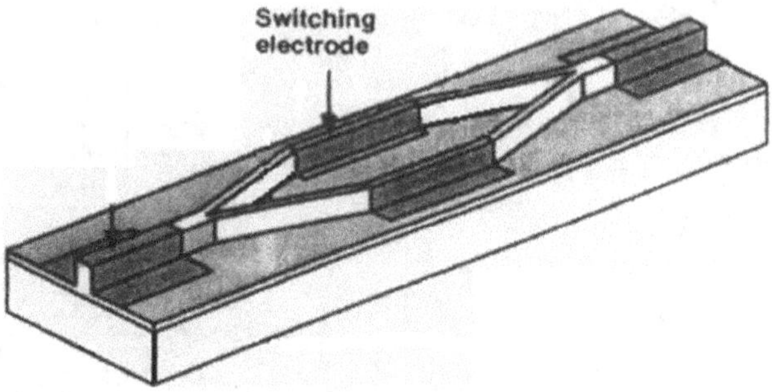

Figure 5.12: The Schematic of MZM. The "Switching electrodes" noted above is used for modulation.

transmission applications. Many modulators have been developed using both bulk and MQW GaAs/AlGaAs and InGaAsP/InP systems. The MZ type modulator consists of two waveguides parallel to each other and two couplers that combine the parallel waveguides, as shown in Figure 5.12. The optical signal is split into two branches. After traveling along the two parallel waveguides, the two branches are recombined. The electric fields or carrier injection applied to the two arms cause refractive index change and thus produce variable phase shift for optical interference at the output. For each arm of the modulator, the change in the optical field due to the effective drive voltage is given by

$$E(t) = E_0 \exp\left[-\left(\frac{\Delta\alpha(t)}{2} + j\Delta\beta(t)\right) L\right] \tag{5.1}$$

where L is the interaction length of the modulator arm, $\Delta\alpha(t)$ is the power attenuation constant, and $\Delta\beta(t)$ is the phase change. Both the electro-optic and free carrier effects are involved in the refractive index changes [32, 33]. The electro-optic effect consists of both the linear electro-optic and the quadratic electro-optic effects. The free carrier effect includes both the plasma and the band-gap shift effects. The band-gap shift effect is a result of both the band filling and band-gap shrinkage effects. The optical properties of semiconductor quantum wells are significantly different from that of bulk materials. The parabolic density of states for electrons and holes in bulk materials change to a step-like density of states for quantum-confined structures. Large absorption coefficient changes have been observed due to QCSE, and the use of QCSE in EA modulator has been described earlier. Associated with these changes are changes in the real part of the refractive index. From the measured EA spectrum, it is possible to calculate the electro-refraction spectrum through the Kramers–Kronig dispersion relation [34, 35]. When the detuning wavelength is increased, the loss modulation decreases, but a relatively large index modulation remains. The attenuation and phase constants of the optical signal are nonlinear functions of the applied voltage due to the QCSE. Although MZM using QCSE have low drive voltages, they have high optical insertion loss and microwave propagation loss which needs to be reduced for commercial applications.

There are two kinds of driving configuration for MZMs: single-sided drive and push–pull drive. In the single side drive, only one arm of the interferometer is phase modulated, the other being held at the same potential as the substrate. In a push–pull drive, the two arms of the interferometer are modulated simultaneously. By using a common electric field to phase modulate the two interfering waves in opposite phase, the drive voltage is reduced. Although the modulation efficiency is doubled in the push–pull drive, so is the capacitance. This may result in reduced bandwidth.

5.2.2.1 *GaAs/AlGaAs MZMs*

$Al_xGa_{1-x}As$ is lattice matched to GaAs for all x values. It is possible to adjust the index of refraction of the $Al_xGa_{1-x}As$ material by adjusting x. Increasing Al composition (x) decreases its index of refraction. By growing such layers epitaxially using techniques such as MBE and MOCVD, it is possible to sandwich a higher index material between two lower index materials. This forms a slab waveguide and provides waveguiding in the vertical direction. The most common approach to providing lateral waveguiding is to etch a rib and fabricate a rib waveguide. The effective index inside the rib is higher than the effective index outside the rib. Among other methods of fabrication are: proton implantation, buried heterostructures, p–n junctions, and disordering. However, rib waveguide is often used due to ease of fabrication, and vertical and lateral index profiles and dimensions can be independently controlled. Typical rib widths are in the 2–4 μm range, and typical rib heights are less than 1 μm.

The MZM using undoped epitaxial layers of AlGaAs/GaAs and push–pull drive has been fabricated with a bandwidth >40 GHz at 1550 nm [36]. The frequency response of these devices is flat up to 20 GHz and starts to roll off gradually and is 2 dB lower at 40 GHz. The extrapolated bandwidth was estimated to be between 50 and 60 GHz. The voltage needed for π phase shift difference between the two arms (V_π) of this MZM (Figure 5.1) is 14 V at 1300 nm and 16.8 V at 1550 nm. These values are too high for a high-speed modulator. The high drive voltage is caused by the poor overlap of the vertical component of the applied electric field with the optical mode. One way to improve the overlap is to use a doped layer for low drive voltage V_π of 4–5 V [37]. However, this results in excessive capacitance, which reduces the fill factor, and so true push–pull operation becomes difficult. A promising approach is to use substrate-removed waveguide and process both sides of the epilayer [38]. This could yield very high-speed modulators with low drive voltage.

5.2.2.2 *InGaAsP/InP MZM*

The monolithic integration of lasers and modulators in InGaAsP/InP semiconductor is a very important area in the research of InGaAsP/InP MZMs. The MZMs can virtually achieve zero-chirp operation and provide more flexibility to control chirp compared with EA modulator [39]. A successful integration of a distributed Bragg Reflector (DBR) laser with an MZM has been demonstrated [40] with an etch and regrowth of the active region and the passive waveguide. In another approach, a DFB laser and an MZM were monolithically integrated using the selective area low-pressure MOVPE growth technique [41]. The modulator achieved a modulation depth of 12 dB with an injection current of 20 mA. The device shows typical roll-off characteristics with a 3 dB bandwidth of 1.2 GHz. The low bandwidth is caused by the low carrier density as well as long carrier lifetime in the modulator. In order to minimize the design constraints on the MZ and DFB active region, an integrated MZ DFB device was fabricated by butt coupling [42]. MZM has been integrated with gain-coupled DFB lasers, and the device can operate at 10 Gb/s [43]. The extinction ratio was 10.8 dB. An MZM integrated with a semiconductor amplifier has also been demonstrated [44]. The active electro-optic material in the modulator is InGaAsP/InP quantum wells with a 1.4 μm band gap, which provides optical phase modulation at an operating wavelength near 1.55 μm. The modulator has 14 dB modulation depth for $V_\pi = 3$ V.

5.3 Semiconductor surface-normal modulator

There is considerable interest in semiconductor surface-normal modulators for application in parallel processing and optical interconnection technology. Since surface-normal modulators can be easily fabricated in a two-dimensional array form, it can be used as spatial light modulator, which is important for switching and signal processing application [45].

For most surface-normal modulators, the major difficulties are in obtaining high on/off ratios and high modulation efficiency since the interaction length with an optical beam is limited to a few micrometers. It is possible to increase the interaction length by adding reflectors at one or both end of the modulator's active region. Figure 5.13 shows the schematics of those modulators. Figure 5.13(a) shows a single backside reflector, which consists of a $\lambda/4$ thick DBR. The thickness of the optical absorption layer is effectively doubled in this structure [46]. In the case of two reflectors, as shown in Figure 5.13(b), the effective interaction length is multiplied by the finesse, F, of the Fabry–Perot (F–P) cavity that is formed, where $F = \pi R/(1 - R)$, and R is the mean mirror reflectivity. This configuration, which consists of asymmetric multireflection layers, can change the reflectivity and enhance the modulation efficiency, even under a low applied voltage. The cavity resonance characteristics is the sharpest and the reflectivity is zero when both DBR reflectivities are the same and no optical absorption exists between the upper

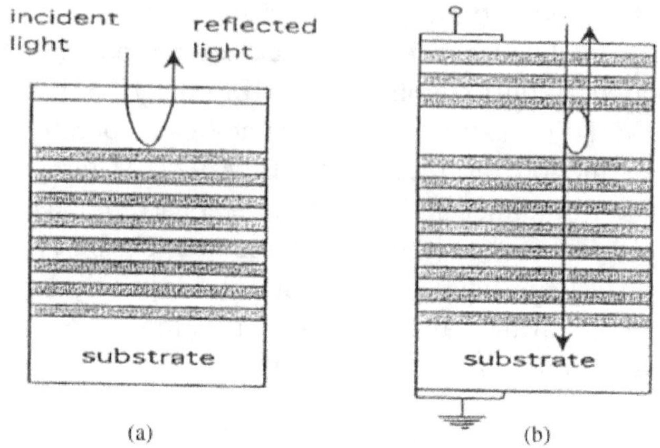

(a) (b)

Figure 5.13: The schematic of surface-normal modulator (a) non-resonant type, and (b) resonant type.

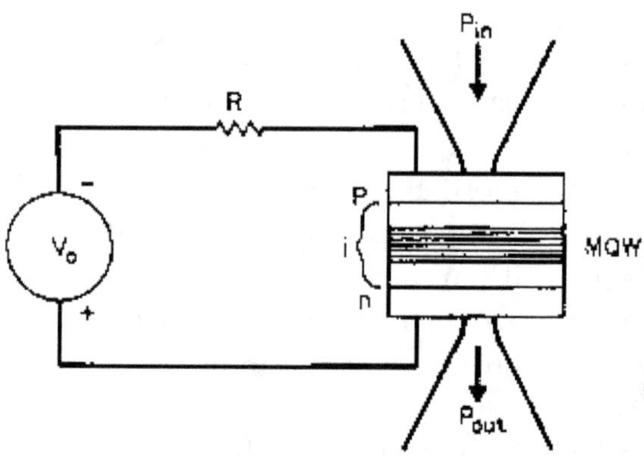

Figure 5.14: The schematic circuit of a simple SEED.

and the lower DBR layers. Contrast ratio of ~100 is possible in this structure. When designing such a structure, both EA and electro-refraction effects must be taken into account.

5.3.1 *GaAs/AlGaAs surface-normal modulator*

Among the first use of GaAs/AlGaAs surface-normal modulators were self-electro-optic effect devices (SEED). Figure 5.14 shows the schematic circuit of a simplest SEED.

A SEED consists of a reverse-biased p-i-n diode containing the quantum wells in its intrinsic region and an electric load. The diode uses the phenomenon of QCSE for the photodetector operation. The operation of the SEED is based on the optoelectronic feedback caused by the simultaneous operation of the quantum well diode as an absorption modulator and a photodetector [47]. For incident wavelength near the zero-field exciton peak, decreasing reverse bias produces increasing absorption and photocurrent. The resulting voltage drop in the load further decreases the bias, increasing the absorption and photocurrent even more. This positive optoelectronic feedback can switch the device to a high-absorption state [47]. Once the low-voltage, high-absorption state is reached, the input power can be decreased to a low critical switch up power, and the device switches back to a high-voltage, low-absorption state. A review of SEED operation is discussed in Ref. [47].

The electrical load in a SEED can be a resistor, or a photodiode (D-SEED or diode-biased SEED) [39], or a transistor (T-SEED or transistor-biased SEED) [48], or a field effect transistor (F-SEED or FET-biased SEED) [49], or another identical quantum well diode (S-SEED or symmetric SEED) [49]. Among those SEEDs, S-SEED is the most promising modulator because of its many desirable features, such as improved low contrast ratio, high speed, and low switching energy [49–54]. Figure 5.15 shows a typical S-SEED structure.

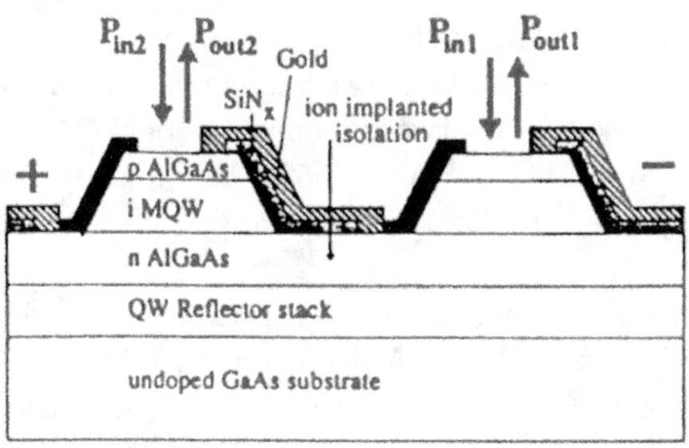

Figure 5.15: A typical S-SEED structure [49].

The switching time of S-SEED is determined by the speed of the quantum well as well as the RC time constant of the circuit. The response time of the quantum well material to the applied electric field is ~100 fs, and the escape time of photoexcited carriers is <10 ps. An RC time constant of 10 ps or less can be obtained by using small area devices. S-SEED switching time of 33 ps has been measured. In another S-SEED design, the voltage swing has been reduced to 2 V and the switching energy to 340 fJ.

A three-terminal optical gate, based on exciton absorption and reflection, has been proposed to improve the low contrast ratio and realize gain in the same device [55]. A high reflection DBR mirror is often introduced in these types of structures to double the interaction length. A high contrast ratio and optical gain have been demonstrated. Another optical switching device called double-heterostructure optoelectronic switch consisting of pnpn thyristors has been developed to realize the surface-normal operation under small operating power [56].

5.3.2 *InGaAsP/InP surface-normal modulator*

InGaAsP/InP materials are also used for surface-normal modulators. When quantum wells are separated by very thin barriers in a periodic structure [superlattice structure (SL)], the discrete energy levels broaden into minibands because of the resonant tunneling effect. Carriers are delocalized and the SL exhibits, to some extent, a three-dimensional like behavior. When a small electric field is applied to SL, the resonance is broken, carriers tend to localize, and the structure recovers its two-dimensional behavior. This is called the Wannier–Stark effect (WSE) [57]. Modulators based on WSE have been reported in the 1550 nm range [58–61] and at 1300 nm range [62].

In order to obtain high extinction ratio at low operating voltage, an asymmetric Fabry–Perot (ASFP) structure is generally employed [63, 64]. Figure 5.16 shows a schematic diagram of the device structure of ASFP modulator. First, a 17.5-pair InP/InGaAsP ($\lambda_g = 1200$ nm) DBR is grown by MOCVD on n-type (100) InP substrate.

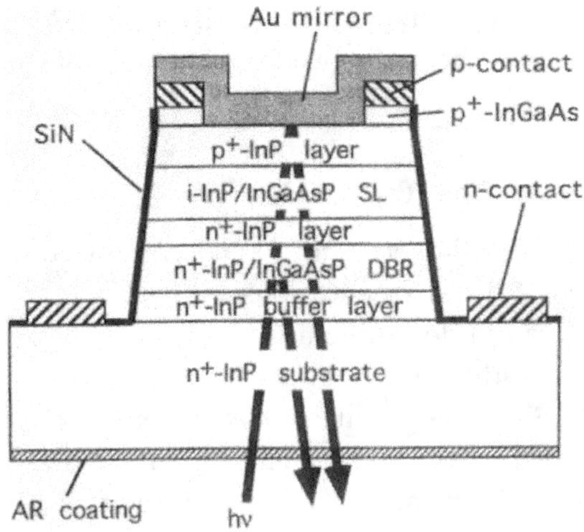

Figure 5.16: A schematic diagram of the device structure of ASFP modulator [63].

Then, a 1.3 μm thick active layer with lattice-matched InP/InGaAsP ($\lambda_g = 1400$ nm) SL is grown by gas-source MBE. An Au mirror with a reflectively of 0.93 at 1300 nm acts as high-reflectively mirror. The absorption coefficient at 1300 nm is 3000 cm^{-1}. This device has an extinction ratio of 11.6 dB at an applied voltage of 14 V and a 3 dB bandwidth of 5.5 GHz.

5.4 $LiNbO_3$ Mach–Zehnder modulator

The guided-wave $LiNbO_3$ interferometers used to modulate laser beams were fabricated as early as 1980 [65]. $LiNbO_3$ has been the material of choice for electro-optic MZM due to its combination of high electro-optic coefficient and high optical transparency in the near-infrared wavelength used for telecommunications. $LiNbO_3$ MZM can operate satisfactorily over a wavelength range of 1300–1550 nm [66]. $LiNbO_3$ modulator has been widely used in today's high-speed digital fiber communication. $LiNbO_3$ MZMs with stable operation over a wide temperature range, with very low bias-voltage drift rates, and with bias-free operation are commercially available. In this section, an overview of $LiNbO_3$ M-ZM is described. In Section 5.4.1, the design and fabrication of $LiNbO_3$ MZM are discussed. In Section 5.4.2, an overview of the high-speed performance of the $LiNbO_3$ MZM is given.

5.4.1 *Design of $LiNbO_3$ MZM*

$LiNbO_3$ wafers with diameters up to 100 mm are commercially available in different crystal orientations (x-, y-, and z-cut). The first choice encountered in designing a $LiNbO_3$ MZM is to choose the orientation of the crystal cuts, and thus the orientation of the waveguide and electrodes. Since the strongest component of the applied electric field must be aligned with the z-axis of the crystal, which has the highest electro-optic coefficient, this requires that the waveguide is placed between the electrodes for an x-cut configuration and beneath the electrodes for z-cut. Thus, crystal orientations affect both modulator efficiency, as characterized by voltage needed for a π phase shift (between the two arms) V_π, and modulation chirp, which is described by the chirp parameter α. Figure 5.17 shows the four most common electrode structures used in MZM [67]. Z-cut devices always require a buffer layer to minimize attenuation of the optical mode due to absorption by metal contacts placed on top of waveguides. Z-cut devices also require conductive buffer layer and charge bleed layers to mitigate dc drift and pyroelectric charge build up, respectively [68]. X-cut devices do not inherently need a buffer layer because the electrodes are not placed directly above the waveguides; however, to achieve multigigahertz operation, broadband x-cut devices do use a buffer layer for velocity matching of the radiofrequency (RF) and optical waves. Finite-element and finite-difference [69] methods have been used to model the applied electric field and electro-optic efficiency of various electrode topologies. Optimum design of electrode and waveguide are required to obtain efficient modulation [70–74].

Chirp-free modulation can be obtained in x-cut topologies as shown in Figures 5.16(a) and 5.16(b) due to the push–pull symmetry of the applied field. Although a z-cut wafer is more difficult to handle because the crystal's piezoelectric and pyroelectric properties produce electric charge accumulation on the z-faces during processing,

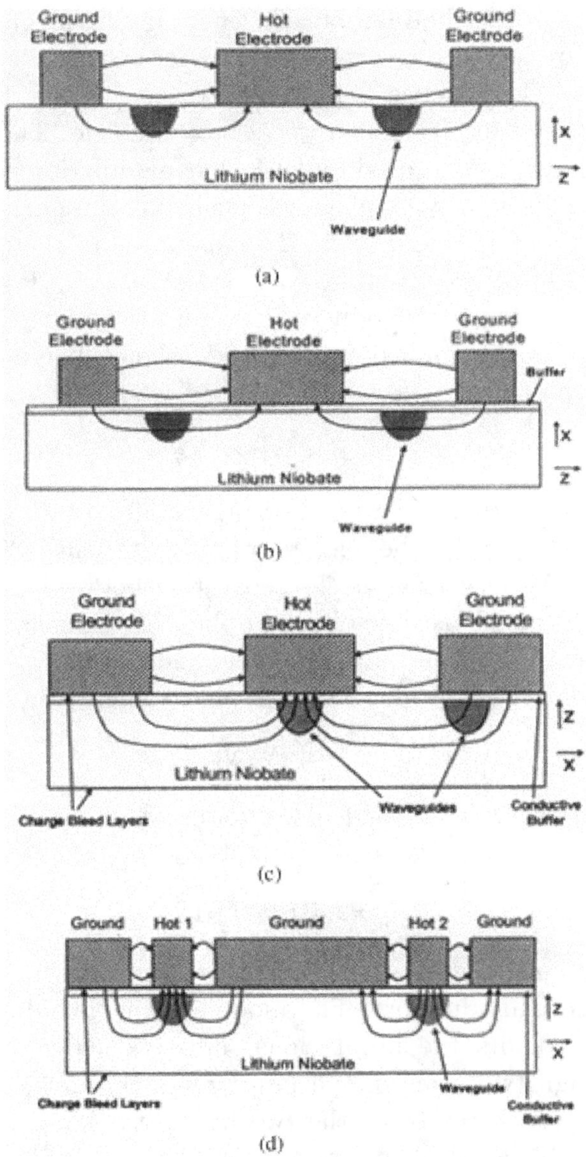

Figure 5.17: Four most common electrode configuration, (a) non-buffered x-cut, (b) buffered x-cut, (c) buffered single-drive z-cut, and (d) buffered dual-drive z-cut [67].

it can improve the modulation efficiency. In the single-drive z-cut device [Figure 5.16(c)], the waveguide under the hot electrode experiences a doubled electric field; however, the waveguide under the ground electrode experiences a weaker electric field. The overall efficiency can thus be improved by 20%. The modulation efficiency can be further improved by employing a push–pull dual drive configuration as shown in Figure 5.16(d). The operation of the modulator is based on the linear electro-optic effect (also known as Pockel's effect) i.e. a change in index is produced in an anisotropic crystal, which is proportional to the applied electric field. The change in index (Δn) as a function of electric field (E) is

$$\Delta n(E) = n_e^3 r_{33} E\Gamma/2 \tag{5.2}$$

where n_e is the extraordinary refractive index of $LiNbO_3$ (~2.15 at 1550 nm), r_{33} is the electro-optic coefficient (30.8×10^{-12} m/V), E (= V/d, where d is the separation between the electrodes (~20–30 μm) and V is the voltage applied) is the electric field, and Γ is the overlap between the optical field and RF field ($\Gamma \sim 0.3-0.5$). The push–pull effect produces a total phase difference (2ϕ) given by:

$$2\phi = \pi V/V_\pi \tag{5.3}$$

where V_π is the voltage needed to produce a phase difference of π. It is given by (in an anisotropic crystal)

$$V_\pi = \lambda d/(2n_e^3 r_{33}\Gamma L) \tag{5.4}$$

where L is the length of the electrode. V_π is typically ~3–5 V for commercial modulators. The modulator has a y-branch at the input, which splits the input signal into two equal parts (which travel through two arms of the interferometer) and a y-branch at the output, which combines the two inputs to this y-branch to an output port. In the ideal situation, when there is no phase difference between the two arms, $P_{\text{in}} = P_{\text{out}}$ and when there is a phase difference of π, then $P_{\text{out}} = 0$. This is illustrated in Figure 5.18. The ratio $P_{\text{out}}/P_{\text{in}}$ as a function of phase difference in units of π is also shown. An important consideration for modulators is its bandwidth.

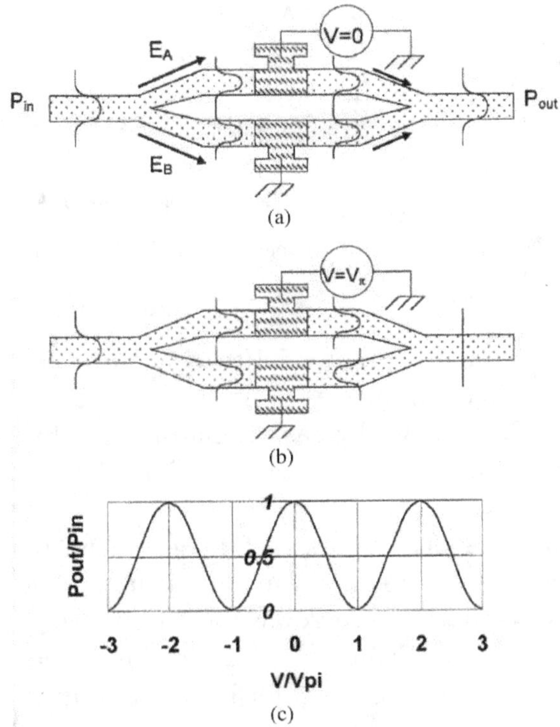

Figure 5.18: Principle of operation of an MZ intensity modulator. The top figure is for V = 0, middle figure is for voltage V = V_π i.e. voltage needed for π phase difference between the two arms. The bottom figure is a schematic of P_{out}/P_{in} ($P_{out}/P_{in} = 1 + \text{Cos}\ (\pi\ V/V_\pi)$) [71].

In order to achieve high bandwidth, velocity matching of RF input and the optical wave along the modulator length is important. The effective refractive index of guided optical wave is 2.15 and that of the microwave is 4.2 as a plane wave. Thus, a velocity mismatch exists between the two waves which limits the optical bandwidth. It is possible to reduce the microwave index by optimizing the buffer layer (SiO_2) thickness and the sizes of the electrodes. A schematic of the waveguides with the buffer layer and electrodes is shown in Figure 5.19.

The microwave index can be calculated using finite element analysis for various values of the electrode dimensions. The calculated

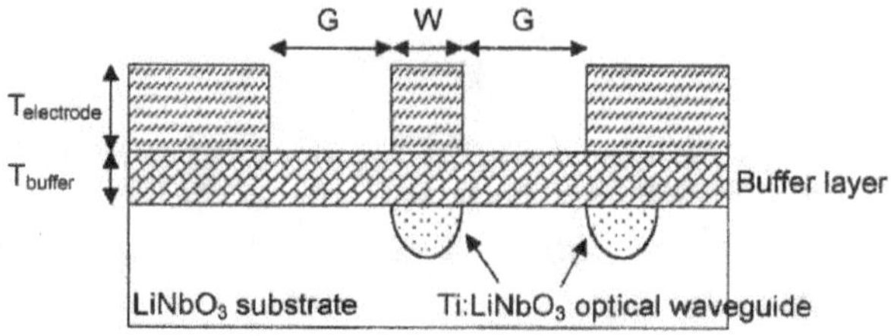

Figure 5.19: A schematic of the waveguides, buffer layers, and electrodes of a $LiNbO_3$ MZM [74].

result (for the design of Figure 5.19) as a function of buffer layer thickness for various thicknesses of Au is shown in Figure 5.20. Note that a value of 2.15 (shown by dashed line) can be obtained. Another parameter of interest is the characteristic impedance (Z) of the electrical line.

The optical response is also a function of microwave attenuation. The optical response as a function of input frequency for a 4 cm long velocity-matched (microwave and optical indices are equal) modulator is shown in Figure 5.21 [71]. The microwave attenuation (a_0) is in cm^{-1}. The result shows for high bandwidth the microwave attenuation must be low.

5.4.2 *Fabrication of* $LiNbO_3$*MZM*

$LiNbO_3$ waveguides have been traditionally fabricated by in-diffusion of Ti at temperature near 1000°C. The process is straightforward, but precautions must be taken to suppress the out-diffusion of Li from the surface. A Li-deficient surface results in an unwanted planar waveguide for z-polarized light and can seriously affect the performance of the modulator. An alternative method for fabricating $LiNbO_3$ waveguide is through the annealed proton exchange (APE) process [75–84]. This process consists of proton exchange from an acid bath and is followed by a high-temperature anneal. It can only be applied to x- and z-cut wafers, since the acid will chemically etch y-cut wafers. Because the modulator can be several centimeters long, preservation of the waveguide cross-sectional dimension along the length of the device is essential. This requires good uniformity in

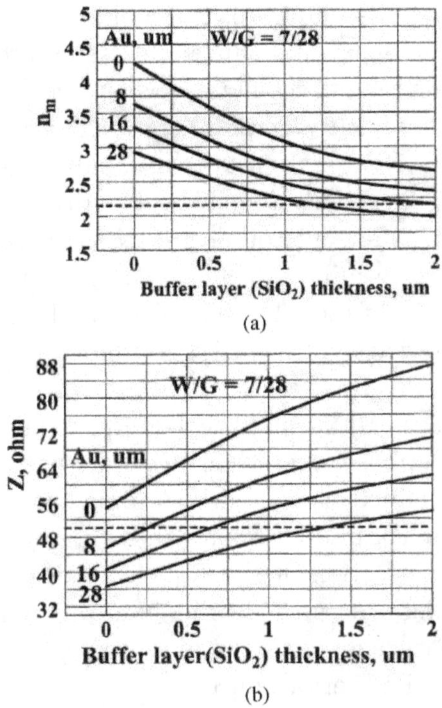

Figure 5.20: (a) Calculated microwave index as a function of buffer layer thickness for W/G is width to gap ratio, (b) impedance as a function of buffer layer thickness [73].

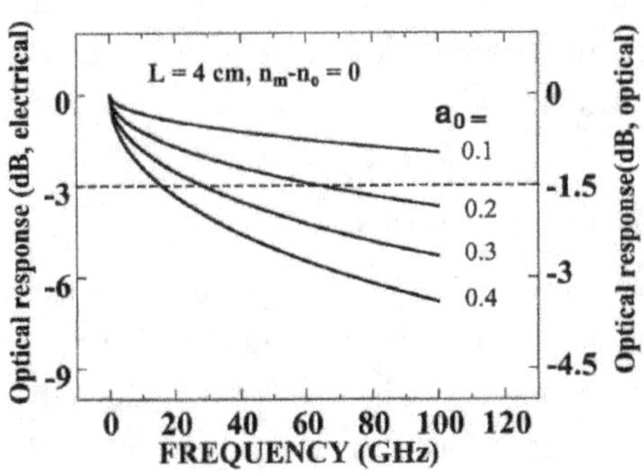

Figure 5.21: Optical response of a velocity-matched modulator for different microwave losses (a_0 in cm^{-1}) [71].

Ti film deposition. Also, uniform thermal zones in furnace used for the diffusion, exchange, and anneal processes are needed. Careful screening of wafers for surface flatness is also very important in minimizing waveguide width variation across the wafer. As modulation speed increases to greater than 20 Gb/s, ridge waveguides on z-cut substrates may be necessary to achieve low drive voltage and reduce waveguide length. A few micron deep $LiNbO_3$ ridge can be obtained by wet-etching or ion-milling techniques. Smooth ridge sidewalls as well as good control of etch depth and sidewall directions are key to achieving low-loss waveguides with good mode confinement.

Gold is generally used as the RF electrode metal. An adhesion layer, such as Ti, is first vacuum-deposited on the wafer, followed by the deposition of a base layer. The electrode pattern is then photolithographically defined. Special lithographic processes are needed to define the thick mask and to ensure that the electroplating process preserves the cross-section of the RF electrode along the length of the devices. Plating processes yielding high-purity metal, small grain size, minimum feature distortion, and reasonable plating rate are key to obtaining good RF performance.

A number of methods have been employed to enhance the electro-optic efficiency. Ridge waveguides can focus the electric flux under the hot electrode and have been used to demonstrate V_π of 3.5 and 5.1 V for z-cut modulator with 30 and 70 GHz bandwidth, respectively [61]. A 15% improvement has also been shown in x-cut device using ridge structure [83]. The overlap between the electric and optical field can also be improved in both x- and z-cut by increasing the waveguide refractive index and narrowing the ground electrode width [84, 85].

5.4.3 *Modulation performance*

Low microwave attenuation (necessary for high bandwidth, Figure 5.21) can be achieved by designing a proper electrode structure. Microwave losses include dielectric loss (inherent loss in the crystal), higher order mode propagation loss, losses due to electrode bends and tapers, stripline conductor loss, connector to contact pad loss, and addition packaging losses. One possible approach

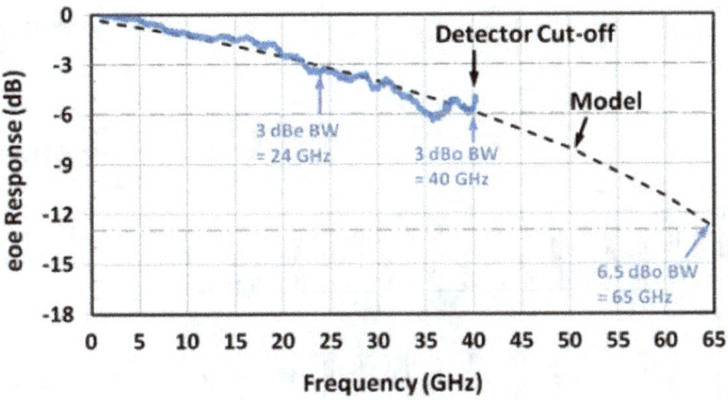

Figure 5.22: Optical response of a modulator [86].

is to use a coplanar waveguide structure [74]. Measured optical response of a modulator with modifications to reduce losses is shown in Figure 5.22. The dashed line is the modeled response. The 3-dB bandwidth is 24 GHz.

5.5 Advanced modulation formats

The digital data coding scheme using a high (1) and low state (0) is also known as OOK. The optical output, which maps the electrical data signal can be produced by both EA and MZM provided the modulators can operate at enough speed. The optical data signal is called non-return to zero (NRZ) () i.e. if the optical signal is present (or high) during the entire duration of electrical bit "1" and it is zero during the entire electrical bit "0". Other types of modulation formats have been studied because they have significant advantages over NRZ format. Some of the simple schemes are illustrated in Figure 5.23.

5.5.1 *Intensity modulation*

For high-speed systems, it is better to have RZ (return to zero) optical data. In this case, the optical signal (high) is not present during the entire electrical bit "1". The effect of chromatic dispersion is lower for RZ signals compared to that for NRZ signals. RZ signals

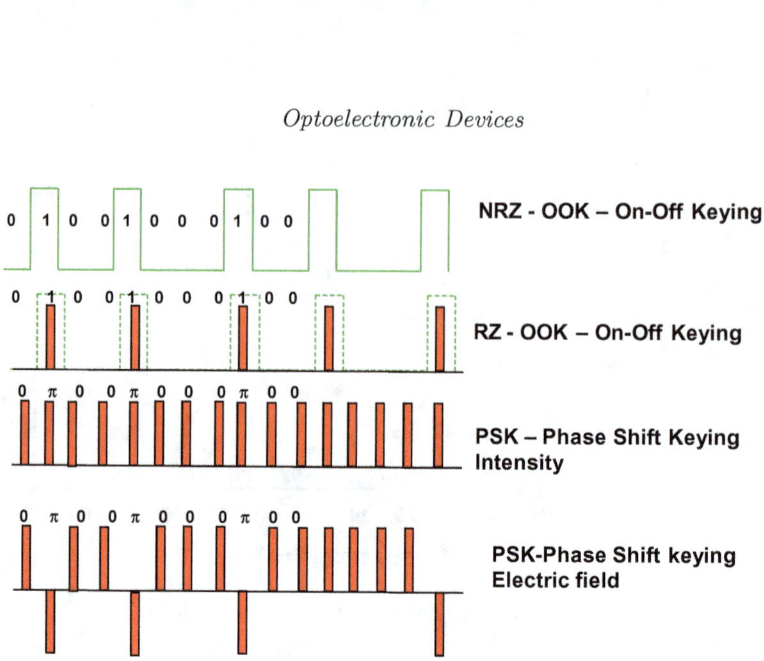

Figure 5.23: Examples of various data formats are shown. NRZ intensity follows the data, RZ intensity duration is less than that of the data, phase shift keying (PSK) — changes phase between 0 and π, the lowest figure shows the electric field for PSK.

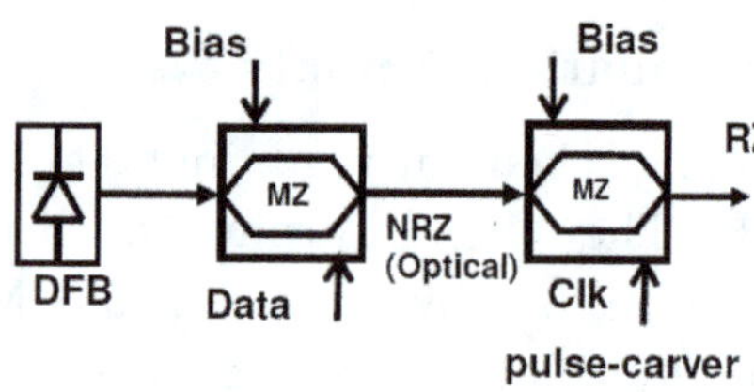

Figure 5.24: A schematic set up for generation of RZ data signals [87, 89].

are produced using the MZMs in series (Figure 5.24). MZMs in this section are generally based on $LiNbO_3$.

The first MZM encodes the data and second is a pulse carver. Optical pulses with a full-width at half-maximum of 50% of the bit duration (a duty cycle of 50%) is obtained by sinusoidally driving an MZM (pulse carver) at the data rate between minimum and maximum transmission. The drive for other duty cycles is discussed in Ref. [89].

Another modulation format is carrier suppress return to zero (CSRZ). It is characterized by a reversal of the phase of the optical

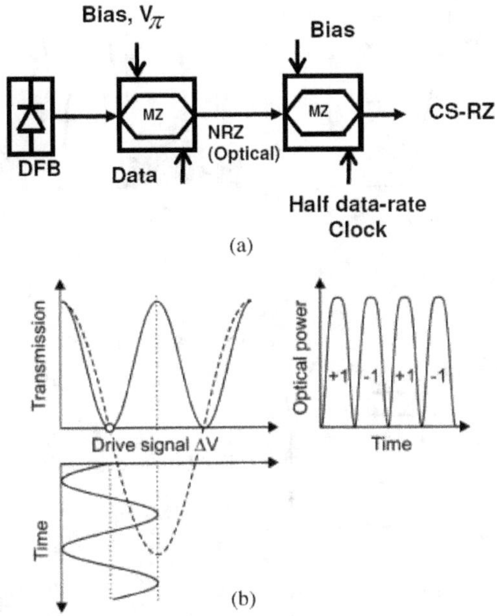

Figure 5.25: The schematic of a CSRZ TX set up. Also shown is the bias setting and output of pulse carver [87, 89].

field at each bit transition. Figure 5.25 shows typical CSRZ TX set up. CSRZ (67% RZ) is generated by driving an MZM pulse carver sinusoidally at half the data rate between its transmission maxima, as shown in Figure 5.25(b) [87]. Phase inversions between adjacent bits are achieved because optical field transfer function of the MZM changes its sign at the transmission minimum. Thus, on average, the optical field of half the 1 bits has positive sign, while the other half has negative sign, resulting in a zero-mean optical field envelope [89]. As a result, the carrier at the optical center frequency is diminished.

Another class of formats known as correlative coding formats introduces a distinct correlation between the optical phase and the data information [89–92]. Optical duo-binary (DB) and alternate mark inversion (AMI) are examples of the general class of correlative coding formats. In optical DB signaling, a phase change

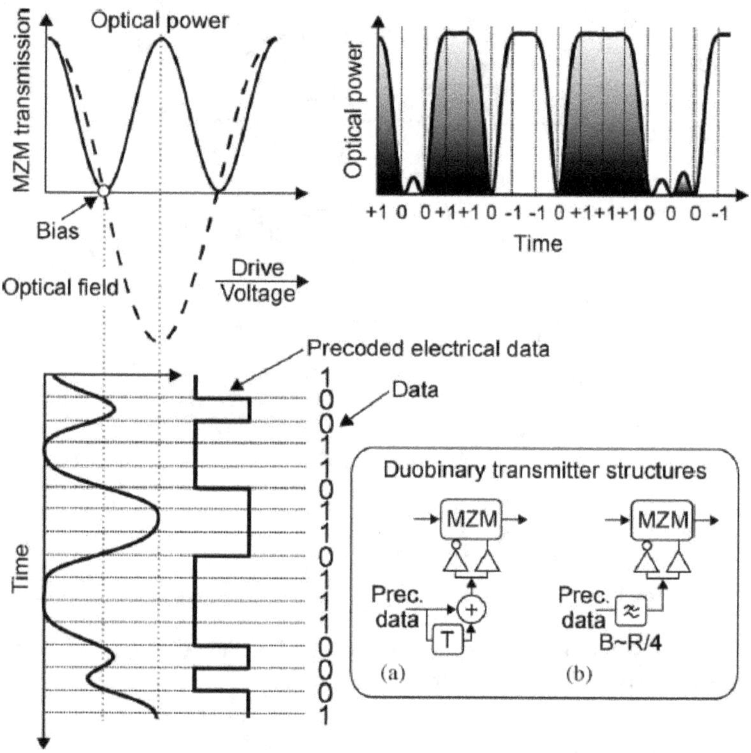

Figure 5.26: The DB signals are generated by driving an MZM around its transmission minimum using a three-level electrical drive signal, generated from the precoded data. signal by either (a) delay-and-add circuit or (b) an appropriate low-pass filter [89].

occurs whenever there is an odd number of 0 bit between two successive 1 bit, whereas for AMI, the phase changes for each 1 bit (even for adjacent 1 bit) is independent of the number of 0 bit in between.

The DB modulation is now discussed. DB modulation is also known under the names PSBT and PASS [89]. The DB signals have higher tolerance (causes less errors) to chromatic dispersion and narrow-band optical filtering compared to binary signaling formats [89]. Figure 5.26 shows the mechanism of a DB TX. The inset shows that DB TXs use a differentially precoded version of the data signal [89] at the input; this precoded data stream exhibits a level change for every 0 bit contained in the original data sequence and prevents error propagation after detection [89]. The precoded sequence is converted into a three-level electrical signal by means of severe electrical low-pass filtering. The low-pass filter can either be implemented digitally, as a delay-and-add filter [Figure 5.26(a)], or as some other type of analog filter with suitable roll-off characteristics and a 3-dB bandwidth of about 25% of the bit rate [Figure 5.26(b)]. Delay-and-add filters typically result in better back-to-back sensitivity, whereas carefully tailored low-pass characteristics better shape the pulse sequence spectrum and yield higher tolerance to dispersion [93]. The three-level electrical signal is then used to drive an MZM between its transmission maxima as shown in Figure 5.26. Alternatively, an MZM designed for one-fourth of the desired data rate may be used to combine the functionality of low-pass filtering and modulation [89]. All of these MZMs can be integrated on the same chip. MZM-based DB modulation has been demonstrated at 107 Gb/s [94]. Passing a binary signal with levels $\{-1, +1\}$, also known as PSK, through a narrow-band optical bandpass filter produces optical DB signal [95].

AMI format is typically implemented in RZ form (RZ-AMI) in optics. The additional phase modulation of AMI can reduce the effects of some fiber nonlinearities. One possible AMI TX implementation is identical to the DB TX in Figure 5.26(a), except that it uses an electrical delay-and-subtract (high-pass) filter instead of the delay-and-add (low-pass) filter used for DB. To obtain RZ-AMI, the data modulator is then followed by an RZ pulse carver.

Most intensity modulation schemes, commonly known as OOK, use two intensity levels (0, 1). Another class of intensity modulation scheme is known as PAM4, it has four intensity levels (0, $\frac{1}{2}$, 1, and $1\frac{1}{2}$). The PAM4 modulation scheme is important for short distance (<1 km) optical communication, such as between computers and memories in data centers. This is discussed in Chapter 11.

5.5.2 *Phase modulation*

Phase modulation-based formats have also been used in optical transmission. MZ intensity modulator is based on phase modulation of the light by the data signal is one arm. Thus, a similar waveguide type device either as single or dual form can be used for modulation between of the input optical electric field into two phase states (phase of $-\pi$ and $+\pi$) or multiple phase states. This is called PSK. In order to detect the information carried by the optical field's phase by a photodiode, phase-to-intensity converting elements (interferometers) have to be inserted into the optical path prior to the photodiode. Due to the absence of an optical phase reference at such a (non-coherent) RX (single photodiode), the phase reference has to be provided by the signal itself: each bit acts as a phase reference for another bit, which is the basis of all differential PSK (DPSK) formats [96].

The commonly known DPSK is a binary phase shift-based scheme. A 1 bit is encoded with a π phase change, whereas a 0 bit is encoded by the absence of a phase change. Like OOK, DPSK can be implemented in RZ and NRZ formats. The main advantage of using DPSK instead of OOK comes from a 3-dB improvement in RX sensitivity [97, 98]. An optical DPSK TX differentially encodes the transmit data prior to modulation to avoid error propagation at the RX [99]. The precoded sequence then modulates the phase of the transmit laser between 0 and π. To perform optical phase modulation, one can use a single-waveguide phase modulator (PM) or an MZM [100]. The speed of phase transitions is limited by the combined bandwidth of the driver amplifier and the waveguide PM, and any overshoot or ringing in the drive waveform appears as phase distortions. An MZM, which is symmetrically driven around zero transmission, always produces exact π phase jumps at the expense of residual optical intensity dips at the locations of phase transitions. Since exact phase modulation is more important for DPSK than a constant optical intensity, practical DPSK TXs are implemented using an MZM. Similar to that for OOK, a pulse carver can be added in series with the data modulator to convert the NRZDPSK signal to RZ-DPSK signal. Since DPSK cannot directly be received using

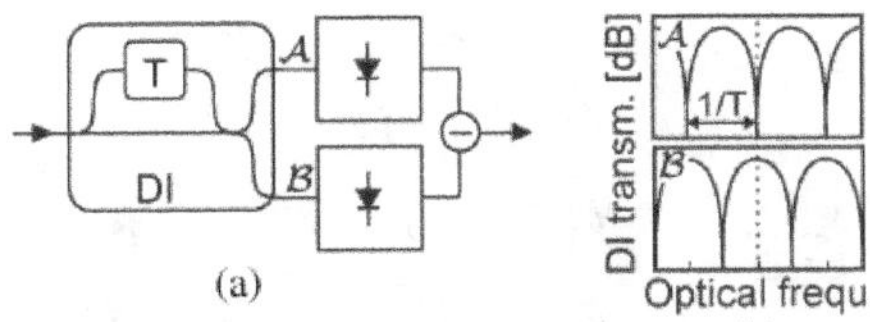

(a) (b)

Figure 5.27: (a) The schematic of balanced DPSK RX using an optical DI to convert phase modulation to intensity modulation. (b) Periodic (with a periodicity of inverse of bit period) DI power transmission characteristics for the destructive port (A) and the constructive port (B) [8, 89].

a photodiode (which detects a change in intensity), a delayed interferometer (DI) is used in the optical path at the RX to convert the differential phase modulation into intensity modulation.

Figure 5.27 shows a DI splitting the phase-modulated signal into two paths, which experience a delay difference equal to the bit duration (25 ps at 40 Gb/s) to let two neighboring bits interfere at the DI output. At the DI output port A (the destructive port), the two optical fields interfere destructively whenever there is no phase change and constructively whenever there is a phase change between subsequent bits, thus converting the phase modulation into intensity modulation. Due to energy conservation, the second DI output port B (the constructive port) yields the logically inverted data pattern. In principle, one of the two DI output ports is sufficient to fully detect the DPSK signal (single-ended detection). However, the 3-dB sensitivity advantage of DPSK is only seen for balanced detection, as shown in Figure 5.26, a balanced RX forms the difference of ports A and B to obtain the electrical decision variable. The reason for the superior performance of balanced detection compared to single-ended detection is the non-Gaussian statistics of square-law detected Gaussian noise.

Another modulation format is differential quadrature PSK (DQPSK) which is based on quadrature phase shift-based scheme. DQPSK is a multilevel modulation format i.e. it has more than 1 bit per symbol. It transmits the four phase shifts $\{0, +\pi/2, -\pi/2, \text{and } \pi\}$ at a symbol rate of half the aggregate bit rate. Figure 5.28 shows the design of a DQPSK TX. The TX is implemented by two nested MZMs

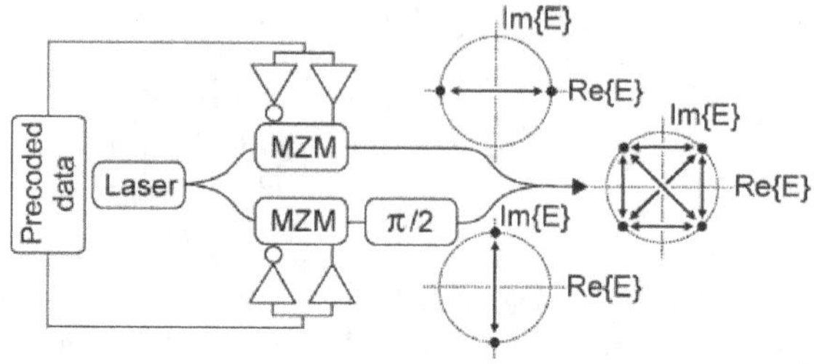

Figure 5.28: DQPSK TX design. Two MZMs are used as phase modulators. The two separately modulated electric fields are then combined with a $\pi/2$ phase shift [89, 101].

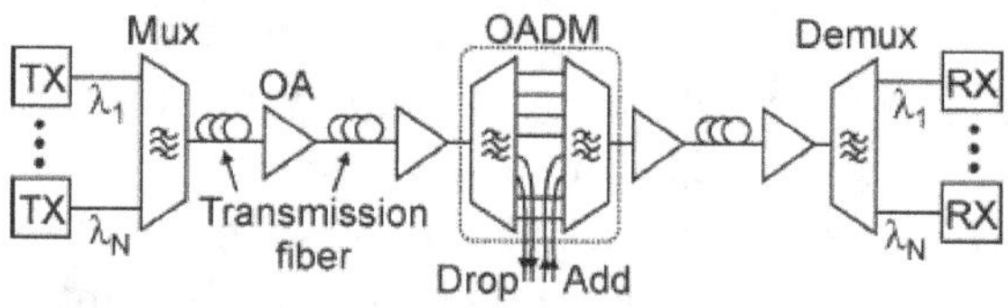

Figure 5.29: Typical design of a WDM optical transmission system, consisting of N TXs and RXs at wavelengths λ_i. Wavelength channels are combined using WDM MUX and separated by WDM DEMUX [89].

operating as phase modulators. Light from a continuously operating laser source is split and divided into two paths of equal intensity, and light in each path goes through an MZM. The output from one of the MZMs goes through an optical $\pi/2$ phase shifter, then the two outputs are combined to produce a single output signal. For this TX design, exact π phase shifts are produced by MZMs, independent of drive signal overshoot and ringing. This is followed by the $\pi/2$ phase shifter in one path, which when combined with the signal in the other path produced the quadrature PSK (QPSK) signal. This TX design requires only binary electronic drive signals, which are much easier to generate at high speeds than multilevel drive waveforms. A pulse carver MZM can be added to the design to produce RZ-DQPSK.

High-capacity optical networks require densely packed wavelength division multiplexed (WDM) channels for propagation over a single optical fiber. The basic structure of such a WDM system is shown in Figure 5.29. The optical signals generated by N TXs at wavelengths λ_i are multiplexed onto a single optical fiber using a

Table 5.1: The TX and RX complexity expressed as number of devices for various modulation formats [89].

Modulation format	TX complexity	RX complexity
NRZ-OOK	1 MZM	1 PD
50% RZ-OOK	1–2 MZMs	1 PD
67% CSRZ-OOK	2 MZMs	1 PD
DB	1 MZM	1 PD
33% RZ-AMI	1–2 MZMs, 1 DI	1 PD
VSB-NRZ-OOK	1 MZM + 1 OF	1 PD
VSB-CSRZ	2 MZMs + 1 OF	1 PD
NRZ-DPSK	1 MZM	1 DI + 2 PDs
50% RZ-DPSK	1–2 MZMs	1 DI + 2 PDs
NRZ-DQPSK	2 nested MZMs	2 DIs + 4 PDs
50% RZ-DQPSK	2 nested MZMs + 1 PC	2 DIs + 4 PDs

Notes: MZM — Mach–Zhender modulator, DI — delayed interferometer, PD — photodiode, PC — pulse carver, OF — optical filter [89].

WDM multiplexer (MUX). At the receiving end, the incident light at multiple wavelengths is separated into individual wavelengths using an optical demultiplexer (DEMUX).

The TX and RX complexity (in terms of devices) for the various modulation formats discussed above is shown in Table 5.1.

Advanced modulation formats are important for increasing the spectral efficiency, dispersion tolerance, and tolerance to nonlinear propagation effects of WDM transmission over long distance in optical fibers. Often, the advantages are bit-rate dependent. For example, at 10 Gb/s, using DB increases dispersion tolerance, whereas at 40 Gb/s, DB is attractive for high spectral efficiency for WDM systems [89]. The TX and RX designs for DB are more complex compared to that for directly modulated lasers. Phase modulation-based (DPSK and DQPSK) systems have received considerable attention recently mainly due to improved RX sensitivity relative to OOK when balanced detection is employed.

Advances in modulator technologies, which includes integration of several MZMs on a single chip for advanced communication formats, and for lower voltage operation will continue.

References

1. W. Franz, *Z. Naturforsch.*, **13a** (1958) 484.
2. L. V. Keldysh, *Zh. Eksp. Teor. Fiz.*, **34** (1958) 1138.
3. R. Williams, *Phys. Rev.*, **117** (1960) 1487.
4. R. Williams, *Phys. Rev.*, **126** (1962) 442.
5. T. S. Moss, *J. Appl. Phys.*, **32** (1961) 2136.
6. K. W. Boer, H. J. Hasche, and U. Kummel, *Z. Phys.*, **155** (1959) 170.
7. N. K. Dutta and N. A. Olsson, *Electron. Lett.*, **20** (1984) 634.
8. Y. Noda, M. Suzuki, Y. Kushiro, and S. Akiba, *Electron. Lett.*, **21** (1985) 1182.
9. D. A. B. Miller, D. S. Chemla, T. C. Damen, A. C. Gossard, W. Wiegmann, T. H. Wood, and C. A. Burrus, *Phys. Rev. Lett.*, **53** (1984) 2174.
10. D. A. B. Miller, D. S. Chemla, T. C. Damen, A. C. Gossard, W. Wiegmann, T. H. Wood, and C. A. Burrus, *Phys. Rev.*, **B32** (1985) 1043.
11. T. G. Beck Mason, Chapter 7 in *WDM Technologies: Active Optical Components*, A. K. Dutta, N. K. Dutta, and M. Fujiwara (eds.), Academic Press, USA, 2002.
12. Y. Furushima, *Electron Lett.*, **34** (1998) 767.
13. M. Bachmann *et al. Electron Lett.*, **32** (1996) 2076.
14. R. J. Deri, C. Caneau, E. Colas, L. M. Schiavone, N. C. Andreadakis, G. H. Song and E. C. M. Pennings, *Appl. Phys. Lett.*, **61** (1992) 652.
15. H. Yamazaki *et al.*, *IEEE J. Sel. Top. Quant. Electron.*, **3** (1997) 1392.
16. K. Satzke, D. Baums, U. Cebulla, H. Haisch, D. Kaiser, E. Lach, E. Kuhn, J. Weber, R. Weinmann, P. Wiedemann, and E. Zielinski, *Electron. Lett.*, **31** (1995) 2030.
17. T. Ido, S. Tanaka, M. Suzuki, M. Koizumi, H. Sano, and H. Inoue, *J. Lightwave Technol*, **14** (1996) 2026.
18. S. Z. Zhang, Y. Chiu, P. Abraham, and J. E. Bowers, *IEEE Photonics Technol. Lett.*, **11(2)** (1999) 191.
19. N. Soda, M. Furutsu, K. Sato, N. Okazaki, S. Yamazaki, H. Nishimoto, and H. Ishikawa, *Electron. Lett.*, **26(1)** (1990) 9.
20. T. Tanbuk-EK, Y. K. Chen, J. A. Grenko, E. K. Byrne, J. E. Johnson, A. Tate, A. M. Sergent, K. W. Wecht, P. F. Sciortino Jr., and S. N. G. Chu, in *7th International Conference on Metalorganic Vapor Phase Epitaxy*, Yokohama, Japan, 1994.
21. T. Sasaki, M. Kitamura, and I. Mito, *J. Cryst. Growth,* **132** (1993) 435.
22. T. Sasaki and K. Kudo, Chapter 10, in WDM Technologies: Active Optical Components, A. K. Dutta, N. K. Dutta, and M. Fujiwara (eds.), Academic Press, USA, 2002.
23. H. Yamazaki, Y. Sakata, M. Yamaguchi, Y. Inomoto, and K. Komatsu, *Electron Lett.* **32** (1996) 109.
24. H. Takeuchi, K. Tsuzuki, K. Sato, K M. Yamamoto, Y. Itaya, A. Sano, M. Yoneyama, and T. Otsuji, *IEEE J. Sel. Top. Quant. Electron.*, **3** (1997) 336.

25. F. Dorgeuille and F. Devaux, *IEEE J. Quantum Electron.*, **30** (1994) 2565.
26. M. Ishizaka, M. Yamaguchi, Y. Sakata, Y. Inomoto, J. Shimizu, and K. Komatsu, *IEEE Photonic Tech. Lett.*, **9** (1997) 1406.
27. M. Suzuki, H. Tanaka, S. Akiba, and Y. Kushiro, *J. Lightwave Technol.*, **6** (1988) 779.
28. J. Yu, C. Rolland, D. Yevick, A. Somani, and S. Bradshaw, *IEEE photon. Technol.*, **8** (1996) 1018.
29. Ph. Delansay, D. Penninckx, S. Artigaud, J. G. Provost, J. P. Hebert, E. Boucherez, J. Y. Emery, C. Fortin, and O. Le Gouezigou, *Electron. Lett.*, **32** (1996) 1820.
30. J. E. Zuker, K. L. Jones, M. A. Newkirk, R. P. Gnall, B. I. Miller, M. G. Young, U. Koren, C. A. Burrus, and B. Tell, *Electron. Lett.*, **28** (1992) 1888.
31. T. Tanbunek, P. F. Scirtino, A. M. Sergent, K. W. Wecht, P. Wisk, Y. K. Chen, C. G. Bethea, and S. K. Sputz, *IEEE photon. Technol. Lett.*, **7** (1995) 1019.
32. D. M. Adams, C. Rolland, N. Puetz, R. S. Moore, F. R. Shepherd, H. B. Kim, and S. Bradshaw, *Electron. Lett.*, **32** (1996) 485.
33. J. Faist and F. K. Reinhart, *J. Appl. Phys.*, **67** (1990) 6998.
34. G. Mendoza-alvalez, L. A. Coldren, A. Alping, R. H. Yan, T. Hausken, K. Lee, and K. Pedrotti, *J. Lightwave Technol.*, **6** (1988) 793.
35. J. S. Weiner, D. A. B. Miller, and D. S. Chemla, *Appl. Phys. Lett.*, **50** (1987) 842.
36. R. Spickermann, S. R. Sakamoto, M. G. Peters, and N. Gagli, *Electron Lett.*, **32** (1996) 1095.
37. R. G. Walker, *IEEE J. Quantum Electron.*, **27** (1991) 654.
38. S. R. Sakamoto, C. Ozturk, Y. T. Byun, J. Ko, and N. Dagli, *IEEE photon. Technol. Lett.*, **10** (1998) 985.
39. J. C. Cartledge, C. R. Rolland, S. Lemerle, and A. Solheim, *IEEE Photon. Technol. Lett.*, **6** (1994) 282.
40. J. E. Zuker, K. L. Jones, M. A. Newkirk, R. P. Gnall, B. I. Miller, M. G. Young, U. Koren, C. A. Burrus, and B. Tell, *Electron. Lett.*, **28** (1992) 1888.
41. T. Tanbun-EK, P. F. Sciortino. Jr., A. M. Sergent, K. W. Wecht, P. Wisk, Y. K. Chen, C. G. Bethea, and S. K. Sputz, *IEEE Photon. Technol. Lett.*, **7** (1995) 1019.
42. D. M. Adams, C. Rolland, N. Puetz, R. S. Moore, F. R. Sheperd, H. B. Kim and S. Bradshaw, *Electron. Lett.*, **32** (1996) 485.
43. C. Rolland, R. S. Moore, F. R. Sheperd, and G. Hiller, *Electron. Lett.*, **29** (1993) 471.
44. E. Zuker, K. L. Jones, B. I. Miller, M. G. Young, U. Koren, J. D. Evankow, and C. A. Burrus, *Appl. Phys. Lett.*, **69** (1992) 277.
45. H. S. Hinton, *IEEE J. Sel. Area Comm.*, **6** (1988) 1209.
46. G. D. Boyd, D. A. B. Miller, D. S. Chemla, S. L. McCall, A. C. Gossard, and J. H. English, *Appl. Phys. Lett.*, **50** (1987) 1119.

47. D. A. B. Miller, J. E. Henry, A. C. Gossard, and J. H. English, *Appl. Phys. Lett.*, **49** (1986) 821.
48. D. A. B. Miller, *Opt. Quantum Electron.*, **22** (1990) 561.
49. T. K. Woodward, L. M. F. Chirovsky, and A. L. Lentine, LEOS'91, Paper **PD-10** (1991).
50. A. L. Lentine, D. A. B. Miller, J. E. Henry, J. E. Cummingham, L. M. F. Chirovsky, and L. A. D'Asaro, *IEEE J. Wuantum Electron.*, **25** (1989) 1928.
51. S. Schmitt-Rink, D. S. Chemla, W. H. Knox, and D. A. B. Miller, *Opt. Lett.*, **15** (1990) 60.
52. G. Livescu, A. M. Fox, D. A. B. Miller, T. Sizer, W. H. Knox, J. E. Cunningham, A. C. Gossard, and J. H. English, *Semicond. Sci. Technol.*, **5** (1990) 549.
53. G. D. Boyd, A. M. Fox, D. A. B. Miller, L. M. F. Chirovsky, L. A. D'asaro, J. M. Kuo, R. F. Kopf, and A. L. Lentine, *Appl. Phys. Lett.*, **57** (1990) 1843.
54. A. L. Lentine, L. M. F. Chirovsky, M. W. Focht, J. Freund, G. D. Guth, R. E. Leibenguth, G. J. pryzylek, and L. E. Smith, *Appl. Phys. Lett.*, **60** (1992) 1809.
55. C. Amano, S. Matsuo, T. Kurokawa, *Photon. Technol. Lett.*, **3** (1991) 736.
56. C. Amano, S. Matsuo, T. Nakahara, and T. Kurokawa, *IEEE J. Quant. Electron.*, **QE-29** (1993) 275.
57. J. Bleuse, G. Bastard, and P. Voisin, *Phys. Rev. Lett.*, **60** (1998) 220.
58. J. Bleuse, P. Voison, M. Allovon, and M. Quillec, *Appl. Phys. Lett.*, **53** (1988) 2632.
59. E. Bigan, M. Allovon, M. Carre, C. Braud, A. Carenco, and P. Voisin, *IEEE J. Quant. Electron.*, **QE-28** (1992) 214.
60. M. Allovon, S. Fouchet, J. C. Hamiand, A. Ougazzaden, B. Rose, A. Gioukhian, and F. Devaux, *IEEE Photon. Tech. Lett.*, **7** (1995) 185.
61. H. C. Neitzert, C. Cacciatore, D. Carnpi, C. Rigo, C. Cmiasso and A. Stano, *IEEE Photon. Tech. Lett.*, **7** (1995) 875.
62. O. Tadanaga, T. Kagawa, Y. Itoh, C. Amano, Y. Matsuoka, Jpn. *J. Appl. Phys.*, **38** (1999) 1278.
63. D. S. Gerber, R. Droopad, and G. N. Maracas, *IEEE Photon. Tecbnol. Lett.* **5** (1993) 55.
64. A. J. Moseley, J. Thompson, M. Q. Keadey, D. 1. Robbins, and M. J. Goodwin, *Electron. Lett.*, **26** (1990) 913.
65. F. J. Leonberger, *Opt. Lett.*, **5** (1980) 312.
66. R. C. Alferness, *IEEE Trans. Microwave Theory Tech.*, **MTT-30**, 8 (1982) 1121.
67. E. L. Wooten, K. M. Kissa, A. Yi-Yan, E. J. Murphy, D. A. Lafaw, P. F. Hallemeier, D. Maack, D. V. Attanasio, D. J. Fritz, G. J. McBrien, and D. E. Bossi, *IEEE J. Sel. Top. Quant. Electron.*, **6** (2000) 69.
68. R. S. Moyer, R. Grencavich, F. F. Judd, R. C. Kershner, W. J. Minford, and R. W. Smith, *IEEE Trans. Comp. Packg. Manufact. Technol.*, **B21** (1998) 130.

69. K. Noguchi, O. Mitomi, and H. Miyazawa, *J. Lightwave Technol.*, **16** (1998) 615.
70. K. Imai, T. Tsuritani, N. Takeda, K. Tanaka, N. Edagawa, and M. Suzui, in *Tech. Dig. Optical Fiber Communication Conf.*, Feb., **PD5**, 1999.
71. R. Madabhushi, Chapter 6 in *WDM Technologies: Active Optical Components* A. K. Dutta, N. K. Dutta, and M. Fujiwara, Academic Press, USA, 2002.
72. K. Noguchi, O. Mitomi, K. Kawano, and M. Yanagibashi, *IEEE Photon Tech. Lett.*, **5** (1993) 52.
73. R. Madabhushi and T. Miyakawa, in *Proc. IOOCWD1-3* (1995).
74. K. Noguchi, O. Mitomi, H. Miyazawa, and S. Seki, *J. Lightwave Tech.*, **13** (1995) 1164.
75. R. C. Alferness, S. K. Korotky, and E. A. J. Marcatilli, *J. Lightwave. Technol.*, **20** (1984) 301.
76. S. J. Chang, C. L. Tsai, Y. B. Lin, J. F. Liu, and W. S. Wang, *J. Lightwave Technol.*, **17**(1999) 843.
77. D. W. Dolfi and T. R. Ranganath, in *Proc. Integrated Photonics Research Conf.*, PD-2 (1992).
78. P. G. Suchoski, T. K. Findakly, and F. J. Leonberger, *Opt. Lett.*, **13** (1988) 1050.
79. J. L. Jackel, C. E. Rice, J. J. Veselka, *Appl. Phys. Lett.*, **41** (1982) 607.
80. K. Noguchi, O. Mitomi, H. Miyazawa, and S. Seki, *J. Lightwave Technol.*, **13** (1995) 1164.
81. F. Laurell, J. Webjorn, G. Arvidsson, and J. Holmberg, *J. Lightwave Technol.*, **10** (1992) 1606.
82. W. L. Chen, R. S. Chen, J. H., Lee, and W. S. Wang, *IEEE Photon. Technol. Lett.*, **7** (1995) 1318.
83. K. Noguchi, O. Mitomi, K. Kawano, and M. Yangagibashi, *IEEE Photon. Technol. Lett.*, **5** (1993) 52.
84. R. Weinmann, D. Baums, U. Cebulla, H. Haisch, D. Kaiser, E. Kuhn, E. Lack, K. Satzke, J. Weber, P. Wiedemann, and E. Zielinski, *IEEE Photon. Technol. Lett.*, **8,** (1996) 891.
85. M. Yoneyama, Y. Miyamoto, T. Otsuji, H. Toba, Y. Yamane, T. Ishibashi, and H. Miyazawa, *J. Lightwave Technol.*, **18** (2000) 34.
86. V. Stenger, J. Toney, A. Pollick, J. Busch, J. Scholl, P. Pontius, S. Sriram, in *Proc. SPIE*, Vol. 8624 86240I-1.
87. M. Haris, PhD. Thesis, Georgia Tech. ECE Dept., 2008.
88. S. Makovejs, PhD. Theis, Univ. College London, EE Dept., 2011.
89. P. J. Winzer and R. J. Essiambre, *J. Lightwave Tech.*, **24** (2006) 4711.
90. J. Conradi, in *Optical Fiber Telecommunications IV B*, I. Kaminow and T. Li (eds). Academic Press, USA, 2002, pp. 862–901.
91. P. J. Winzer and R.-J. Essiambre, *Proc. IEEE*, **94** (2006) 952.
92. M. Ohm and J. Speidel, *IEEE Photon. Technol. Lett.*, **15** (2003) 159.
93. S. Walklin and J. Conradi, *IEEE Photon. Technol. Lett.*, **9** (1997) 1005.

94. P. J. Winzer, G. Raybon, and M. Duelk, *Euro. Conf. Optical Commun. (ECOC)*, Glasgow, UK, 2005, Paper Th4.1.1.
95. D. Penninckx, H. Bissessur, P. Brindel, E. Gohin, and F. Bakhti, *Euro. Conf. Optical Commun.* (*ECOC*), Amsterdam, The Netherlands, 2001, Paper We.P.40.
96. A. H. Gnauck and P. J. Winzer, *J. Lightwave Technol.*, **23** (2005), pp. 115.
97. G. Jacobsen, *Noise in Digital Optical Transmission Systems.* Norwood, MA: Artech House, 1994.
98. P. A. Humblet and M. Azizoglu, *J. Lightwave Technol.*, **9** (1991) 1576.
99. P. Kabal and S. Pasupathy, *IEEE Trans. Commun.*, **COM-23** (1975) 921.
100. T. Chikama, S. Watanabe, T. Naito, H. Onaka, T. Kiyonaga, Y. Onoda, H. Miyata, M. Suyama, M. Seino, and H. Kuwahara, *J. Lightwave Technol.*, **8** (1990) 309.
101. R. A. Griffin and A. C. Carter, *Optical Fiber Commun. Conf. (OFC)*, Anaheim, CA, 2002, Paper WX6.

Chapter 6

Photodetectors

6.1 Introduction

A photodetector is a device which is used to detect the light signal. It converts the optical signal to an electrical signal which is then processed (such as amplification, gating, and decision making) to extract the relevant information in the optical signal. For early review see Refs. [1–9]. For optical transmission systems, the subsystem used to capture the information from the incoming digital optical system is called a receiver circuit. A receiver circuit consists of a photodetector followed by an amplifier, a clock recovery circuit, and a decision circuit.

Semiconductor photodiodes fabricated using InP-based materials are used in receiver circuits for long wavelength (1300 and 1500 nm) optical transmission systems. Si-based photodiodes are often used for short wavelength (850 nm) and low-speed optical transmission systems. The two primary optical detectors used in fiber optic communications are the PIN photodiode and avalanche photodiode (APD). The PIN photodiode has reverse biased p–n junction near the light absorbing region, which is generally undoped. The APD in addition has a high-field region where the photogenerated carriers (electrons and holes) undergo multiplication by impact ionization and hence the device has internal gain. The principles of operation, fabrication, and performance of InP-based photodetectors are discussed in this chapter.

6.2 PIN photodiode

Photodiodes have been fabricated using many material systems. There are Si, Ge, GaAs, and InP material-based photodiodes. Si-based devices are generally used in the visible region of the spectrum and InP-based devices are generally used in the near infrared region of the spectrum, for example, for optical transmission systems. PbS and HgCdTe-based photodetectors are important for >5 μm region of the spectrum [1–14].

6.2.1 *Principle of operation*

The mechanism responsible for the photodetection process is the absorption of light producing electron–hole pairs. If P_{in} is the optical power incident on an absorbing material of thickness w, the transmitted power P_t is given by $P_t = P_{\text{in}} \exp(-\alpha w)$, where α is the absorption coefficient. The absorbed power is $P_a = P_0 - P_{\text{in}}$. The photocurrent ($I_p$) produced is given by, assuming every absorbed photon produces an electron–hole pair which is collected in the external circuit (i.e. unity internal quantum efficiency),

$$I_p = P_a/h\nu = P_{\text{in}}(1 - e^{-\alpha w})/h\nu \tag{6.1}$$

where $h\nu$ is the energy of the incident photons. Often the surface of the photodiode has a reflectivity R, in this case, the input light to the absorbing material becomes $P_{\text{in}}(1 - R)$ and the above equation becomes

$$I_p = P_{\text{in}}(1 - R)(1 - e^{-\alpha w})/h\nu \tag{6.2}$$

The ratio I_p/P_{in} expressed in A/W is often called the responsivity of the photodetector. For many semiconductor materials, the absorption coefficient approaches 10^5 cm^{-1} for incident photon energies >100 meV larger than the direct band gap. Thus, it is feasible to have absorptions in the 80–90% range using only a few micron thick absorbing layer. The schematic of absorption mechanism for both direct gap (e.g. InGaAs) and indirect gap (e.g. Si) is shown in Figure 6.1. For indirect-gap semiconductor, the absorption process involves the emission of a phonon. The absorption coefficient depends

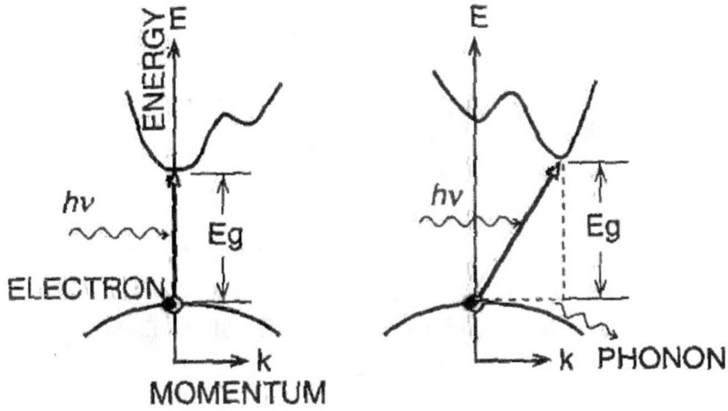

Figure 6.1: Optical transition in direct- and indirect-gap semiconductors.

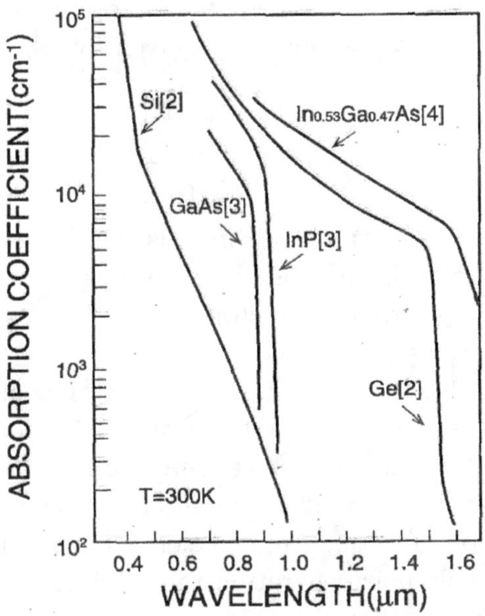

Figure 6.2: The dependence of absorption coefficient on wavelength for various semiconductors.

on the wavelength of the incident light. This dependence is shown in Figure 6.2 for various semiconductors.

The schematic of a PIN photodiode under reverse bias is shown in Figure 6.3. The photogenerated carriers are produced in the I

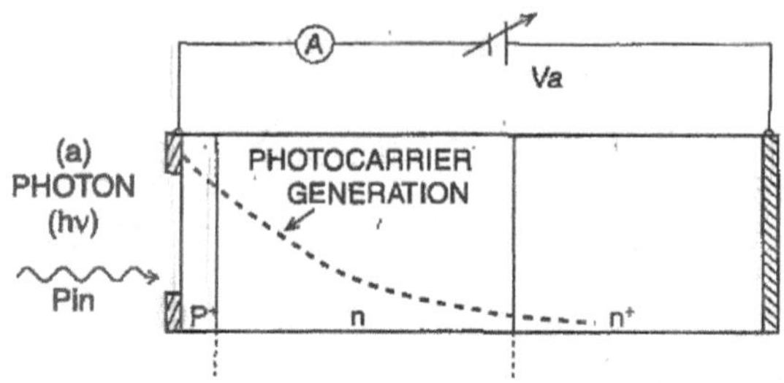

Figure 6.3: Schematic of a PIN photodiode [5].

(low-doped) region. The electrons diffuse (and also accelerate in the drift field through the depletion region) to the N-region where they are collected as current flow in the external circuit. For an applied bias voltage V_a, the relationship between the depletion layer width W and the junction capacitance C is given by [5]

$$V_a = \left(\frac{e}{2\epsilon}\right) N_d W^2$$

$$C = A\sqrt{e\epsilon N_d/2V_a} \tag{6.3}$$

where e is the charge of an electron, ε is the dielectric constant, N_d is the doping density in the absorbing layer, and A is the area of the diode. For low-capacitance detectors (needed for high-speed optical transmission systems), the doping density of InGaAs absorbing layer must be low ($< 2 \times 10^{14}$ cm^{-3}). The capacitance of the photodiode is <10 pF for use in 10 Gb/s transmission systems. In addition, low-doping density results in fewer inter subband states and hence low-dark current (i.e. current in the absence of incident light) due to thermal excitation. The quantum efficiency (η) is defined as the ratio of the electron–hole pair generation rate (which produces photocurrent) to the incident photon incidence rate, i.e. $\eta = (I_p/e)\ (P_{\text{in}}/h\nu)$, where $h\nu$ is the photon energy. Figure 6.4 shows the typical quantum efficiency of commercial photodiodes.

6.2.2 *Materials and device structure*

Initially, the photodetectors were fabricated using liquid-phase epitaxy (LPE) growth technique. With the development of metal

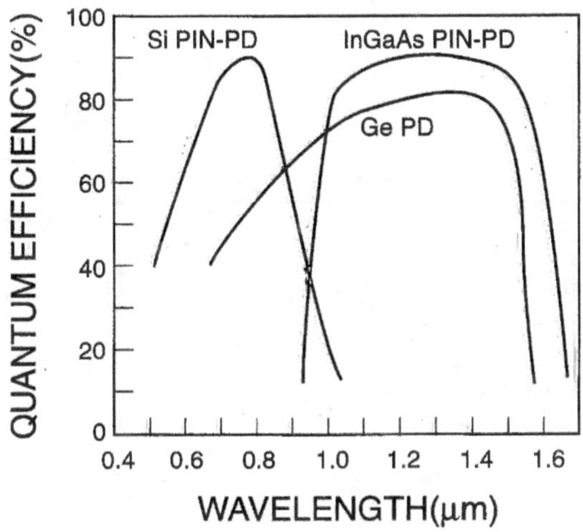

Figure 6.4: Quantum efficiency of commercial photodiodes [5].

organic chemical vapor deposition (MOCVD) and vapor-phase epitaxy (VPE) (see Chapter 2) growth techniques in the 1980s, the reproducibility of the fabrication process and performance have considerably improved. A measure of the performance of photodiode is the dark current, i.e. current produced in the absence of light due to thermal generation of carriers. Very low background doping is needed to have sufficiently low dark current for optical transmission system applications. Photodiodes with such low dark current have been manufactured by MOCVD and VPE growth techniques. The epitaxial growth of InP-based materials is described in Chapter 2. The layer structure schematic of a typical InGaAs photodiode is shown in Figure 6.5. The undoped InGaAs absorbing layer is sandwiched between n-InP and p-InP higher band-gap layers. The heavily doped p-InP layer (labeled p+ in the figure) is produced by Zn diffusion into the top n-InP layer. The rest of the layers are grown over an n-InP substrate. An annular contact was made on the p-InP layer for reverse bias and photocurrent collection. The noise in InGaAs photodiode is determined by the dark current. The dark current depends on the recombination of thermally generated electron–hole pair in the depletion region (generation-recombination current). The dark

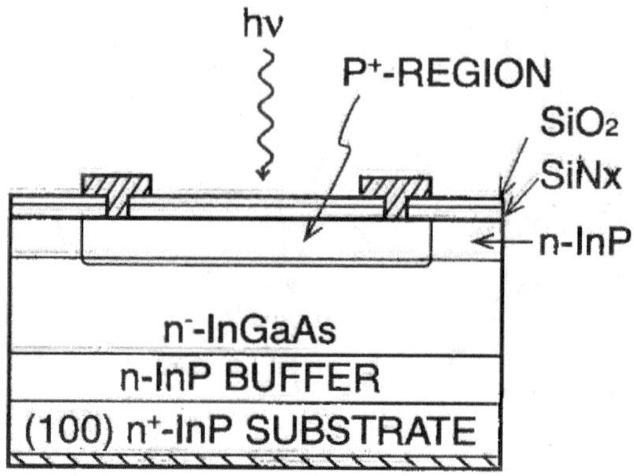

Figure 6.5: Schematic of the layer structure of an InGaAs photodiode [9].

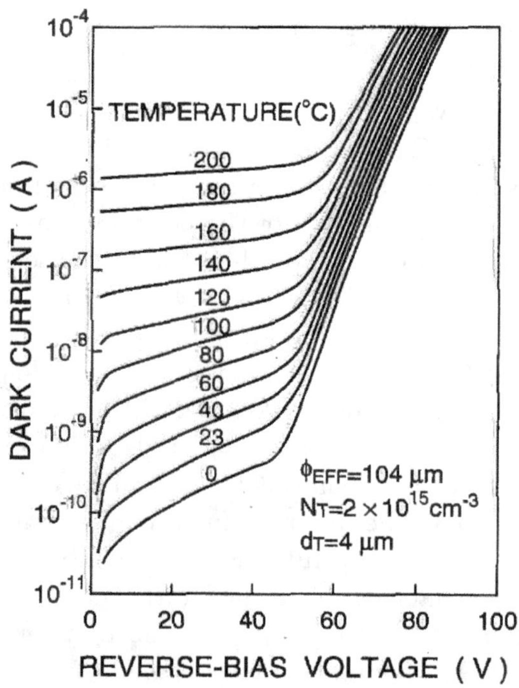

Figure 6.6: Measured dark current at different temperatures. The diameter of absorbing region is 104 μm, doping density 2×10^{15} cm^{-3}, and thickness 4 μm [5].

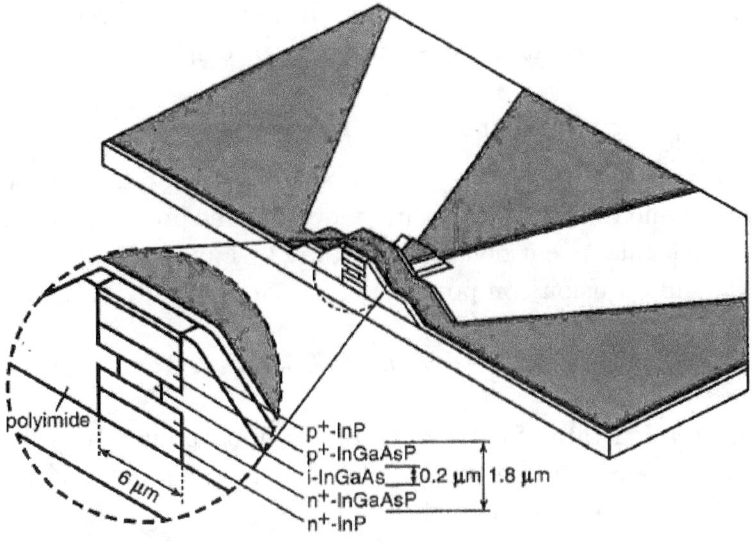

Figure 6.7: Schematic of an InGaAs waveguide photodiode [15].

current increase rapidly with increasing temperature. The data are shown in Figure 6.6.

So far a photodiode design where the light is incident normal to the surface has been discussed. A waveguide-based photodetector has been fabricated. In this case, the absorbing region (InGaAs) is sandwiched between higher gap InGaAsP ($\lambda \sim 1300$ nm) layers. These layers are further confined between n-InP and p-InP layers. The entire layer structure is grown over an n-InP substrate. The absorbing InGaAs layer is preferentially etched to produce a 5 μm wide absorbing layer. The waveguide photodiode shown in Figure 6.7 is designed for high-speed operation. The length is 10 μm and the capacitance is 0.1 pF. A 3-dB bandwidth of 110 GHz is observed for the device structure shown in Figure 6.7. Waveguide photodiodes are important for photonic integrated circuits (see Chapter 8).

Another class of photodiodes is metal–semiconductor–metal (MSM) photodiodes. These photodiodes rely on Schottky barrier in metal–semiconductor contacts. They are generally fabricated with interdigitated contacts. The planar geometry of these photodiodes makes them suitable for integration with electronic circuits (amplifiers) fabricated on the same semiconductor chip.

6.2.3 *Modulation response and noise*

The electron and hole pairs generated in the photodiode absorbing layer drift and diffuse to the contact layers in the presence of a reverse bias. The total current density generated by a photodiode is the sum of drift and diffusion components:

$$J = J_{\mathrm{dr}} + J_{\mathrm{diff}} \tag{6.4}$$

The generation rate of carriers $G(x)$ per unit volume (at a distance x from the surface) is given by photon flux (photons/cm^2/s) at x, multiplied by absorption coefficient

$$G(x) = \alpha\phi(x) = \alpha\phi_0 e^{-\alpha x} \tag{6.5}$$

where $\frac{d\phi}{dx} = -\alpha\phi$, or $\phi(x) = \phi_0 e^{-\alpha x}$ and ϕ_0 is the incident optical flux (photons/cm^2/s). If W is the width of the absorption region, the total drift current

$$J_{\mathrm{dr}} = -e\int_0^W G(x)dx = -e\phi_0(1 - e^{-\alpha W}) \tag{6.6}$$

where e is unit charge, $e = 1.6\times10^{-19}$ Coulomb. The diffusion component of the photocurrent is now considered. This current is produced by the diffusion of holes produced in the n^+ layer beyond the i- layer (or n-layer in Figure 6.5). The production of carriers in the top (p^+) layer is negligible. The diffusion equation for hole density (p) is

$$D_h\frac{d^2p}{dx^2} - \frac{p - p_0}{\tau_h} + G(x) = 0 \tag{6.7}$$

with the boundary condition $p = p_0$ at $x \to \infty$ and $p = 0$ at $x = W$. The second condition arrives from the fact that the minority carrier density at the edge of the depletion region of a reverse biased diode is zero. The solution of above is

$$p = p_0 - (p_0 + Ce^{-\alpha W})e^{(W-x)/L_h} + Ce^{-\alpha x} \tag{6.8}$$

where $L_h = \sqrt{D_h \tau_h}$ is the hole diffusion length and

$$C = \frac{\phi_0 \alpha L_h^2}{D_h(1 - \alpha^2 L_h^2)} \tag{6.9}$$

The hole diffusion current is then given by

$$J_h = -eD_h \left.\frac{\partial p}{\partial x}\right|_{x=W} = -e\phi_0 \frac{\alpha L_h}{1 + \alpha L_h} e^{-\alpha W} - ep_0 \frac{D_h}{L_h} \tag{6.10}$$

Adding (6.6) and (6.10), the total current is

$$J = -e\phi_0 \left(1 - \frac{e^{-\alpha W}}{1 + \alpha L_h}\right) - ep_0 \frac{D_h}{L_h} \tag{6.11}$$

Since p_0 is very small, the second term in (6.11) is small compared to the first term. Then $J \propto \phi_0$, which is the expected result. If R is the reflectivity of the semiconductor, the above expression is multiplied by $1 - R$ of ϕ_0 is the incident field. Also, for small L_h

$$J \to (1 - R)\phi_0(1 - e^{-\alpha W}) \tag{6.12}$$

Small signal response is now considered. In general, we need to consider both hole and electron densities,

$$\begin{aligned} \frac{\partial p}{\partial t} &= -\frac{p - p_0}{\tau} + G - \frac{1}{q}\nabla \cdot \mathbf{J}_h \\ \frac{\partial n}{\partial t} &= -\frac{n - n_0}{\tau} + G + \frac{1}{q}\nabla \cdot \mathbf{J}_e \end{aligned} \tag{6.13}$$

where $\mathbf{J}_h$ and $\mathbf{J}_e$ are hole and electron currents.

$$\begin{aligned} J_h &= e\mu_h pE - eD_H \frac{\partial p}{\partial x} \\ J_e &= e\mu_e nE + eD_e \frac{\partial n}{\partial x} \end{aligned} \tag{6.14}$$

The total current is given by

$$J = J_h + J_e + \varepsilon_s \frac{\partial E}{\partial t} \tag{6.15}$$

where p_0, n_0 are equilibrium hole and electron densities, τ is the recombination time, E is the electric field, ε_s is the dielectric constant, and μ_h and μ_e are hole and electron mobility. The following assumptions are usually valid: (i) transit time is much shorter than recombination lifetime, so that recombination term (first term on the right) in Eq. (6.13) can be neglected and (ii) electric field is large so that diffusion term in Eq. (6.14) is neglected. Hence, Eq. (6.14) becomes

$$\begin{aligned} J_h &= ev_h p \\ J_e &= ev_e n \end{aligned} \tag{6.16}$$

where $v_h = \mu_h E$ and $v_e = \mu_e E$ are the saturation velocities of holes and electrons, respectively. Equation (6.13) using Eq. (6.16) becomes

$$\begin{aligned} \frac{\partial p}{\partial t} &= G - \frac{1}{e}\frac{\partial J_h}{\partial x} \Rightarrow \frac{1}{ev_h}\frac{\partial J_h}{\partial t} = G - \frac{1}{e}\frac{\partial J_h}{\partial x} \\ \frac{\partial n}{\partial t} &= G + \frac{1}{e}\frac{\partial J_e}{\partial x} \Rightarrow \frac{1}{ev_e}\frac{\partial J_e}{\partial t} = G + \frac{1}{e}\frac{\partial J_e}{\partial x} \end{aligned} \tag{6.17}$$

6.2.3.1 *Modulation response: small signal analysis*

For small signal analysis, the incident photon flux is assumed to be of the form

$$\mathbf{\Phi}_{\text{inc}} = \mathbf{\Phi}_0(1 + me^{i\omega t}) \tag{6.18}$$

where $m \ll 1$. Then the generation rate and carrier densities are given by

$$\begin{aligned} G(x,t) &= G_0(x) + G_1(x)e^{i\omega t} \\ p(x,t) &= p_{dc}(x) + p_1(x)e^{i\omega t} \\ n(x,t) &= n_{dc}(x) + n_1(x)e^{i\omega t} \end{aligned} \tag{6.19}$$

where the subscript 1 denotes a.c. quantities. Equation (6.14) then becomes

$$\frac{\partial J_{e1}}{\partial x} - i\omega\frac{J_{e1}}{v_e} = -eG_1$$
$$\frac{\partial J_{h1}}{\partial x} - i\omega\frac{J_{h1}}{v_h} = eG_1 \tag{6.20}$$

where $G_1 = \alpha m\mathbf{\Phi}_0 e^{-\alpha x}$. Solving for J_{e1} and J_{h1} using the boundary condition $J_{e1}(x = 0) = 0$ and $J_{h1}(x = W) = 0$, we get

$$J_{h1}(x) = -\alpha q m\mathbf{\Phi}_0\left[\frac{e^{-\alpha x} - e^{-\alpha W + i\omega\frac{(W-x)}{v_h}}}{\alpha - \frac{i\omega}{v_h}}\right]$$
$$J_{e1}(x) = -\alpha q m\mathbf{\Phi}_0\left[\frac{e^{\frac{i\omega x}{v_e}} - e^{-\alpha x}}{\alpha + \frac{i\omega}{v_e}}\right] \tag{6.21}$$

The total short circuit (where the terms contain the applied bias zero) the total a.c. current is

$$J(w) = \frac{1}{w}\int_0^w (J_{e1}(x) + J_{h1}(x))dx \tag{6.22}$$

It is important to consider a photodiode in conjunction with its intrinsic capacitance, resistance, and bond wire inductance for evaluating the measured response. The equivalent circuit of a p–i–n photodiode is shown in Figure 6.8.

The typical values are [16]: $C_j = 80$ fF, $L_s = 60$ pH, $R_D = 100$ MΩ, $C_p = 15$ fF, $R_s = 10$ Ω, $R_L = 50$ Ω, $J_L(\omega) = J(\omega)H(\omega)$, where $J_L(\omega)$ is the output current across the load and $H(\omega)$ is the transfer function of the equivalent circuit [16]

$$H(\omega) = \frac{R_D}{a + i\omega(b - c\omega^2) - d\omega^2} \tag{6.23}$$

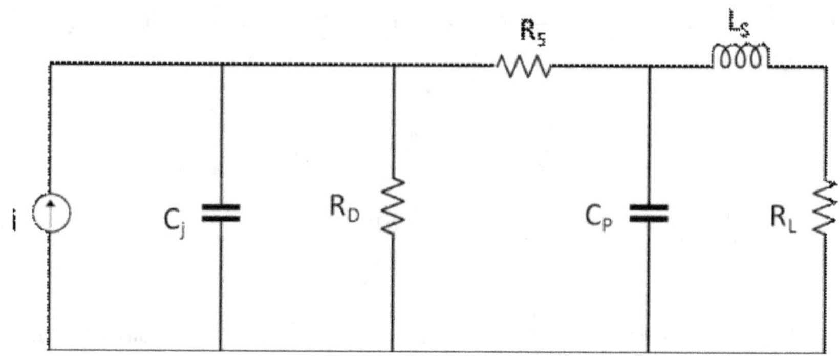

Figure 6.8: Equivalent circuit of a p–i–n photodiode [16].

a, b, c, d are related to the elements in the equivalent circuit as follows:

$$\begin{aligned}
a &= R_S + R_L + R_D \\
b &= R_S R_L C_P + L_S + (R_S + R_L) R_D C_j + R_L R_D C_P \\
c &= R_S L_S C_P R_D C_j \\
d &= (R_S R_L C_p + L_S) R_D C_j + R_S L_S C_P + R_D C_P L_S
\end{aligned} \tag{6.24}$$

6.2.3.2 *Noise in photodiode*

The noise sources in photodiode are shot noise, dark current, and Johnson noise in the electrical circuit. Shot noise is a noise from the random nature of the electron–hole generation process. The average root-mean-square (r.m.s.) value of shot noise current (i_s) is given by

$$\bar{i}_s^2 = 2e(I_{\text{ph}} + I_D)B \tag{6.25}$$

where I_{ph} is the photocurrent, I_D is the dark current, B is the bandwidth, and q is the charge of an electron. Johnson noise is caused by resistances in the diode circuit. The average r.m.s. value of Johnson noise current (i_j) is given by

$$\bar{i}_j^2 = \frac{4kTB}{R_{\text{eq}}} \tag{6.26}$$

where $R_{\text{eq}} = (\frac{1}{R_D} + \frac{1}{R_L} + \frac{1}{R_i})^{-1}$, where R_i is the resistance of the i-layer.

Assume a modulated incident power given by $P(\omega) = P_{\text{inc}}$ incident on the photodiode. The r.m.s. photocurrent is

$$i_{\text{ph}} = \frac{e\eta P_{\text{inc}}}{h\nu} \tag{6.27}$$

Hence, the signal power is

$$S = i_{\text{ph}}^2 R_{\text{eq}} \tag{6.28}$$

and the noise power is $N = (\bar{i}_s^2 + \bar{i}_j^2)R_{\text{eq}}$. The total r.m.s. noise current is $i_N^2 = \bar{i}_s^2 + \bar{i}_j^2$.

The ratio of signal to noise power is then given by:

$$\frac{S}{N} = \frac{\left(\frac{e\eta P_{\text{inc}}}{h\nu}\right)^2}{2e(I_{\text{ph}} + I_{\text{D}})B + \frac{4kTB}{R_{\text{eq}}}} \tag{6.29}$$

When $S/N = 1$, signal power equals noise power. Noise equivalent power (NEP) is defined as the power P_{inc} at which the r.m.s. signal current equals the noise current.

$$\text{NEP} = \frac{h\nu}{e\eta}\left[2e(I_{\text{ph}} + I_{\text{D}}) + \frac{4kT}{R_{\text{eq}}}\right]^{1/2} B^{1/2} \tag{6.30}$$

NEP is a measure of the minimum detected power. At $\lambda = 1300$ nm, NEP $= 2.5 \times 10^{-13}$ WHz^{-1}.

6.3 Avalanche photodiode

APD is a photodiode with internal gain which results in higher sensitivity or ability to detect lower levels of optical signals than the regular p-i-n photodiode [17–28]. The device is designed to have an electric field of (∼30–50 V/μm) in the structure. The internal gain arises from impact ionization where an electron (or hole) with sufficiently high-energy (gained by accelerating in the electric field) collides with a hole (or electron) to produce an additional electron–hole pair. The newly produced electron (or hole) undergoes acceleration

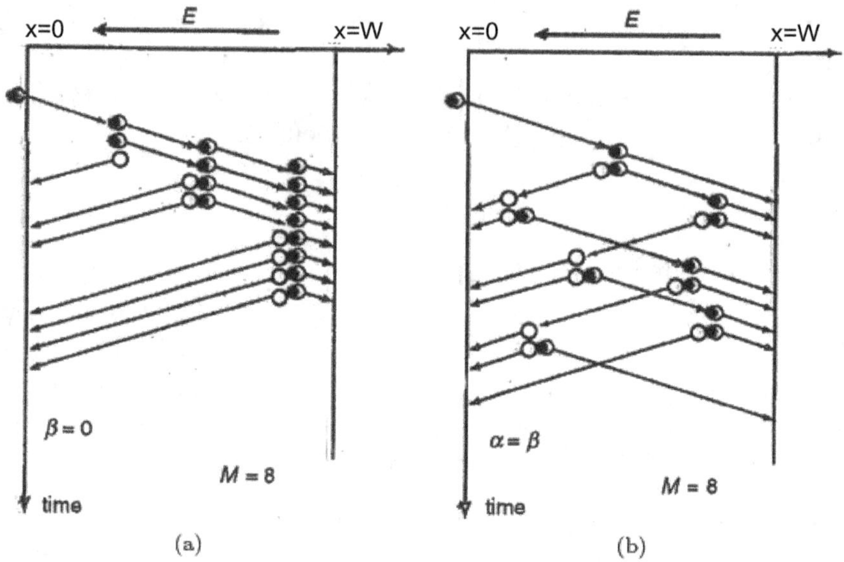

Figure 6.9: Illustration of multiplication of electrons and holes in the high field region. The electrons are represented by solid circles and holes are represented by open circles.

in the same electric field to produce more pairs. This leads to an avalanche process and the original photocurrent is multiplied by this avalanche process. The optimum multiplication factor is typically ~10 for practical APDs.

6.3.1 *Principle of operation*

The APD was first fabricated using Ge and Si. Ge-based APDs were originally used in various optical transmission and other studies. With the development of InGAs APDs with superior performance than Ge-APDs in the early 1980s, most of the current work is carried out using these devices. The avalanche process involving electrons (solid circles) and holes (open circles) are shown in Figure 6.9. The figure shows the photogenerated electrons and holes (labeled 1) producing more electrons and holes as they get accelerated (and travels) through the electric field until they reach the end of the field region. The same avalanche process also effects thermally generated carriers (electrons and holes) and hence results in noise higher than

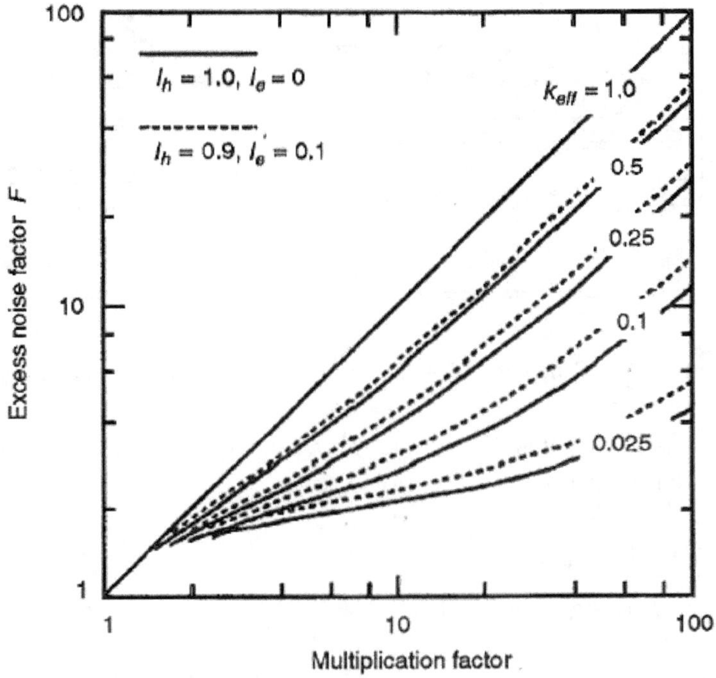

Figure 6.10: Excess noise as a function of multiplication factor for various k_{eff} values [17].

that for a photodiode. This noise is referred to as excess noise. Key to the design and operation of APDs is to reduce excess noise in the presence of high multiplication factor.

Electron and hole ionization coefficients α and β influence both the multiplication factor and excess noise of APDs. There are generally given by the following equation:

$$\alpha(E) = A_e \exp(-B_e/E) \quad \text{and} \quad \beta(E) = A_h \exp(-B_h/E) \qquad (6.31)$$

where A_e, A_h, B_e, B_h are generally experimentally determined constants. A quantity k_{eff}, which is the ratio of the electron and hole ionization coefficients (i.e. $k_{\text{eff}} = \alpha/\beta$), is important for reducing excess noise. k_{eff} depends on both temperature and electric field. Figure 6.10 shows the excess noise factor as a function of multiplication factor

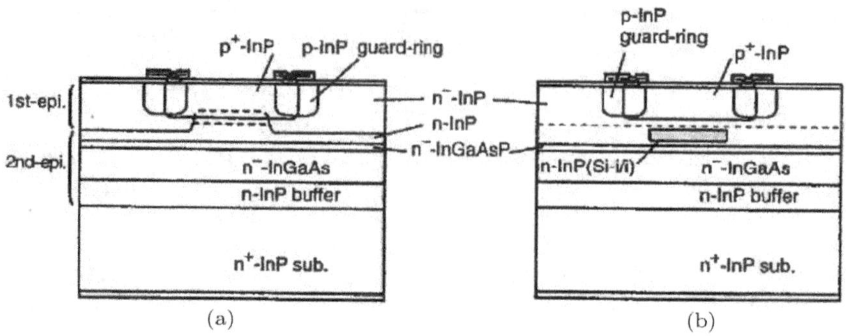

Figure 6.11: Schematic of APDs with deep and shallow guard rings [17].

for various k_{eff} values. The figure shows a low k_{eff} value is needed for high multiplication with low noise.

6.3.2 *Device structure*

Two APD designs with InGaAs absorbing region are shown in Figure 6.11. The devices (generally known as SAM-APD) utilize a separate absorption and multiplication (SAM) region [12, 18–21]. The narrow gap InGaAs layer cannot support a large enough electric field to achieve high avalanche multiplication without excessive tunneling leakage current, so the multiplication is achieved in the wide band gap InP layer grown next to the absorbing layer. The p^+InP region is produced by Be ion implantation in the n-InP (most commonly used technique). It can also be produced using Zn diffusion or by a second epitaxial growth. The entire set of layers is grown by a VPE growth technique on an n-InP substrate.

Typical multiplication vs. reverse voltage of an SAM-APD is shown in Figure 6.12. The multiplication factor (M) increases with increasing reverse voltage, but the dark current, and hence the multiplied dark current, also increases. This increases noise of the APD. Thus an optimum value for reverse bias is generally chosen with $M \sim 10$.

6.3.3 *Multiplication in APD*

The electrons drifting towards n^+ layer produce a current J_e and the holes drifting towards p^+ region produce a current J_h. The

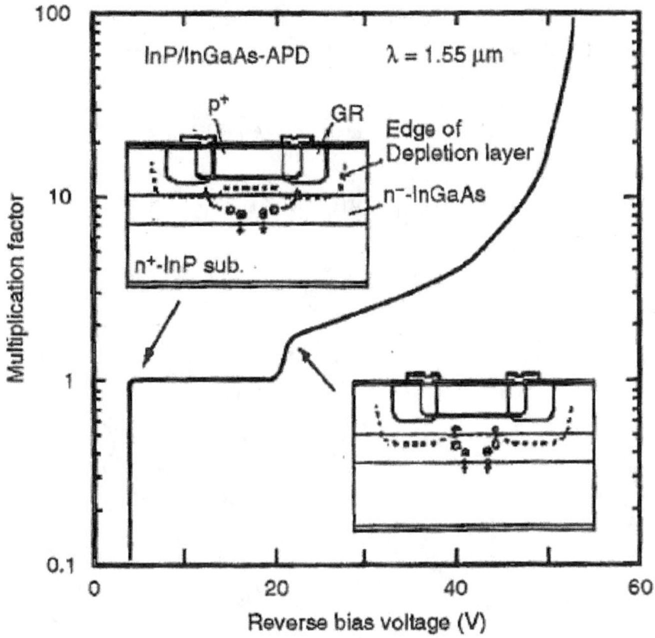

Figure 6.12: Multiplication factor as a function of reverse voltage [17].

total current $J_T = J_e + J_h$. For electrons $M_e(E) = \frac{J_e(w)}{J_{ei}}$ is the multiplication factor due to avalanche, where J_{ei} and $J_e(w)$ are the electron currents at $x = 0$ and $x = w$, respectively. Consider a slice Δx of material in the multiplication region

$$\Delta J_e(x) = J_e(x)\alpha_e \Delta x + J_h(x)\alpha_h \Delta x \tag{6.32}$$

where $\Delta J_e(x)$ is the increase in electron current, the first term is due to electrons generated by electron ionization and the second term due to electrons generated by hole ionization. From Eq. (6.32)

$$\frac{dJ_e}{dx} = J_e\alpha_e + J_h\alpha_h = J_e\alpha_e + (J_T - J_e)\alpha_h$$

$$\frac{dJ_e}{dx} + (\alpha_h - \alpha_e)J_e = \alpha_h J_T \tag{6.33}$$

The electron and hole ionization coefficients α_e, α_h are function of the field (E) which depend on x, and J_T (the sum of electron and hole currents) is independent of x. Consider the case $\alpha_e = \alpha_h$, the solution of (6.33) is

$$J_e(w) = J_T(w) \int_0^w \alpha_e dx + J_{\mathrm{ei}}$$

Using $J_e(w) = J_T$ it follows from the above

$$J_e(w) = J_e(w) \int_0^w \alpha_e dx + J_{\mathrm{ei}} \tag{6.34}$$

Using $M_e(E) = J_e(w)/J_{ei}$ (defined earlier) it follows from (6.34)

$$M_e = M_e \int_0^W \alpha_e dx + 1 \tag{6.35}$$

or

$$M_e = \frac{1}{1 - \int_0^W \alpha_e dx} \tag{6.36}$$

Thus, the breakdown condition (when the current becomes infinite, i.e. $M_e \to \infty$) is

$$\int_0^W \alpha_e dx = 1 \tag{6.37}$$

For $\alpha_e(E) \neq \alpha_h(E)$, the breakdown condition, if the avalanche process is initiated by electrons, becomes

$$\int_0^W \alpha_h \exp\left[\int_0^x (\alpha_h - \alpha_e) dx'\right] dx = 1 \tag{6.38}$$

If the avalanche process is initiated by electrons, $M_e \to \infty$ at breakdown.

6.3.4 *Noise in APD*

APDs have noise over that of p–i–n photodiode due to the fact that the avalanche process is probabilistic, i.e. there are random fluctuations in the actual distance between successive ionizing collisions [23]. These variations give rise to fluctuations in the total number of carriers generated by the primary carrier and have an excess noise. The excess noise factor for electron initiated avalanche process is

$$F_e = M_e[1 - (1 - k_p)[(M_e - 1)/M_e]^2] \tag{6.39}$$

where $k_p = \alpha_h/\alpha_e$. For hole initiated avalanche process, the excess noise factor is

$$F_h = M_h \left\{1 - \left(1 - \frac{1}{k_p}\right)[(M_h - 1)/M_h]^2\right\} \tag{6.40}$$

The S/N of APD is very similar to that of p–i–n photodiode

$$i_{\text{ph}} = \frac{q\eta P_{\text{inc}}}{h\nu}\Gamma_G \tag{6.41}$$

where Γ_G is the gain of the APD. The shot noise is given by

$$\overline{i_s^2} = 2q(I_{\text{ph}} + I_D)\Gamma_G^2 FB \tag{6.42}$$

where F is the excess noise factor. Thus the signal power-to-noise power ratio is

$$\left(\frac{S}{N}\right)_{\text{power}} = \frac{\left(\frac{q\eta P_{\text{inc}}\Gamma_G}{h\nu}\right)^2}{2q(I_{\text{ph}} + I_D)F\Gamma_G^2 B + \frac{4kTB}{R_{\text{eq}}}} \tag{6.43}$$

Then the avalanche gain can improve the (S/N) of the device and hence reduce the NEP if Johnson noise (second term in denominator) is the dominant noise.

6.4 Solar cells

Solar cells and photodiodes are similar devices, i.e. in both of these devices, the incident light creates electron–hole pairs which produces a photocurrent in a closed circuit (for photodiodes) or generates a voltage (photovoltaic effect for solar cells). The photovoltaic device was built first in 1939 using a Si p–n junction. Photodiodes generally work in a narrow range of wavelengths and its performance metric is quantum efficiency (i.e. the number of electron–hole pairs generated per incident photon). Solar cells work in a large wavelength range (solar spectrum) and its metric is power conversion efficiency, i.e. ratio of the power delivered (in W) to an external circuit to Watts of incident solar power. The external circuit is often optimized to deliver maximum power performance. Several reviews of solar cells exit [29–31].[1]

6.4.1 *Principle of operation*

A typical solar cell is a p–n junction diode. The n region is heavily doped and thin so that the light can penetrate through it easily. The p region is lightly doped so that most of the depletion region lies in the p side. Light absorption and electron–hole pair generation take place mainly in the depletion region, and due to the built-in potential, electrons move to the n region and the holes to the p region. When an external load is applied, the excess electrons travel through the load to recombine with the excess holes. Electrons and holes are also generated in the p and n regions, as seen from Figure 6.13(a).

The typical solar cell is built on a substrate with a p–n junction on top. The light incident side generally has finger electrodes (metal contacts) connected to a thicker bus electrode so that the light can penetrate to the semiconductor. This is shown in Figure 6.13(b). Similar to a photodiode, solar cell generates a photocurrent (I_{ph}) in an external load (R_L). The equivalent circuit of a solar cell connected to a load resistor is shown in Figure 6.14.

The solar cell is a p–n junction diode. The current vs. voltage (I–V) characteristics of a diode (in the absence of light) is similar to that of a p–n junction and is given by

$$I = I_s \left[\exp\left(\frac{eV}{k_B T} \right) - 1 \right] \tag{6.44}$$

where V is the voltage, I is the current and I_s is the saturation current, T is the temperature, and k_B is the Boltzmann constant. In the presence of light, a photocurrent (I_{ph}) is generated, which

[1] For a review see "Basic Photovoltaic Principles and Methods", Feb. 1982 National Technical Information Service U.S. Department of Commerce 5285 Port Royal Road Springfield, VA 22161 Stock Number: SERIISP-290-1448.

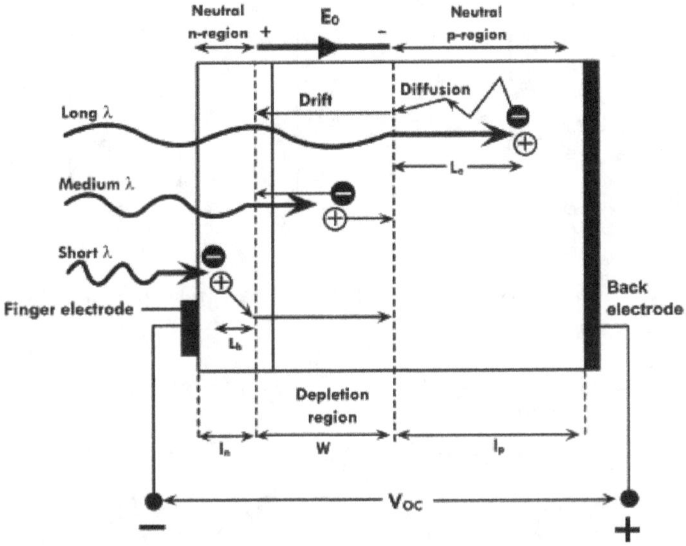

(a)

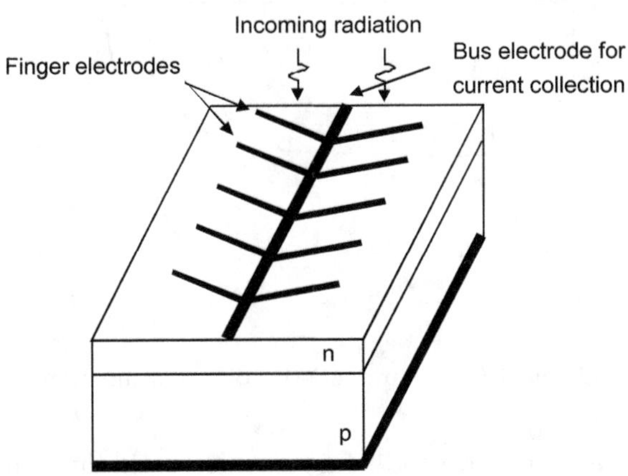

(b)

Figure 6.13: (a) Schematic of light absorption and electron–hole pair generation in a solar cell. Long wavelength radiation penetrates far into the material due to lower absorption. (b). Schematic of a p–n junction solar cell with finger electrodes [29].

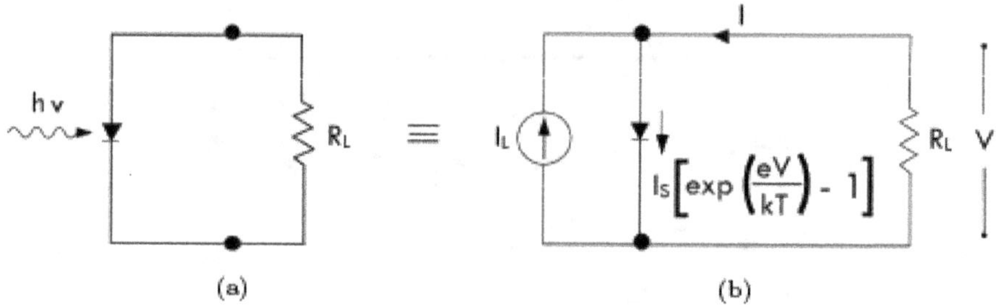

Figure 6.14: Solar cell connected to a load resistor and its equivalent circuit [30].

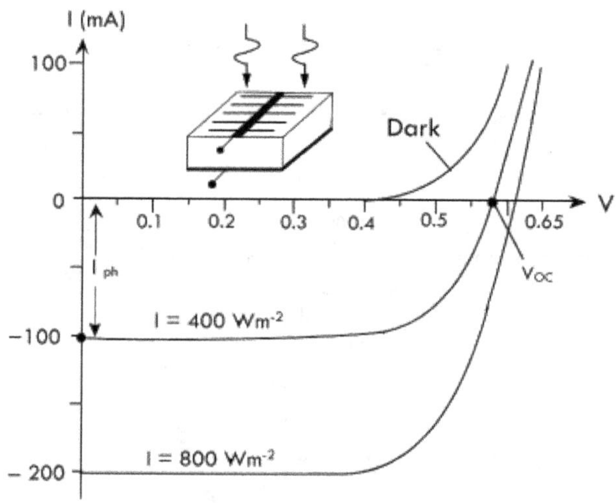

Figure 6.15: Typical I–V characteristics of a p–n junction solar cell [29].

opposes a forward biased diode current, the total current is given by

$$I = -I_{\text{ph}} + I_s \left[\exp\left(\frac{eV}{k_B T}\right) - 1\right] \tag{6.45}$$

with $I_{\text{ph}} = \kappa I_w$, where κ is a constant and I_w is the incident power in Watts. The I–V characteristics of a diode with and without illumination are shown in Figure 6.15. The effect of light is to shift the I–V characteristics of the diode. It is possible to define an open circuit voltage, V_{oc}, which is the voltage when the net current in the circuit is zero. Using Eq. (6.45), the following equation is obtained for V_{oc}. The photocurrent (I_{ph}) for this open circuit voltage is also noted in Eq. (6.46).

$$\begin{aligned} V_{\text{oc}} &\simeq \frac{k_B T}{e} \ln\left(\frac{I_{\text{ph}}}{I_s}\right) \\ I_{\text{ph}} &= I_s \left[\exp\left(\frac{eV_{\text{oc}}}{k_B T}\right) - 1\right] \end{aligned} \tag{6.46}$$

For high photon flux, I_{ph} is high and hence V_{oc} is high. Similarly, lower I_s can also cause higher V_{oc}. I_s can be lowered using a material with a higher band gap. But this will also reduce the range of wavelengths that can be absorbed by the material, which will reduce I_{ph}. The choice of commercial solar cell material is determined by many other factors such as availability, cost, and reliability. The output power (P) of a solar cell is given by

$$P = IV = I_s V \left[\exp\left(\frac{eV}{k_B T}\right) - 1\right] - I_{\text{ph}} V \tag{6.47}$$

The total power as a function of voltage can be maximized using $dP/dV = 0$. This results in the following relationships between the optimum voltage (V_m) and the optimum current (I_m) for maximum power:

$$\begin{aligned} I_m &\simeq I_{\text{ph}} \left(1 - \frac{k_B T}{eV_m}\right) \\ V_m &\simeq V_{\text{oc}} - \frac{k_B T}{e} \ln\left(1 + \frac{eV_m}{k_B T}\right) \\ P_m = I_m V_m &\simeq I_{\text{ph}} \left[V_{\text{oc}} - \frac{k_B T}{e} \ln\left(1 + \frac{eV_m}{k_B T}\right) - \frac{k_B T}{e}\right] \end{aligned} \tag{6.48}$$

Thus, the maximum power can be increased by increasing V_{oc}, i.e. by decreasing I_s (from Eq. (6.46)).

6.4.2 *Material and design*

Silicon is the most common material used for commercial solar cells. Typical solar cells are made of p–n homojunction although some heterojunction materials have been studied in research laboratories. The absorption coefficient of a typical semiconductor is low for photons with energies lower than its band gap. Thus silicon with a band gap near 1 μm does not absorb solar light for wavelengths longer than 1 μm. This phenomenon fundamentally limits the efficiency of silicon cells. Solar radiation spectrum is shown in Figure 6.16. The spectrum at sea level is affected by absorption lines of various molecular species in the atmosphere. Figure 6.17 shows the efficiency of various materials for solar cell. The lines are drawn at the band gaps of the materials indicated. The solar spectrum is also shown.

Conventional solar cells are made of Si single crystal and have an efficiency of around 22–24%, while polycrystalline Si cells (cheaper to manufacture) have an efficiency of 18% due to defects in the microstructure that can trap carriers. Amorphous solar cells have an even lower efficiency, but can be grown directly on glass substrates by

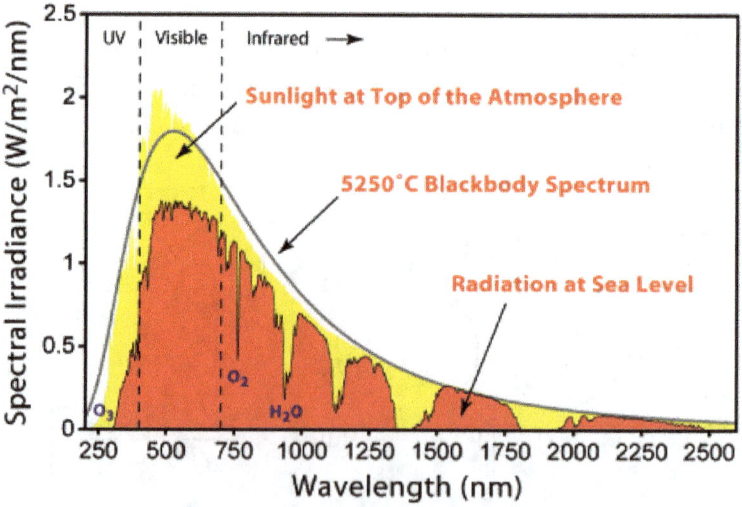

Figure 6.16: Spectrum of solar radiation above the atmosphere and at sea level. Modified from http://en.wikipedia.org/wiki/Sunlight.

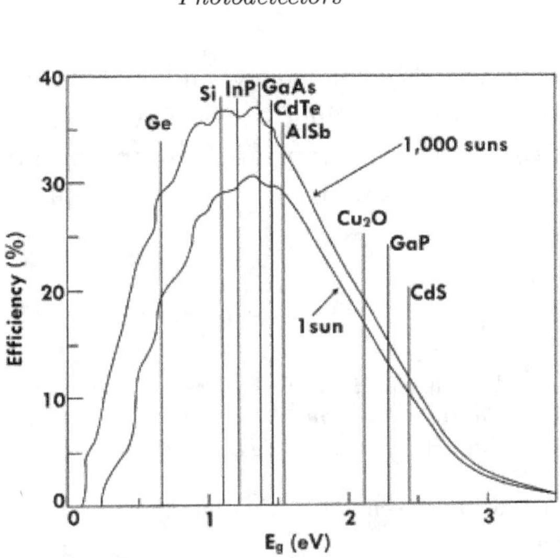

Figure 6.17: Solar cell efficiency of various materials. The lines indicate the band gap of the respective materials [30].

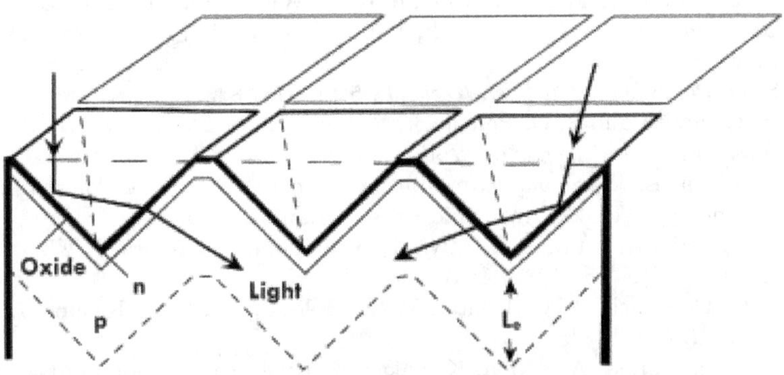

Figure 6.18: Example of a corrugated structure used for increasing efficiency. An efficiency of 24% was achieved using this design. A textured top surface is useful for trapping light, and thereby increasing the efficiency [30].

techniques like sputtering so that the overall cost of manufacturing is lowered. Design improvements which can effectively increase the surface area to incident radiation have been used to increase efficiency. An example is shown in Figure 6.18.

Photodiodes with high-efficiency and high-speed operation for communication systems continue to be developed. Solar cells represent one of the most important application areas of photodiodes. High- efficiency solar cells with high-power capability are important research areas.

References

1. H. Melchior, M. B. Fischer, and F. R. Arams,*Proc. IEEE*, **58** (1970) 1466.
2. J. Conradi, *Solid State Electron.*, **17** (1974) 99.
3. B. R. Bennett and R. A. Soref, *IEEE J. Quant. Electron.,* **QE-23** (1987) 2159.
4. F. R. Becher, J. S. Blakemore, J. T. Ebner, and J. R. Arthur, *Phys. Rev.*, **B37** (1988) 2551.
5. K. Taguchi, Chapter 8, in A. K. Dutta, N. K. Dutta, and M. Fujiwara (eds.), *WDM Technologies: Active Optical Components*, Academic Press, New York, 2002.
6. H. Melchoir, *J. Luminescence*, **7** (1973) 393.
7. G. E. Stillman and C. M. Wolfe, in R. K. Willardson and A. C. Beer (eds.), *Semiconductor and Semimetals*, Vol. 12, Academic Press, New York, 1977, p. 291.
8. S. D. Personick, *Bell Syst. Tech. J.*, **52** (1973) 843.
9. R. G. Smith and S. D. Personick, *Semiconductor Devices for Optical Communication*, Springer, New York, 1980.
10. S. R. Forrest, *Semiconductors and Semimetals*, Lightwave Communications Technology, Vol. 22, Academic Press, Orlando, 1985.
11. J. Yamada, A. Kawana, T. Miya, L. Nagai, and T. Kimura, *IEEE J. Quant. Electron.*, **QE-18** (1982) 1537.
12. J. C. Campbell, A. G. Dentai, W. S. Holden, and B. L. Kasper, *Electron. Lett.*, **18** (1983) 818.
13. Y. Matsushima, A. Akiba, K. Sakai, K. Kushiro, Y. Noda, and K. Utaka, *Electron. Lett.*, **18** (1982) 945.
14. D. R. Smith, M. C. Brain, and R. C. Cooper, *Telecom J.*, **48** (1981) 670.
15. K. Kato, A. Kozen, Y. Muramoto, Y. Itaya, T. Nagatsuma, and M. Yaita, *IEEE Photon Tech. Lett.*, **6** (1994) 719.
16. P. Bhattacharya, Chapter 8, *Semiconductor Optoelectronic Devices*, Prentice Hall, Upper Saddle River, 1997.
17. M. Kobayashi and T. Mikawa, Chapter 9 in A. K. Dutta, N. K. Dutta, and M. Fujiwara (eds.), WDM Technologies: Active Optical Components, Academic Press, New York, 2002.
18. K. Nishi, K. Taguchi, and Y. Matsumoto, *Appl. Phys. Lett.*, **35** (1979) 251.
19. H. Kanbe, N. Susa, H. Nakagome, and H. Ando, *Electron Lett.*, **16** (1980) 155.

20. J. C. Campbell, A. G. Dentai, W. S. Holden, and B. L. Kasper, *Electron. Lett.*, **19** (1983) 818.
21. J. C. Campbell, W. S. Holden, G. J. Qua, and A. G. Dentai, *IEEE J. Quant. Electron.*, **QE-21** (1985) 1743.
22. G. E. Stillman, Chapter 11, , in E. E. Basch and H. W. Sams (eds.), *Optical Fiber Transmission,* Howard W. Sams & Co., USA, 1986.
23. R. J. McIntyre, *IEEE Trans. Elec. Devices,* **ED-13** (1966) 164; **ED-19** (1972) 703.
24. R. P. Webb, R. J. McIntyre, and J. Conradi, *RCA Re.*,**35** (1974) 234.
25. S. R. Forrest, O. K. Kim, and R. G. Smith, *Appl. Phys. Lett.*, **41** (1982) 95.
26. L. W. Cook, G. E. Bulman, and G. E. Stillman, *Appl. Phys. Lett.*, **40** (1982) 589.
27. F. Capasso, A. Y. Cho, and P. W. Fox, *Electron Lett.*, **20** (1984) 635.
28. O. Mikami, H. Ando, H. Kanbe, T. Mikawa, T. Kaneda, and Y. Toyama, *IEEE J. Quant. Electron.*, **QE-17** (1981) 210.
29. T. Markvart and L. Castañer, Chapter 1A,in S. O. Kasap (ed.), *Principles of Electronic Materials.*
30. S. M. Sze, *Physics of Semiconductor Devices,* John Wiley, NY.
31. M. Zeman, Introduction to Solar Energy.

Chapter 7

Semiconductor Optical Amplifier

7.1 Introduction

Optical amplifier, as the name implies, is a device that amplifies an input optical signal. The amplification factor or gain can be higher than 1000 (>30 dB) in some devices. There are two principal types of optical amplifiers. They are the semiconductor optical amplifier (SOA) and the fiber optical amplifier. The latter is described in Chapter 9. Fiber amplifiers generally have higher gain and can amplify the signal to a higher power level compared to an SOA. In addition to light amplification, SOA is an important device for photonic integrated circuits and for optical logic.

SOA [1–7] is a device very similar to a semiconductor laser. Optical amplifiers amplify light through stimulated emission which is responsible for gain. Hence, their operating principles, fabrication, and design are also similar. The semiconductor amplifier is electrically pumped (current injection) to produce gain. For details on lasers please see Chapters 3 and 4. In an SOA, amplification of light takes place when it propagates through a semiconductor medium. The amplifier medium is sandwiched between semiconductor layers with higher band gap and lower refractive index. The three layers form a waveguide through which the light travels and undergoes amplification. SOAs are typically less than few mm in length. The operating principles, design, fabrication, and performance characteristics of InP-based semiconductor amplifiers are described in this chapter.

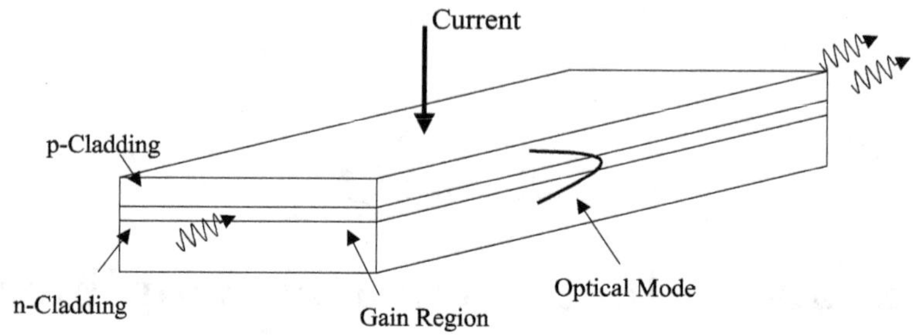

Figure 7.1: Schematic of a semiconductor optical amplifier with no waveguide in the lateral (along the plane of p–n junction) direction.

A schematic of an SOA is shown in Figure 7.1. It shows a slab waveguide of the SOA. The layer in the middle is responsible for optical gain. The layers on top and bottom of the gain region along with the gain region form a waveguide for the optical mode similar to that for a laser (see Chapter 3). The SOA chip has antireflection (AR) coating on both facets to reduce reflection from the cleaved mirror facets. The figure also shows a typical SOA application where light is injected at one facet and an amplified signal emerges from the other facet. Semiconductor lasers and amplifiers are described in several books [8–14].

7.2 Optical gain

The optical gain depends on the frequency (or wavelength) of the incident signals and also on the local intensity of the signal. For simplicity, the gain medium can be modeled as a homogeneously broadened two-level system (similar to that often done for a two-level laser). The gain coefficient (gain per unit length) of such a medium can be written as

$$g(\omega) = \frac{g_0}{1 + (\omega - \omega_0)^2 T_2^2 + P/P_s} \tag{7.1}$$

where g_0 is the maximum (peak) value of the gain (g_0 includes the confinement factor — Chapter 3 and Appendix 1), ω is the optical frequency of the incident signal, ω_0 is the electronic transition frequency, and P is the optical power of the signal being amplified. The quantity

P_s is known as the saturation power of the gain medium. It depends on the parameters of the gain medium such as radiative recombination time and transition cross-section (Chapter 3 and Appendix 5). The parameter T_2 in Eq. (7.1) is known as the dipole relaxation time with a typical value in the 0.1–1 ps range. For radiative transitions, there is another characteristic — lifetime (T_1). This radiative lifetime is often called the fluorescence lifetime or the population relaxation time. The value of T_1 can vary from 100 ps to 10 ms depending on the gain medium of the amplifier. For semiconductor amplifiers, Eq. (7.1) is a useful approximation of the actual gain spectrum. For semiconductors, $T_2 \sim 0.1$ ps and $T_1 \sim 200$ ps to 1 ns. Equation (7.1) is now used to discuss important amplifier characteristics such as gain bandwidth, net amplification, and output power.

7.2.1 *Gain spectrum and bandwidth*

For low powers, i.e. $P/P_s \ll 1$, the gain coefficient is given by

$$g(\omega) = \frac{g_0}{1 + (\omega - \omega_0)^2 T_2^2} \tag{7.2}$$

The equation shows the gain is maximum when the incident frequency $\omega = \omega_0$. The gain spectrum of Eq. (7.2) is given by a Lorentzian profile that is characteristic of a two-level system. As shown in Chapter 3, the actual gain spectrum can deviate from this Lorentzian profile. The gain bandwidth is defined as the full width at half maximum (FWHM) of the gain spectrum $g(\omega)$. For the spectrum of Eq. (7.2), the gain bandwidth $\Delta\omega_g$ is given by $\Delta\omega_g = 2/T_2$ or equivalently by $\Delta\nu_g = \Delta\omega_g/2\pi = 1/\pi T_2$. For a typical semiconductor amplifier $T_2 \sim$0.1 ps, this results in $\Delta\nu_g \sim 3$ THz. Amplifiers with large bandwidth are desirable for multiwavelength amplification such as would be the case for a wavelength division multiplexed (WDM) optical transmission system. $\Delta\nu_g$ is the bandwidth of the gain spectrum. A related concept is the gain bandwidth. The difference between the two becomes clear when the amplifier is considered as a single element with a gain G defined as

$$G = P_{\text{out}}/P_{\text{in}} \tag{7.3}$$

where P_{in} and P_{out} are the optical input and output power of the amplifier. P_{in} and P_{out} are related by the amplifier equation

$$dP/dz = gP \tag{7.4}$$

where $P(z)$ is the optical power at a distance z from the input end ($z = 0$). For a constant g, Eq. (7.4) has the solution

$$P(z) = P_{\text{in}} \exp(gz) \tag{7.5}$$

For an amplifier medium of length L, using $P(L) = P_{\text{out}}$, it follows from the above

$$G(\omega) = \exp[g(\omega)L] \tag{7.6}$$

The amplifier gain $G(\omega)$ is maximum for $\omega = \omega_0$, i.e. when $g(\omega)$ is maximum, and decreases when ω departs from ω_0. However, $G(\omega)$ decreases faster than $g(\omega)$ because of the exponential dependence. Using Eq. (7.6), the FWHM of $G(\omega)$, $\Delta\nu_a$ can be related to $\Delta\nu_g$ by

$$\Delta\nu_a = \Delta\nu_g \left(\frac{\ln 2}{g_0 L - \ln 2} \right) \tag{7.7}$$

$\Delta\nu_a$ is known as the amplifier bandwidth. As expected, $\Delta\nu_a$ is smaller than $\Delta\nu_g$, the value of $\Delta\nu_a$ depends on the amplifier gain.

7.2.2 *Gain saturation*

Equation (7.1) shows that the material gain depends on the optical power (P). For low optical powers, the gain is higher than that for high optical powers. This is the origin of gain saturation. Since g is reduced when P becomes comparable to P_s, the amplifier gain G also decreases when P becomes large. For simplicity, consider the case when the frequency of the incident light $\omega = \omega_0$, in this case, using Eqs. (7.1) and (7.4) it follows

$$\frac{dP}{dz} = \frac{g_0 P}{1 + P/P_s} \tag{7.8}$$

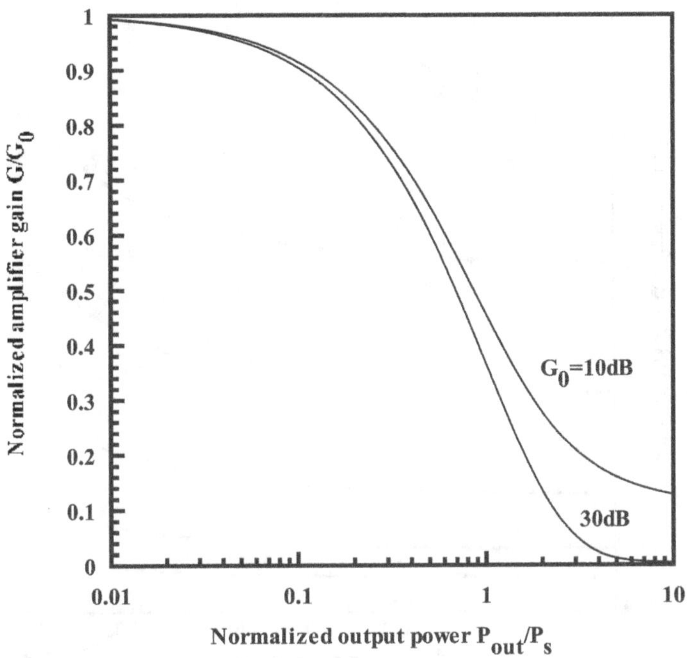

Figure 7.2: Normalized saturated amplifier gain G/G_0 as a function of the normalized output power.

The above equation can be solved using the boundary conditions $P(0) = P_{\text{in}}$ and $P(L) = P_{\text{out}}$, where L is the length of the amplifier. The amplifier gain $G = P_{\text{out}}/P_{\text{in}}$. The following relation is then obtained for the large signal amplifier gain:

$$G = G_0 \exp\left(-\frac{G-1}{G}\frac{P_{\text{out}}}{P_s}\right) \tag{7.9}$$

where $G_0 = \exp(g_0 L)$ is the unsaturated gain of the amplifier, i.e. gain for $P \ll P_{\text{s}}$. Equation (7.9) shows that the gain G decreases from G_0 when P_{out} becomes comparable to P_s. Figure 7.2 shows the saturation characteristics under two different unsaturated gain G_0 values.

A parameter known as output saturation power $P_{s\text{out}}$ is defined as the output power at which the saturated gain decreases to half of its unsaturated value, i.e. value of P_{out} for which $G = G_0/2$. Using

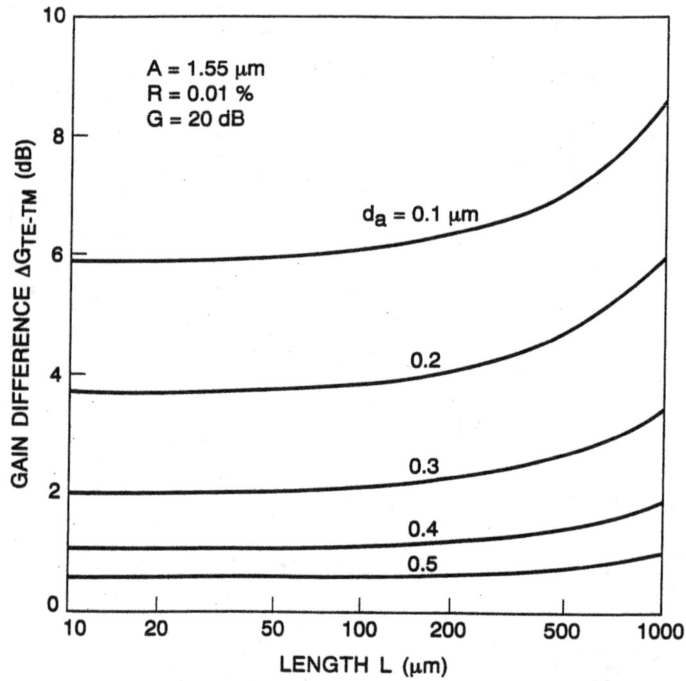

Figure 7.3: The optical gain difference between the TE and TM modes of a semiconductor amplifier is plotted as a function of device length for different active layer thicknesses [7]. For the plot, the amplifier gain for the TE mode is 20 dB. The reflectivity of each facet is 0.01%.

Eq. (7.9) P_{sout} is given by

$$P_{sout} = \frac{G_0 \ln 2}{G_0 - 2} P_s \tag{7.10}$$

P_{sout} is smaller than P_s. For amplifiers with large gain ($G_0 > 100$), P_{sout} is nearly independent of G_0. For a typical amplifier with 30 dB gain ($G_0 = 1000$), $P_{sout} \sim 0.69\ P_s$.

7.2.3 *Dielectric waveguide*

Stimulated emission in a semiconductor amplifier arises from electron–hole radiative recombination in the active region (gain producing region), and the light generated is confined and guided by a dielectric waveguide very similar to that of a laser (Figure 3.3

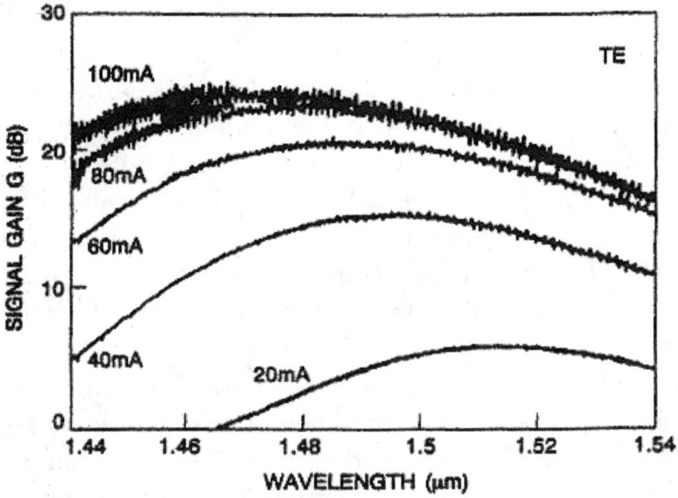

Figure 7.4: Measured gain spectrum of a traveling wave (TW) type semiconductor optical amplifier at several current levels.

of Chapter 3). Similar to that for a laser, the active region has a slightly higher index than the p-type and n-type cladding layers, and the three layers form a dielectric waveguide. The fraction of the optical mode confined in the active region is called the confinement factor (Γ). The optical gain of a signal in the fundamental mode traveling through the amplifier is given by

$$g_m = \Gamma_g \tag{7.11}$$

Since the transverse electric (TE) and transverse magnetic (TM) modes have different confinement factors the optical gain for the TE and TM modes are different (Chapter 3), i.e. the SOA generally exhibits a polarization dependence of gain. This is generally not desirable. Figure 7.3 (plotted using Figure 3.4 of Chapter 3) shows that amplifiers with thick active regions will have low polarization dependence of gain due to similar Γ for both TE and TM modes. If P_s is the saturation power of the gain medium, the observed saturation power is given by

$$P_s(\text{observed}) = P_s/\Gamma \tag{7.12}$$

Thus, amplifiers with small mode confinement factor (small Γ), such as quantum well amplifiers, exhibit high saturation power. The amplifiers used in lightwave system applications, either as preamplifiers in front of a receiver or as in-line amplifiers as a replacement of regenerators, must exhibit nearly equal optical gain for all polarizations of the input light. The calculated TE/TM gain difference for an amplifier operating in the 1550 nm wavelength range is shown in Figure 7.3. The active region consists of InGaAsP active layer with p-InP and n-InP cladding layers. The condition for amplification is similar to that of a laser and is described in Chapter 3.

The active region thickness of an optical amplifier is typically in the range of 0.2–0.4 μm. Amplifiers with thick active regions are needed for polarization independent gain. The semiconductor materials used for semiconductor laser fabrication (Chapters 2 and 3) can be used to fabricate SOAs. For SOAs, the differences are typically thick active region for polarization independent gain or longer lengths for higher gain. Optical amplifiers with quantum well active regions have low confinement factor which, in general, leads to high saturation power and are therefore suitable for integration with other optical devices on a single chip. Such photonic integrated circuits are discussed in Chapter 8.

7.3 CW amplifier characteristics

Extensive work on optical amplifiers was carried out in the 1960s using the AlGaAs material system [1–3]. Much of the recent experimental work on SOAs have been carried out using the InGaAsP material system, with the optical gain centered around 1300 or 1550 nm. This is primarily due to their potential use in all-optical networks.

SOAs can be classified into two categories, the Fabry–Perot (FP) amplifier and the traveling wave (TW) amplifier. An FP amplifier has considerable reflectivity at the input and output ends, which results in resonant amplification at FP modes. Thus, an FP amplifier exhibits a very large optical gain at wavelengths corresponding to the longitudinal modes of the cavity, and the gain is small in between the cavity modes. This modulation in gain is not desirable if the amplifier is used in optical networks. The TW amplifier, by contrast, has negligible reflectivity at each end which results in signal amplification during a single pass. The optical gain spectrum of a TW amplifier is quite broad and corresponds to that of the semiconductor gain medium. Most practical TW amplifiers exhibit some small ripple in the gain spectrum which arises from residual facet reflectivity. TW amplifiers are more suitable for system applications. An example of the gain spectrum of a TW amplifier with AR coated cleaved facets is shown in Figure 7.4. The output exhibits modulations at longitudinal modes of the cavity because the optical gain is slightly higher at the modes than in between the modes. Both TE and TM fundamental modes can propagate in a TW amplifier. These modes have slightly different effective indices due to slightly different confinement factors and hence the precise AR coating needed for these modes are different. Thus, the residual modulation in TE and TM mode gain in TW amplifiers due to non-zero facet reflectivity are spectrally displaced. This problem is reduced considerably for amplifiers with thick active regions. For plane waves, a quarter wave thick (thickness $= \lambda_0/4n$, where n is the index of the dielectric layer) single layer dielectric coating with an index $n = n_m^{1/2}$, where n_m is the index of the semiconductor material, is sufficient to create a perfect (zero reflectivity) AR coating. The waves propagating in the amplifier are not plane waves and the above therefore serves only as an approximation. For low gain modulation (also known as gain ripple), the facet reflectivity of a TW amplifier must be <0.01%. Much effort has been devoted to fabricating amplifiers with very low effective facet reflectivity. Such amplifier structures either utilize special low effective reflectivity dielectric coatings or have tilted or buried facets [8].

Schematic cross-section of a buried-facet optical amplifier is shown in Figure 7.5. Current confinement in this structure is provided by semi-insulating Fe-doped InP layers grown by the metal organic chemical vapor deposition (MOCVD) growth technique. Fabrication of this device involves a procedure similar to that used for lasers. The

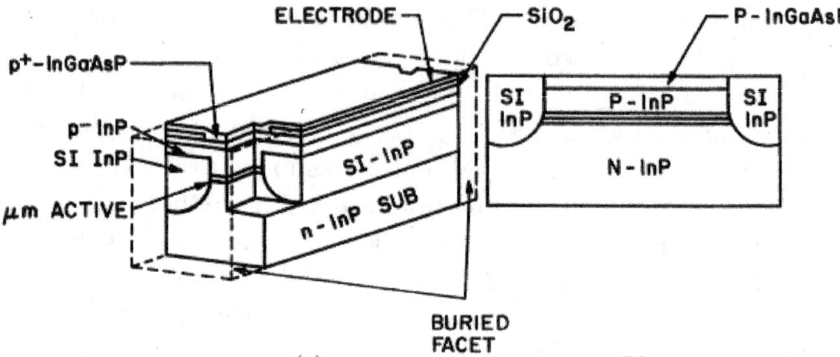

Figure 7.5: Schematic of a buried facet optical amplifier [8].

first four layers are grown on a (100)-oriented n-InP substrate by MOCVD. These layers are (i) an n-InP buffer layer, (ii) an undoped InGaAsP ($\lambda \sim 1550$ nm) active layer, (iii) a p-InP cladding layer, and (iv)a p-InGaAs ($\lambda \sim 1650$ nm) layer. Mesas are then etched on the wafer along the [110] direction with 15 μm wide channels normal to the mesa direction using a SiO_2 mask. The latter is needed for buried-facet formation. Semi-insulating Fe-doped InP layers are then grown around the mesas by MOCVD growth process with the oxide mask in place. The oxide mask and the p-InGaAs layer is removed and a p-InP cladding layer and p-InGaAs ($\lambda \sim 1650$ nm) contact layer are then grown over the entire wafer by the MOCVD growth technique. The wafer is processed using standard lithography methods, metal depositions, and cleaved to produce 0.5–3-mm-long buried-facet chips with $\sim$7-μm-long buried facets at each end. The facets of the chips are then AR-coated using a single-layer film of ZrO_2. Fabrication of cleaved-facet devices follows the same procedure as described above, except that the mesas are continuous with no channels separating them. The latter is needed for defining the buried-facet regions. The semi-insulating layer, in both types of devices, provides current confinement and lateral index guiding. For buried-facet devices it also provides the buried-facet region where beam expansion takes place. Detailed fabrication and performance of low reflectivity designs are described in Chapter 5 of Ref. [8].

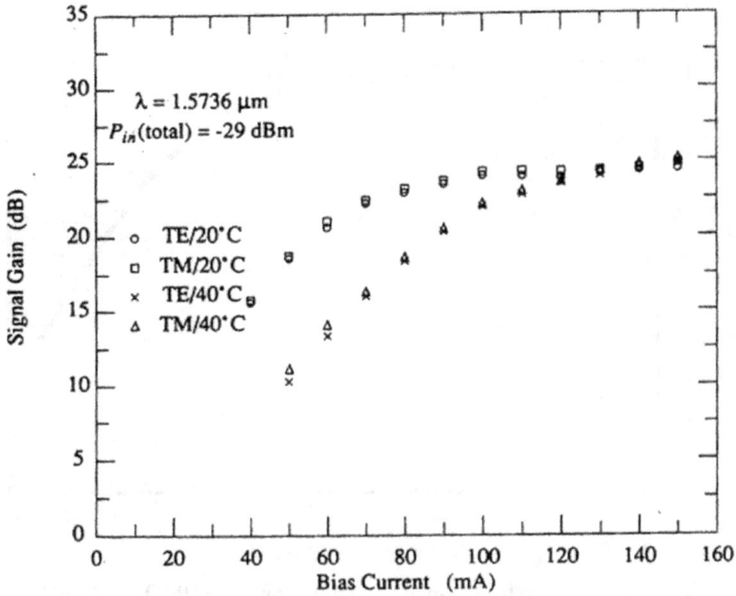

Figure 7.6: Gain as a function of current of a semiconductor amplifier. The input power is -29 dBm ($\sim$1 μW). The device has a buried heterostructure design.

A parameter of interest for amplifier characteristics is gain as a function of current. The optical gain is found to increase with increasing current, then it saturates. Typical data for these characteristics are shown in Figure 7.6. The optical gain decreases with increasing temperature at a given current. This is due to several factors. Electrons are distributed over a wider energy range at high temperature, and hence the number of electrons available for participating in optical transition (gain) at a given energy is fewer. Non-radiative recombination which increases with increasing temperature for long wavelength materials ($\sim$1550 nm) causes a reduction in gain for a given current. In addition, many semiconductor amplifiers are fabricated using buried heterostructure designs (for good mode stability). These designs rely on reverse biased junctions for confinement of the current to the active region and therefore necessarily have higher leakage current (current going around the active region) at high temperatures. This is discussed in Chapter 3.

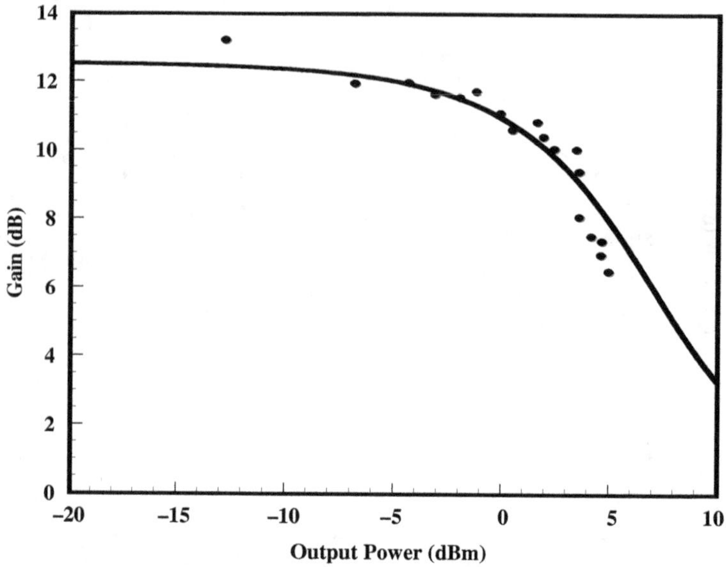

Figure 7.7: Gain as a function of output power. Solid line is plotted using Eq. (7.9), with $G_0 = 18$, $P_s = 2.5$ mW. Dots are measured gain.

As discussed earlier, the optical gain decreases with increasing optical input power. This phenomenon also exhibits itself as saturation in gain as a function of current. With increasing current, the gain increases and hence the signal acquires high enough intensity to reduce its gain before reaching the output facet. Since the optical power in a TW amplifier increases as the signal propagates through the amplifier, the maximum gain saturation occurs near the output port. The gain saturation characteristics of an amplifier are generally plotted as the measured gain as a function of output power (Figure 7.7). The saturation output power is the power at which the gain is reduced by a factor of 2 (3 dB) from that for small output powers. When an amplifier is operated in the saturated regime with multiple input signals (for example at different wavelengths), a cross-talk (i.e. transfer of modulation from one signal to another) occurs. This cross-talk limits the use of amplifiers in multichannel transmission systems.

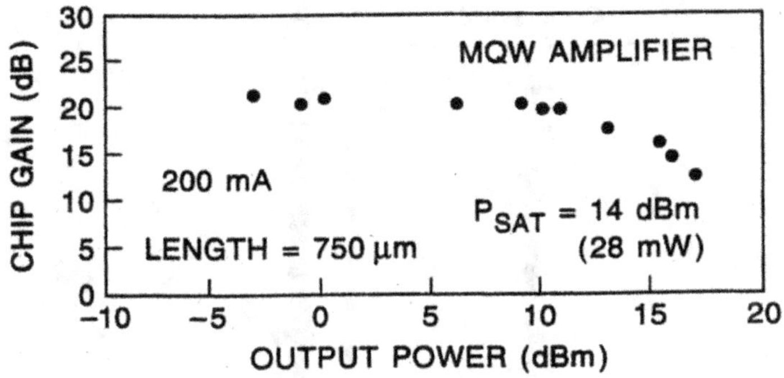

Figure 7.8: Gain as a function of output power for multiquantum well amplifiers.

7.3.1 *Multiquantum well amplifiers*

Multiquantum well (MQW) lasers are described in several articles [15–19]. Semiconductor amplifiers with an MQW active region (see Chapter 3 for details) have high gain in TE polarization. The gain in TM polarization is considerably lower. In addition, these amplifiers have high saturation power due to small confinement factor. Typical TE-mode gain as a function of output power is shown in Figure 7.8. The saturation power of this amplifier is 28 mW. This amplifier had a four-quantum well active region. Because of high saturation power and high TE-mode gain, MQW amplifiers are attractive candidates for on-chip integration with lasers (which also emit in TE mode) as power boosters. An important reason for higher saturation power of MQW amplifiers is low confinement factor of the optical mode (Eq. (7.12)).

MQW amplifiers fabricated using lattice-matched semiconductors have high gain in the TE mode. The TM mode gain is significantly smaller. This is because TM mode transitions involve light holes and the light hole population is smaller than the heavy hole population.

However, polarization independent gain in an MQW SOA with tensile strained active region has been reported [16, 17] (see Chapters 3 and 4 for strained layer growth).

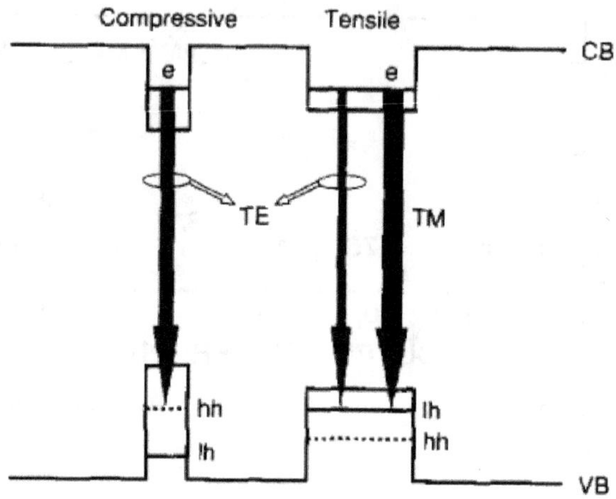

Figure 7.9: Operating principle of polarization insensitive multiquantum well amplifier [18]. High gain is indicated by thick line for the transition.

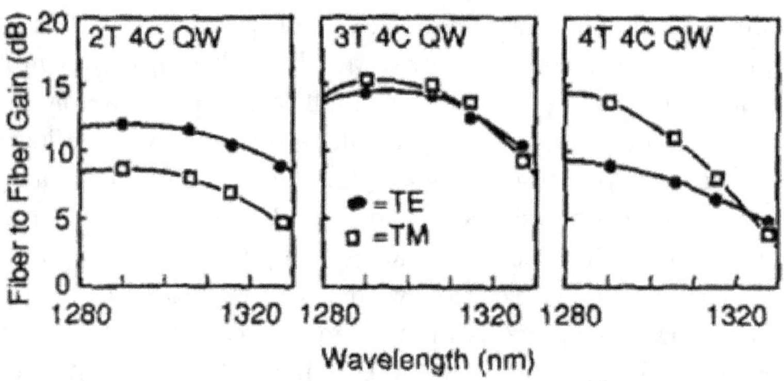

Figure 7.10: TE and TM polarization resolved gain spectrum for four compressively strain (4C) and two, three, and four tensile strained (2T, 3T, and 4T) MQW amplifiers, respectively [18].

The tensile strain results in nearly degenerate valence-band maximum and thus equal transition probability for light and heavy holes. A more versatile design has been reported using both compressive and strained quantum well layers in the MQW structure [18, 19]. In this design, the quantum well layers are tailored (strain, composition, or well width) in order to equalize the band gaps as shown in

Figure 7.9. The compressively strained quantum well structures only contribute to TE polarized gain, and the tensile strained quantum wells contribute mainly to TM gain and to TE gain to some extent. The ratio of TE to TM gain (given by ratio of the matrix elements) for electron–light hole recombination (in tensile strained material) is about 1–4. The ratio of the number of both types (compressive and tensile) strained layers and their widths are optimized in the structure so as to yield polarization insensitive gain. Figure 7.10 shows the polarization resolved (fiber-to-fiber) gain spectra of an MQW SOA with four 1% compressively (4C) strained $In_{0.83}Ga_{0.17}As_{0.67}P_{0.33}$ (width = 4.5 nm) and two, three, or four 1% tensile (2T, 3T, and 4T) strained $In_{0.5}Ga_{0.5}As_{0.78}P_{0.22}$ (width = 11 nm) wells. The amplifier is driven at 100 mA and the amplifier length is 750 μm. Nearly polarization independent gain is obtained for the 3T/4C design.

Amplifiers using quantum dot (QD) active regions have also been studied. These amplifiers are discussed in Section 7.9.

7.4 Amplifier rate equations for pulse propagation

The dynamics of SOA performance is discussed in this section. A TW amplifier with very low facet reflectivity (~0.01%) is a promising candidate for amplifier applications. The amplifier has a large bandwidth (~5 THz), this allows amplification of pulses as short as few ps. If the amplifier is operated in the linear regime, the amplified pulse is a replica of the input pulse. However, in most cases the amplifier operates near saturation, particularly near the peak of the pulse, resulting in time-dependent optical gain. Thus, the amplified pulse undergoes a change in shape.

7.4.1 *Amplifier rate equations*

In this section, the basic rate equations for pulse propagation in SOAs are developed [20–27]. The amplitude and the phase change of an optical signal propagating through the SOA is a result of gain saturation. The propagation of an electromagnetic field inside the amplifier is governed by the wave equation

$$\nabla^2 \vec{E} - \frac{\varepsilon}{c^2}\frac{\partial^2 \vec{E}}{\partial t^2} = 0 \tag{7.13}$$

where $\vec{E}(x, y, z, t)$ is the electric field vector of the wave, c is the velocity light, and ε is the dielectric constant of the amplifier medium. The dielectric constant is given by

$$\varepsilon = n_b^2 + \chi \tag{7.14}$$

where n_b is the background refractive index which takes into account the dielectric waveguiding in the semiconductor medium. It is, in general, a function of x, y which represent the transverse coordinates of the waveguide medium. The susceptibility χ represents the effect of injected carriers on the dielectric constant. It depends on the injected carrier density n. The exact dependence of χ on n is quite complicated. However, a phenomenological model is found to be quite effective. In this model, χ is given by [23]

$$\chi(n) = -\frac{\bar{n}c}{\omega_0}(\alpha + i)g(n) \tag{7.15}$$

where $\bar{n}$ is the effective mode index, ω_0 is the frequency, $g(n)$ is the optical gain, and α is the linewidth enhancement factor. As discussed in Chapter 3, the optical gain (g) approximately varies as

$$g = a(n - n_0) \tag{7.16}$$

where a is the gain constant, n is the injected carrier density, and n_0 is the carrier density needed for transparency. The carrier induced change in index is accounted for using the linewidth enhancement factor, α, which is the ratio of the change in real part of the index to change in the imaginary part of the index. The latter is proportional to optical gain. The typical values of α are in the range of 3–8. It is useful to make some simplifying approximation of the wave equation. Assume a TW amplifier with a single propagating optical mode whose electric field distribution is $\phi(x, y)$. Assume the light is linearly polarized and it remains linearly polarized in the amplifier. The electric field $\vec{E}(x,y,z,t)$ can then be written as

$$\vec{E}(x,y,z,t) = \hat{\varepsilon}\frac{1}{2}\{\phi(x,y)A(z,t)\exp[i(k_0 z - \omega_0 t)] + c.c.\} \tag{7.17}$$

where $\hat{\varepsilon}$ is the polarization unit vector, $k_0 = \bar{n}\omega_0/c$, ω_0 is the photon angular frequency, and $A(z,t)$ is the slowly varying amplitude of the propagating wave. Using Eqs. (7.13) and (7.17), and neglecting second derivatives of $A(z,t)$ with respect to t and z and integrating over the transverse dimensions x, y, the following is obtained:

$$\frac{\partial^2\phi}{\partial x^2} + \frac{\partial^2\phi}{\partial y^2} + (n_b^2 - \bar{n}^2)\frac{\omega_0^2}{c^2}\phi = 0 \tag{7.18}$$

$$\frac{\partial A}{\partial z} + \frac{1}{v_g}\frac{\partial A}{\partial t} = \frac{i\omega_0\Gamma}{2\bar{n}c}\chi A - \frac{1}{2}\alpha_{\text{int}} A \tag{7.19}$$

where the group velocity $v_g = c/n_g$, the group index n_g is given by

$$n_g = \bar{n} + \omega_0\left(\frac{\partial \bar{n}}{\partial \omega}\right) \tag{7.20}$$

Γ is the confinement factor of the optical mode. The group velocity dispersion has been neglected in the above Eq. (7.19) since the amplifier lengths are typically small (~1 mm). The evolution of carrier density (n) can be described by the following Eq. (7.21) [8, 23]. The equation assumes that the carrier density is nearly uniform in the transverse direction and hence an average value can be used. Also, carrier diffusion has been neglected since the amplifier length is much larger than the diffusion length, and the width and thickness are much smaller than the diffusion length. In Eq. (7.21), V is the volume of the active region, I is the injected current, e is the charge of the electron, τ_c is the carrier lifetime, and σ_m is the cross-section of the active region.

$$\frac{\partial n}{\partial t} = \frac{I}{eV} - \frac{n}{\tau_c} - \frac{\Gamma g(n)}{\hbar\omega_0\sigma_m}|A|^2 \tag{7.21}$$

For pulse propagation, Eqs. (7.19) and (7.21) can be further simplified by using the retarded time frame:

$$\tau = t - z/v_g \tag{7.22}$$

Then, the following is obtained:

$$\frac{\partial n}{\partial \tau} = \frac{I}{eV} - \frac{n}{\tau_c} - \frac{\Gamma g(n)}{\hbar\omega_0\sigma_m}|A|^2 \tag{7.23}$$

$$\frac{\partial A}{\partial z} = \frac{i\omega_0\Gamma}{2\bar{n}c}\chi A - \frac{1}{2}\alpha_{\text{int}} A \tag{7.24}$$

Assume

$$A = \sqrt{P}\exp(i\phi) \tag{7.25}$$

where $P(z,\tau)$ and $\phi(z,\tau)$ are the instantaneous power and the phase of the propagating pulse, respectively. Using Eqs. (7.15)–(7.25), the following equations are obtained [8, 23]:

$$\frac{\partial A}{\partial z} = \frac{g}{2}(1 + i\alpha)A \tag{7.26}$$

$$\frac{\partial g}{\partial \tau} = -\frac{g - g_0}{\tau_c} - \frac{gP}{E_s} \tag{7.27}$$

$$\frac{\partial P}{\partial z} = (g - \alpha_{\text{int}})P \tag{7.28}$$

$$\frac{\partial \varphi}{\partial z} = -\frac{1}{2}\alpha g \tag{7.29}$$

Equation (7.26) is the equation for the complex amplitude A. The quantity $E_s = \tau_c P_s$. P_s is the saturation power of the amplifier, which is given by

$$P_s = h\omega_0\sigma_m/(a\Gamma\tau_c) \tag{7.30}$$

g_0 is the small signal gain, which is given by

$$g_0 = \Gamma a(I\tau_c/eV - n_0) \tag{7.31}$$

and $\sigma_m \cong wd$ (w and d are the width and thickness of the active layer).

7.5 Pulse amplification results

For many applications, a series of intensity modulated signals are injected into the amplifier. Such intensity modulated pulses are short in duration for high-speed communication systems. Amplification of pulses of light is considered in this section. Using the formulation of the previous section, a net gain $g(z,\tau)$ can be defined as

$$g(z,\tau) = g - \alpha_{\text{int}} \tag{7.32}$$

Equation (7.28) can be integrated over the amplifier length to provide the following expression for output power P_{out} (τ):

$$P_{\text{out}}(\tau) = P_{\text{in}}(\tau)\exp[h(\tau)] \tag{7.33}$$

where P_{in} (τ) is the input power and $h(\tau)$ is the total integrated net gain defined as

$$h(\tau) = \int_0^L g(z,\tau)dz \tag{7.34}$$

Under the assumption that the pulse width $\tau_0 \ll \tau_c$, the first term in Eq. (7.27) can be neglected. Then Eq. (7.27) can be integrated over the amplifier length L to provide the following equation for $h(\tau)$.

$$\frac{dh}{d\tau} = -\frac{P_{\text{in}}(\tau)}{E_s}(e^h - 1) \tag{7.35}$$

Equation (7.35) can be integrated to obtain $h(\tau)$ for different input pulse shapes $P_{\text{in}}(\tau)$. The parameter of interest is the evolution of total gain $G(t)$ as a function of time. From Eq. (7.35) it is given by [23]

$$G(\tau) = \exp(h(\tau)) = \frac{G_0}{G_0 - (G_0 - 1)\exp[-E_0(\tau)/E_s]} \tag{7.36}$$

where G_0 is the unsaturated gain of the amplifier and the quantity $E_0(\tau)$ is given by

$$E_0(\tau) = \int_{-\infty}^{\tau} P_{\text{in}}(\tau)d\tau \tag{7.37}$$

Equation (7.36) shows the gain is time dependent, i.e. different portions of the pulse experience different amounts of gain. This leads to a modification of the shape of the pulse after transmission through the amplifier. For the purpose of general discussion, a super-Gaussian pulse is used as the input.

$$P_{\text{in}}(\tau) = P_0 \exp[-(\tau/\tau_p)^{2m}] \tag{7.38}$$

where m is the shape parameter. For $m = 1$, the pulses are Gaussian and it becomes flat topped as m increases. Figure 7.11 shows the output pulse shape for a Gaussian input pulse for several values of unsaturated gain G_0. The pulse is given by

$$P_{\text{in}}(\tau) = P_0 \exp[-(\tau/\tau_0)^2] \tag{7.39}$$

Figure 7.11 is obtained by solving the above equations. The calculated pulse shapes for several values of gain are shown. For higher gain, the output pulse is asymmetric, i.e. the leading edge of the pulse has a faster rise time than the trailing edge. The optical gain is accompanied by a change in phase. The gain changes the carrier density in the material. The phase of the propagating pulses changes because the real index varies with the carrier density. The total phase change for the pulse propagating through the amplifier is given by integrating Eq. (7.29) [23]

$$\phi(\tau) = -\frac{1}{2}\alpha \int_0^L g(z,\tau)d\tau = -\frac{1}{2}\alpha h(\tau) \tag{7.40}$$

where α is the linewidth enhancement factor. The value of α depends on the wavelength of the input light. It is typically in the 3–8 range for regular double heterostructure gain region and is smaller (~2–5) for MQW gain region.

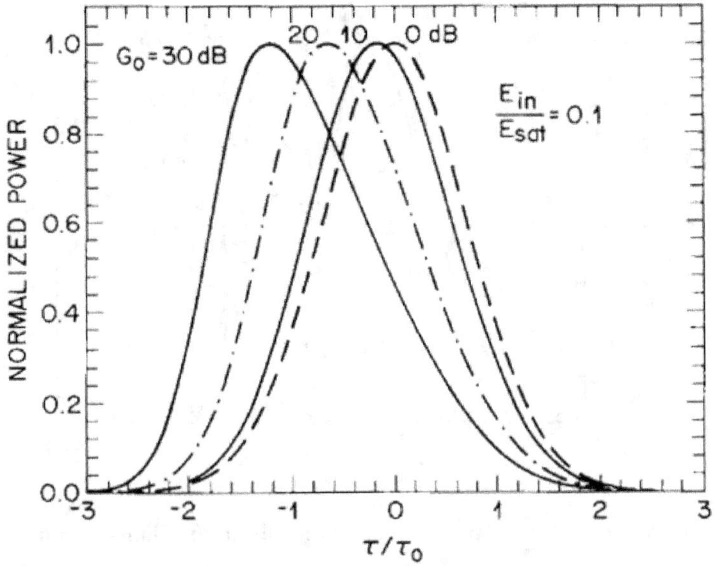

Figure 7.11: Output amplified pulse shape for several values of τ/τ_0 when a Gaussian pulse of width τ_0 is amplified by an amplifier. The output pulse is shown for several values of small signal gain [23].

A time-dependent phase change leads to a variation in optical frequency of the pulse. The instantaneous variation in frequency, known as the frequency chirp $\Delta\nu(\tau)$ is given by

$$\Delta\nu(\tau) = -\frac{1}{2\pi}\frac{d\phi}{d\tau} = \frac{\alpha}{4\pi}\frac{dh}{d\tau} \tag{7.41}$$

The chirp of the output pulse $\Delta\nu_{\text{out}}(\tau)$ is given by $\Delta\nu_{\text{out}}(\tau) = \Delta\nu_{\text{in}}(\tau) + \Delta\nu(\tau)$, where $\Delta\nu_{\text{in}}(\tau)$ is the chirp of the input pulse. Figure 7.12 shows the pulse spectrum during the evolution of a Gaussian pulse for various amplifier gains. The linewidth enhancement factor, $\alpha = 5$ is used in the calculation. For high gain, the spectrum has multiple peaks. With increasing gain, the main peak in the spectrum shifts to the low-frequency side (red shift). The shift is larger than the spectral width of the pulse. For 10 ps wide pulses, the shift can exceed 100 GHz.

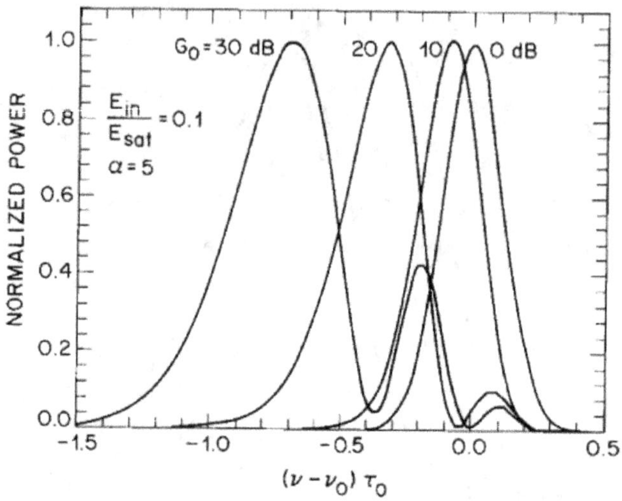

Figure 7.12: Output pulse spectrum for an input Gaussian pulse [23].

The phenomenon of multiple peaks in the output spectrum and the associated change in output phase is called self-phase modulation (SPM). Such SPM induced broadening has been observed in the amplifier output spectrum. Figure 7.13 shows the shape of the output pulse for a super-Gaussian input pulse of width τ_0.

The complete Eq. (7.27) for the temporal evolution of gain is used in this calculation [23]. The pulse shape is calculated for various values of τ_0/τ_c. The x-axis is in units of τ_0. The pulse energy is chosen so that $E_{\text{in}}/E_{\text{s}} = 0.1$. The amplifier has a 30 dB small signal gain. The calculated pulse distortion is larger for short pulses (small τ_0/τ_c). As before, a numerical solution of Eq. (7.41) gives $\Delta\nu\ (\tau)$ as a function of time for super-Gaussian pulses. Figure 7.14 shows the frequency chirp of an amplified pulse for various values of pulse energies. The frequency chirp is larger for higher energy pulses because the gain saturation occurs earlier for such pulses. A value of five for α is used in the calculation. In addition to the above frequency chirp, pulses propagating through an SOA also exhibit self-phase modulation. This happens for high enough energies for which the index varies with the intensity of the pulse.

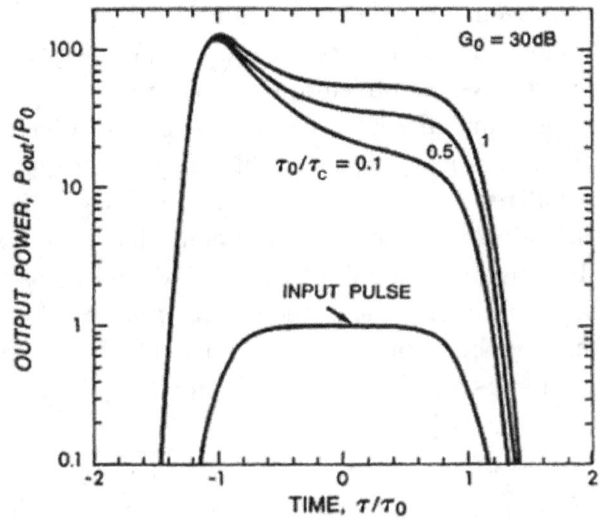

Figure 7.13: Amplified pulse shape for several values of τ_0/τ_c when a super-Gaussian pulse of width τ_0 is amplified by an amplifier with a small signal gain of 30 dB [8, 28].

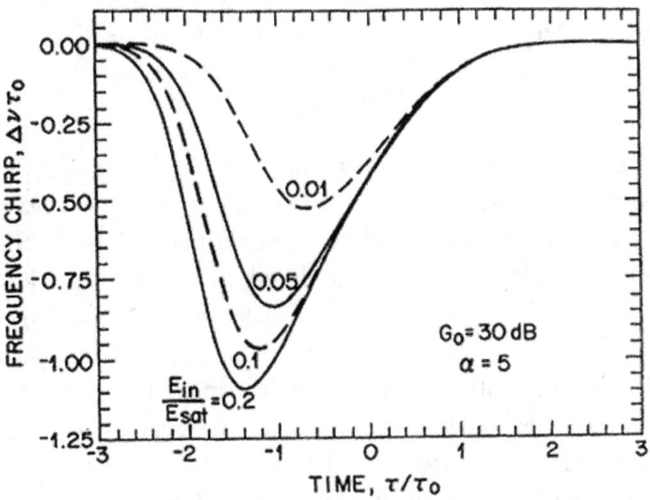

Figure 7.14: Frequency chirp as a function of time during the pulse duration for various pulse energies [8, 28].

SOAs have been used to simultaneously amplify many signals at different wavelengths such as that used in a WDM communication system. An error-free amplification is possible if there is no cross-talk between light at different wavelengths. Cross-talk can arise from two mechanisms, viz. gain saturation and carrier density modulation. Gain saturation results in cross-talk because presence of one signal saturates the gain of the other. If the amplifier is operated in the unsaturated regime, the gain saturation induced cross-talk can be made small. Experimental and theoretical studies of multichannel amplification have been published [29–37]. Results of optical transmission experiments are described in Chapter 11.

7.6 Amplifier noise

Amplifiers degrade the signal-to-noise ratio (SNR) of the amplified signal primarily because of amplified spontaneous emission (ASE) which adds noise to the signal [28]. The effect of spontaneous emission is to add fluctuations in the signal which adds noise to the detected photocurrent. The SNR degradation is generally characterized by a quantity called noise figure, F, which is defined as

$$F = \frac{(SNR)_{\text{in}}}{(SNR)_{\text{out}}} \tag{7.42}$$

where SNR refers to the signal-to-noise ratio of the electrical power generated when the signal is converted to current by a photodetector. The SNR of the photodetector is determined by both the thermal noise and the shot noise. For an ideal detector, considered below, shot noise is the only noise. In this case, simple expressions can be obtained.

Consider an amplifier with an input optical power P_{in}, amplifier gain G, a detector with a responsivity R with an electrical bandwidth B. The responsivity R is the ratio of the photocurrent (Amp) per unit power (W). For an ideal detector $R = (e/h\nu)$, where e is the charge of the electron and ν is the frequency of the incident optical signal. The photocurrent generated is given by

$$I_i = RP_{\text{in}} \tag{7.43}$$

and the r.m.s. (root mean square value of the photocurrent) due to shot noise is

$$\sigma^2 = 2e(RP_{\text{in}})B$$

Thus, the input SNR for ideal detector is given by

$$(SNR)_{\text{in}} = \frac{\langle I_i \rangle^2}{\sigma^2} = \frac{P_{\text{in}}}{2h\nu B} \tag{7.44}$$

In order to determine the output SNR, the spontaneous emission must be added. The spectral density of spontaneous emission induced noise (white noise) is given by

$$S(\nu) = (G-1)N_{\text{sp}}h\nu \tag{7.45}$$

where N_{sp} is the spontaneous emission factor or the population inversion factor. For a two-level atom, it is given by

$$N_{\text{sp}} = \frac{N_2}{N_2 - N_1} \tag{7.46}$$

where N_2 and N_1 are the number of atoms in the upper and lower state, respectively. Thus, for high population inversion $N_2 \gg N_1$, $N_{\text{sp}} \sim 1$. However, for low population inversion, N_{sp} could be much larger. For the discussion here only the dominant contribution to noise is considered, i.e. the variance of the photocurrent arises mainly from the beating of the spontaneous emission with the signal. The spontaneous emission mixes coherently with the signal at the photodetector and produces a fluctuating photocurrent within the bandwidth B. The signal and the variance of the photocurrent after amplification is given by

$$I = RGP_{\text{in}} \tag{7.47}$$

$$\sigma^2 = 2qIB + 4IRSB \tag{7.48}$$

The variance in Eq. (7.48) is the sum of the shot noise and spontaneous-signal beat noise. All other noise contributions have been neglected (see Ref. [8]). Thus, the SNR at the output is given by

$$(SNR)_{\text{out}} = \frac{\langle I \rangle^2}{\sigma^2} \approx \frac{GP_{\text{in}}}{4SB} \tag{7.49}$$

where the last relation is obtained by neglecting shot noise and assuming $G \gg 1$. The amplifier excess noise figure can then be written as

$$\begin{aligned} F &= 2N_{\text{sp}}(G-1)/G \\ &\approx 2N_{\text{sp}} \end{aligned} \tag{7.50}$$

The equation shows that the SNR of the amplifier is degraded by a factor of 2 (3 dB) for the ideal case of $N_{\text{sp}} = 1$. For most practical amplifiers the SNR degradation is in the range of 5–7 dB.

7.7 Mach–Zhender interferometer using SOA

SOAs incorporated into Mach–Zhender interferometer (MZI) or Michelsen interferometer configurations have been used for wavelength conversion and to demultiplex high-speed time division multiplexed optical signals [8, Chapter 9]. They are also important for all-optical logic studies. The schematic of the device is shown in Figure 7.15. The input signal beam is split into two beams by an input y-junction, which propagate through semiconductor amplifiers positioned in the upper and lower arms of the interferometer. These beams then merge and interfere at an output y-junction. The phase of the interference is altered by injecting another signal in one of the other ports.

The ratio of the output (P_o) to the input signal (P_i) is given by

$$\frac{P_o}{P_i} = \frac{1}{8}\{G_1 + G_2 - 2\sqrt{G_1G_2}\cos(\phi_1 - \phi_2)\} \tag{7.51}$$

where G_1, G_2 are the gains of the amplifiers and ϕ_1, ϕ_2 are phase changes induced by nonlinear effects in the two amplifiers. The phase

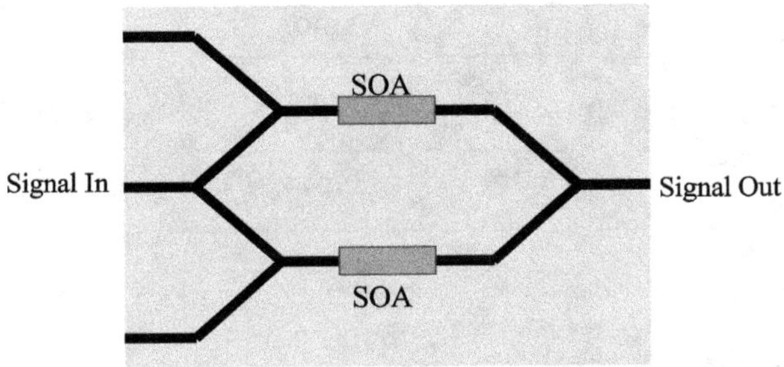

Figure 7.15: Schematic of a Mach–Zhender interferometer with integrated SOA's.

difference, $\Delta\phi = \phi_1 - \phi_2$, at the output is given by:

$$\Delta\phi = \frac{\alpha}{2P_s}\left\{-\frac{hv}{e}(I_1 - I_2) + \frac{P_c}{2}(G_1 - 1)\right\} \tag{7.52}$$

where $\alpha = \delta n_R/\delta n_{\mathrm{Im}}$ is the ratio of the carrier induced change in the real and imaginary part of the index, also known as the linewidth enhancement factor. P_c is the injected power in one of the amplifiers, P_s is the saturation power in the amplifier, and I_1, I_2 are the currents through the two amplifiers. For $I_1 \cong I_2$ and $G_1 \cong G_2$, the above reduces to

$$\frac{P_o}{P_i} = \frac{G}{2}\sin^2\left(\frac{\Delta\phi}{2}\right), \quad \Delta\phi = \frac{\alpha}{4}\frac{P_c}{P_s}(G - 1) \tag{7.53}$$

Using typical values; $\alpha = 3$, $P_s = 10$ mW, and experimental saturated gains $G \cong 40$, we get $P_c \cong 1$ mW for $\Delta\phi = \pi$. Thus, the optical power that needs to be injected in one of the other ports for π-phase shift can be easily achieved. There may be some thermally induced refractive index changes, but these are considerably smaller than the carrier induced changes considered above and can be ignored. Thus, MZI devices operate on the principle of optically induced phase shift on a carrier signal. The Michelsen interferometer (MI) operates on a similar principle. Both MZI and MI devices have been used in wavelength conversion, optical demultiplexing, and photonic logic systems [8, Chapters 9 and 10].

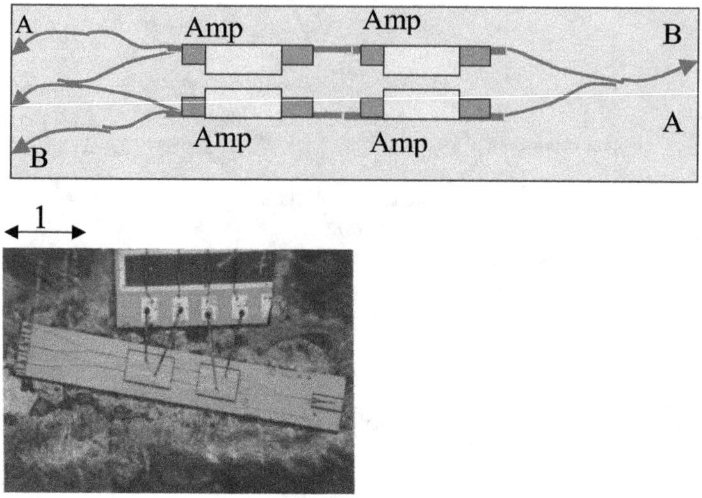

Figure 7.16: Photograph of MZI chip and a schematic of a fabricated device.

The schematic of a MZI device and bonded version of a similar device used in operation are shown in Figure 7.16. The MZI device is fabricated with tilted waveguides (to reduce residual reflections) at the facets and beam expanders at input and output for high-coupling efficiency to single mode fibers which are used to couple light to the semiconductor waveguides. The devices have long amplifiers ($\sim$2 mm long) for high gain/phase change. The applications of the SOA-based interferometer devices, such as Mach–Zhender and Michelsen interferometers are discussed in Ref. [8].

7.8 Gain dynamics in SOA

The gain and phase dynamics in an SOA are important for many SOA applications [38–59]. For example, for SOA-based Mach–Zhender interferometers, which work on the principle of gain and phase change imposed on a probe signal in the presence of a series of pulsed signals. For simplicity, the gain, and phase change of a CW probe signal in the presence of a pulsed signal is considered first. The schematic of an experiment is shown in Figure 7.17 where the wavelength of both

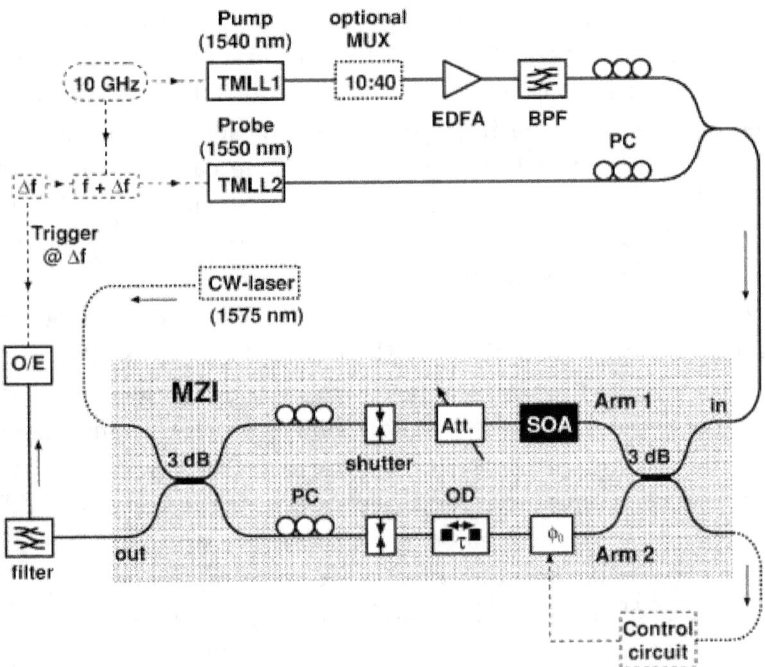

Figure 7.17: Schematic of a gain recovery (top) and phase recovery (bottom) measurement [48].

the CW probe beam and that of the pulsed beam are within the gain bandwidth of the amplifier.

The high-power pulsed signal saturates the gain of the amplifier. This results in a decrease in gain (decrease in transmission) for the CW signal following which the gain recovers due to redistribution of carriers and injection current. Thus, the gain dynamics are inferred from the measured transmission of the probe as a function of time. The phase change of the probe beam as a function of time can be obtained if the SOA is used in a Mach–Zhender configuration. Then, the output of the Mach–Zhender is determined by the phase change in the arm which has the SOA.

For SOAs with bulk or MQW active regions, the mechanisms for gain recovery are current injection, carrier heating, and spectral hole burning. The amplifier is biased with a certain CW current which provides carriers. Thus, the carriers depleted by the pulse are continuously being replaced by the current carrying carriers. The time scale of this process is given by the carrier lifetime (τ_c). The quantity τ_c decreases with increasing injected current density. For high current densities, it is in the 100–200 ps range. The carrier heating results from a thermalization of carriers in the entire energy band following the pulse. This is fast process occurring is time scale of ~1–0.7 ps. The injected pulse reduces the gain at the photon energy of this pulse, i.e. in the gain spectrum it burns a hole. The process is known as spectral hole burning. Following the pulse, the gain spectrum returns to its original shape through redistribution of carriers. The process is fast occurring in a time scale of ~0.6–0.3 ps. The phase recovery is related to the gain recovery by the α-factor (Eq. (7.29)). Separate measurements of gain and phase recovery have shown different α-factor values are associated with the three processes. For carrier injection, the value of α depends on the wavelength and it is typically in the range of $\alpha \sim 4-7$. For carrier heating, $\alpha \sim 1$ and for spectral hole burning, $\alpha = 0$. The small α-factor for fast processes suggests that phase recovery is slower than the gain recovery. In addition, in order to observe the fast gain recovery, sufficiently short (~1 ps wide) pulses must be used. The measured gain and phase dynamics of a CW probe signal following a series of pulses at 10 and 40 GHz are shown in Figure 7.18. The phase dynamics is obtained using an interferometer setup. The fast recovery observed for gain recovery is not present in the phase recovery data. Also, for higher repetition rate (40 GHz), the gain modulation and phase modulation produced by the pulses are smaller than that for low repetition rate (10 GHz) due to shorter recovery time between pulses.

The data of Figure 7.18 illustrates the presence of different response times for gain and phase recovery. The gain recovery following a short pulse takes place due to several processes. They are: current injection, carrier heating, and spectral hole burning. It also may include carrier relaxation from a higher energy state in a semiconductor layer next to active region. Since such relaxation times are fast (~1–3 ps), this layer serves as a carrier reservoir. The carrier

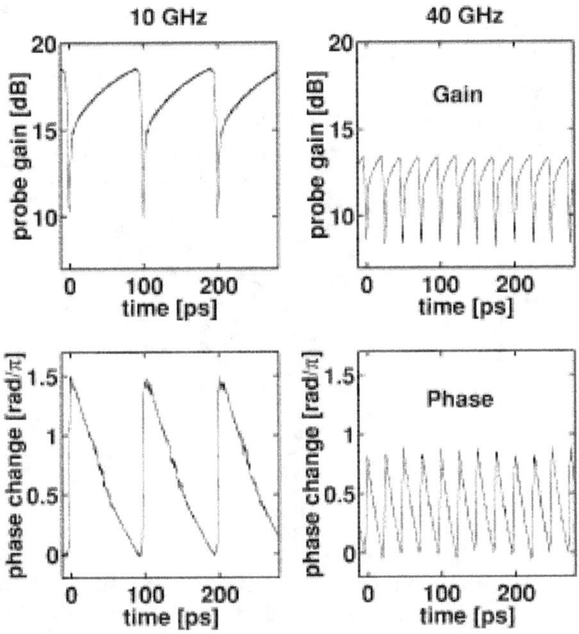

Figure 7.18: Measured gain and phase dynamics for 10 and 40 GHz repetition rate. The amplifier length is 0.75 mm. The pump pulses are 1.6 ps wide with an average power of 8.4 dBm [48].

reservoir allows a fast replacement of carriers depleted by the short pulse [59]. The gain saturation energies for different pulse widths have been measured. The result is shown in Figure 7.19. In the experiment, the optical pulses were obtained from a Ti-sapphire laser. The pulse repetition rate is 250 KHz. The gain saturation (decrease of gain with increasing optical power) takes place at lower energy for short pulses.

Fast gain recovery is important for high-speed applications of SOAs. This is generally achieved by increasing the injection current or by increasing the input optical power. Increasing injection current leads to heating effects which may not be desirable. Another technique for decreasing the gain recovery is by injection of an optical signal into the SOA [71–74]. This optical signal generally has a wavelength shorter than the system operating wavelength (e.g. 1480 nm wavelength light is injected into a 1550 nm wavelength-band SOA).

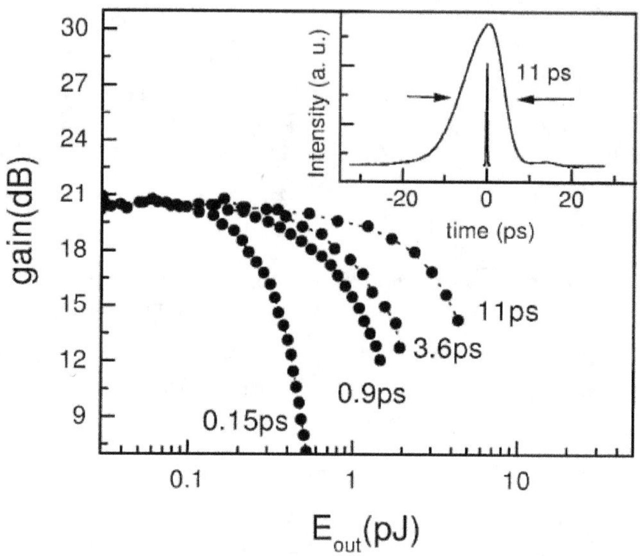

Figure 7.19: Measured gain as a function of output energy for different input pulse widths. The SOA is 0.55 mm long and the bias current is 200 mA. The inset shows the autocorrelation traces of the longest (11 ps) and the shortest (0.15 ps) pulses [50].

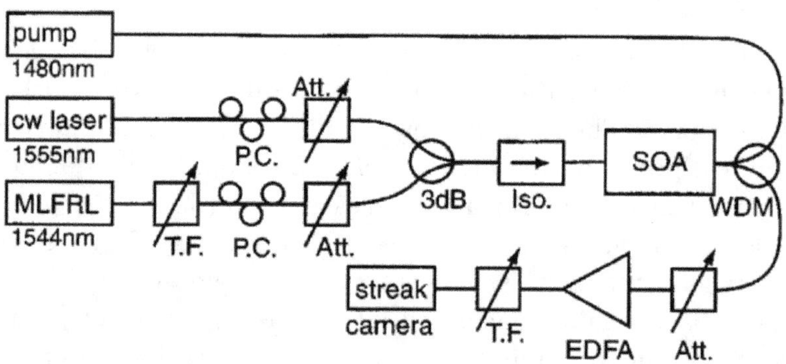

Figure 7.20: Experimental setup used for gain recovery measurements [51].

The absorbed optical signal creates carriers in the conduction and valence band. These carriers relax to the bottom of the band to replace the carriers depleted by the input pulse. The experiment setup used for the measurement is shown in Figure 7.20. The CW laser light on which the gain recovery is measured has a wavelength of 1555 nm. The wavelength of the CW pump light used for gain recovery is 1480 nm and the mode-locked fiber ring laser (MLFRL) which depletes the carriers operates at a wavelength of 1544 nm. The laser produces 3 ps wide pulses at 2.5 GHz repetition rate. An optical filter is used prior to streak camera to detect the 1555 nm signal.

The streak camera measurements of the 1555 nm light as a function of time is shown in Figure 7.21 (top traces). The bottom part of Figure 7.21 shows the 3-dB gain recovery times plotted as a function of the power of the 1480 nm wavelength light. Gain recovery times ~27 ps are obtained with 80 mW input power. This method of decreasing the gain recovery time using a holding beam has been used in several transmission system demonstrations. The long gain recovery times cause pattern effects when an SOA is used in an optical system for wavelength conversion, i.e. response of a single pulse "1" is different from that of a long string of 1's or 0's. One way to examine this is by plotting the eye pattern. Experiments have been carried out where an 80 Gb/s pulse train is amplified using an SOA with or without holding beam (also known as assist light). The amplified output of 80 Gb/s pattern @ 1555 nm wavelength is found to be much clearer with the holding beam (10 mW @ 1480 nm) than in its absence.

7.9 Quantum dot-SOA

The quantum dot semiconductor optical amplifiers (QD-SOA) have some advantages over conventional bulk or quantum well devices because of higher saturation power, faster gain recovery, and large amplification bandwidth [60–69]. The QD material for SOA is similar to that for a laser and has been discussed in Chapter 4. The QD active region (i.e. QDs and the embedding layers) for SOA are often surrounded by a set of separate confinement layers which have n-type and p-type cladding layers on either side similar to that for a regular laser (Chapter 4). The lateral confinement in this heterostructure could be either strongly index guided (buried heterostructure) or weakly index guided (e.g. ridge waveguide type), i.e. the lateral

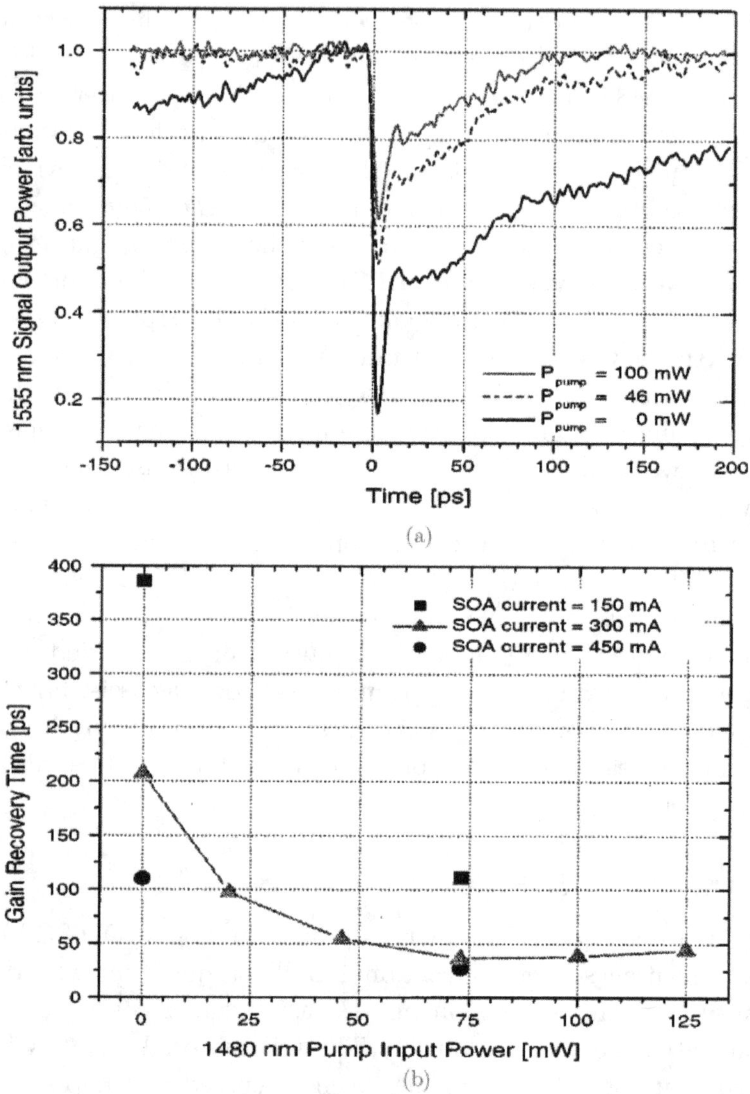

Figure 7.21: (a) Transmitted intensity of 1555 nm signal as a function of time for various levels of 1480 nm pump. (b) 3-dB gain recovery time plotted as a function of power of 1480 nm pump [51].

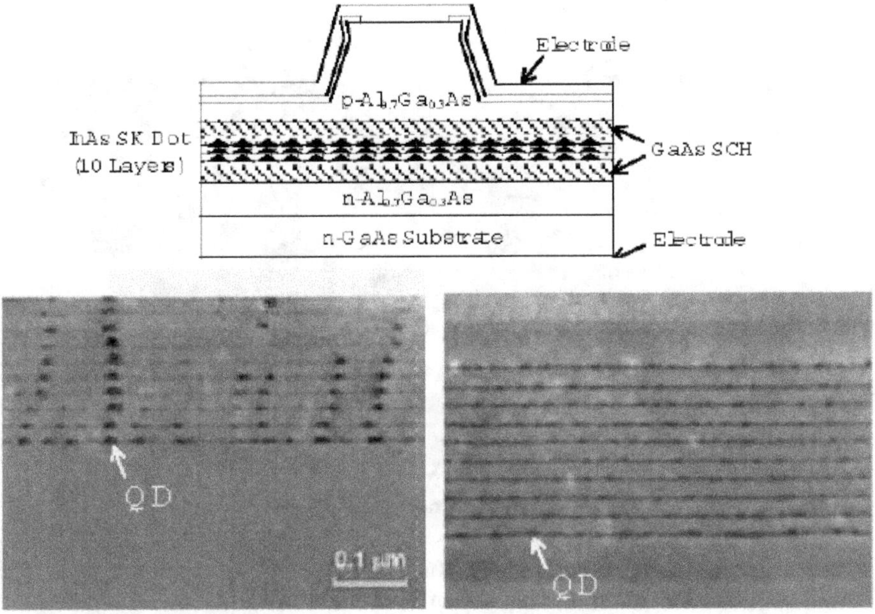

Figure 7.22: Structure of ridge waveguide type quantum dot (QD) SOA. GaAs separate confinement heterostructure (SCH) is also shown (a) aligned QDs (b) non-aligned QDs [66].

index and current confinement follows the same methods as described in Chapters 3 and 4.

The structure of a QD-SOA along with the scanning electron photomicrographs of the QD active region are shown in Figure 7.22 [66]. The QD layers consist of InAs QDs (grown by molecular beam epitaxy) embedded in $In_{0.17}Ga_{0.83}As$. The QD material in this case has a GS emission peak of ∼1300 nm. The embedding layer affects the emission wavelength [67]. In order to obtain sufficient gain from the QDs, the QD layers are generally repeated a few times (10 in Figure 7.22). Generally, close stacking of QD layers causes vertical lining of dots (Figure 7.22(a)), which leads to strain accumulation. The researchers in Refs. [66–68] developed a technique to eliminate lining up (Figure 7.22(b)) and hence strain accumulation. This leads to improved crystal quality. This material was processed into a ridge waveguide structure 10 μm wide and 2.5 mm long. The

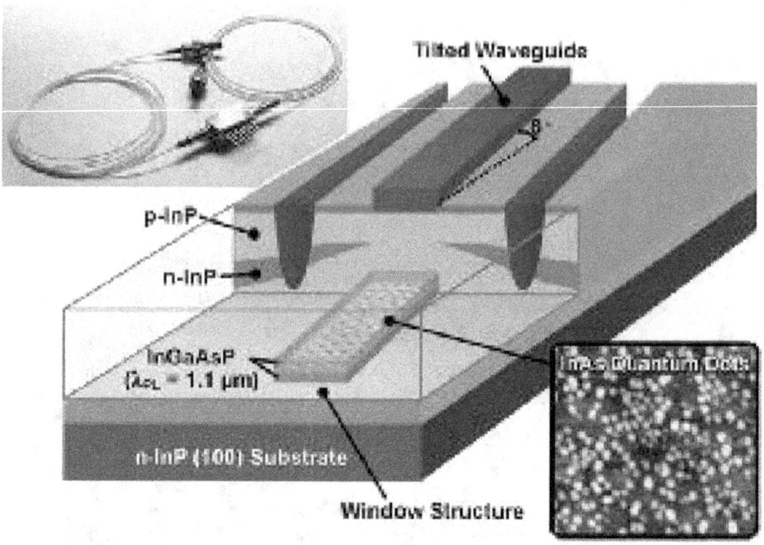

Figure 7.23: Structure of an InAs/InGaAsP/InP SOA fabricated on an InP substrate [67]. The inset is a picture of fiber-pigtailed butterfly module (which has a QD-SOA) with temperature controller.

facet reflectivity was reduced by applying AR coating. The entire structure is grown on an n-GaAs substrate.

The structure of an index guided QD amplifier is shown in Figure 7.23. In this structure, the InAs QDs are buried in InGaAsP. The final cladding layers are p-InP. The structure is laterally buried in a current confining (InP-based) structure and the entire set of layers are grown on an n-InP substrate. The QD waveguide is tilted with respect to the facet to reduce facet reflection, and the facets are further coated with AR films.

A modification of the QD-SOA is a columnar QD-SOA (CQD-SOA) [71]. A tensile strained barrier layer is used to control the position of the QDs in this case. The structure has 22 stacks of QD layers, each sandwiched between 160 nm thick tensile strained InGaAsP separate confinement heterostructure layers, as shown in Figure 7.24. This type of dot configuration can significantly increase the modal gain of the SOA and has polarization-independent gain [71]. The planar density of the dots is 9×10^{10} cm^{-2}, the height of the

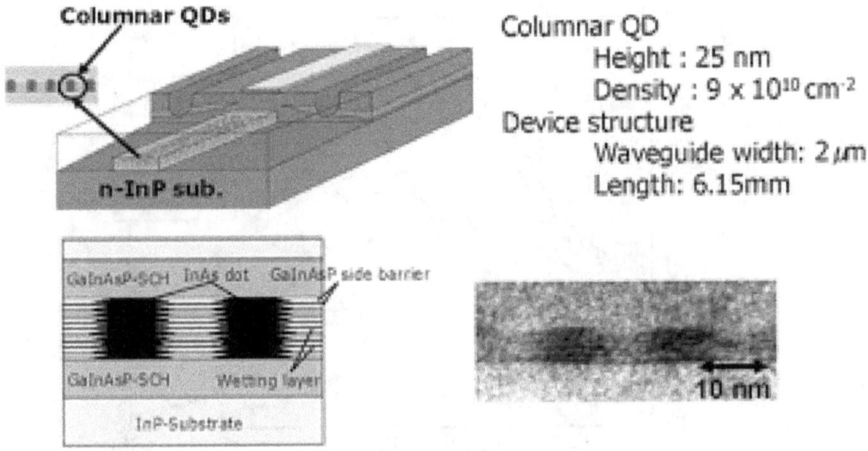

Figure 7.24: Structure of columnar quantum dot (CQD) with a number of layers, *Notes*: SCH: separate confinement heterostructure [71].

dots is 25 nm, and the total active region thickness is about 350 nm. This QD amplifier operates near 1500 nm. The CQD-SOA, shown in Figure 7.24 is a 6.15-mm-long device with an 8° tilted waveguide (to reduce facet reflection) and is AR-coated.

7.9.1 *Quantum dot amplifier performance*

The QDs have a discrete set of energy levels in the conduction band and valence band. Depending on the size of the dots and the band gap of the wetting layer (the layer in which the dots are embedded), one or two of these discrete energy levels are distinct, i.e. separate from the wetting layer. The lowest energy level is called ground state (GS) here and the higher energy level is called excited state (ES) level here. Optical transitions take place from both of these levels. The higher energy levels merge with the wetting layer. These differences are evident in the ASE spectrum of the QD-SOA.

The ASE from an InAs/InGaAs/InP QD-SOA is shown in Figure 7.25 [71].

At low currents, the conduction band GS of the QD is populated, and the spectrum peak is near 1540 nm. The broadening of the spectrum is believed to be due to distribution in the dimension of the

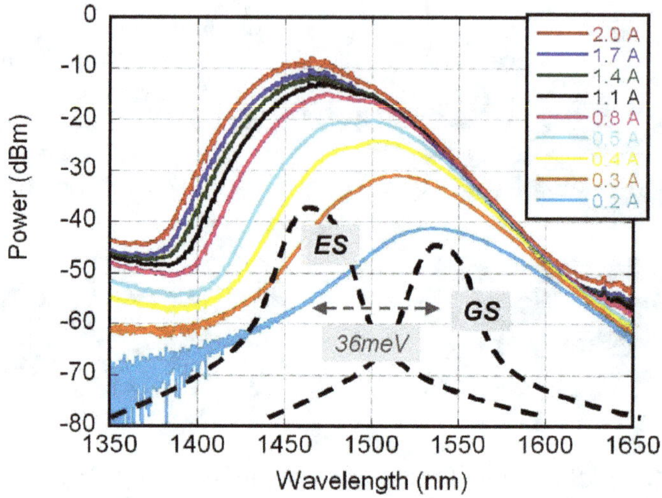

Figure 7.25: The amplified spontaneous emission (ASE) spectra of a quantum dot (QD) amplifier at different bias current. Also, shown in the figure are the extrapolated QD ground and excited state emission profiles. This spectrum is for the device shown in Figure 7.24 [71].

dots, i.e. not all dots are of the same size. At high currents, a second peak in the spectrum (near 1470 nm) emerges due to transitions from the ES in the conduction band to the valence band. In Figure 7.25, the extrapolated QD ES and GS ASE profiles are also shown. The ~70 nm difference between the peaks of the spectrum (ES and GS transitions) corresponds to an energy spacing of ~36 meV, which agrees well with the calculation using the dot dimensions. The measured gain as a function of wavelength of the QD-SOA in Figure 7.24 is shown in Figure 7.26. Note that a fiber-to-fiber gain of >30 dB is feasible. The 3-dB gain saturation power is ~17 dB.

The QD-SOA structure shown in Figure 7.23 has been characterized. The performance shown (Figure 7.27) was measured with a device having a tilted waveguide (8° off angle) and window structure to suppress lasing action. The waveguide length, stripe width, number of QD layers, and density of QDs of the QD-SOA are 6.15 mm, 2.2 μm, 5, and 4×10^{10} cm^{-2}, respectively [70].

Nonlinear effects in SOAs such as four-wave mixing and cross-gain modulation are described in Appendix 13.

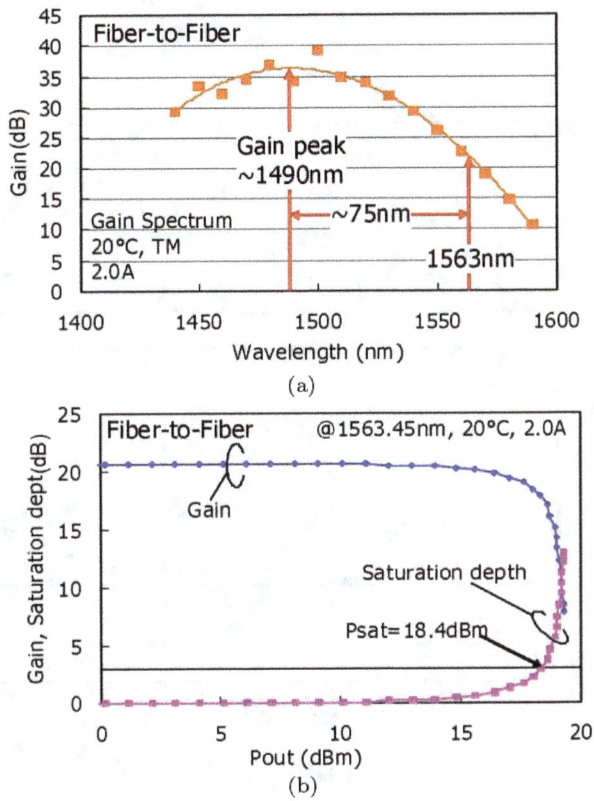

Figure 7.26: (a) The measured gain spectra of a quantum dot (QD) amplifier at 20°C. The device was put in a packaged module. (b) The measured gain as a function of output power. This data is for the device shown in Figure 7.24 [71].

7.10 Broadband SOA source

SOA has a broad emission spectrum. The emission arises from the ASE in the active region of the device [72–84]. The width of the emission spectrum is typically 40–60 nm and can be larger if ES emission in QD material is considered. This broad emission spectrum can be made of higher power using a reflective SOA (RSOA) type structure or a superluminescent diode (SLD) type structure. The conventional TW SOA, the reflective SOA, and the SLD have essentially similar active region structure, they differ in facet reflectivity (facet coating used) and may differ in transverse mode confinement depending on

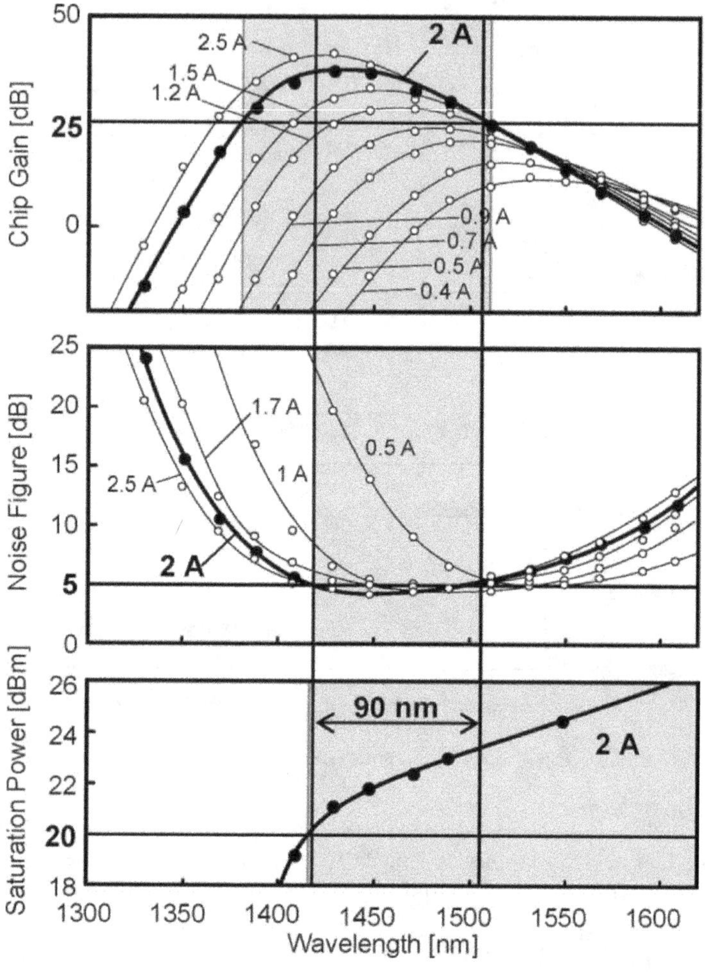

Figure 7.27: Wavelength dependence gain, noise figure, and saturation output power is shown for the QD-SOA structure of Figure 7.23 [70].

the application of the device. A high power broadband SOA type source is generally called a superluminescent diode (SLD). The SLD device structure is optimized for high power and broad spectral width and has been fabricated using GaAs/AlGaAs, GaAs/InGaAs, and InP/InGaAsP material systems. These systems generally produce SLDs in the wavelength ranges of 800, 1000, and 1300–1600 nm,

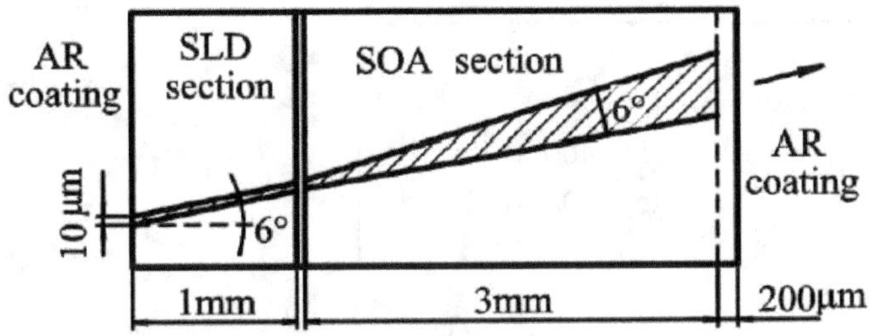

Figure 7.28: Schematic of the integrated SLD and SOA structure [84].

respectively. The output power of the SLD (with a zero reflectivity at the output facet) is approximately given by [73]

$$P = R_{\text{sp}}[(e^{(g-\alpha)L} - 1)/(g - \alpha)] \tag{7.54}$$

where R_{sp} is the spontaneous emission per unit length into the guided mode, g is the optical gain, α is the optical absorption, and L is the length of the device. Equation (7.54) is obtained by integrating the net gain $(g - \alpha)$ over the length of the device. Since the optical gain increases with increasing current, the output of the SLD increases rapidly with increasing current until it saturates due to heating and other effects such as leakage current in the device (Chapter 3). For high power output and broad emission spectrum, a QD-SLD monolithically integrated with an SOA has been reported. The active region consists of 10 InAs-QD layers separated from each other by GaAs spacer layers. The QDs are formed by depositing a monolayer InAs at 480°C and covering by a 2-nm thick $In_{0.15}Ga_{0.85}As$. 10 QD layers plus the GaAs waveguide layers form the whole active region which is sandwiched between n-type and p-type $Al_{0.5}Ga_{0.5}As$ cladding layers followed by a p-GaAs contact layer. The entire structure is grown over an n-GaAs substrate. The wafer was processed into 10 μm wide ridge waveguide devices in a form where the SLD is monolithically integrated with the SOA (Figure 7.28) [84]. The SLD is 1 mm long and the SOA is 3 mm long. The light vs. SOA current and spectrum of the device under different conditions are shown in Figure 7.29.

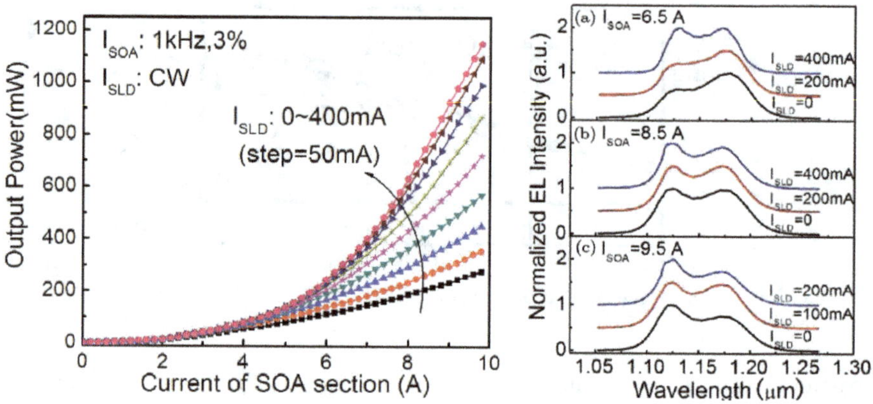

Figure 7.29: The measured light as a function of SOA current and spectrum at various currents are shown [84].

Many of the advances in the SOA development and functional applications (such as wavelength conversion for WDM systems, optical demultiplexing, and optical logic) would not have been possible without the advances in materials and processing technology. The challenges of much of the current SOA research is intimately linked with the challenges in materials growth, which include not only the investigation of new material systems, but also improvements in existing technologies, such as quantum dots, to make them more reproducible and predictable.

References

1. M. J. Coupland, K. G. Mambleton, and C. Hilsum, *Phys. Lett.*, **7** (1963) 231.
2. J. W. Crowe and R. M. Craig, Jr., *Appl. Phys. Lett.*, **4** (1964) 57.
3. W. F. Kosnocky and R. H. Cornely, *IEEE J. Quant. Electron.*, **QE-4** (1968) 225.
4. T. Saitoh and T. Mukai, Chapter 7, in Y. Yamamoto (ed.), Coherence, *Amplification and Quantum Effects in Semiconductor Lasers*, Wiley, New York, 1991.
5. M. Nakamura and S. Tsuji, *IEEE J. Quant. Electron.*, **QE-17** (1981) 994.
6. T. Saitoh and T. Mukai, J. *Lightwave Technol.*, **6** (1988) 1656.
7. M. O'Mahony, *J. Lightwave Technol.*, **5** (1988) 531.
8. N. K. Dutta, *Semiconductor Optical Amplifier*, World Scientific, Singapore, 2012.

9. H. Kressel and J. K. Butler, *Semiconductor lasers and Heterojunction LEDs*, Academic Press, New York, 1977.
10. H. C. Casey, Jr. and M. B. Panish, *Heterostructure Lasers*, Academic Press, New York, 1978.
11. G. P. Agrawal and N. K. Dutta, *Semiconductor Lasers*, Van Nostrand Reinhold, NY, 1986, 2nd Edn., 1992.
12. L. A. Coldren and S. W. Corzine, *Diode Lasers and Photonic Integrated Circuits*, John Wiley, NY, 1995.
13. G. H. B. Thompson, *Physics of Semiconductor Laser Devices,* John Wiley & Sons, New York, 1980.
14. G. P. Agrawal and N. K. Dutta, *Semiconductor Lasers*, Section 2.5, Van Nostrand Reinhold, NY, 1992.
15. P. Zory (Ed.), *Quantum Well Lasers*, John Wiley, New York, 1992.
16. K. Magari, M. Okamoto, and Y. Noguchi, *IEEE Photon. Tech. Letts.*, **3** (1991) 998.
17. M. Joma, H. Horokawa, C. Q. Xu, K. Yamada, Y. Katoh, and T. Kamijoh, *Appl. Phys. Lett.*, **62** (1993) 121.
18. P. J. A. Thijs, L. F. Tiemeijer, J. J. M. Binsma, and T. van Dongen, *IEEE J. Quant. Electron.*, **QE-90** (1994) 477.
19. L. F. Tiemeijer, P. J. A. Thijs, T. van Dongen, R. W. M. Slootweg, J. M. M. van der Heijden, J. J. M. Binsma, and M. P. C. M. Krijn, *Appl. Phys. Lett.*, **62** (1993) 826.
20. G. P. Agrawal and N. A. Olsson, *Opt. Lett.*, **14** (1989) 500.
21. N. A. Olsson and G. P. Agrawal, *Appl. Phys. Lett.*, **55** (1989) 13.
22. N. A. Olsson, G. P. Agrawal, and K. W. Wecht, *Electron. Lett.*, **25** (1989) 603.
23. G. P. Agrawal and N. A. Olsson, *IEEE J. Quant. Electron.*, **QE-25** (1989) 2297.
24. I. W. Marshall, D. M. Spirit, and M. J. O'Mahony, *Electron. Lett.*, **23** (1987) 818.
25. J. M. Wiesenfeld, G. Eisenstein, R. S. Tucker, G. Raybon, and P. B. Hansen, *Appl. Phys. Lett.*, **53** (1988) 1239.
26. T. Saitoh, H. Itoh, Y. Noguchi, S. Sudo, and T. Mukai, *IEEE Photon. Technol. Lett.*, **1** (1989) 297.
27. G. P. Agrawal, *Electron. Lett.*, **27** (1991) 620; *IEEE J. Quant. Electron.*, **QE-27** (1991) 1843.
28. G. P. Agrawal and N. K. Dutta, *Semiconductor Lasers*, Chapter 11, van Nostrand Reinhold, 1992.
29. G. Grosskopf, R. Ludwig, and H. G. Weber, *Electron. Lett.*, **22** (1986) 900.
30. T. Mukai, K. Inoue, and T. Saitoh. *Electron. Lett.*, **23** (1987) 396.
31. R. M. Jopson, K. L. Hall, G. Eisenstein, G. Raybon, and M. S. Whalen, *Electron. Lett.*, **25** (1987) 510.
32. G. P. Agrawal, *Electron. Lett.*, **23** (1987) 1175.
33. K. Inoue, *Electron. Lett.*, **23** (1987) 1293.

34. R. M. Jopson, T. E. Darcie, K. T. Gayliard, R. T. Ku, R. E. Tench, T. C. Rice, and N. A. Olsson, *Electron. Lett.*, **23** (1987) 1394.
35. I. M. I. Habbab and G. P. Agrawal, *J. Lightwave Technol.*, **LT-7** (1989) 1351.
36. S. Ryu, K. Mochizuki, and H. Wakabayashi, *J. Lightwave Technol.*, **LT-7** (1989) 1525.
37. G. P. Agrawal and I. M. I. Habbab, *IEEE J. Quant. Electron.*, **QE-26** (1990) 501.
38. K. L. Hall and K. A. Rauschenbach, *Electron. Lett.*, **32** (1996) 1214.
39. A. J. Poustie, K. J. Blow, R. J. Manning, and A. E. Kelly, *Opt. Commun.*, **159** (1999) 208.
40. T. Fjelde, A. Kloch, D. Wolfson, B. Dagens, A. Coquelin, I. Guillemot, F. Gaborit, F. Poingt, and M. Renaud, *IEEE Photon. Technol. Lett.*, **13** (2001) 750.
41. N. S. Patel, K. L. Hall, and K. A. Rauschenbach, *Appl. Opt.*, **37** (1998) 2831.
42. M. Jinno and T. Matsumoto, *Opt. Lett.*, **16** (1991) 220.
43. T. Houbavlis, K. Zoiros, A. Hatziefremidis, H. Avramopoulos, L. Occhi, G. Guekos, S. Hansmann, H. Burkhard, and R. Dall'Ara, *Electron. Lett.*, **35** (1999) 1650.
44. C. Bintjas, M. Kalyvas, G. Theophilopoulos, T. Stathopoulos, H. Avramopoulos, L. Occhi, L. Schares, G. Guekos, S. Hansmann, and R. Dall'Ara, *IEEE Photon. Technol. Lett.*, **12** (2000) 834.
45. T. Fjelde, D. Wolfson, A. Kloch, B. Dagens, A. Coquelin, I. Guillemot, F. Gaborit, F. Poingt, and M. Renaud, *Electron. Lett.*, **36** (2000) 1863.
46. Q. Wang, G. Zhu, H. Chen, J. Jaques, J. Leuthold, A. B. Piccirilli, and N. K. Dutta, *IEEE J. Quant. Electron.*, **40** (2004) 703.
47. H. Chen, G. Zhu, J. Jaques, J. Leuthold, A. B. Piccirilli, and N. K. Dutta, *Electron. Lett.*, **38** (2002) 1271.
48. L. Schares, C. Shubert, C. Schmidt, H. G. Weber, L. Occhi, and G. Guekos, *IEEE, J. Quant. Electron.*, **39** (2003) 1394.
49. R. Gutierrez-Castrejon, L. Occhi, L. Schares, and G. Guekos, *Opt. Commun.*, **195** (2001) 167.
50. P. Borri, S. Scaffetti, J. Mork, W. Langbein, J. M. Hvam, A. Mecozzi, and F. Martelli, *Opt. Comm.*, **164** (1999) 51.
51. J. Pleumeekers, M. Kauer, K. Dreyer, C. Burrus, A. G. Dentai, S. Shunk, J. Leuthold, and C. H. Joyner, *IEEE Photon. Tech. Lett.*, **14**, (2002) 12.
52. M. Usami, T. Tsurusawa, and Y. Matsushima, *Appl. Phys. Lett.*, **72** (1998) 2657.
53. M. A. Dupertuis, J. L. Pleumeekers, T. P. Hessler, P. E. Selbmann, B. Deveaud, B. Dagens, and J. Y. Emery, *IEEE Photon. Tech. Lett.*, **12** (2000) 1453.
54. R. J. Manning and D. A. O. Davies, *Opt. Lett.*, **19** (1994) 889.
55. J. Mark and J. Mørk, *Appl. Phys. Lett.*, **61** (1992) 2281.
56. A. Mecozzi and J. Mørk, *IEEE J. Sel. Top. Quant. Electron.*, **3** (1997) 1190.

57. A. Mecozzi and J. Mork, *J. Opt. Soc. Am. B*, **14** (1997) 761.
58. J. M. Tang and K. A. Shore, *IEEE, J. Quant. Electron.*, **35** (1999) 1704.
59. G. Eisenstein, J. M. Wiesenfeld, M. Wegener, G. Sucha, D. S. Chemla, S. Weiss, G. Raybon, and U. Koren, *Appl. Phys. Lett.*, **58** (1991) 158.
60. T. Akiyama, H. Kuwatsuka, T. Simoyama, Y. Nakata, K. Mukai, M. Sugawara, O. Wada, and H. Ishikawa, *J. Quantum. Electron.*, **37** (2001) 1059.
61. P. Borri, W. Langbein, J. Hvam, F. Heinrichsdorff, M-. Mao, and D. Bimberg, *Phys. Stat. Sol.*, **224** (2001) 419.
62. P. Borri, W. Langbein, J. M. Hvam, F. Heinrichsdorff, M.-H. Mao, and D. Bimberg, *IEEE Photon. Technol. Lett.*, **12** (2000) 594.
63. M. Sugawara, N. Hatori, T. Akiyama, Y. Nakata, and H. Ishikawa, *Jpn. J. Appl. Phys.*, **40** (2001) L488.
64. M. Sugawara, N. Hatori, T. Akiyama, and Y. Nakata, *Proc. Indium Phos. Rel. Mat.*, **01** (2001) 358.
65. A. Sakamoto and M. Sugawara, *Photon. Tech. Lett.*, **12** (2000) 107.
66. M. Sugawara, T. Akiyama, N. Hatori, Y. Nakata, Y. Otsubo, and H. Ebe, *Proc. SPIE*, **4905** (2002).
67. M. Sugawara, H. Ebe, N. Hatori, M. Ishida, Y. Arakawa, T. Akiyama, K. Otsubo, and Y. Nakata, *Phys. Rev., B.* **69** (2004) Art. No. 235332.
68. M. Sugawara, T. Akiyama, N. Hatori, Y. Nakata, H. Ebe, and H. Ishikawa, *Meas. Sci. Technol.*, **13** (2002) 1683.
69. A. Sakamoto and M. Sugawara, *IEEE Photon.Technol. Lett.*, **12** 107 (2000) 107.
70. T. Akiyama, M. Sugawara, and Y. Arakawa, *Proc. IEEE*, **95** (2007) 1757.
71. G. Contestabile, A. Maruta, and K. Kitayama, *IEEE Photonic Tech. Lett.*, **22** (2010) 987.
72. G. A. Alphonse, D. B. Gilbert, M. G. Harvey, and M. Ettenberg, *IEEE J. Quant. Electron.*, **24** (1988) 2454.
73. N. K. Dutta and P. P. Deimel, *IEEE J. Quantum Electron.*, **QE-19** (1983) 496.
74. N. S. K. Kwong, K. Y. Lau, and N. Bar-Chaim, *IEEE J. Quant. Electron.*, **25**, (1989) 696.
75. K. Tateoka, H. Naito, M. Yuri, M. Kume, K. Hamada, H. Shimizu, M. Kazumura, and I. Teramoto, *IEEE J. Quant. Electron.*, **27** (1991) 1568.
76. L. Goldberg and D. Mehuys, *Electron. Lett.*, **30** (1994) 1682.
77. O. Imafuji, T. Takayama, H. Sugiura, M. Yuri, H. Naito, M. Kume, and K. Itoh, *IEEE J. Quant. Electron.*, **29** (1993) 1889.
78. T. Takayama, O. Imafuji, H. Sugiura, M. Yuri, H. Naito, M. Kume, and K. Itoh, *Jpn. J. Appl. Phys.*, **34** (1995) 3533.
79. T. Takayama, O. Imafuji, M. Yuri, H. Naito, M. Kume, A. Yoshikawa, and K. Itoh, *IEEE J. Select. Topics Quant. Electron.*, **1** (1995) 562.
80. T. Takayama, O. Imafuji, Y. Kouchi, M. Yuri, A. Yoshikawa, and K. Itoh, *IEEE J. Quant. Electron.*, **32** (1996) 1981.

81. N. Liu, P. Jin, and Z. G. Wang, *Electron. Lett.*, **41** (2005) 25.
82. E. V. Andreeva, A. E. Zhukov, V. V. Prokhorov, V. M. Ustinov, and S. D. Yakubovich, *Quant. Electron.*, **36** (2006) 527.
83. T. K. Ong, M. Yin, Z. Yu, Y. Chuen Chan, and Y. L. Lam, *Meas. Sci. Tech.*, **15**, (2004) 1591.
84. X. Li, P. Jin, Q. An, Z. Wang, X. Lv, H. Wei, J.Wu, J. Wu, and Z. Wang, *Nano Res. Lett.*, **6** (2011) 625.

Chapter 8

Photonic Integrated Circuits

8.1 Introduction

There have been a significant number of developments in the technology of integration of semiconductor lasers, amplifiers, modulators, and other related devices on the same chip. These chips allow higher levels of functionality than those achieved using single devices. The name photonic integrated circuits (PIC) are generally used when all the integrated components are photonic devices, e.g. lasers, detectors, amplifiers, modulators, and couplers. Several review articles and books have been published on PICs [1–6]. The classes of PICs which use amplifiers are being investigated for photonic logic systems. The design and performance of these PICs are described in Ref. [7].

Many PICs use SOAs integrated with another photonic device such as a laser. Since the light from the laser is TE polarized, the parameter of importance for these SOAs is high TE mode gain and high saturation power. Since multiquantum well (MQW) materials generally have high TE gain than regular double heterostructure materials, MQW active regions are generally used for amplifiers which are integrated in a photonic circuit. The saturation output power, i.e. the power at which the gain decreases by 3 dB from its value at low power is higher for MQW amplifiers. Saturation power of as high as 200 mW has been reported for MQW amplifiers.

An important epitaxial growth technique for PIC fabrication is selective area growth because this allows interconnections with low optical loss.

8.2 Selective area growth

During the fabrication of photonic integrated circuits, it is important to fabricate materials of different band gap on the same plane. This is important for building low-loss interconnection between active/active and active/passive regions. Such interconnections are needed for integrated chips consisting of many optical devices, such as lasers, amplifiers, and photodetectors. The low-pressure metal organic chemical vapor deposition (MOCVD)-based selective area growth (SAG) process [8–10] is used for making the interconnections. The key reason for using the SAG process is that it produces a smooth transition between materials of different composition. The SAG process development has been carried out for the InGaAsP/InP material system.

The fundamental concept in SAG is the following: Consider an InP substrate with prefabricated oxide (SiO_2)-based masks deposited on it (Figure 8.1). When such a substrate is put inside a MOCVD growth chamber, the growth precursors trimethylindium (TMIn),

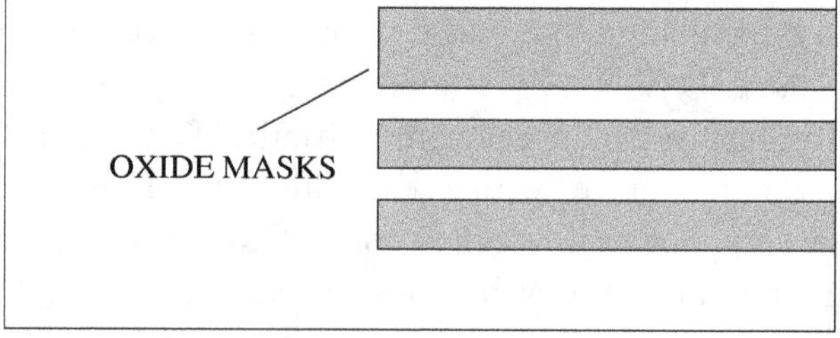

Figure 8.1: Schematic of a wafer with oxide masks. The shaded regions represent masks of different widths.

trimethylgallium (TMGa), arsenic (As), and phosphorus (P) are incident on the wafer.

However, they do not react to form the compounds on the oxide and hence must diffuse to the exposed regions for growth. Thus, the exposed regions in between the oxide masks have more materials compared to regions further away from the oxide masks. This results in variation in composition and thickness of the material deposited near the oxide masks relative to those far from the masks. By a suitable design of the oxide masks it is then possible to deposit films of desired thickness, composition, and strain selectively on different parts of the wafer. These films could then be active regions of different devices (e.g. laser, amplifiers, and modulator) or they could form waveguides for interconnection. For a laser-amplifier integrated device, the active region of the laser and the amplifier are grown in the same epitaxial growth, but with the amplifier region grown between oxide masks so that the composition of the amplifier active region has lower band gap compared to that of the laser region. This is important since the amplifier gain shifts to shorter wavelength with increasing current. In addition, the laser area can be grown over a grating so that it emits at a specific wavelength determined by the periodicity of the grating. The advantage of the SAG process is that both the amplifier and laser active regions are vertically self-aligned. A similar SAG process is important for laser-modulator [electroabsorption (EA)] integrated devices (see Section 5.2.1.1 of Chapter 5).

A set of SAG-based devices and corresponding masks are shown schematically in Figure 8.2. Figure 8.2(a) shows a laser-amplifier integrated device. Figure 8.2(d) shows a type of mask needed for the fabrication of an integrated laser and amplifier. Figure 8.2(b) shows the case of an expanded beam amplifier for low coupling losses to fibers. In this case, a waveguide is integrated with the amplifier output and input so that the mode at the facet is expanded for better coupling to optical fibers. For this adiabatic spot size conversion (expansion) to take place the waveguide must change its thickness. Such a vertically tapered waveguide could be formed using a set of oxide masks whose separation varies with the distance. Figure 8.2(e) shows such a mask.

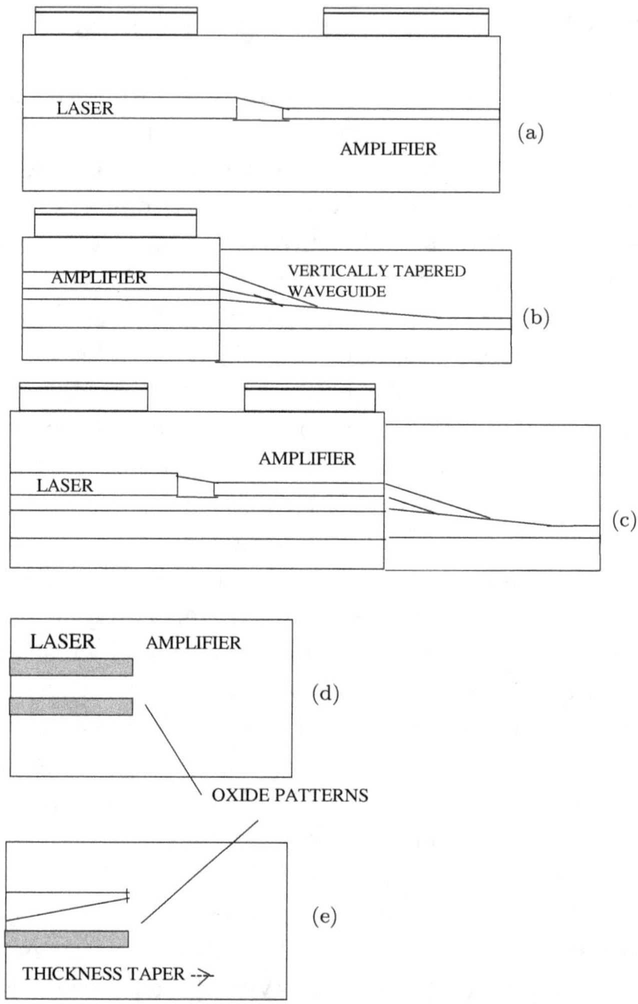

Figure 8.2: (a) Laser-amplifier integrated chip, (b) expanded beam amplifier for low coupling losses, (c) expanded beam laser–amplifier, (d) oxide mask for laser-amplifier, laser active region is fabricated between masks, and (e) oxide mask for expanded beam amplifier.

The tapering in the lateral direction could be carried by standard photolithography and etching techniques. Figure 8.2(c) shows the case of an expanded beam laser-amplifier integrated device with an expanded beam region.

8.2.1 *Model of SAG*

High-performance integrated devices require precise control of composition and layer thickness. A model of the SAG process would be very useful in the design of oxide masks needed for selective growth in various parts of the wafer. Such a model should be able to predict the composition, thickness, and strain of the grown layers for different oxide pattern configurations.

A modeling of the selective area growth process by solving the diffusion equation in conjunction with the reaction rate equation for various reacting species (TMIn, TMGa, As, and P) has been developed [9–10]. An iterative procedure using both analysis and experimental results is employed to determine the optimum conditions for materials growth that are applicable to various devices. The modeling is important for the design of masks for specific device fabrication. Figure 8.3 shows an example of the modeling results.

8.2.2 *Materials growth using SAG*

Semiconductor layer growth using the SAG process in an MOCVD reactor has been carried out for various types of integrated device fabrication. For growth studies, a series of MQW layers with fixed lattice constant is grown over InP substrates with different SiO_2 mask patterns. The electroluminescence (EL) wavelength of the material grown between the masks is measured. Figure 8.4 shows the measured EL peak wavelength as a function of the width of the dielectric SiO_2 mask. By simply varying the width, the EL wavelength can be changed considerably. For the data shown in Figure 8.4, the epitaxial growth was carried out at 650°C and at a reactor pressure of 80 mbar. TMIn, TMGa, AsH_3, and PH_3 were used as precursors.

Although the SAG process was initially developed for EA modulated laser (EML) device fabrication (Section 5.2.1.1 of Chapter 5), it is very important for amplifiers with low coupling loss or in general for other amplifier-based devices such as Mach–Zehnder modulators that also need to have low coupling losses. A major importance of SAG process for SOA-based devices is for the fabrication of mode-matched beam modifiers.

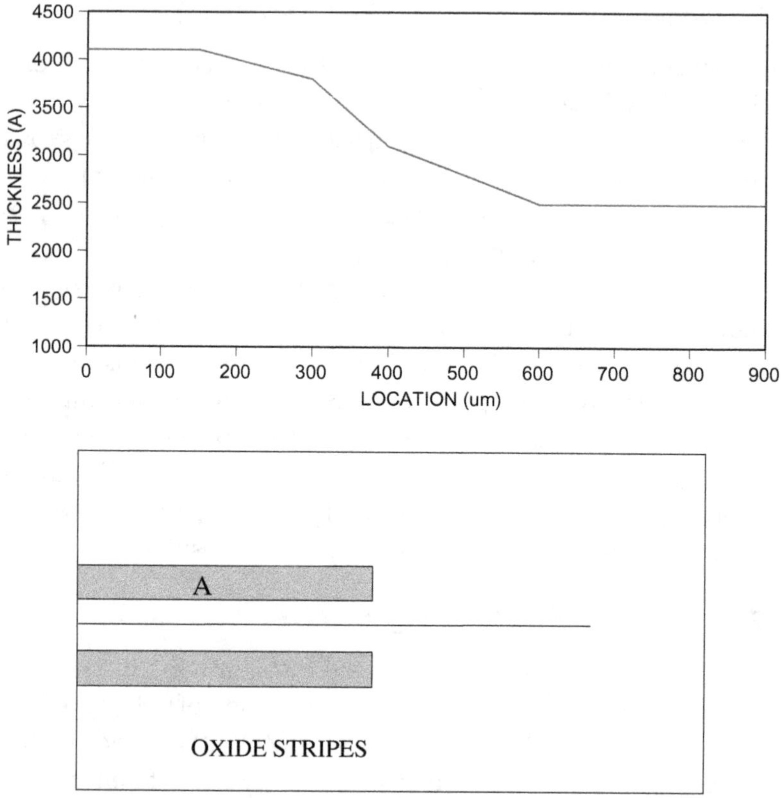

Figure 8.3: Calculated variation in thickness of the grown layer as a function of distance from the pattern. The lower figure shows the geometry used in the calculation. The location refers to position along the line A. A location of ∼400 μm corresponds to the place where the masks end. The oxide masks are separated by 10 μm and have a width of 10 μm. Experimental results on variation of peak wavelength are shown in Figure 5.2.7 for a different configuration of masks.

8.3 Integrated laser and amplifier

Since MQW amplifier is ideally suited for amplifying the output power of a semiconductor laser, it is useful to combine the amplifier and laser on a single chip and thereby produce a compact device with low coupling losses. MQW amplifiers have been integrated with both distributed feedback (DFB) and distributed Bragg reflector (DBR) lasers. The schematic of such an integrated device with an MQW

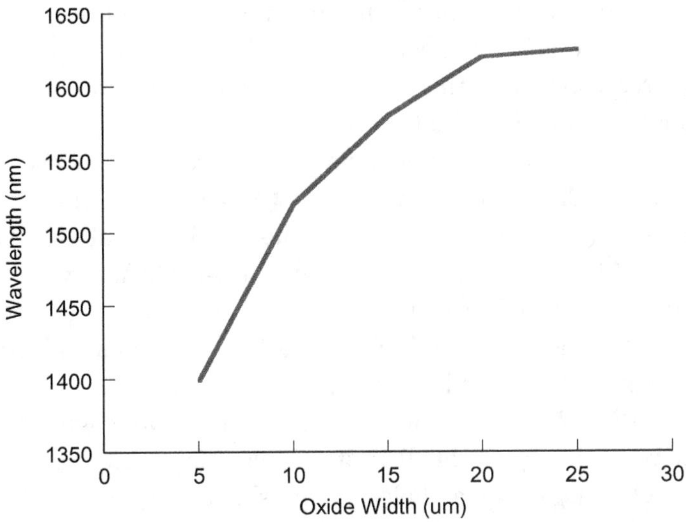

Figure 8.4: Variation of electroluminescence peak wavelength as a function of the width of the oxide masks. The separation between the masks is 10 μm. Experimental results on variation of peak wavelength for a different configuration are also shown in Figure 5.7.

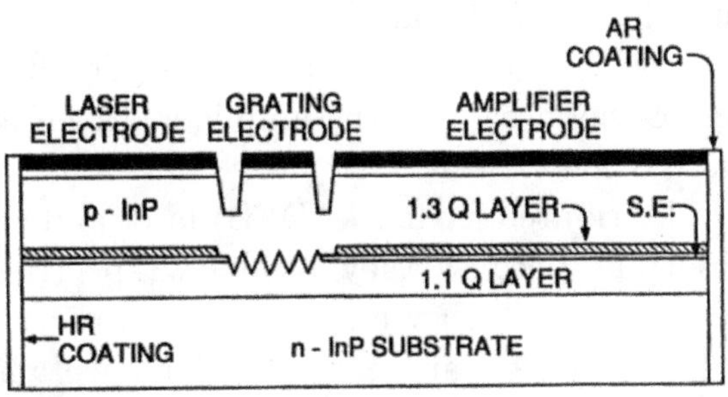

Figure 8.5: Schematic of a photoni58c integrated circuit with a DBR (distributed Bragg reflector) laser and amplifier [11].

amplifier integrated with a DBR laser is shown in Figure 8.5. The integrated device has a laser whose active region has a band gap of 1300 nm, a waveguide layer which is present throughout the length of the chip with a band gap of ~1100 nm, and an amplifier active region with a band gap of 1300 nm. The waveguide layer couples the laser and the amplifier. The grating with a periodicity of 205 nm provides frequency selective feedback so that the laser emits in a single frequency. For the structure shown, both the laser and the amplifier have an MQW active region.

The fabrication of a typical similar device involves the following steps: An n-InGaAsP waveguide layer ($\lambda \sim 1100$ nm InGaAsP, ~0.5 μm thick), n-InP layer (~0.05 μm thick), a InGaAsP MQW region and a p-InP layer (~1 μm thick), and a p-InGaAs layer (~0.5 μm thick) are successively grown over a n-InP substrate using MOCVD growth technique. The InGaAsP MQW region has four InGaAsP active regions (~8 nm thick) and InP barrier layers (~10 nm thick). The thin InP layer serves as a stop etch layer for wet chemical etching. The active region and the InP stop etch layers are removed using selective etching using a dielectric mask. The grating is produced in these regions using wet chemical etching and a phase mask. The wafer is then fabricated into buried heterostructure devices using an active stripe definition by etching followed by second MOCVD growth of current confining layers as described in Chapters 3 and 4. This is followed by metal deposition and cleaving. The cleaved facets are antireflection (AR) and high-reflection (HR) coated using deposition of dielectrics of suitable thickness. The light vs. current characteristics of the laser with the amplifier driven at 170 mA are shown in Figure 8.6.

Integrated distributed feedback (DFB) laser and amplifier structures are important because they have low spectral width under modulation [12, 13]. The schematic of such a structure is shown in Figure 8.7. The DFB laser and the amplifier have an MQW active region. Semi-insulating Fe-doped InP layers are used for both current confinement and for isolation of DFB laser and the amplifier. The DFB laser region has a grating etched under the active region which provides frequency-selective feedback.

For low chirp operation, the laser is modulated with low current (~5 mA) which produces small chirp and the amplifier is used to amplify the output to the required power levels (~10–30 mW). Another interesting laser-amplifier integrated device has a series of

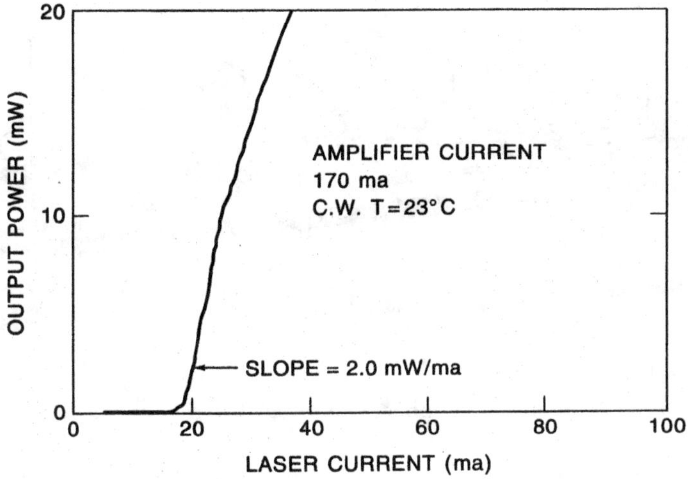

Figure 8.6: Power output as a function of laser current of the device shown in Figure 7.2.1 [11].

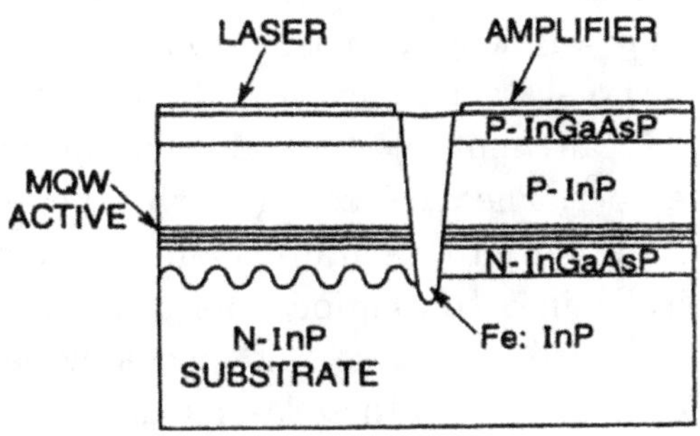

Figure 8.7: Schematic of a distributed feedback (DFB) laser and amplifier structure [12].

amplifiers connected in tandem with a laser (Figure 8.8) [14]. The light is emitted normal to the chip in this device. The first active region-grating combination acts as a laser oscillator. A fraction of the output of this laser propagates along the waveguide and is amplified by the next active region (amplifier). This propagation continues throughout the chip. The device emits in a series of beams normal to the grating. The emission is in a single wavelength and all the emitters emit in phase.

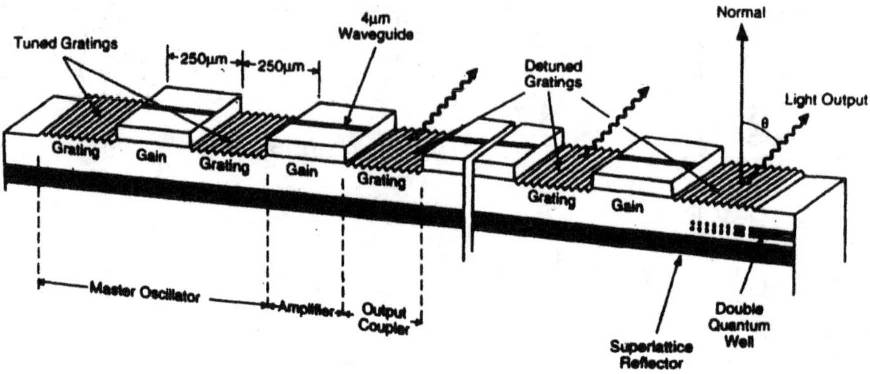

Figure 8.8: A number of amplifiers and a DBR laser connected in series [14].

8.3.1 *Multichannel WDM sources with amplifiers*

An alternative to single channel high-speed (>20 Gb/s) data transmission for increasing the transmission capacity is multichannel transmission using wavelength division multiplexing (WDM) technology. In the WDM system, many (4, 8, 16, or 32) wavelengths carrying data are optically multiplexed and simultaneously transmitted through a single fiber. The received signal with many wavelengths is optically demultiplexed into separate channels which are then processed electronically in a conventional form. Such a WDM system needs transmitters with many lasers at specific wavelengths.

It is desirable to have all of these laser sources on a single chip for compactness and ease of fabrication like electronic integrated circuits. Figure 8.9 shows the schematic of a photonic integrated circuit with multiple lasers for a WDM source [13]. This chip has four individually addressable DFB lasers, the output of which are combined using a waveguide-based multiplexer. Since the waveguide multiplexer has an optical loss of ~8 dB, the output of the chip is further amplified using a semiconductor amplifier. The laser output in the waveguide is TE polarized and hence an amplifier with a multiquantum well gain region which has a high-saturation power is integrated in this chip. Many similar devices which use SOA for output amplification have been fabricated [4, 5].

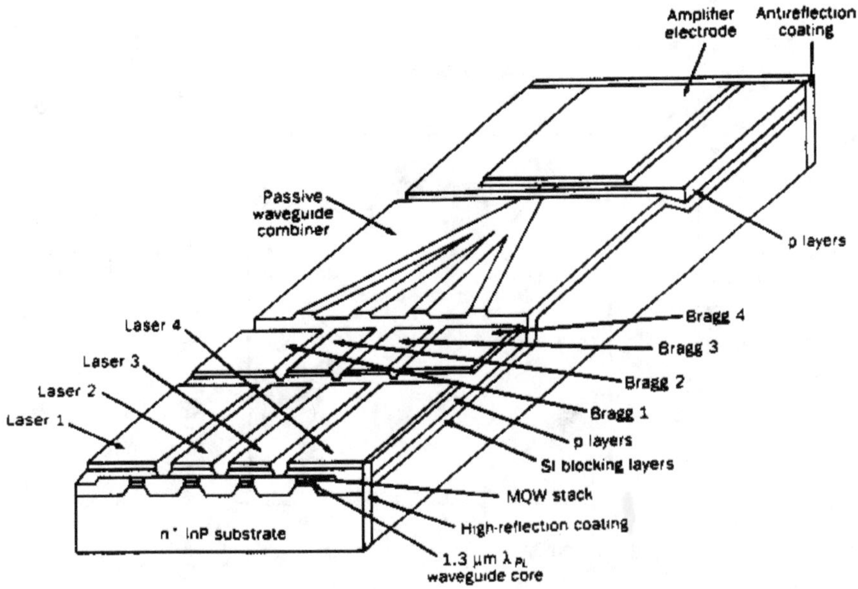

Figure 8.9: Schematic of a photonic integrated circuit with multiple lasers for a WDM source [13].

8.4 Transmitter photonic integrated circuit (PIC)

For combining the output of several DFB or DBR lasers into a single mode fiber, a more suitable approach is an arrayed waveguide (AWG) router. The AWG combines the output of several sources into a single output. It can be fabricated using planar optical waveguides. AWGs using InGaAsP/InP waveguides have been fabricated. The schematic of an AWG is shown in Figure 8.10.

The boxes labeled 1st and 2nd slabs are multimode interference couplers (MMI). The sketch above shows the device as a demultiplexer [17]. An AWG typically consists of one input and output multimode interference coupler (MMI) [15–17] connected by a series of waveguides. This array of waveguides along with the MMI couplers is responsible for the multiplexing (combining signals from multiple sources to one) and demultiplexing (separating the output of one port to multiple ports) property of AWG. An MMI has the property of coupling the light from one input to multiple outputs with

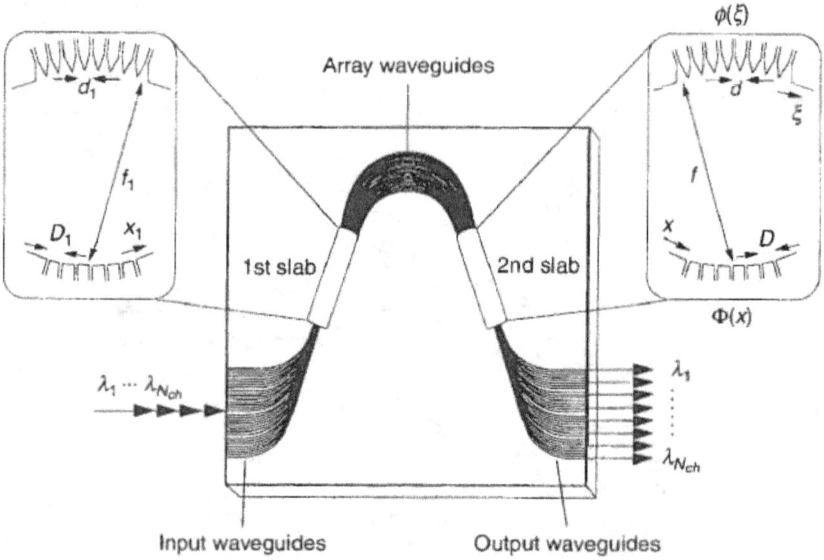

Figure 8.10: Schematic of an arrayed waveguide router (AGW).

negligible excess loss. As shown in Figure 8.10, a set of n input ports can bring in n input signals. The input nXm (n input ports and m output ports) MMI splits the signals into m ports with equal amplitude. The arrayed waveguides modify the phases of these m outputs before they reach the output MMI coupler. The interference of these signals in the mXn output MMI is a function of the lengths of the AWGs. Thus, by suitable design, the output n ports can be made wavelength specific. For multiplexing application in a WDM transmission, the arrayed waveguides are designed so that all inputs at different wavelengths arrive at the same output port of output MMI. For a demultiplexing application, all wavelength channels are present at one input port of input MMI and the demultiplexed channels arrive at separate output ports of output MMI. The AWG (both MMIs and arrayed waveguides) has been fabricated on Si substrate using Si/SiO_2-based planar waveguides and also on InP substrates using InP/InGaAsP planar waveguides. The latter is more suitable for integrated optical circuits which include lasers, modulators, and detectors on one substrate.

Integrated laser modulator devices (EML) are described in Chapter 5. They consist of a distributed feedback laser integrated on the chip with an EA modulator. The laser and the modulator are fabricated in the same epitaxial growth step by the selective area growth process described in Section 8.2. Six such EML devices (emitting at different wavelengths 1.6 nm apart) along with coupler waveguides, an MMI coupler, and an SOA have been fabricated on the same chip. This is shown in the top of Figure 8.11. The PIC in Figure 8.11 uses SOAs (as power amplifier) to increase the output power of multiwavelength sources. When an SOA is used as a power amplifier, a key requirement is high saturation power. As discussed in Chapter 7, SOA with MQW active region has high saturation power. Such SOA has a strong polarization dependent gain, i.e. TE mode gain is considerably higher than TM mode gain. Since the laser light is TE polarized, this is not a concern. For both PIC designs shown in Figure 8.11, the amplifier is used to compensate for the losses from the coupler. Such integrated chips are suitable as transmitters for WDM systems.

8.5 Photoreceiver PIC

A photonic integrated circuit with an SOA and photodiode has been reported [19]. The device uses a traveling wave type photodiode. The input light is injected into a dual section SOA. The front section has high optical confinement for high gain and the back section is offset to produce low confinement and hence high saturation power. The structure is shown in Figure 8.12. The front section is labeled c-MQW (c-centered) and the back section is labeled o-MQW (o-offset). The SOAs are followed by the photodiode, which has an InGaAs active region. The wafer is fabricated using the InP/InGaAsP material system using multiple MOCVD growth steps and is processed into ridge waveguide type devices for lateral confinement. The maximum chip gain of the SOA region is 28 dB. The saturation output power of SOA is 18 dB. The internal quantum efficiency of the photodiodes are determined to be 28 and 32% for the 30 and 40 μm long photodiodes,

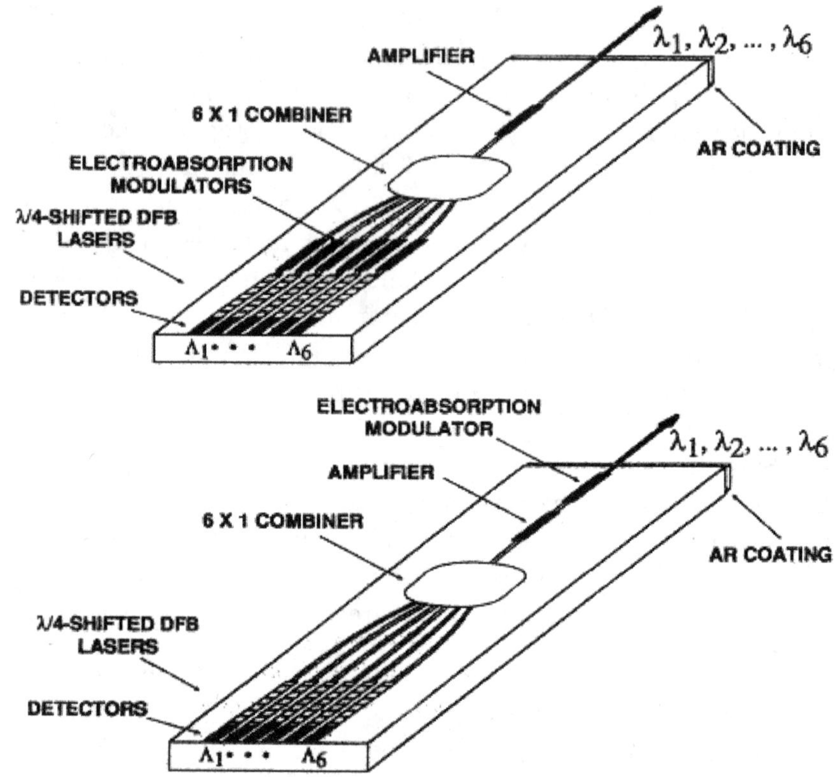

Figure 8.11: The figure shows two configurations of distributed feedback laser array as sources for transmission in WDM systems. (a) Each laser can be separately modulated at high speed for simultaneous multiwavelength transmission and the output is combined to a single port using an AWG. Electroabsorption modulators are integrated in this PIC (b) The lasers' output is combined using an AWG for single wavelength, wavelength selectable transmission. The lasers can be modulated by modulating current input to laser [18]. Detectors monitor the back facet output power of lasers.

respectively [19]. These measurements were carried out at an input light wavelength of 1550 nm.

The device was used as a photoreceiver in a 40 Gb/s transmission experiment. The bit-error-rate (BER) as a function of received power for two designs with different SOA lengths is shown in Figure 8.13. The measured receiver sensitivity (received power for 10^{-9} BER) for the two designs are -17 and -20 dBm, respectively.

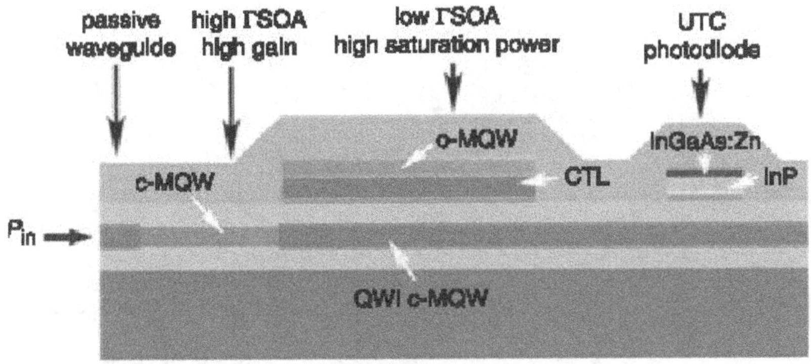

Figure 8.12: Integrated photodiode and SOA preamplifier structure [19].

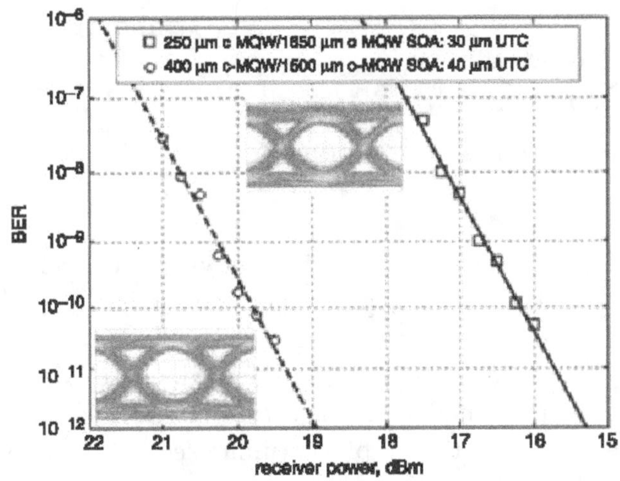

Figure 8.13: Bit-error-rate (BER) as a function of received power [19].

Semiconductor optical amplifiers continue to be used as power amplifiers and preamplifiers in various photonic integrated circuits. The power amplifier applications are primarily in wavelength division multiplexing circuits to compensate for the optical loss of the multiplexer. The preamplifier applications are for a photo-receiver in conjunction with a photodiode. The latter is suitable as an alternative to avalanche photodiode-based receiver which requires high voltage (~50–100 V) for operation.

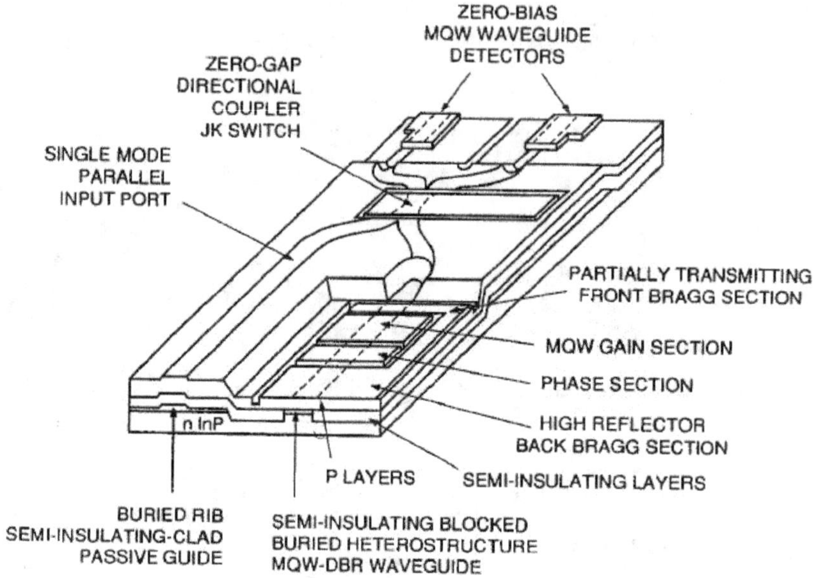

Figure 8.14: Schematic of a heterodyne receiver photonic integrated circuit [20].

For coherent transmission, a heterodyne receiver PIC has been demonstrated (Figure 8.14) [20]. The receiver has a local oscillator laser, waveguides for mixing the input signal with the local oscillator, and a set of balanced waveguide-based photodiodes. The local oscillator laser is a tunable multi-section distributed Bragg reflector laser. It has a gain section, a phase tuning section, and a wavelength tuning (DBR) section.

8.6 Transmitters and receiver PICs for WDM systems

PICs, for multichannel WDM transmission systems, both as transmitter and receiver integrated circuits (ICs) have been reported [21–25]. Schematic architecture of a 10-channel transmitter PIC is shown in Figure 8.15. It has 10 tunable lasers which can be set on a 200 GHz spacing required by International Telecommunication Union (ITU) for a WDM system. Each laser is followed by an EA modulator, a SOA, a back facet power monitor for laser, and a variable optical

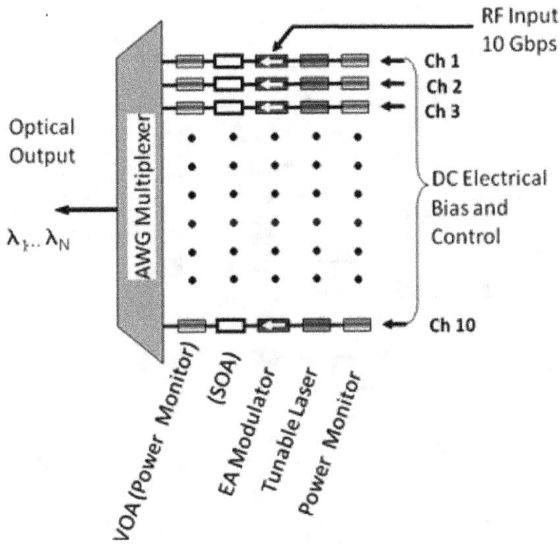

Figure 8.15: Schematic architecture of a 10 channel WDM transmitter PIC [21].

attenuator (VOA) on each laser path. The latter is used to equalize the power output for each wavelength signal out of the AWG. The AWG multiplexes all modulated light signals on to a single fiber. The entire PIC is built using InP/InGaAsP device fabrication technology. The layers are grown using the MOCVD growth process.

The 10-channel receiver PIC is relatively simpler (Figure 8.16). It consists of 10 photodiodes fabricated as an array each of which is waveguide coupled to an AWG which serves as a demultiplexer, i.e. the input to the AWG has 10 wavelength channels after transmission and the AWG demultiplexes the input into 10 separate wavelengths each of which arrives at a photodetector. The photodetectors convert the optical signals to electrical signals which are then separately processed (amplification, clock recovery, and decision circuit) in electronic chips. Good error-free performance has been reported. Also, the reliability of the PICs is good for system applications. Transmitter and Receiver PICs with higher total transmission capacity have been reported using polarization multiplexing and quadrature phase shift keying (QPSK) modulation scheme [22]. The transmitter

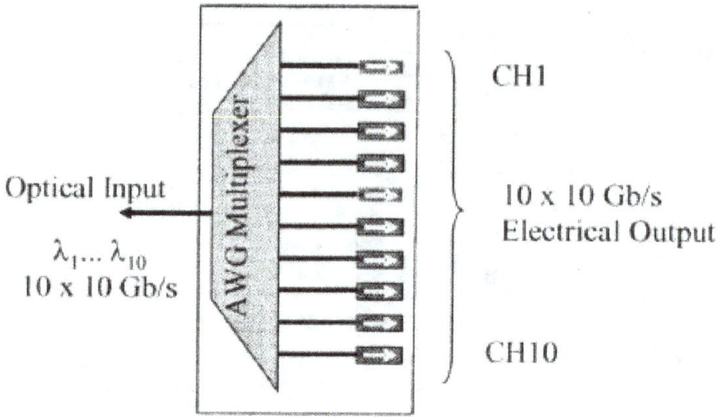

Figure 8.16: Schematic architecture of a 10-channel WDM receiver PIC [23].

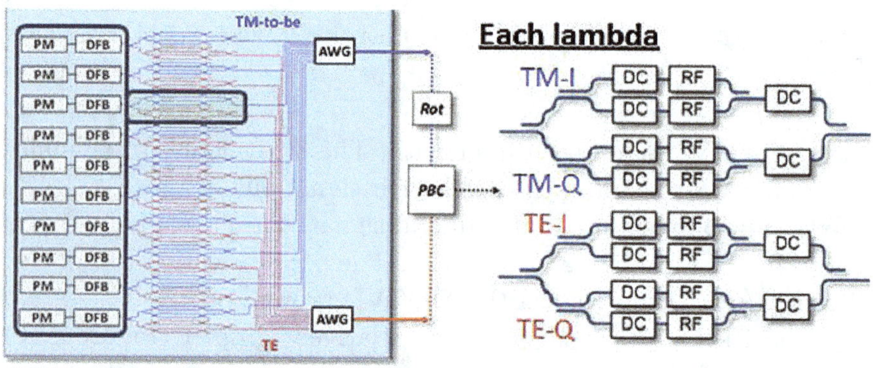

Figure 8.17: Schematic architecture of 10-channel polarization multiplexed WDM QPSK transmitter [22].

generates 10 wavelengths using tunable DFB lasers in the ITU grid. The light from each laser is split into two, one of which is called TE and the other TM-to-be (Figure 8.17).

Each laser was a back face power monitor. The two sets go through different Mach–Zhender interferometers (MZI) which serve as phase modulators and impress different phase modulated data on the two split signals. The light from DFB laser is in the TE mode. Hence, one of the split lights is rotated to an orthogonal polarization

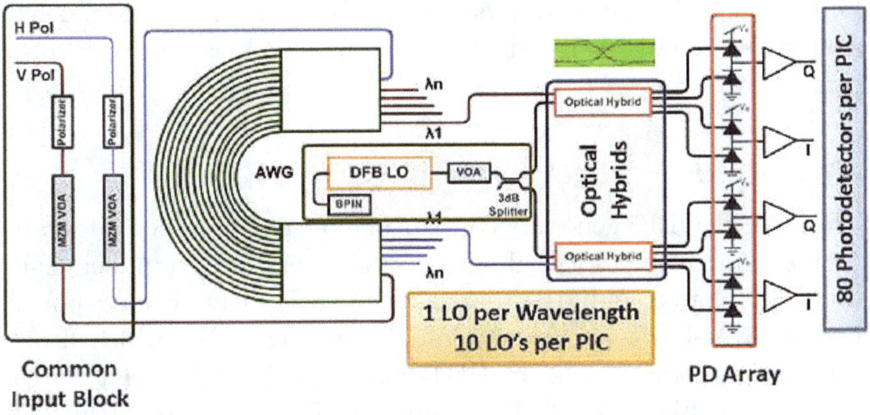

Figure 8.18: Schematic architecture of one-channel polarization of receiver IC. Each channel has its own local oscillator [21].

prior to transmission — this light is labeled TM-to-be. The TE and TM-to-be signals are multiplexed using separate AWGs.

The polarization rotator and polarization combiner (prior to fiber input) are implemented off-chip using discrete components. The transmitter is capable of transmitting at 28Gbaud per channel. The schematic architecture of one of the channels of receiver IC is shown in Figure 8.18. At the input of the receiver, the TE (H pol) and TM (V pol) are split off-chip using a polarization beam splitter. They then become input to the AWGs which are designed to demultiplex the input to different wavelengths at its output. The different wavelengths are then processed using standard coherent receiver technique, i.e. the input signal is mixed with a local oscillators at 90° optical hybrids, the output of which go to four balanced photodetectors. The photodetector output is followed by amplifiers. The 10-channel transmitter–receiver PIC is capable of a total capacity of 1.12 Tb/s.

8.7 Optoelectronic integrated circuit (OEIC)

So far, PICs where all the elements are optical elements (such as lasers, photodiodes, optical waveguides, etc.) have been described. ICs known as optoelectronic circuits (OEIC) have also been reported

[5, 6, 26–27]. In these circuits, an optical device is integrated with simple electronic circuits, such as amplifiers and laser drivers on the same.

8.7.1 *Receiver OEIC*

The receiver OEIC generally consists of a p-i-n photodiode integrated with an amplifier fabricated on the same substrate. Amplifiers using both FET-based and HBT-based technologies have been reported. Figure 8.19 shows the schematic of a receiver OEIC which has an InGaAs photodiode and an HBT-based amplifier [26]. The circuit is shown in Figure 8.20. The amplifier circuit is also shown. The detected photocurrent is amplified by transistors labeled T_1 and T_2. The OEIC is fabricated on a semi-insulating (Fe:InP) substrate to provide electrical isolation between photodiode and the transistor circuit.

8.7.2 *Transmitter OEIC*

Both AlGaAs/GaAs and InGaAsP/InP-based transmitter OEICs have been reported. They generally consist of a laser and a laser driver [27]. The laser driver produces the current needed to drive the laser from a voltage source.

The laser driver uses MESFET technology for GaAs-based drivers and HBT technology for InP-based drivers. An InGaAsP/InP-based OEIC is shown in Figure 8.20. The epitaxial layers are grown by MOCVD growth technique on a semi-insulating InP (Fe:InP) substrate. The HBTs driving the laser are arranged in a differential pair. The circuit operates at 5 Gb/s.

Photonic and optoelectronic integrated circuits continue to be developed for transmitter and receiver applications. GaAs-based surface emitting laser arrays and photodiode arrays with ~250 μm spacing have been developed for optical data link applications. The reliability of these circuits is suitable for commercial applications.

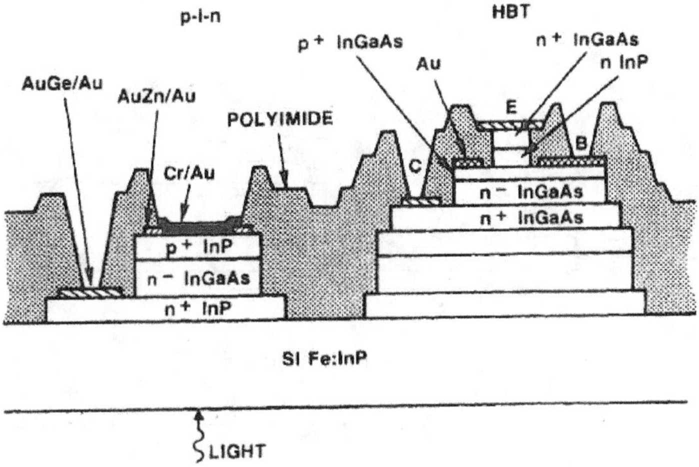

Figure 8.19: Schematic of the layer structure of a p-i-n photodiode and HBT amplifier.

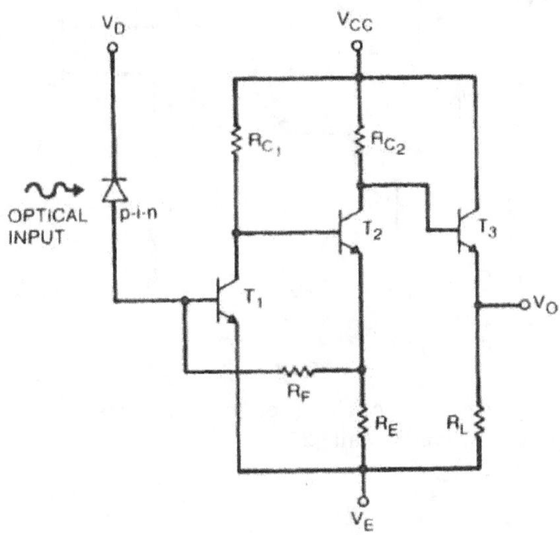

Figure 8.20: Layout of the HBT in the amplifier and photoreceiver circuit. T_1 and T_2 amplify the signal and T_3 acts as an output buffer [26].

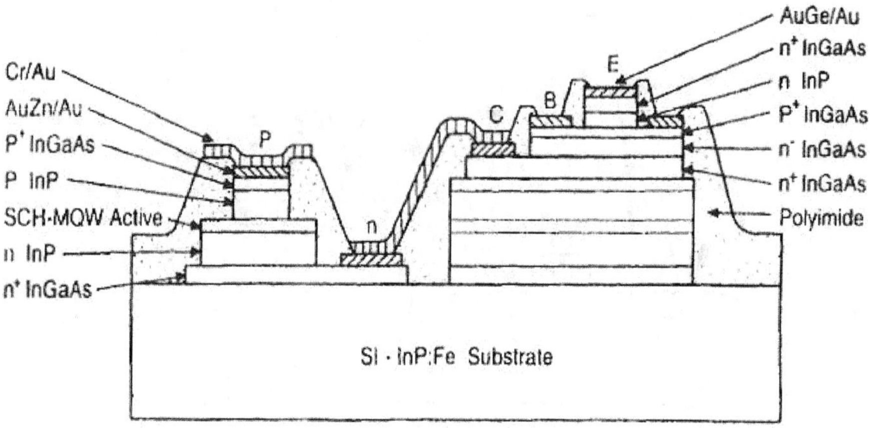

(a)

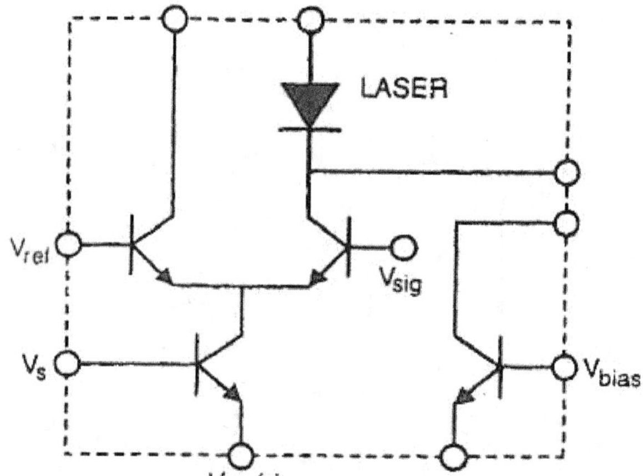

(b)

Figure 8.21: (a) Schematic of the layers of a laser transmitter OEIC. (b) The schematic of the laser driver circuit [27].

References

1. K. Kobayashi, Chapter 15, in S. E. Miller and I. P. Kaminow (eds.), *Optical Fiber Telecommunications II*, . San Diego, Academic Press, USA, 1988.
2. T. L. Koch and U. Koren, *IEEE J. Quant. Electron.*, **QE-27** (1991) 641.
3. O. Wada, T. Sakurai, and T. Nakagami. *IEEE J. Quant. Electron.*, **QE-22** (1986) 850.
4. S. R. Forrest, *J. Lightwave Technol.*, **LT-3** (1985) 1248.

5. M. Dagenais (ed.), *Integrated Optoelectronics*, John Wiley, NJ, USA, 1995.
6. O. Wada (ed.), *Optoelectronic Integration: Physics Technology and Applications*, Kluwer Academic, USA, 1994.
7. N. K. Dutta, *Semiconductor Optical Amplifiers*, World Scientific, Singapore, 2012.
8. R. Bhat, *J. Crys. Growth*, **120** (1992) 362.
9. R. P. Gale, R. W. McClelland, J. C. C. Fan, and C. O. Bozler, *Appl. Phys. Lett.*, **41** (1982) 545.
10. C. Joyner, Chapter 5, in I. P. Kaminow and T. L. Koch (eds.), *Optical Fiber Communications* , III B, Academic Press, USA, 1997.
11. T. L. Koch, U. Koren, R. P. Gnall, F. S. Choa, F. Hernandez-Gil, C. A. Burrus, M. G. Young, M. Oron, and B. I. Miller, *Electron. Lett.* **25** (1989) 1621.
12. N. K. Dutta, J. Lopata, R. Logan, and T. Tanbun-Ek, *Appl. Phys. Lett.*, **51**, (1991) 1676.
13. U. Koren, B. I. Miller, G. Raybon, M. Oron, M. G. Young, and T. L. Koch, *Appl. Phys. Lett.*, **57** (1990) 1375.
14. D. F. Welch, D. Mehuys, R. Parke, R. Waarts, D. Scifres, and W. Streifer, *Electron. Lett.*, **17** (1990) 1327.
15. C. Dragone, C. A. Edwards, and R. C. Kistler, *Photon Tech. Lett.*, **3** (1991) 896.
16. H. Takahashi, S. Suzuki, K. Kato, and I. Nishi, *Electron. Lett.*, **26** (1990) 87.
17. K. Okamoto, Chapter 2, in A. K. Dutta, N. K. Dutta, and M. Fujiwara (eds.) *WDM Technologies, Passive Optical Components*, Academic Press, USA, 2003.
18. T. L. Koch, Chapter 4, in I. P. Kaminow and T. L. Koch (eds.), *Optical Fiber Communication IIIB.*, Academic Press, USA, 1997.
19. J. W. Raring, L. A. Johansson, E. J. Skogen, M. N. Sysak, H. N. Poulsen, S. P. DenBaars, and L. A. Coldren, *Electron. Lett.*, **42** (2006) 16.
20. T. L. Koch, U. Koren, R. P. Gnall, F. S. Choa, F. Hernandez-Gil, C. A. Burrus, M. G. Young, M. Oron, and B. I. Miller, *Electron. Lett.*, **25** (1989) 1621.
21. F. A. Kish *et al.*, *IEEE J. Sel. Top. Quant. Electron.*, **17** (2011) 1470.
22. R. Nagarajan, M. Kato, D. Lambert, P. Evans, S. Corzine, V. Lal, J. Rahn, A. Nilsson, M. Fisher, M. Kuntz, J. Pleumeekers, A. Dentai, H. Tsai, D. Krause, H. Sun, K. Wu, M. Ziari, T. Butrie, M. Reffle, M. Mitchell, F. Kish, and D. Welch, *Semicond. Sci. Tech.*, **27** (2012) 94003.
23. F. A. Kish *et al. Conf. Lasers and Electro-optics (CLEO),* 2005, paper CMGG3.
24. R. Nagarajan, M. Kato, J. Pleumeekers, P. Evans, D. Lambert, A. Chen, V. Dominic, A. Mathur, P. Chavarkar, M. J. Missey, A. Dentai, S. Hurtt, J. Back, R. Muthiah, S. Murthy, R. A. Salvatore, S. Grubb, C. Joyner, J. Rossi, R. Schneider, M. Ziari, F. Kish, and D. Welch, *Electron. Lett.,* **42**, (2006) 771.

25. S. Corzine, P. Evans, M. Fisher, J. Gheorma, M. Kato, V. Dominic, P. Samra, A. Nilsson, J. Rahn, I. Lyubomirsky, A. Dentai, P. Studenkov, M. Missey, D. Lambert, A. Spannagel, R. Muthiah,R. Salvatore, S. Murthy, E. Strzelecka, J. Pleumeekers, A. Chen, R. Schneider, R.Nagarajan,M. Ziari, J. Stewart, C. Joyner, F. Kish, and D. Welch, *IEEE Photon. Technol. Lett.*, **22** (2010) 1015.
26. G. K. Chang, W. P. Hong, J. L. Gimlett, R. Bhat, C. K. Nguyen, G. Sasaki, and G. C. Young, *IEEE Photon. Technol. Lett.*, **2** (1990) 197.
27. S. Hata, M. Ikeda, S. Kondo, and Y. Noguchi, *Tech. Digest. TECE Conf.*, Vol. 85, Japan, 1985.

Chapter 9

Optical Fiber and Fiber-Based Devices

9.1 Introduction

The optical fiber used for transmission has a core about 10 μm in diameter which is surrounded by a cladding layer about 100 μm in diameter. The core region has a slightly higher index ($\sim$0.3%) which allows the light to be guided in the core. The cladding is further surrounded by a polymer jacket which provides both protection from the environment and strength. The concentration of rare earth dopants in rare earth-doped fiber (used for optical amplification) can vary considerably. It can be as much as $\sim$1000 ppm (parts per million) in multicomponent glasses, to as small as $<$100 ppm in erbium-doped fibers. In erbium-doped fibers used for distributed amplification (over the entire transmission path), it can be a few ppm.

Optical fibers can be fabricated by several methods. The principal among them are modified chemical vapor deposition (MCVD), plasma-assisted chemical vapor deposition (PCVD), axial vapor deposition (VAD), and outside vapor deposition (OVD). Rare earth dopants for silica fiber are often introduced during the growth of the fiber itself. Thus, the fabrication of rare-earth-doped fiber is very much linked to the fabrication of the regular low-loss transmission fiber. Typical rare earth dopant level is ~100 ppm.

9.1.1 *Laser fundamentals*

In Sections 3.1.1 and 3.4.1.1 of Chapter 3, the conditions for gain in a semiconductor laser medium are discussed. In this section, the fundamentals of gain and laser operation in rare-earth-doped glass are described. The considerations, such as population inversion, for rare-earth-doped glass are similar to that for lasers based on atomic or molecular transition in gaseous medium. Examples of the latter class of lasers are Ar^+ lasers, He–Ne lasers, CO_2 laser, etc. For simplicity, a two-level system (shown in Figure 9.1) is considered. The realistic cases of three-level and four-level systems are discussed in Appendix 4.

Figure 9.1 shows the two relevant energy levels involved in the laser transition, i.e. the atom makes a transition from the higher-energy level (2) to lower-energy level (1) with the emission of a photon. Let N_2 and N_1 be the population density of levels 2 and 1, respectively, and σ is the cross-section of emission for transition from 2 to 1. The cross-section for absorption for transition from 1 to 2 is also σ. The typical σ values are in the 3×10^{-21} to $3 - 10^{-20}$ cm^2 range. Then, the following equations are obtained for an incident

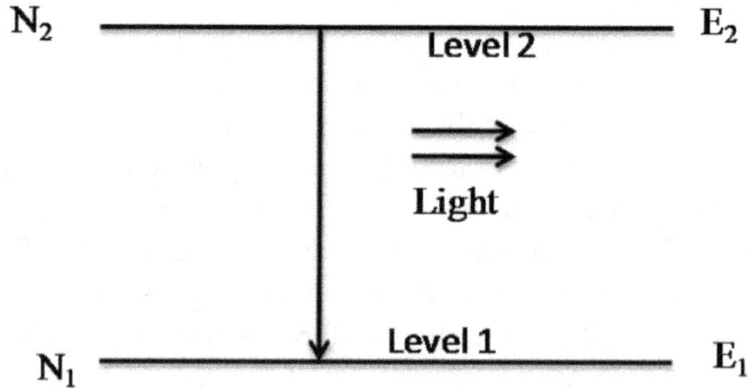

Figure 9.1: The two relevant energy levels (noted as E_2 and E_1) of a two-level laser is shown. N_2 and N_1 are the population densities of levels 2 and 1, respectively.

light at intensity I (expressed in number of photons/cm^2/s),

$$\begin{aligned}
&\text{Stimulated emission rate} = N_2\sigma I \\
&\text{Absorption rate} = N_1\sigma I \\
&\text{Net stimulated emission rate} = (N_2 - N_1)\sigma I
\end{aligned} \tag{9.1}$$

Thus, for net stimulated emission, N_2 must be larger than N_1, i.e. the population density of the high-energy state must be higher than that for the lower-energy state. This condition is known as "population inversion" since under normal thermal equilibrium the population of the higher-energy state is lower than that of the lower-energy state. The population inversion is created by a pump source which produces atoms in the higher-energy state. For rare-earth-doped atoms (e.g. in fiber lasers), the pump source is another laser. For He–Ne laser, the pump source is an energy transfer mechanism by collision of excited He atom to Ne atom, which produces Ne atoms in a high-energy state. The quantity

$$g = (N_2 - N_1)\sigma \tag{9.2}$$

is the optical gain of the laser medium. In addition to stimulated emission, spontaneous emission (and other transfers) also cause a decrease in population of high-energy level. Assuming τ is the lifetime of level 2 due to these processes, the rate equation for N_2 is given by:

$$\frac{dN_2}{dt} = P - (N_2 - N_1)\sigma I - \frac{N_2}{\tau} \tag{9.3}$$

where P is the pump rate and $N_2 + N_1 = N$, where N is the total number density of atoms. Typical τ values are in 0.5–3 ms range. In steady state, the following equations for N_2 and gain (g) are obtained:

$$\begin{aligned}
N_2 &= \frac{P\tau + NI/2I_S}{1 + I/I_S} \\
g &= \frac{g_0}{1 + I/I_S}
\end{aligned} \tag{9.4}$$

with $g_0 = (2P\tau - N)\sigma$ and $I_S = 1/2\tau\sigma$, where g_0 is known as the unsaturated gain and I_s is known as the saturation intensity. The rate equation for intensity (I) is given by

$$\frac{dI}{dt} = gI - \alpha_t I \tag{9.5}$$

where α_t is the total optical loss. In the context of Section 3.1.1, $\alpha_t = \alpha + \alpha_m$, i.e. it is sum of scattering and mirror losses (see Section 3.1.1). In steady state, gain = loss, i.e. $g = \alpha_t$, this leads to the following equation for I:

$$\begin{aligned} I &= I_S(g_0/\alpha_t - 1) \\ &= \frac{1}{\alpha_t}(P - P_0) \end{aligned} \tag{9.6}$$

where $P_0 = N/2\tau + \alpha_t/2\tau\sigma$.

The equation shows stimulated intensity, $I > 0$ for $P > P_0$. The quantity P_0 is called the threshold pump rate. At pump rates higher than threshold pump rate, the laser intensity increases linearly with pump power. This is similar to the result in Section 3.1.1 where it is shown that the semiconductor laser output increases linearly with current for currents higher than the threshold current.

9.2 Fiber for optical transmission

The optical fiber used for transmission has a core (Ge-doped SiO_2) about 10 μm in diameter which is surrounded by a cladding (SiO_2) about 100 μm in diameter. Addition of Ge in SiO_2 increases the index. Optical fiber waveguide mode is discussed in Appendix 2 and optical pulse propagation in fibers is discussed in Appendix 3.

9.2.1 *Optical fiber fabrication*

The optical fiber used in optical transmission systems is based on silica (SiO_2) material. Low loss silica fibers were first proposed in 1966 [1] and by early ~1970s, low loss fibers had an attenuation of 20 dB/km. Since then the manufacturing processes have been refined and developed so that transmission fibers have typical losses of <0.15 dB/km in the 1550 nm region of the optical spectrum [1–8].

The principal methods of optical fiber fabrication are (i) MCVD, (ii) PCVD (iii) VAD, and (iv) OVD, [2–16]. The MCVD method is described in the next section. The methods for the creation of SiO_2 are hydrolysis (reaction with H_2O) or oxidation (reaction with O_2) of $SiCl_4$. The hydrolysis and oxygen reactions are:

$$SiCl_4 + 2H_2O = SiO_2 + 4HCl \tag{9.7}$$

$$SiCl_4 + O_2 = SiO_2 + 2Cl_2 \tag{9.8}$$

The hydrolysis method is accomplished by flowing $SiCl_4$ vapor into a hydrogen flame and the resulting silica particles get deposited on a rotating target. The oxidation method involves a reaction of oxygen and chloride inside a substrate tube heated to 1200°C. Other materials such as GeO_2 can be incorporated in the silica by using a reaction of $GeCl_4$ with the hydrogen flame or oxygen.

9.2.2 *MCVD process*

The MCVD process was invented around 1980 [3]. It involves vapor deposition of high purity material on the inner surface of a tube. The tube may provide part of the cladding, and additional materials for the cladding and core get deposited on the inner surface of the tube. In the MCVD process, first, a tube of suitable dimension and purity is chosen. The tube is mounted on a lathe, which rotates the tube. It has a chemical (chlorides) delivery system at one end and the other end is the exhaust. A heat torch traverses along the tube which causes the entrants to react forming SiO_2 and depositing it on the walls of the tube.

The deposition process is high-temperature gas phase oxidation of chlorides which creates SiO_2 (oxides) and deposition of these oxides as submicron particles on the tube [4, 7, 8]. These particles are converted to a clear glass film by a process called fusion. The entrants into the tube are controlled amounts of chemical reagents which are generated by passing (O_2 or He) through liquid sources such as $SiCl_4$, $GeCl_4$, or $POCl_3$. Sometimes gaseous sources such as SiF_4 are also mixed. In MCVD, first, a high-quality cladding is deposited, then the core is deposited. The total number of deposited layers and their

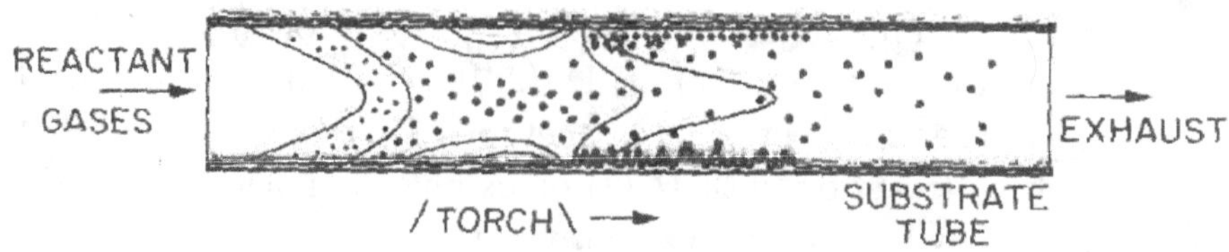

Figure 9.2: Schematic of an MCVD reactor [4].

composition depends on fiber design. The schematic of an MCVD process is shown in Figure 9.2.

9.2.3 *Doping and fiber drawing*

It is necessary to add dopants to the primary glass component SiO_2, to change its index. This allows fabrication of different waveguide designs. Index-raising dopants are germanium, phosphorus, and aluminum and index-lowering dopants are boron and fluorine. They are introduced as halides or gases in the reaction chamber. The halides are carried by oxygen and inert gas. The oxidation process is controlled by the thermodynamic equilibrium established during the doping reaction, deposition, and sintering. Typical fiber index profiles that have been produced are shown in Figure 9.3. Different index profiles are used for fabrication of fibers with different chromatic dispersion characteristics.

A major step in the fabrication of optical fiber is fiber drawing and the associated fiber coating with polymer [4, 18, 19]. This step is important in determining the resultant optical, dimensional, and mechanical properties of the final optical fiber. Figure 9.4 shows the essential features of a fiber drawing facility for drawing silica-based fiber from preforms. The preform is essentially the collapsed tube in the MCVD process after the deposition of the core and cladding materials. A tall structure is used for supporting the control equipment. The glass preform is fed into a high-temperature furnace by a preform feed mechanism where it is aligned and centered with respect to the furnace. For silica fiber drawing, the temperature of the furnace is set in the 1950–2300°C range. This allows proper viscosity for drawing. The tip of the preform softens as it is fed to the

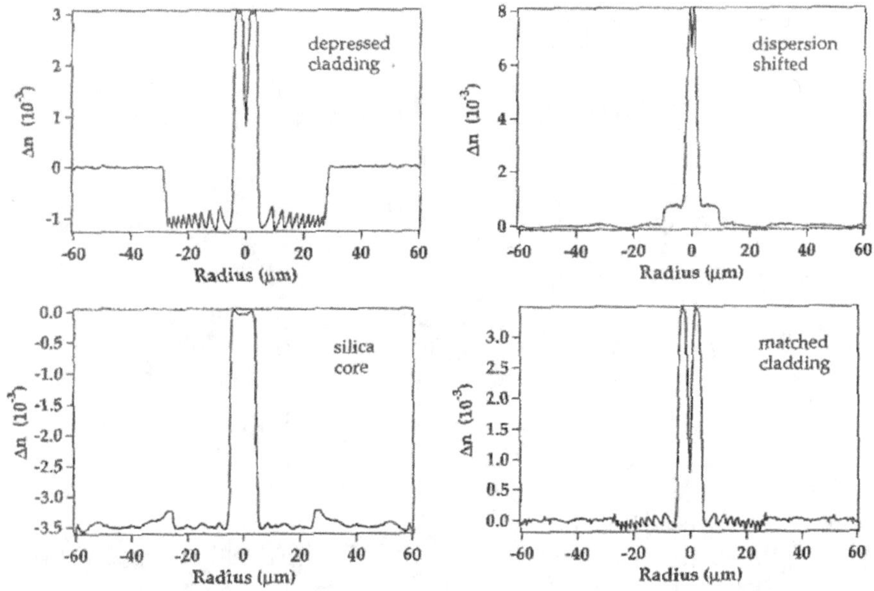

Figure 9.3: Schematic of various index profiles that have been fabricated. Plotted is change in index from SiO_2 as a function of distance from the center.

high-temperature region, and both gravity and an applied tensile force cause it to shrink to small diameter. A controlled tensile force is provided by a fiber pull mechanism (capstan). The preform feed rate and the capstan rotation rate determine the ratio of the drawn down radius of fiber to preform. Typically, preforms with diameter in the 10–70 mm range are drawn down to a fiber 100–225 μm in diameter. 125 μm is the diameter of a typical transmission fiber. In line coating is a key step in fiber drawing process. Both glass fiber diameter and coated fiber diameter are measured and controlled by feedback loops.

The modern optical fiber used in transmission has zero dispersion near 1300 nm and lowest loss ($\sim$0.15 dB/km) at 1550 nm. However, using suitable cladding geometries such as that shown in Figure 9.3, the dispersion of a fiber can be tailored for a specific wavelength range. The dispersion is the sum of material dispersion and waveguide dispersion. The latter is varied by varying the cladding layer

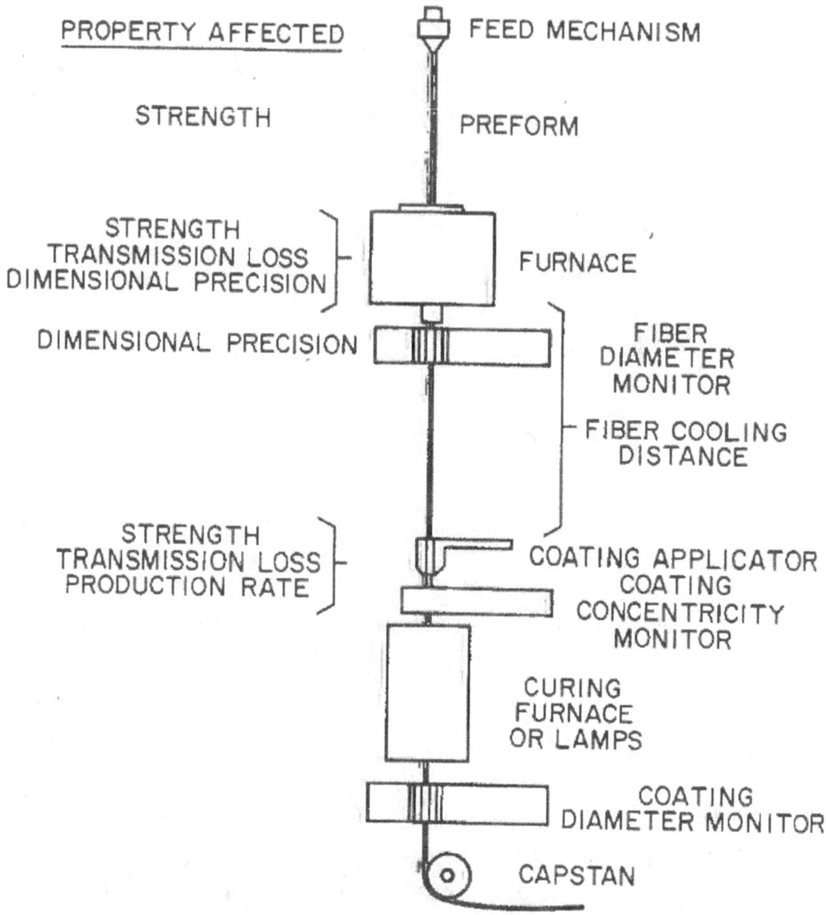

Figure 9.4: Schematic of a fiber drawing and coating apparatus [4].

geometry. The chromatic dispersion as a function of wavelength of three types of fiber are shown in Figure 9.5. The conventional fiber has zero dispersion at 1300 nm. The dispersion shifted fiber has zero dispersion at 1550 nm. The true wave fiber has a dispersion of $\sim$1 ps/km nm near 1550 nm, this is important for reducing four-wave mixing-based nonlinear effects in multiwavelength (wavelength division multiplexed — WDM, see Chapter 11) transmission. Measured optical loss of a single mode fiber is shown in Figure 9.6.

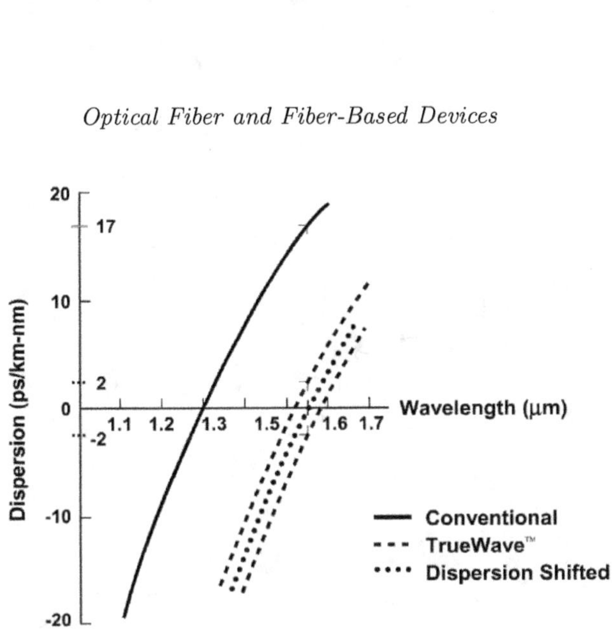

Figure 9.5: Chromatic dispersion as a function of wavelength for three fiber types [20].

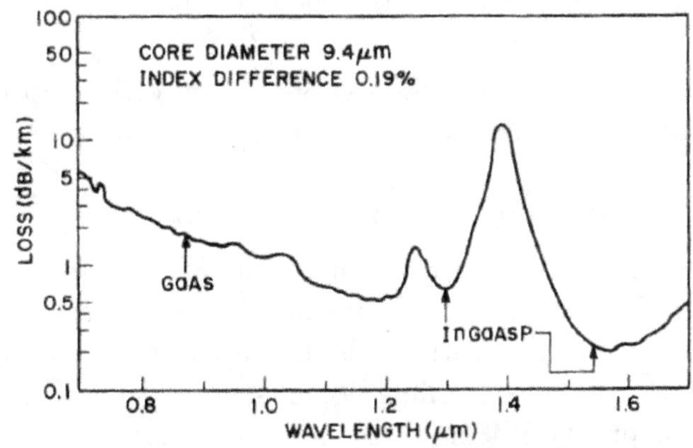

Figure 9.6: Measured optical loss of a single-mode fiber as a function of wavelength. Typical wavelength for optical communication along with the semiconductor material for lasers is indicated. The wavelengths are 850, 1300, and 1550 nm, respectively.

9.3 Fiber amplifiers

Fiber amplifiers are similar to regular optical fibers except that they are doped with rare earth dopants. These are typically Er, Yb, Nd, Er–Yb, etc.

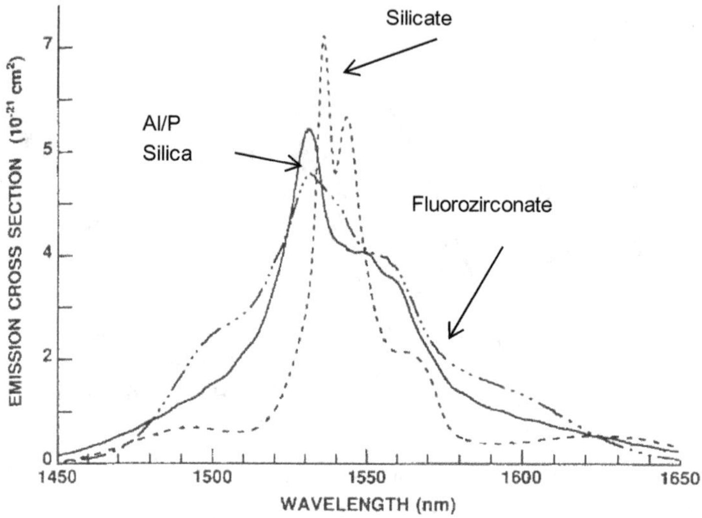

Figure 9.7: Emission cross-section spectrum of Er near 1500 nm for different glass hosts [21]. Fluorozirconate is also known as ZBLAN.

As an example, an emission cross-section of Er-doped material for different glass host types is shown in Figure 9.7.

9.3.1 *Rare-earth-doped fiber fabrication*

Methods of incorporating rare earth dopants in the preform have been developed for all of the perform fabrication techniques stated earlier [21–23]. The rare earth dopants are sent to the reaction region along with the index-controlling dopants. Rare earths generally have a low vapor pressure. This is usually compensated by placing the vapor source close to the reaction zone or by delivering the material as an aerosol or high vapor pressure organic compound (Figure 9.8). The heated frit source is made by soaking previously deposited porous soot with a rare earth chloride solution (Figure 9.8). From the heated source, the rare earth chloride is directly injected (after dehydration) close to the reaction chamber (B). An attractive feature of the heated source injector (C) is that the rare earth source (in the small ampule) is isolated from potential unwanted reaction with other chlorides. The aerosol delivery method does not require

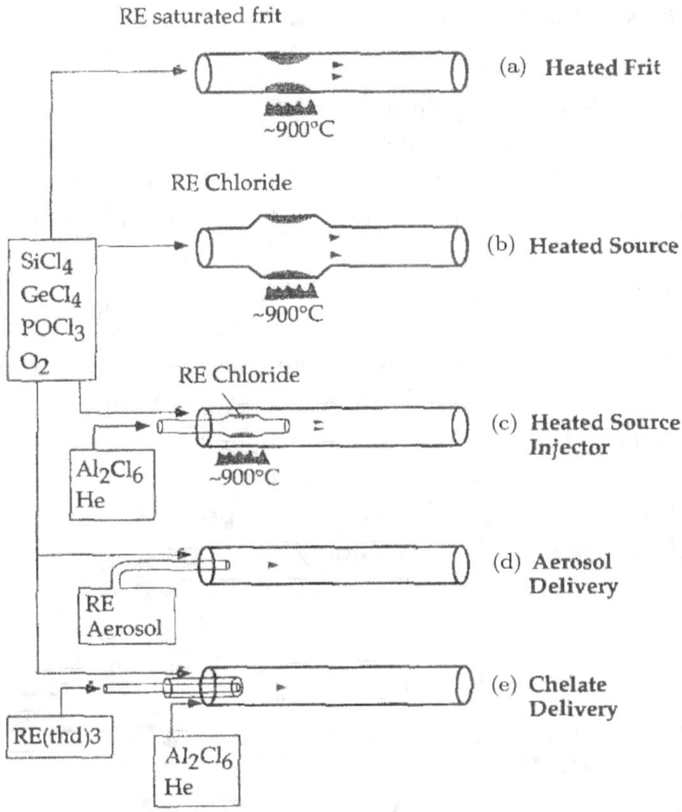

Figure 9.8: Low-vapor pressure dopant (such as rare earth) delivery methods for MCVD [21]. *Note*: RE: rare earth element.

heated source compounds because it generates vapor at the reaction site (D). Organic compound of rare earths (E) provides higher vapor pressure for rare earths than the chlorides, so this is another potential delivery method.

So far, silica-based fiber fabrication has been described. There are several reasons (for example, lower loss than silica at long wavelengths) to consider other types of glass hosts. These glass hosts are selenides, tellurites, germinates, fluorides, chlorides, iodides, and sulfides [24–26]. The glass host influences the transition rates and hence the gain.

9.3.2 *Fiber amplifier characteristics*

Optical fiber amplifiers (OFAs) are typically ~10–30 m in length. The schematic of a fiber amplifier is shown in Figure 9.9. The small signal gain for two signal wavelengths as a function of pump power for a single-mode fiber amplifier is shown in Figure 9.10. The gain increases rapidly at pump powers near threshold and increases slowly at high-pump powers where almost all the erbium ions along the length of the fiber are inverted.

Single-mode amplifiers are often tailored to specific performance characteristics by using different pump configurations and amplifier lengths. They are generally limited to ~500 mW output power.

An approach to a high-power fiber amplifier design is to use a single-mode guide for the signal surrounded by a multimode pump guide (Figure 9.11) [28]. The core is doped with rare earth ions. Pump light is launched at one end into the undoped cladding, where it propagates with a fraction of the pump confined in the doped core,

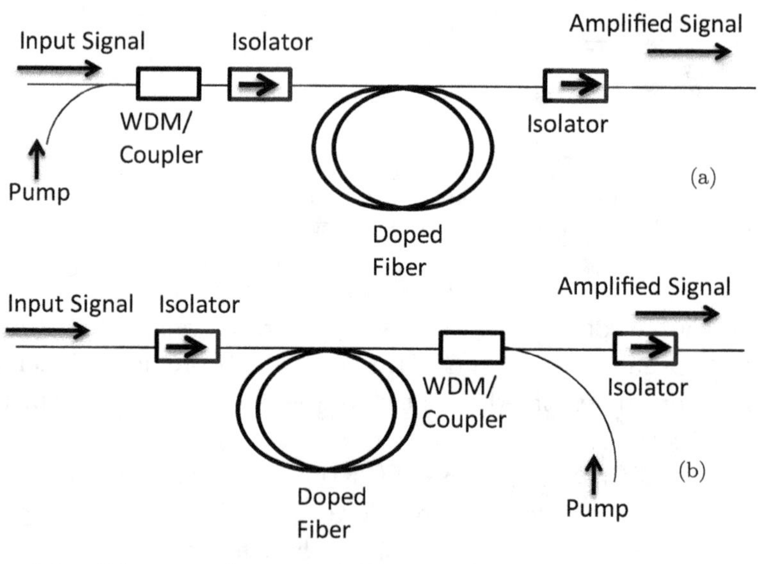

Figure 9.9: Schematic of an optical fiber amplifier. The isolators are needed for reducing the effect of amplified spontaneous emission (ASE) noise. (a) Forward pumping or co-propagating pump, (b) backward pumping, or counter-propagating pump. Bidirectional pumping is also used.

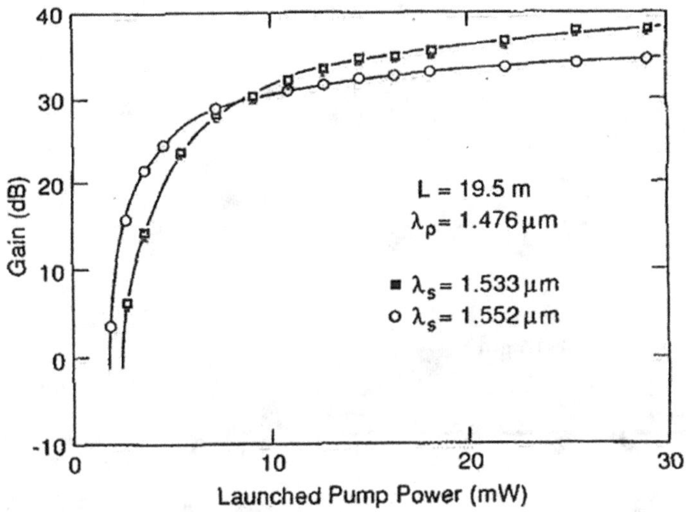

Figure 9.10: Gain as a function of input pump power [27].

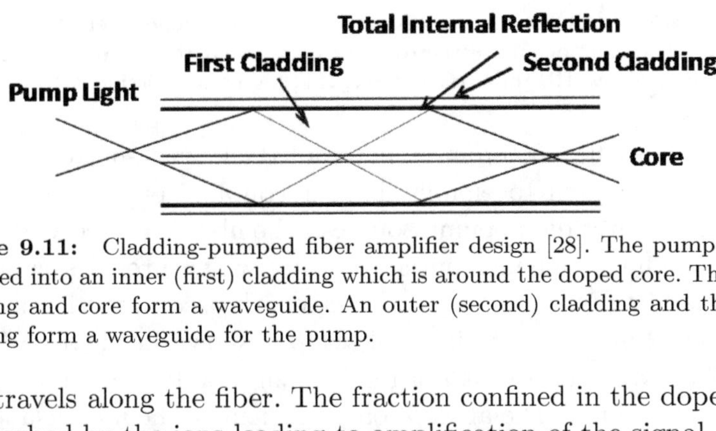

Figure 9.11: Cladding-pumped fiber amplifier design [28]. The pump light is launched into an inner (first) cladding which is around the doped core. This inner cladding and core form a waveguide. An outer (second) cladding and the inner cladding form a waveguide for the pump.

as it travels along the fiber. The fraction confined in the doped core is absorbed by the ions leading to amplification of the signal.

Results on a large core area Yb-doped clad fiber laser are described below. This double-clad fiber has been designed to produce high-power levels (>1 kW). It is drawn from a preform that is fabricated by the MCVD process and solution-doping technique. The fiber has a 40-μm diameter Yb^{3+}-doped alumino-silicate-based glass core with an NA of 0.05. The mode field area for the fundamental mode is estimated to be $\sim 0.9 \times 10^3$ μm^2. The D-shaped inner

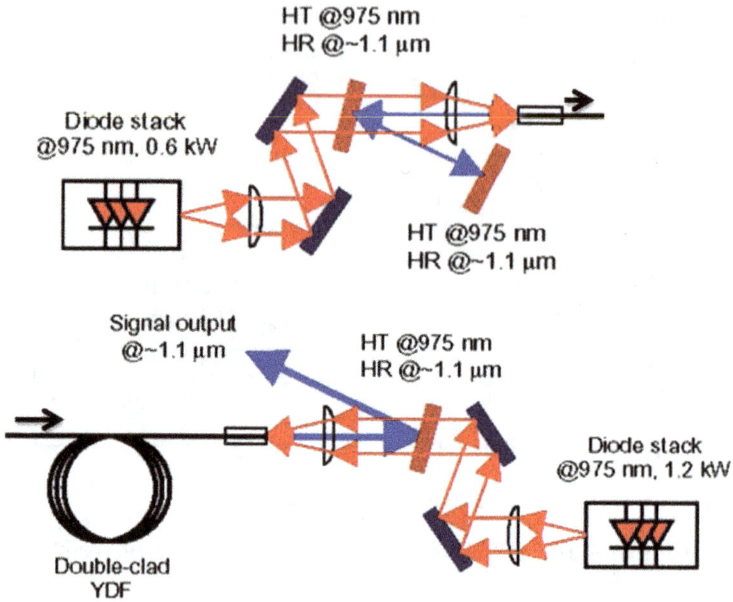

Figure 9.12: YDCF experimental setup. Note to diode laser stacks for end pumping. The components in the top figure are input to the components in the lower figure. *Notes*: HR, high reflectivity; HT, high transmission [29].

cladding has a 650/600-μm diameter for the longer/shorter axis. This diameter is chosen to enable efficient coupling of the large beams from high-power diode pump sources. The fiber is coated with a low-refractive-index polymer outer cladding which provides a nominal inner-cladding NA of 0.48. The fiber length used in the laser experiments is 12 m. The experimental setup is shown in Figure 9.12 [29]. The fiber was pumped at both ends using diode laser stacks which emit near 975 nm. The absorption coefficient in the inner cladding is $\sim$1.5 dB/m at the pump wavelength of 975 nm. This corresponds to an Yb^{3+}-concentration of $\sim$6000 ppm by weight. $\sim$1.3 kW of power was obtained using $\sim$1.8 kW of pump power.

9.4 Fiber Bragg gratings

Fiber Bragg gratings (FBGs) are an important component of fiber lasers, particularly fiber Raman lasers. They are used for providing feedback to the gain region, as both input and output couplers. For a multilayered planar dielectric medium with a small refractive index difference between the layers, reflections take place at every interface for a normally incident light. For a significant reflection from the dielectric stack, the reflections at each interface must add up in phase. Similar phenomenon must happen for an index grating close to a waveguide. For the schematic shown in Figure 9.13 (an index grating near a dielectric layer is a periodic region of alternating high and low index), the small reflections of the optical field outside the waveguide from the "crests" of the grating must add up in phase to produce a significant reflectivity.

This constructive interference occurs if the period of the grating (Λ) obeys the following equation

$$2n\Lambda = m\lambda \tag{9.9}$$

where n is the index of the material, ($n \sim 1.5$ for silica glass), λ is the wavelength in free space, and m is an integer. Equation (9.9) is called the Bragg condition. For $m = 1$, the grating is called a first-order grating. For a first-order grating, the Bragg wavelength λ_B for a grating with a period Λ is $\lambda_B = 2n\Lambda$. Thus, for reflection at $\lambda \sim 1500$ nm, the grating period is ~500 nm for first-order gratings in silica ($n = 1.5$).

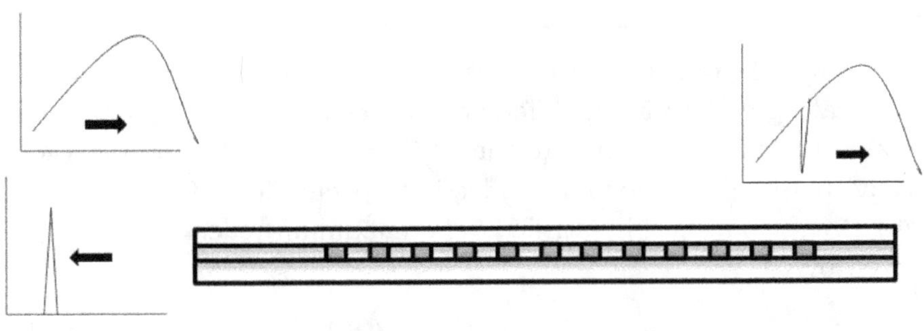

Figure 9.13: Schematic of an index grating and its operation. The grating reflects a specific wavelength. The grating period (in silica glass) is ~500 nm for reflection near 1550 nm. The reflected light spectrum and transmitted light spectrum are sketched for an incident light signal with the spectrum shown.

9.4.1 *Theory and design*

Typically, first-order gratings are used in FBG applications. In a laser cavity, the light travels both in the forward and backward direction. Thus, for an FBG in a laser cavity, the FBG reflection couples the forward and backward traveling waves. The reflectivity of FBG has been derived using coupled mode equations. The result for reflectivity, R is [30–32]

$$R = \frac{K^2 \text{Sinh}^2(|S|L)}{\delta\beta^2 \text{Sinh}^2(|S|L) + |S|^2 \text{Cosh}^2(|S|L)} \quad \text{for } K^2 > \delta\beta^2$$

$$= -\frac{K^2 \text{Sinh}^2(|S|L)}{\delta\beta^2 - K^2 \text{Cosh}^2(|S|L)} \quad \text{for } K^2 < \delta\beta^2$$

with

$$S = \sqrt{[\delta\beta^2 - K^2]} \tag{9.10}$$

In these equations, $\delta\beta = 2\pi n_e/\lambda - m\pi/\Lambda$, where n_e is the effective index of the waveguide. The quantity $\delta\beta$ is a measure of the mismatch between Bragg condition and the actual free space wavelength (λ). The quantity K is the coupling constant between the forward and backward traveling waves and L is the length of the grating region. For uniform sinusoidal modulation of the grating index

$$K = (\pi\delta n/\lambda_B)\eta \tag{9.11}$$

where η is the overlap of the electric field in the waveguide core with the grating and δn is the difference between the maximum and minimum index in the grating region. If the grating is far from the core, K decreases. The maximum reflectivity occurs for $\delta\beta = 0$, i.e. when the Bragg condition is satisfied. The maximum reflectivity, R_{max}, is given by

$$R_{\text{max}} = \tanh^2(KL) \tag{9.12}$$

For $KL > 3$, the reflectivity approaches 100%. For a significant reflectivity, few mm of grating length is needed. Another measure of an FBG is its bandwidth. A measure of this quantity is the wavelength difference ($\Delta\lambda$) between the two minimum on either side of the maximum reflection peak. It is given by (for $\lambda \sim \lambda_B$),

$$\Delta\lambda = \frac{\lambda_B^2}{\pi n_e L}[(kL)^2 + \pi^2]^{1/2} \tag{9.13}$$

Thus, the reflection peak position, the width of the reflection spectrum, and the maximum reflectivity can be varied by varying the grating pitch and the grating length.

9.4.2 *Grating fabrication*

There are several techniques for writing gratings. The principals among them are the dual beam holographic technique and the phase mask technique. These techniques are also used for fabricating grating on a semiconductor substrate for distributed feedback laser fabrication. In the holographic technique, the light from a UV laser is split up into two beams using a beam splitter. The beams are focused back using lenses on a fiber with the cladding removed. Cylindrical lenses are used to elongate the beams along the fiber length. The interfering light on the fiber produces an intensity pattern. Figure 9.14 shows the schematic of the holographic method.

The interference pattern writes a grating on the fiber. This is due to changes in absorption in fiber when exposed to UV light. An ideal silica fiber would have SiO_2 bonds. When dopants, such as

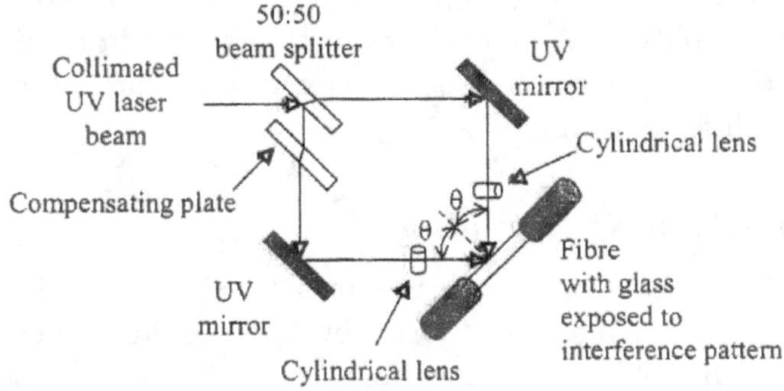

Figure 9.14: Schematic of grating writing using a dual beam holographic method [30].

Ge are introduced to change the index to form the fiber waveguide, defects are introduced which can take the form of Si–Si, Si–Ge, and Ge–Ge bonds. In the presence of UV light (~244 nm) these bonds break increasing the absorption. An index change is then caused through the Kramers–Kronig relationship. Thus, a periodic index change is produced by the periodic interference pattern of UV light. The grating period is given by

$$\Lambda = \frac{n_{\text{eff}}\lambda_{\text{laser}}}{n_{\text{laser}}\sin\theta} \tag{9.14}$$

where λ_{laser} is the wavelength of laser, n_{laser} is the index at laser wavelength in the fiber, n_{eff} is the index of fiber near Bragg wavelength, and θ is the angle shown in Figure 9.14. The quality of the grating can be improved by increasing the photosensitivity of the fiber using different dopants. The dual beam holographic technique requires a highly coherent laser source, a stable grating writing system over a large area (few cm), and low alignment tolerances. These disadvantages are overcome using a phase mask technique [30]. The performance of fiber gratings and their use are discussed in later sections.

9.5 CW fiber lasers

Fiber laser is a device very similar to a fiber amplifier [31–40]. Fiber amplifiers amplify light using a rare-earth-doped gain medium. Fiber lasers also use a rare-earth-doped fiber gain medium, and in addition have optical feedback.

9.5.1 *Single mode fiber laser*

A typical Er-doped fiber laser is shown in Figure 9.15 [38]. In this design, the gain length is reduced by high doping, ~600 ppm vs. a typical value of ~10–30 ppm for a typical erbium-doped amplifier. The Ge-doped core is co-doped with Al to reduce concentration quenching effects.

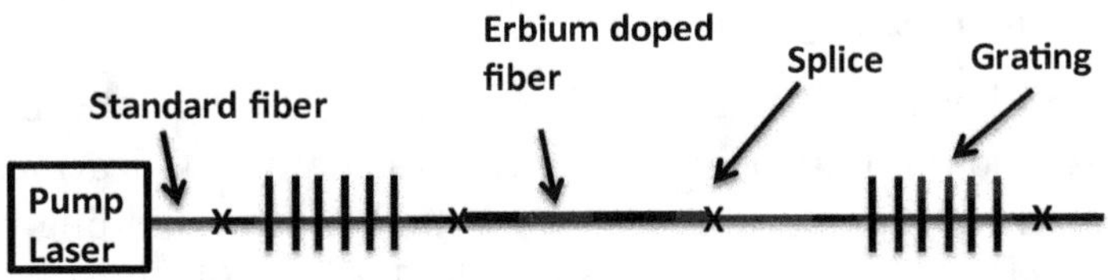

Figure 9.15: Schematic of an Erbium-doped fiber laser with fiber Bragg gratings (FBGs).

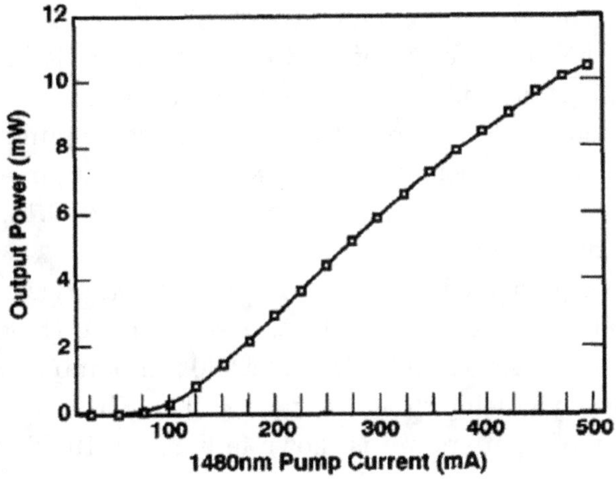

Figure 9.16: Laser output as a function of 1480 nm pump laser current [38].

The core diameter of the fiber is ~2.6 μm. The Er-doped fiber is spliced to two FBGs. FBGs are refractive index gratings generated in the fiber by optical interference of two light (UV) beams (Section 9.4). They can have very high reflectivity at a specific wavelength. The output coupler grating and the rear reflector have reflectivity of 98% and >99%, respectively. Er-doped fiber lasers of the type shown in Figure 9.15 have been pumped using both 1480 and 980 nm pump lasers. The laser output as a function of pump power for the laser design of Figure 9.15 is shown in Figure 9.16. The laser emits near 1539 nm. The amplifier fiber length is 2 cm. These types of fiber laser have been used as sources in optical transmission experiments.

9.5.2 *High-power fiber laser*

Yb-doped fiber lasers, which emit near 1100 nm, have been studied for a high-power laser source, which has many applications. As a result, many Yb-doped fiber lasers are made in the cladding-pumped geometry [41–46]. The fiber used is a double clad fiber with an inner cladding diameter of 240 μm and an outer cladding diameter of 355 μm. The Yb core is 10 μm in diameter with an NA of 0.06. The Yb doping is 1.4×10^{21} cm^{-3} (about 12 wt.% of Yb_2O_3).

The Yb concentration is about 10 times higher than typical values used. The active core, offset from the inner cladding and the outer cladding, is made of phosphate glass. An air hole of 10 μm diameter is added 30 μm from the core to increase pump absorption (Figure 9.17). The schematic of the experimental setup and a microscopic image of the fiber end are shown in Figure 9.18. The pump laser emits at 940 nm.

The fiber gain medium is 71 cm long and the laser has a high reflector (bulk) at one end and the cleaved fiber end (reflectivity ~3–4%) serves as the second reflector. The 940 nm pump light is coupled at the cleaved end using a dichroic mirror. The Yb laser output power as a function of pump power is shown in Figure 9.19. The laser emits

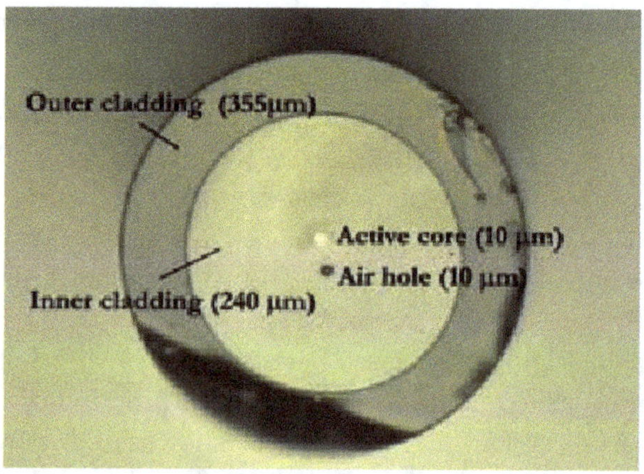

Figure 9.17: Image of the Yb-doped fiber end. The core has a diameter of 10 μm, inner cladding has a diameter of 240 μm, and the outer cladding has a diameter of 355 μm [27].

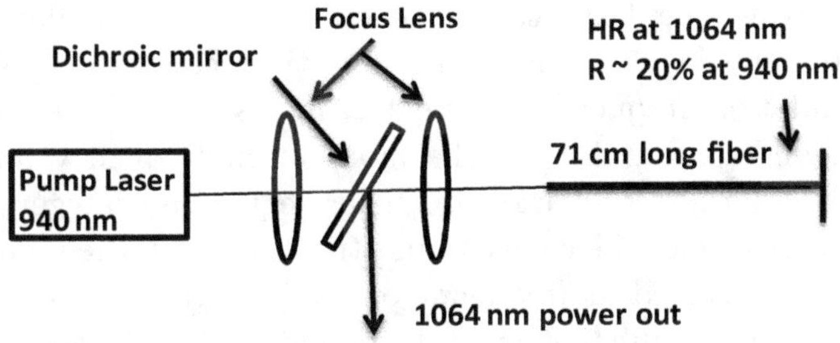

Figure 9.18: Schematic of the experimental setup for Yb-doped fiber [46].

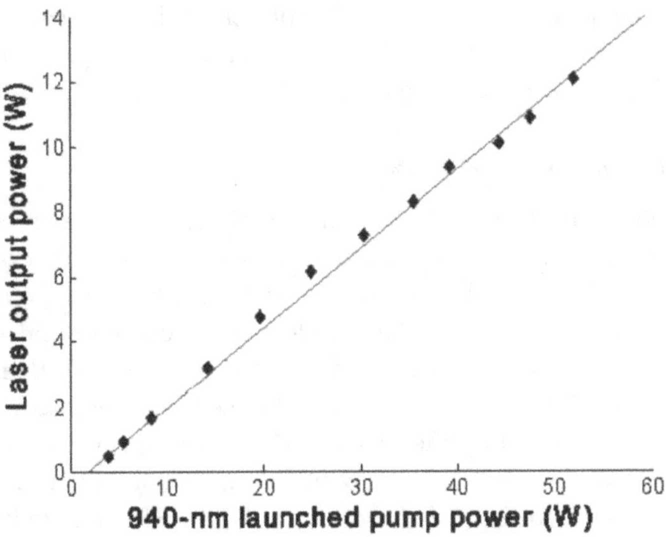

Figure 9.19: Yb laser output as a function of pump power [46].

near 1062 nm. The measured laser slope efficiency against launched power and absorbed power are 24% and 37%, respectively. Tunable fiber lasers have been demonstrated using ring geometry [47].

9.6 Mode-locked fiber lasers

Mode-locked pulses have been reported using rare-earth-doped lasers using both active and passive mode locking schemes. For active mode, locking the laser cavity generally has a modulator which modulates the loss or phase in the optical cavity [48–55]. When the modulation frequency equals the inverse of cavity round-trip time, mode-locked pulses at the modulation frequency are generated. Also, if the modulation frequency equals an integer multiple (n) of the inverse of the cavity round-trip time (f_c), the mode-locked pulses are generated at frequency of f_m ($= nf_c$). This phenomenon is called harmonic mode locking. If the modulation frequency equals $(n+1/p)f_c$, i.e. $f_m = (n+1/p)f_c$, where p is an integer, mode-locked pulses are also generated at a frequency of pf_m, i.e. at a multiple of the modulation frequency. This phenomenon is called rational harmonic mode locking and it is useful for generating optical pulses at a high-data rate. The theory of harmonic and rational harmonic mode locking is discussed in Appendix 9.

9.6.1 *Harmonic mode locking*

The experimental set-up for a harmonic mode-locking scheme is shown in Figure 9.20. The fiber laser has a 23-m long Er-doped fiber (EDF), a WDM coupler, a 10% output coupler, two polarization-independent isolators, a $LiNbO_3$ electro-optical modulator with a bandwidth of 11 GHz, and a wavelength-tunable optical filter with a 3 dB bandwidth of 1.0 nm. A synthesizer is used to generate the RF signal which is amplified to 30 dBm to drive the modulator. All components are connected by a regular single-mode fiber. The fundamental cavity frequency f_c ($= c/nL$) is 2.58 MHz. The gain in EDF is created using a single-mode laser emitting at 980 nm as a pump. The active harmonic mode locking in this fiber laser is obtained when the modulation signal at frequency $f_m = nf_c$ in the 1–10 GHz range is applied. The shortest and most stable optical pulses are obtained by careful tuning of the modulation frequency f_m, and the polarization of the circulating light, using the polarization controller. The pulse width of mode-locked pulse trains is measured by a sampling oscilloscope or by an autocorrelator.

Figure 9.21 shows that for harmonic mode-locking case $f_m = nf_c$, pulse width is proportional to $f_m^{-1/2}$.

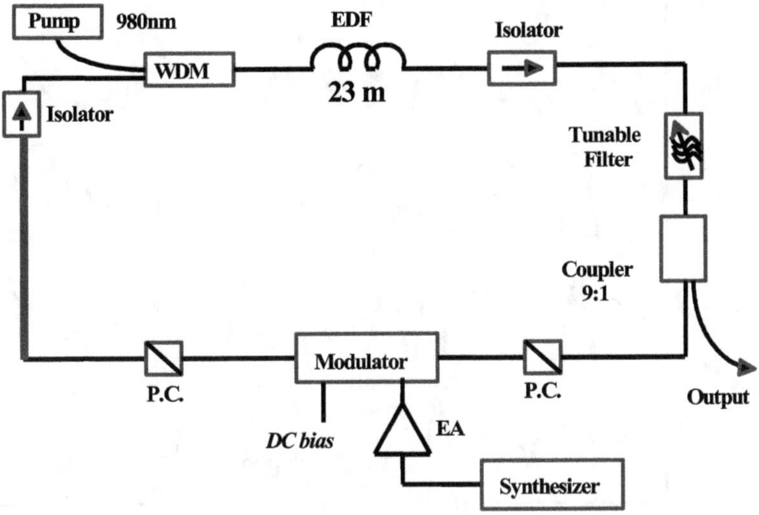

Figure 9.20: Schematic of the fiber laser [52]. *Notes*: EDF, erbium-doped fiber; PC, polarization controller.

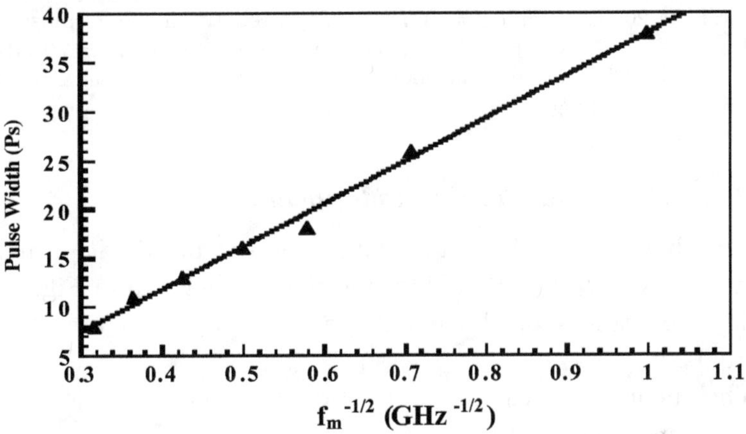

Figure 9.21: Pulse width vs. inverse square root of the modulation frequency for harmonic mode-locking case. The corresponding frequency changes from 1 to 10 GHz [52].

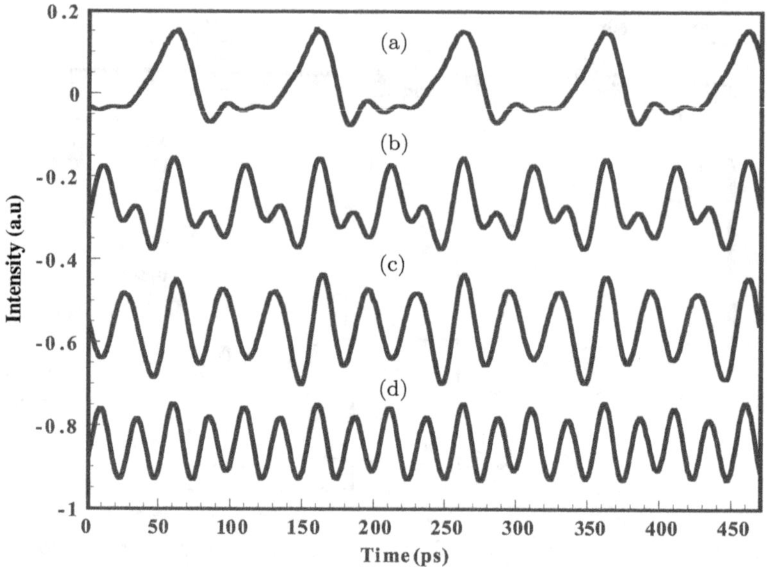

Figure 9.22: Rational harmonic mode-locking pulse trains using $f_m \approx 10$ GHz (observed on sampling oscilloscope). (a) f_m = 9992.5129, MHz = 3872 f_c; repetition rate of pulse train =10 GHz, (b) $f_m = (3872 + 1/2)f_c$; repetition rate of pulse train = 20 GHz, (c) $f_m = (3872 + 1/3)f_c$; repetition rate of pulse train = 30 GHz, (d) $f_m = (3872 + 1/4)f_c$; repetition rate of pulse train = 40 GHz [52]. The pulse width for rational harmonic mode locking is almost same as that for harmonic mode locking.

9.6.2 *Rational harmonic mode-locking*

Rational harmonic pulse trains have been produced with a modulation frequency (f_m) of 10 GHz using the laser shown in Figure 9.20. Oscilloscope traces of the result (for p = 2, 3, 4) are shown in Figure 9.22.

The theory of harmonic and rational harmonic mode-locking is discussed in Appendix 9.

9.6.3 *Passive mode-locking*

Several types of saturable absorbers have been used for passive mode-locking of fiber lasers. Over the last few years, carbon nanotubes (CNT) have been used as saturable absorbers [56–63]. The CNT

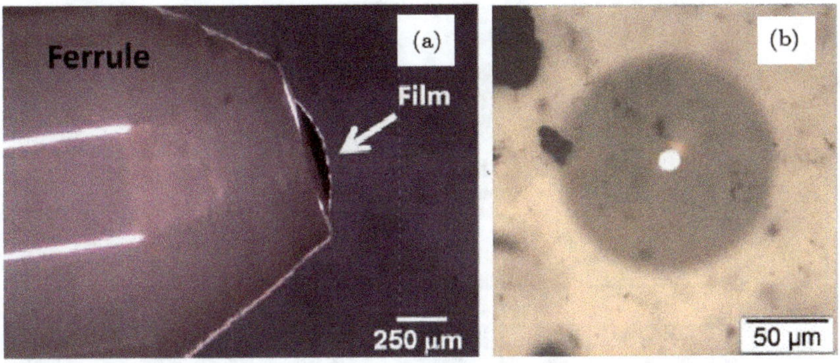

Figure 9.23: Side view (a) and front view (b) of a droplet containing carbon nanotubes at the end of an FC/APC fiber connector [63].

can be deposited at the end of a single-mode fiber. For film fabrication, 1.0 ± 0.2-nm diameter CNTs made by the high-pressure carbon monoxide (HiPCO) method are suspended in a UV-curable, commercial optical polymeric adhesive based on urethane. This adhesive is chosen for its transparency and good adherence to silica. It has a refractive index of 1.56 and a transparency of 90% at 1550 nm. For a homogeneous CNT/polymer dispersion, the samples are mixed in an ultrasonic bath for 90 min. The weight concentrations of CNT are in the 20–50 mg/ml range. A droplet of CNT/polymer suspension is collected with a microtip and transferred to the surface of the ferrule of an FC/APC fiber connector. The microtip has a 1-μm diameter tip and is fabricated by tapering a 1-mm diameter glass capillary. Sometimes the CNT droplet is positioned using a micropositioner. Figure 9.23 shows the microscopic image of the side view and front view of a CNT droplet at the end of a fiber connector (FC/APC type).

The connector with the CNT droplet is included in a standard ring fiber laser cavity. It has an Er-doped fiber, a polarization controller, an isolator, and a WDM for pump laser light coupling. The EDF is pumped by a pump laser emitting near 1480 nm to produce gain (Figure 9.24). The pump laser power is 47 mW and the rest of the fiber laser has 8 m long standard telecommunication fiber. The EDF length is 0.8 m. The shortest pulse width is obtained for a 20 μm

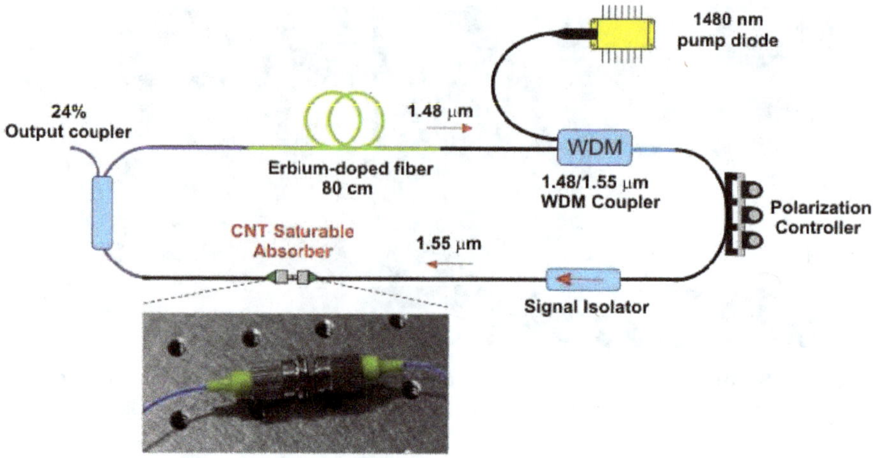

Figure 9.24: Schematic of a fiber laser setup with a photograph of the fiber connector containing CNT [63].

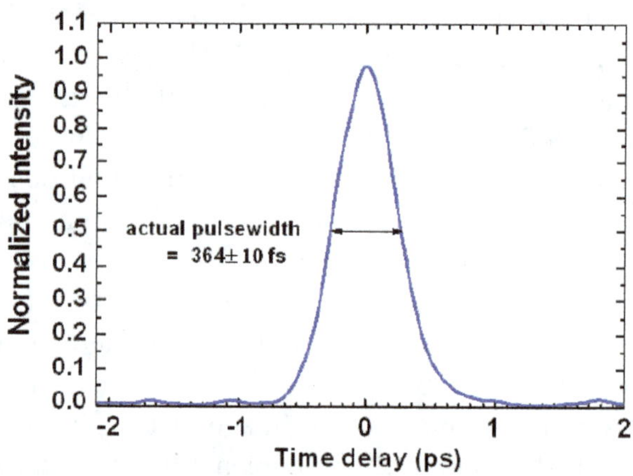

Figure 9.25: Autocorrelation trace of output pulses using carbon nano tube (CNT) saturable absorber [63].

thick CNT sample with a concentration of 40 mg/ml. The autocorrelation trace showing ~370 fs pulse width is shown in Figure 9.25.

Among other saturable absorbers (SA) that have been studied is graphene [64–66]. In the experiment described below, a few layers of graphene deposited on a fused silica substrate are used as the

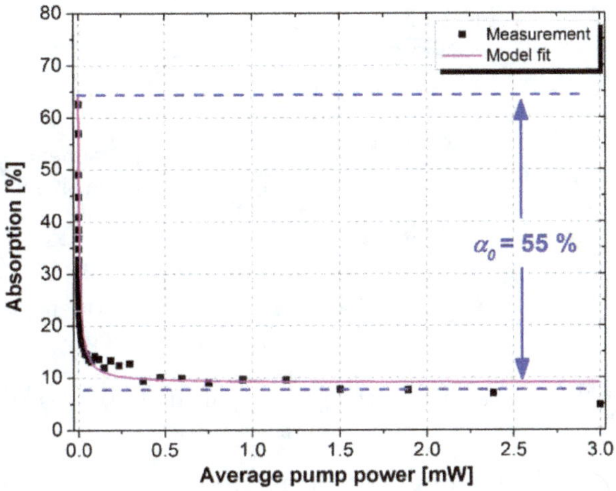

Figure 9.26: Optical transmission as a function of power of a film of graphene [66].

saturable absorber. The performance of the SA is determined using a power-dependent transmission measurement. A mode-locked laser operating at 1556 nm at a repetition rate of 169 MHz with 150 fs pulse width (at 1560 nm) and 3 mW average output power is used as a source for transmission experiment. The signal is transmitted through a variable optical attenuator and then to the SA. The transmission as a function of average power is shown in Figure 9.26. The transmission data can be expressed by the equation:

$$\alpha(I) = \alpha_0/(1 + I/I_s) + \alpha_{\text{ns}} \tag{9.15}$$

where $\alpha(I)$ is the power dependent absorption coefficient, I is the intensity, I_s is the saturation intensity, and α_{ns} is the unsaturable absorption. Based on the results of Figure 9.26, a large modulation in transmission (55%) can be obtained with only 8% unsaturable absorption losses. The mode-locked laser setup using graphene as a saturable absorber is shown in Figure 9.27. The Er fiber is 30-cm long and is pumped by a 980-nm pump laser. Two collimators on XYZ translation stages are used for transmitting the light through the graphene film. The total optical loss between the collimators is 0.9 dB. The total length of the fiber laser loop is 3.1 m.

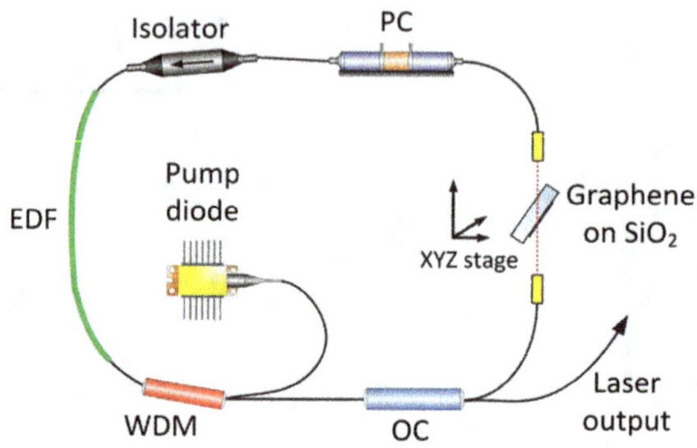

Figure 9.27: Schematic of the graphene mode-locked laser setup [66].

The threshold for laser action is 15 mW of pump power. At 60 mW of pump power, mode-locked operation is observed after adjusting the polarizer. The pulses are equally spaced by ~17 ns which corresponds to 58 MHz repetition rate. The repetition rate agrees well with the inverse of cavity round-trip time. The autocorrelation trace of the pulses is measured and is found to be 315 fs. The pulses have 1.9 mW average power, 104 W of peak power, and pulse energy of 33 pJ. The time-bandwidth product of the pulses is 0.44.

9.7 Fiber raman lasers and amplifiers

Fiber Raman lasers are based on the nonlinear process of stimulated Raman scattering (SRS) [67–71]. When a sufficiently intense light beam is incident on a material, light at a new wavelength is generated by a process called Raman scattering. Typically, the energy of the photons ($h\nu_p$) at the incident wavelength is larger than the energy of the Raman process-generated photon ($h\nu_s$) by an amount equal to the separation between two energy levels (δE) in the material, i.e. $h\nu_p - h\nu_s = \delta E$. The photons at the lower energy are generated by both spontaneous and stimulated emission (Figure 9.28). This is further discussed in Appendix 12.

The SRS produces Raman gain and hence fiber Raman lasers. The energy levels in the material include electronic, vibration, and rotational energy. For typical fiber Raman lasers, the excitations are

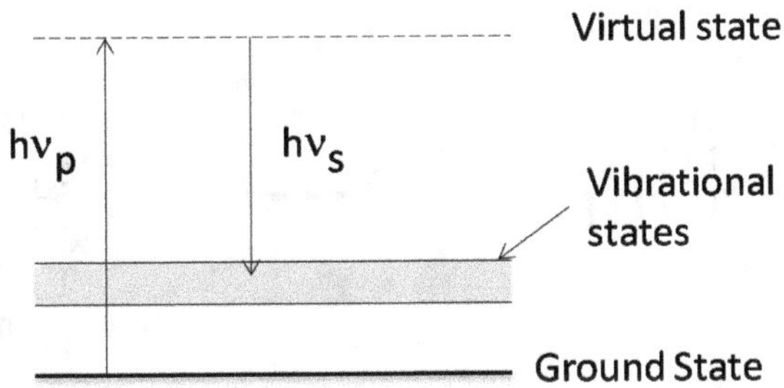

Figure 9.28: Schematic of quantum mechanical transition in a Raman process. The figure shows that a molecule is excited to a virtual state when a pump photon is incident on it and it quickly decays to a different energy level emitting a photon at a lower energy.

vibration energy in the SiO_2 glass matrix, i.e. vibration energy of the Si–O bond. The Raman gain is proportional to the input pump power. The vibration energy is ~467 cm^{-1} and the Raman gain 0.8 cm^{-1}/GW. Since Raman fiber lasers are based on SRS (which produces gain at the energy of the Raman scattered light), the incident light (known as the pump) needs to be sufficiently intense. The Raman gain for the generated Raman signal is proportional to $P_p \sigma N$, where P_p is the intensity of the incident pump, σ is the stimulated Raman cross-section, and N is the number of participating molecules per unit volume. The quantity $\sigma N = g_R$ is often called the Raman gain coefficient. The difference in energy between the pump and signal photons (Raman scattering produced or injected) is determined by the molecular vibration of the host material. This frequency difference is called the Stokes shift. The amorphous nature of silica makes the Raman gain curve quite broad in optical fiber. The simplest Raman fiber laser consists of a single mode fiber to which pump light is injected.

The single-mode fiber has FBG attached at both ends (Figure 9.29(a)). The FBG has the unique property of being fabricated so that it reflects at a very narrow band of wavelengths (by proper choice of the periodicity of the grating) and can be easily spliced to a single-mode fiber with low loss. The grating reflectivity can also be varied by varying the length of the FBG. At the input pump end, the FBG reflects the Raman signal light but lets

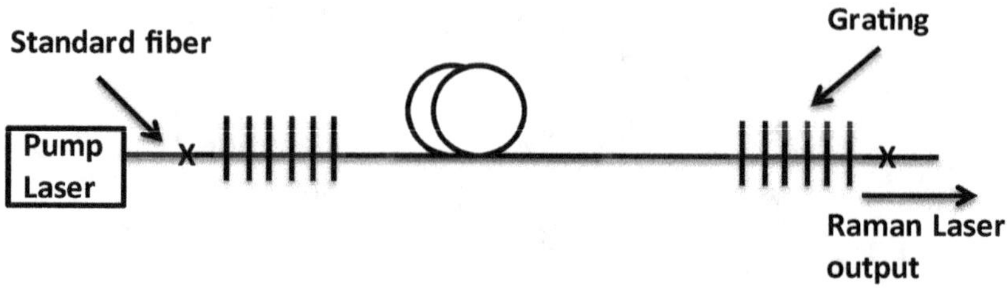

(a)

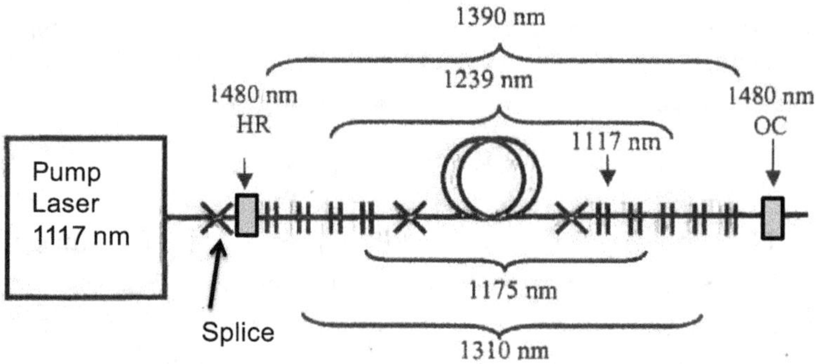

(b)

Figure 9.29: (a) Typical schematic of a fiber Raman laser with fiber Bragg gratings (FBGs). (b) A cascaded Raman laser design [72]. The laser emits at 1480 nm using 1117 nm pump laser. *Notes*: HR, high reflection; OC, output coupler.

the pump light through. At the output coupler, it allows some of the signal light to escape (reflectivity ~90%). FBGs have been used for providing feedback for both fiber lasers and fiber Raman lasers. Using a series of gratings at appropriately chosen wavelength, a cascaded Raman laser (Figure 9.28(b)) has been demonstrated [72, 73]. In this design, four nested pairs of gratings (with reflectivity close to 100%) at appropriate Stokes wavelengths are used to build up the pump light for each Stokes scattering. The pump light is at 1117 nm. The first Stokes scattering produces stimulated emission at 1239 nm, which acts as pump for the second Stokes scattering producing stimulated emission at 1390 nm and this acts as a pump for the third Stokes scattering which produces stimulated emission at 1480 nm. The 1480 nm light could be used as a pump source for Er-doped fibers. Thus, the cascaded Raman process utilizing FBGs produces stimulated emission at 1480 nm using 1117 nm pump light.

9.7.1 *Raman laser*

Raman lasers with higher output power have been studied and demonstrated [74–77]. The experimental setup (Figure 9.30) using Yb-doped double-clad pump lasers is similar to that shown in Figure 9.29. The Raman gain fiber is a 75 m long commercial grade fiber with 10 μm core diameter. The FBG has a reflectivity of 99% (HR FBG) at the pump input end and has a reflectivity of 50% at the output coupler end. The 3-dB bandwidth of the HR FBG is 1.49 nm, and that of the output coupler is 0.52 nm. The high power from Yb-doped fiber laser (at 1120 nm) is delivered by a regular fiber with core diameter of 10 μm. A 99%/1% coupler is spliced before the pump to monitor the stability of pump source.

The Raman laser output at 1173 nm as a function of injected pump has been measured. Raman laser power of 114 W has been obtained with 144 W of pump power. The conversion efficiency is 82%. The Raman laser spectrum has a full width at half maximum of ~2 nm.

For shorter Raman gain fiber lengths, the output power is lower. One importance of high-power output near 1230 nm is its use as pump source for a fiber Raman amplifier or fiber Raman laser emitting near 1300 nm. First Stokes shift of 1230 nm from a silica fiber produces Raman output near 1300 nm. Thus, a Raman amplifier for

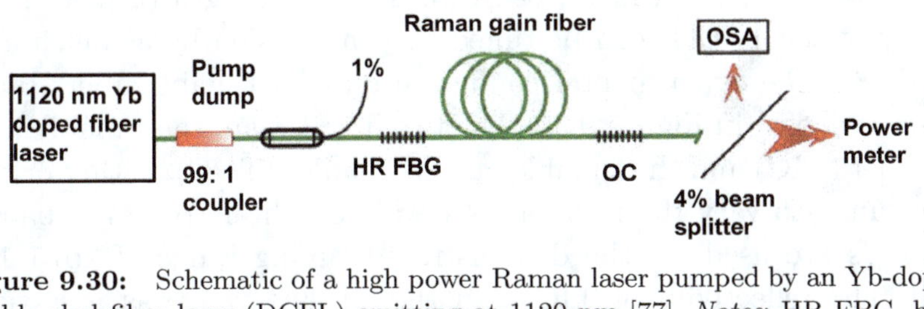

Figure 9.30: Schematic of a high power Raman laser pumped by an Yb-doped double-clad fiber laser (DCFL) emitting at 1120 nm [77]. *Notes*: HR FBG, high reflection fiber Bragg grating; OC, output coupler; OSA, optical spectrum analyzer.

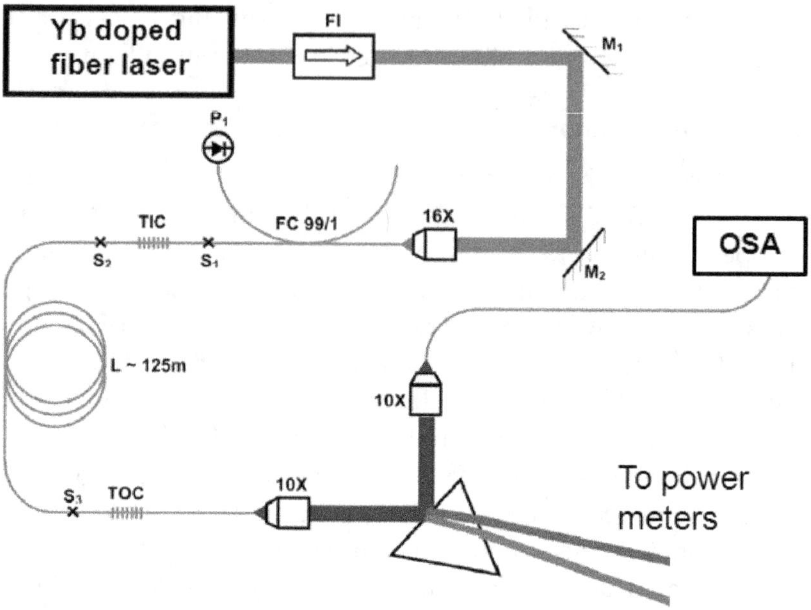

Figure 9.31: Schematic of a tunable Raman fiber laser. It uses tunable fiber Bragg gratings (TIC and TOC) [78]. *Notes*: OSA, optical spectrum analyzer; S, splices.

1300 nm spectral region, which is an important wavelength for optical communication, can be built using this process. Such an amplifier has been demonstrated.

Using essentially similar schemes as described earlier, tunable fiber Raman lasers have been demonstrated [78]. They involve tunable fiber Bragg gratings (TFBGs). The wavelength of peak reflectivity of the TFBGs can be tuned by compression by as much as 60 nm [78]. The experimental configuration of a tunable fiber Raman laser is shown in Figure 9.31. The two TFBGs are the tunable input coupler (TIC) and the tunable output coupler (TOC). Either or both are tuned to vary the Raman laser wavelength. Two sets of pairs of TFBGs are used for the demonstrated tuning range of 1075–1135 nm. The reflectivity of TIC is in the 90–99% range as it is tuned and that of the TOC is in the 60–54 % range as it is tuned. The Raman fiber is a 125 m long fiber with a core diameter of 4.1 μm.

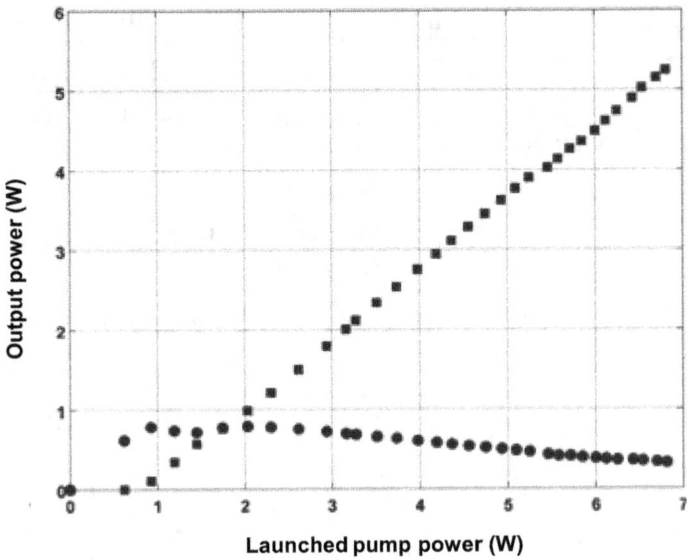

Figure 9.32: Raman laser output as a function of launched power (squares), residual pump power (circles) [78].

Yb-doped fiber laser emitting at 1064 nm is used as a pump. The Raman laser output power as a function of launched pump power is shown in Figure 9.32. The Raman laser emission is at 1109 nm. The residual pump power is also shown. The Stokes spectrum is centered around 1109 nm. It has one peak for 0.9 W Stokes power. As the power increases, the spectrum broadens and can have two peaks.

9.7.2 *Raman amplifier*

Silica fiber-based Raman amplifiers have been extensively studied for use as broadband amplifiers in WDM fiber communication systems [79–84]. If a single pump laser is used, the Raman gain would be downshifted by about 14 THz. The width of Raman gain spectrum is about 6 THz. The schematic of a typical Raman amplifier is shown in Figure 9.33. It has two pumps at about the same wavelength for SRS [84]. The Raman gain spectrum of optical fiber exhibits a broad continuum shape due to the amorphous nature of glass. The magnitude of the Raman gain is determined by gain coefficient and

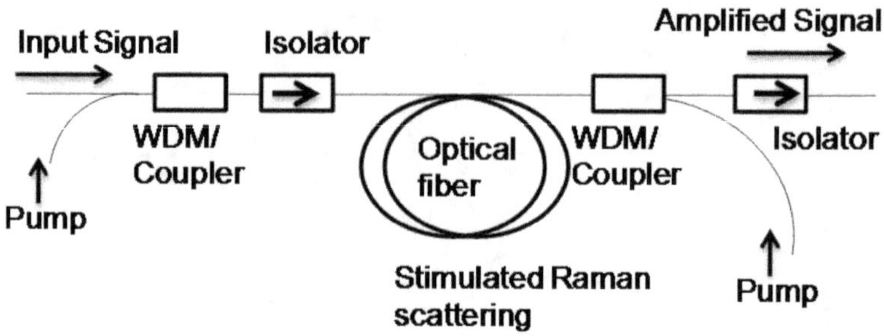

Figure 9.33: Schematic of a typical Raman amplifier [84].

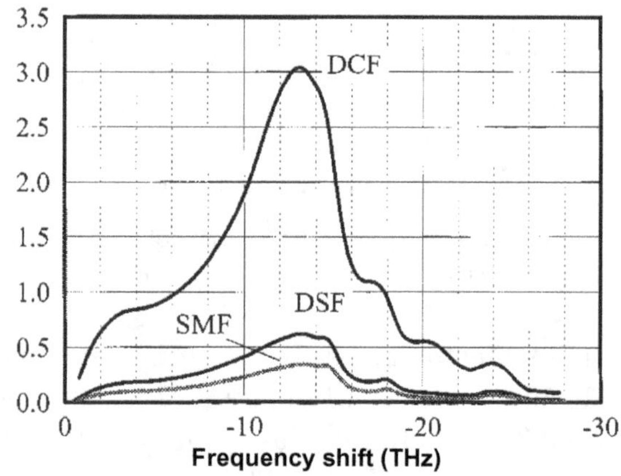

Figure 9.34: Relative Raman gain in different fiber types. They are single mode fiber (SMF), dispersion shifted fiber (DSF), and dispersion compensating fiber (DCF). The differences are due to different germanium oxide concentrations [84].

nonlinearity, and the gain profile is influenced by the fractions of GeO_2 and SiO_2 [82, 83]. The Raman gain for three different fiber types is shown in Figure 9.34. They are SMF: regular single-mode fiber, DSF: dispersion shifted fiber, and DCF: dispersion compensating fiber. The concentration of germanium oxide becomes larger in the order of SMF, DSF, and DCF. It is useful to compare Raman amplifier with Er-doped fiber amplifier (EDFA). The polarization state of light is maintained during SRS, while the gain is almost unpolarized for EDFA. Another difference is the stimulated emission process in Raman amplification occurs through virtual energy states.

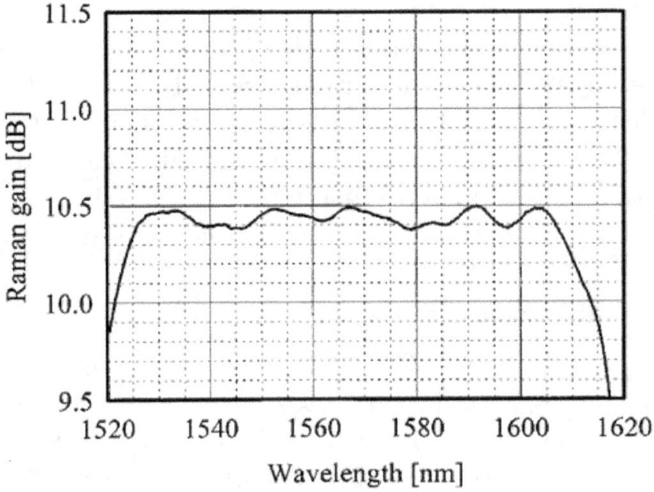

Figure 9.35: Flat Raman gain spectrum using 12 pump lasers, ~1 THz apart, and a 25 km long dispersion shifted fiber [84]. A frequency of 200 THz corresponds to a wavelength of 1500 nm.

The luminescence lifetime of Raman is $\sim 10^{-15}$ s, while that of EDFA is $\sim 10^{-3}$ s.

Using multiple pump lasers at different wavelength, it is possible to increase the width of the gain spectrum significantly. By choosing multiple pump wavelengths separated by approximately the Raman gain spectrum width and by combining them into a single fiber, a wide-width Raman gain spectrum has been demonstrated. Figure 9.35 shows the measured Raman gain profile of an experiment using a 25-km long DSF as the Raman fiber. The pump lasers are used in a counter-propagating configuration. Twelve pump lasers are used with a frequency spacing of ~1 THz. The pump frequencies span from 212 to 199 THz. They are combined using a silica-based planar waveguide combiner. The average Raman gain is almost 10.5 dB and the gain flatness is about 0.1 dB over the wavelength range of 1527–1607 nm. This corresponds to the entire C-band and L-band of EDFA. Since the wavelengths near the middle have overlapping gains from adjacent pump wavelengths and the end regions get gain mostly from the end pumps, for obtaining the flat gain spectrum it is important to optimize both the pump wavelengths and pump powers appropriately. The maximum available Raman gain in this scheme is determined by the maximum output power of the longest and shortest wavelength pumps. It is also feasible to tailor the Raman gain spectrum to any desired shape using appropriately chosen pump laser wavelength and power. Supercontinuum generation using nonlinear effects and Raman effects in optical fiber is discussed in Appendix 7.

9.8 Photonic crystal fiber

Photonic crystal fibers (PCF) also known as microstructure fibers have been studied since the mid-1990s [85–89]. Nonlinear optics studies such as supercontinuum generation and amplification using rare earth-doping have been reported. Although much of these studies have been carried out using silica-based fiber, the PCF fabrication (described in the next paragraph) is, in principle, applicable to a wide range of materials.

9.8.1 *Photonic crystal fiber fabrication*

The fabrication of PCF involves a process by which an arranged macroscopic structure is multiple thinned using high temperatures. In the first step, individual capillaries are created using standard

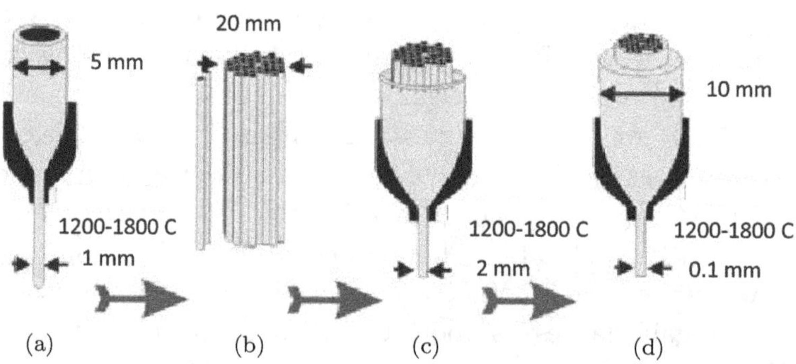

Figure 9.36: Schematic of the microstructure fiber fabrication. (a) Creation of individual capillaries, (b) perform creation, (c) drawing of preform with cladding — intermediate preform, (d) drawing of microstructured fiber or PCF [89].

glass capillary fabrication procedures. Then, they are manually organized (arranged) to form a multicapillary structure of required symmetry. In the middle there is a glass rod in which the light propagates, or there could be a hole for fibers with photonic band gap. Capillaries with different aspect ratios (ratio of inner to outer diameter) can be chosen. The rods are positioned in the structure. Extra glass rods may be added to create a structure of sufficient mechanical integrity and desired dimensions. The structure is then fused in a furnace to create an intermediate preform, a few mm in diameter. The preform is fused and drawn using a fiber drawing tower (described previously) to create the PCF. Figure 9.36 shows the schematic of the process. The fiber cross-section for various types of PCFs is shown in Figure 9.37. One of the capillaries could be rare earth-doped for amplifier fiber fabrication. The process is very general and could be applicable to wide range of materials choices. In principle, many types of glasses (silica, fluorides, germinates, etc.) can be used. PCFs have high nonlinearity due to small waveguide size and hence are useful for a study of nonlinear optical processes.

9.9 Fiber sensors

Several authors have reported results on fiber optic sensors which utilize an FBG and a broadband light source [90–94]. The light sources in many studies have been the amplified spontaneous emission (ASE) from an SOA (or SLD) type source and in some cases it has been the ASE from a doped-fiber amplifier. The sensing element generally is the FBG whose reflection or transmission property changes (i.e. it reflects or transmits a slightly different wavelength) due to a change in its environment. Such changes could be temperature, pressure, vibration, or any mechanical or electromechanical changes that change the effective periodicity of the grating through a change in grating period or a change in refractive index of the grating medium.

The operating principle of an FBG-based sensor system is to monitor the shift in wavelength of the returned reflected signal with the changes in the physical properties of the grating or its environment (e.g. strain, temperature). The Bragg wavelength

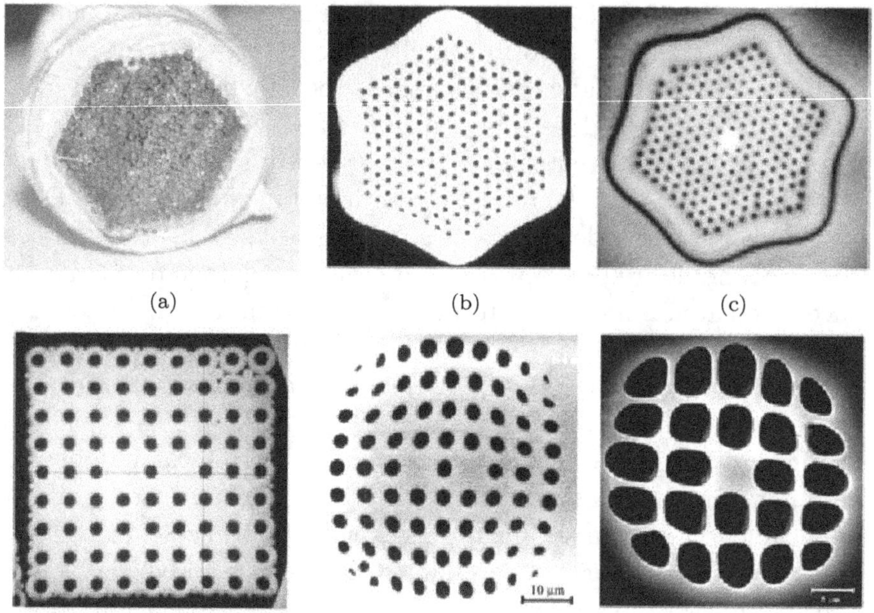

(a) (b) (c) (d) (e) (f)

Figure 9.37: Fabrication of photonic crystal fibers: preforms, intermediate preforms and final fibers: (a) the preform of fiber with holes spaced 1 mm with hexagonal lattice; (b) the intermediate preform with air holes spaced 250 μm with hexagonal lattice; (c) the fiber diameter of 120 μm, air holes diameter $d = 3$ μm; (d) the intermediate preform of double core fiber with square lattice; (e) a double core fiber with a square lattice, diameter of 250 μm, air holes diameter $d = 2.5$ μm; (f) a multimode fiber with a square lattice; diameter of the fiber 160 μm, air holes diameter 3 μm [89].

(the wavelength of the reflected signal, λ_B) is given by [90]

$$\lambda_B = 2n\Lambda \tag{9.16}$$

where Λ is the grating pitch and n is the effective index of the core. When a spectrally broadband source of light is injected into the fiber with an FBG, a narrowband spectral component at the Bragg wavelength is reflected by the grating (Figure 9.38). In the transmitted light, this spectral component is missing. The temperature response of the sensor arises primarily through a change in index (n in Eq. (9.16)) and the strain response is due to a physical change in length

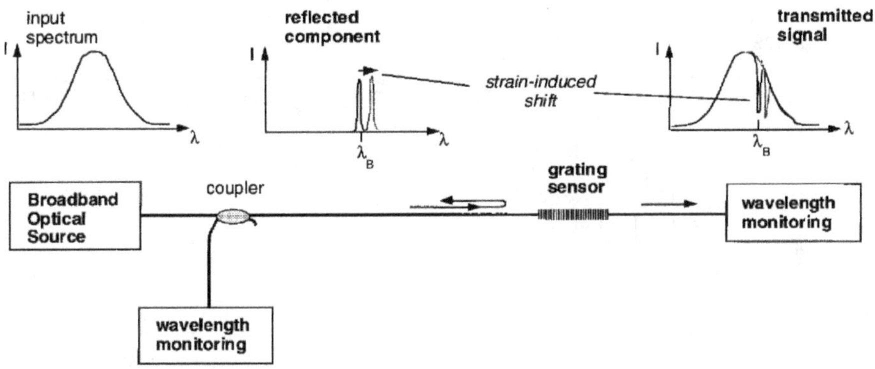

Figure 9.38: Schematic of a fiber Bragg grating (FBG)based sensor [94].

of sensor (i.e. small change in grating pitch Λ). The shift in Bragg wavelength (λ_B)due to a change in strain or temperature can be expressed by [94]

$$\Delta\lambda_B = 2n\Lambda[\{1-(n^2/2)(P_{12}-\nu(P_{11}+P_{12}))\}e+\{\alpha+(dn/dT)/n\}\Delta T] \tag{9.17}$$

where P_{ij} is the Pockel's (piezo) coefficients of the stress-optic tensor, ν is Poisson's ratio, e is the applied strain, α is the coefficient of thermal expansion (CTE) of the fiber material (e.g. silica), and ΔT is the temperature change. The quantity $(n^2/2)(P_{12} - \nu(P_{11} + P_{12}))$ has a numerical value of $\sim$0.22.

The strain response of an FBG at constant temperature has been measured. It is found to be [94]

$$(1/\lambda_B)(\delta\lambda_B/\delta\varepsilon) = 0.78 \times 10^{-6}\mu\varepsilon^{-1} \tag{9.18}$$

($\mu\varepsilon$ is microstrain). This response gives a "rule-of-thumb" measure of the grating reflectivity peak shift with strain of 1 nm per 1000 $\mu\varepsilon$ at 1300 nm wavelength. In silica fibers, the thermal response is the dominant effect, which accounts for 95% of the observed shift. The normalized thermal response at constant strain is [94]

$$(1/\lambda_B)(\delta\lambda_B/\delta T) = 6.7 \times 10^{-6}\mathrm{C}^{-1} \tag{9.19}$$

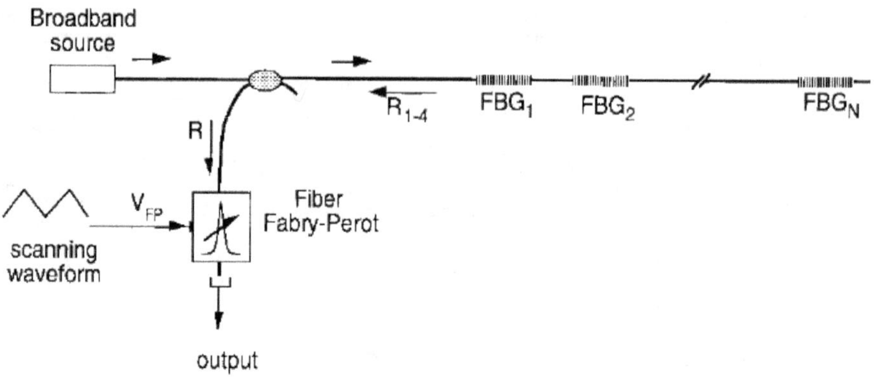

Figure 9.39: Scanning filter for monitoring FBG output [94].

A wavelength resolution of ~1 pm (0.001 nm) is required (at ~1300 nm) to resolve a temperature change of ~0.1°C or a strain change of 1 μ strain.

One of the advantages of Bragg gratings is that the sensed information is encoded directly into the wavelength, which is an absolute parameter. The output wavelength does not depend directly on the total light levels, losses in the connecting fibers and couplers, or source power. The FBG- based sensor system allows wavelength division multiplexing, i.e. each sensor is assigned a different wavelength band of the broadband source spectrum. This allows quasi-distributed sensing of strain, temperature, or other effects by associating each spectral band with a particular spatial location. The concept is illustrated in Figure 9.39. A scanning filter (by tuning voltage) is used to examine the reflected output. From the voltages (or wavelengths) at which the signal is observed the FBGs can be monitored.

Temperature and strain sensors using FBG are used in many applications where remote monitoring is required.

Optical fibers and optical fiber-based devices continue to be developed for many applications including optical fiber communications, high-power lasers, and FBG-based sensors. New materials development for mid-infrared fibers and PCFs with low-optical loss are important research areas.

References

1. C. K. Kao and G. A. Hockmann, *Proc. IEEE*, **133** (1966) 1151.
2. D. B. Keck, *IEEE Comm. Mag.*, **23** (1985) 17.
3. J. B. MacChesney and P. B. O'Connor, US Patent 4,217,027 (1980).
4. S. R. Nagel, Chapter 4, in S. E. Miller and I. P. Kaminow (eds.), *Optical Fiber Telecommunications*, 2nd Edn.
5. K. J. Beales and C. R. Day, *Phys. Chem. Glasses*, **21** (1980) 5.
6. K. J. Beales, S. F. Carter, P.W. France, and S. Partington *J. Non-crystalline Solids*, **70** (1985) 253.
7. K. L. Walker, G. M. Hornsy and F. T. Geyling, *J. Colloid. Interface Sci.*, **69** (1979) 138.
8. K. L. Walker, F. T. Geyling, and S. R. Nagel, *J. Amer Cer. Soc.*, **63** (1980) 552.
9. J. Koenings, D. Kuppers, H. Lydtin, and H. Wilson, *Proc. 5th Int. Conf. CVD*, 270 (1975).
10. P. Geittner, H. J. Hagemann, J. Warnier, and H. Wilson, *J. Lightwave Tech.*, **LT-4** (1986) 818.
11. H. Lydtin, *J. Lightwave Tech.*, **LT-4** (1986) 1034.
12. D. B. Keck, P.C. Shultz, and F. Zimar, US patent 3,737,292 (1973).
13. D. B. Keck, R. D. Maurer, and P. C. Shultz, *Appl. Phys. Lett.*, **22** (1973) 307.
14. P. C. Shultz, *Appl. Opt.*, **18** (1979) 3684.
15. T. Izawa, T. Miyashita, and F. Hanawa, US Patent 4,062, 655 (1977).
16. T. Izawa and N. Inagaki, *Proc. IEEE*, **68** (1980) 1184.
17. F. Hanawa, S. Sudo, M. Kawachi, and M. Nakahara, *Electron Lett.*, **16** (1980) 299.
18. L. L. Blyler and F. V. DiMarcello, *Proc. IEEE*, **68** (1980).
19. F. V. DiMarcello, C. R. Kurkjian, and J. C. Williams, Chapter 4 Optical Drawing and Strength Properties, in T. Li (ed.), *Optical Fiber Communications*, Vol. 1, AcademicPress, NY. 1985.
20. A. H. Gnauck, R. W. Tkach, A. R. Chraplyvy, and T. Li, *IEEE J. Lightwave Tech.*, Vol. 26, 1032, 2008.
21. N. K. Dutta, *Fiber Amplifiers and Fiber Lasers*, World Scientific, Singapore, 2014; P.C. Becker, N. A. Olsson, and J. R. Simpson, Erbium Doped Fiber Amplifiers, Academic Press, New York, 1997.
22. J. Simpson, Fiber Laser Sources and Amplifiers, in M. J. F. Digonnet (ed.), *Proc. SPIE*, **1171** (1989) 2; D. J. DiGiovanni, Fiber Laser Sources and Amplifiers, in M. J. F. Digonnet (ed.), *Proc. SPIE*, **1373** (1990) 2.
23. B. J. Ainslie, *J. Lightwave Tech.*, **9** (1991) 220.
24. P. W. France, S. F. Carter, M. W. Moore, and C. R. Day, *Br. Telecom Tech. J.*, **5** (1987) 28.
25. S. Takanashi and H. Iwasaki, Chapter 5, in I. D. Agrawal and G. Lu (eds.), *Flouride Glass Fiber Optics*, Academic Press, New York, 1991.
26. Y. Ohishi, S. Mitachi, and S. Takahashi, *J. Lightwave Tech.*, **LT-2** (1984) 593.

27. G. P. Agrawal and N. K. Dutta, *Semiconductor Lasers*, Van Nostrand Reinhold, NY, 1992.
28. A. Tunnermann and H. Zellmer, Chapter B4.2, in C. Webb and J. Jones (eds.), *Handbook of Laser Technology and Applications*, Institute of Physics, 2004.
29. Y. Jeong, J. K. Sahu, D. N. Payne, and J. Nilsson, *Opt. Exp.*, **12** (2004) 6088.
30. A. Othonos and K. Kalli, Fiber Bragg Gratings, Artech House, Boston, 1999, Chapter 5.
31. D. Hanna, Chapter B4.1, in C. Webb, and J. Jones (eds.), *Handbook of Laser Technology and Applications*, Institute of Physics, 2004.
32. A. Tunnermann and H. Zellmer, Chapter B4.2, in C. Webb and J. Jones (eds.), *Handbook of Laser Technology and Applications,* Institute of Physics, 2004.
33. G. A. Ball and W. W. Morey, *IEEE Photon. Technol. Lett.*, **3** (1991) 1077.
34. J. Y. Alain, J. F. Bayon, P. Bernage, T. Georges, M. Monerie, and P. Niay, in *Topical Meet on Optical Ampl$ers and Their Applications*, Vol. 14, Tech. Dig., Washington, DC, Optical Society of America, 1993, p. 57.
35. J. T. Kringlebotn, J. L. Archambault, L. Reekie, J. E. Townsend, G. G. Vienne, and D. N. Payne, *Electron. Lett.*, **30** (1994) 972.
36. S. V. Chernikov, J. R. Taylor, and R. Kashyap, *Opt. Lett.*, **18** (1993) 2023.
37. G. A. Ball, G. Meltz, and W. W. Morey, *Opt. Lett.*, **18** (1993) 1976.
38. V. Mizrahi, D. J. DiGiovanni, R. M. Atkins, S. G. Grubb, Y. K. Park, and J. M. Delavaux, *J. Lightwave Tech.*, **11** (1993) 2021.
39. J. L. Zyskind, V. Mizrahi, D. J. DiGiovanni, and J. W. Sulhoff, *Electron. Lett.*, **28** (1992) 1385.
40. J. L. Zyskind, J. W. Sulhoff, P. D. Magill, K. C. Reichmann, and D. J. DiGiovanni, *Electron. Lett.*, **29** (1993) 1105.
41. Y. Jeong, J. K. Sahu, D. N. Payne, and J. Nilsson, *Opt. Exp.*, **12** (2004) 6088.
42. A. Liem, J. Limpert, H. Zellmer, A. Tünnermann, V. Reichel, K. Mörl, S. Jetschke, S. Unger, H.-R. Müller, J. Kirchhof, T. Sandrock, and A. Harschak, in *Conference on Lasers and Electro-Optics,* May 2004, Postdeadline paper CPDD2, 2004.
43. V. P. Gapontsev, D.V. Gapontsev, N. S. Platonov, O. Shkurikhin, V. Fomin, A. Mashkin, M. Abramov, and S. Ferin, in *European Conference on Lasers and Electro-Optics*, July 2005, Paper CJ1-1, 2005.
44. N. S. Platonov, D. V. Gapontsev, V. P. Gapontsev, and V. Shumilin, in *Proc. CLEO 2002*, May 2002 Postdeadline paper CPDC3, 2002.
45. J. Limpert, A. Liem, H. Zellmer, and A. Tunnermann, *Electron. Lett.*, **39** (2003) 645.
46. Y. W. Lee, S. Sinha, M. J. F. Digonnet, and R. L. Byer, in *Proc. CLEO 2006*, Paper CTuI5, 2006.
47. Z. Chen, S. Ma, and N. K. Dutta, *Opt. Exp.*, **17** (2009) 1234.
48. Z. Ahmed and N. Onodera, *Electron. Lett.*, **32** (1996) 455.
49. M. Y. Jeon, H. K. Lee, J. T. Ahn, D. S. Lim, H. Y. Kim, K. H. Kim, and E-H Lee, *Electron. Lett.*, **34** (1998) 182.

50. S. Li, C. Lou, and K.T.Chan, *Electron. Lett.*, **34** (1998) 375.
51. P. Das, W. Kaechele, J. P. Theimer, and A. R. Pirich, *SPIE*, **3075** (1997) 21.
52. C. Wu and N. K. Dutta, *IEEE J. Quant. Electron.*, **36** (2000) 145.
53. T. Pfeiffer and G. Veith, *Electron. Lett.*, **29** (1993) 1849.
54. M. Nakazawa, E. Yoshida, and Y. Kimura, *Electron. Lett.*, **30** (1994) 1603.
55. E. Yoshida, and M. Nakazawa, *Electron. Lett.*, 1996, **32** (1996) 1370.
56. J. W. Nicholson and D. J. DiGiovanni, *IEEE Photon. Technol. Lett*, **20** (2008) 2123.
57. J. W. Nicholson, R. S. Windeler, and D. J. DiGiovanni, *Opt. Exp*, **15** (2007) 9176.
58. Y. W. Song, S. Yamashita, C. S. Goh, and S. Y. Set, *Opt. Lett*, **32** (2007) 148.
59. K. Kieu and M. Mansuripur, *Opt. Lett*, **32** (2007) 2242.
60. N. Nishizawa, *et al*, *Opt. Exp*, **16** (2008) 9429.
61. Z. Sun, *et al*, *Appl. Phys. Lett.*, **93** (2008) 061114-1.
62. A. G. Rozhin, Y. Sakakibara, S. Namiki, M. Tokumoto, H. Kataura, and Y. Achiba, *Appl. Phys. Lett.*, **88** (2006) 051118-1.
63. R. M. Gerosa, D. Steinberg, H. G. Rosa, C. Barros, C. J. S. de Matos, and E. A. T. de Souza, *IEEE Photonics Tech. Lett.*, **25** (2013) 1007.
64. H. Zhang, D. Tang, R. J. Knize, L. Zhao, Q. Bao, and K. P. Loh, *Appl. Phys. Lett.*, **96** (2010) 111112.
65. P. L. Huang, S. Lin, C. Yeh, H. Kuo, S. Huang, G. Lin, L. Li, C. Su, and W. Cheng, *Opt. Exp.*, **20** (2012) 2460.
66. G. Sobon, J. Sotor, I. Pasternak, K. Grodecki, P. Paletko, W. Strupinski, Z. Jankiewicz, and K. M. Abramski, *IEEE J. Lightwave Tech.*, **30** (2012) 2770.
67. G. P. Agrawal, Nonlinear Fiber Optics, 4th edn. Academic Press, NewYork, 2007.
68. R. G. Smith, *Appl. Opt.*, **11** (1972) 2489.
69. R. H. Stolen, Chapter 10, in S. E. Miller and A. G. Chynoweth (eds.), *Optical Fiber Telecommunications,* Academic Press, New York, 1979.
70. R. H. Stolen and J. E. Bjorkholm, *IEEE J. Quant. Electron.*, **QE-18** (1982) 1062.
71. N. Bloembergen, The stimulated Raman effect, *Am. J. Phys.*, **35** (1967) 989.
72. S. Grubb, *Proc. Optical Amplifiers and Their Applications*, Paper PD3-1, 1994.
73. S. Grubb, *Proc. Optical Amplifiers and Their Applications*, Paper SA-4, 197, 1995.
74. Y. Feng, L. R. Taylor, and D. B. Calia, *Opt. Exp*, **17** (2009) 23678.
75. C. A. Codemard, J. Ji, J. K. Sahu, and J. Nilsson, in *Proc. SPIE*, San Francisco, CA, USA, **7580** (2010) 75801.
76. J. AuYeung and A. Yariv, *J. Opt. Soc. Amer*, **69** (1979) 803.
77. H. Zhang, H. Xiao, P. Zhou, X. Wang, and X. Xu, *IEEE Photon J.*, **5** (2003) 1501706.

78. E. Bélanger, M. Bernier, D. Faucher, D. Côté, and R. Vallée *IEEE Photon J, J. Lightwave Tech.*, **26** (2008) 1696.
79. A. J. Stentz, T. Nielsen, S. G. Grubb, T. A. Strasser, and J. R. Pedrazzani, in *Optical Fiber Communication Conf. OSA Tech. Dig.* Washington, DC, 1996.
80. C. R. S. Fludger and V. Henderek, in *Optical Fiber Communication Conf. OSA Tech. Dig.*, Washington, DC, 2001.
81. F. Galeener, J. Mikkelsen, R. Geils, and W. Mosby,*Appl. Phys.Lett.*, **32** (1978) 34.
82. M. Nissov, C. R. Davidson, K. Rottwitt, R. Menges, P. C. Corbett, D. Innis, and N. S. Bergano, in *European Conference Optical Commun.*, 1997, PD Paper, p. 9.
83. M. N. Islam, *IEEE J. Sel. Top. Quant. Electron.*, **8** (2002) 248.
84. S. Namiki and Y. Emori, *IEEE J. Sel. Top. Quant. Electron.*, **7** (2001) 3.
85. J. C. Knight, T. A. Birks, P. St. J. Russell, and D. M. Atkin, *Opt. Lett.*, **21** (1996) 1547.
86. J. C. Knight, J. Broeng, T. A. Birks, and P. S. Russell, *Science,* **282** (1998) 1476.
87. P. S. Russell, *Science,* **299** (2003) 358.
88. R. F. Cregan, B. J. Mangan, J. C. Knight, T. A. Birks, P. S. Russell, P. J. Roberts, D. C. Allan, *Science,* **285** (1999) 1537.
89. R. Buczynski, *Acta Physica Polonica,* **106** (2004) 141.
90. W. W. Morey, J. R. Dunphy, and G. Meltz, in *Proc. SPIE Distributed and Multiplexed Fiber Optic Sensors*, Boston, MA, **1586** (1991) 216.
91. S. M. Melle, K. Liu, and R. M. Measures, *IEEE Photon. Technol. Lett.*, **4** (1992) 516.
92. M. A. Davis and A. D. Kersey, *Electron. Lett.*, **30** (1994) 75.
93. M. A. Davis, D. G. Bellemore, and A. D. Kersey, in *Proc. SPIE 1995 North American Conf. Smart Structures Materials*, San Diego, CA, **2446** (1995) 227.
94. A. D. Kersey, M. A. Davis, H. J. Patrick, M. LeBlanc, K. P. Koo, C. G. Askins, M. A. Putnam, and E. J. Friebele, *J. Lightwave Tech.*, **15** (1997) 1442.

Chapter 10

Planar Waveguide Amplifiers and Lasers

10.1 Introduction

Rare earth-doped optical fiber amplifiers and lasers are important devices for both terrestrial and submarine optical communication systems, for sensor systems, and as laser sources at various wavelengths for various applications including high power applications. Optical fiber amplifiers are typically 10–30 m in length. The core and cladding of the fiber form a waveguide through which the light travels and undergoes amplification. The core is typically doped with the rare earth dopant. The length of a planar waveguide amplifier (PWA) is typically limited to tens of centimeters both due to fabrication limitation and higher optical attenuation in planar waveguides. In order to achieve high gain in ~10 cm long planar waveguides, the rare earth concentration is typically two orders of magnitude higher than that for fibers. This may introduce non-radiative mechanisms in some rare earths such as Er due to up-conversion.

PWAs have been fabricated in many glass materials by several methods [1–8]. These include modifications of refractive index at the surface by ion exchange [9], ion implantation [11], or ion diffusion [12]. Also, a waveguide can be formed by growth or deposition of one material on top of another of a different refractive index [13, 14]. Usually, the waveguides formed by these techniques are in the form of a planar thin film which may be patterned into an array of channel waveguides using photolithographic techniques. The optical confinement of the pump light in the waveguide results in low threshold for net gain or laser action for rare-earth-doped planar waveguide devices. However, key to achieving low-threshold operation is low optical loss in the material. The schematics of typical planar and channel waveguide structures are shown in Figure 10.1.

Efficient coupling of high power diode pump sources is important for low pump threshold. Methods such as optical fiber-based coupler for single mode waveguides, specialized lens optics, and diode beam shapers have been used for efficient coupling. In general, the pump light can be coupled from the end faces of the waveguide, from the side or from the surface. This is shown schematically in Figure 10.2.

Diode lasers are suitable for end pumping because the laser mode is laterally spread like the waveguide allowing good overlap between the gain region in the waveguide and the diode laser mode. A typical configuration for end pumping of waveguide lasers is shown in Figure 10.3(a). Also, shown is the asymmetric lasing and pump beam profile in the waveguide. Dichroic mirrors with high transmission at pump wavelength and high reflectivity at laser wavelength are often used before the fiber end. For double clad structures needed for high power laser fabrication, diode bars (which emit many Ws of power) are used for pumping. The diode bar can be placed next to the

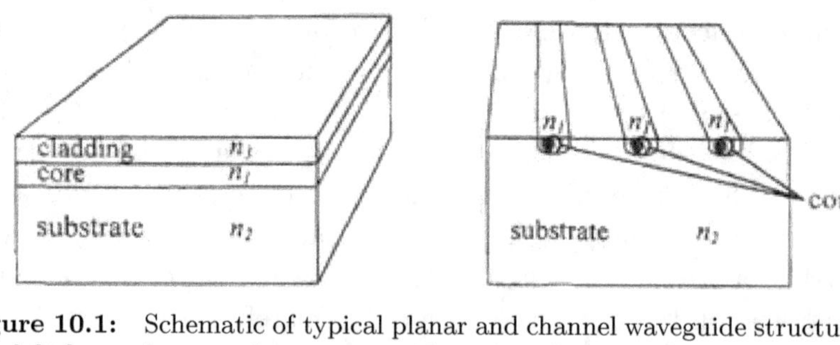

Figure 10.1: Schematic of typical planar and channel waveguide structures [8]. The left figure shows a planar waveguide and to the right is channel waveguide. The core index n_1 is larger than cladding index, n_2.

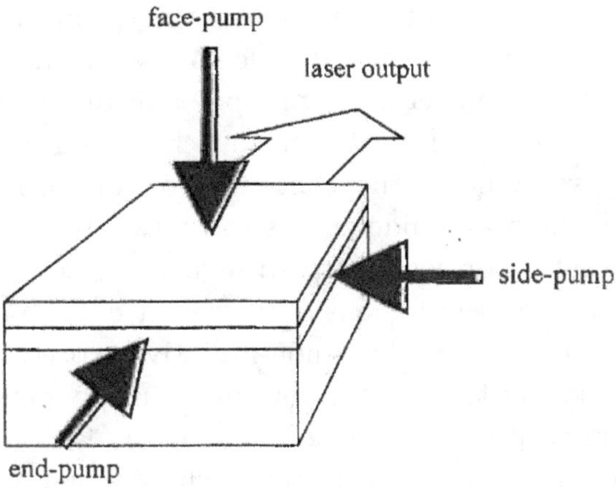

Figure 10.2: Schematic of pump laser coupling into the waveguide [8].

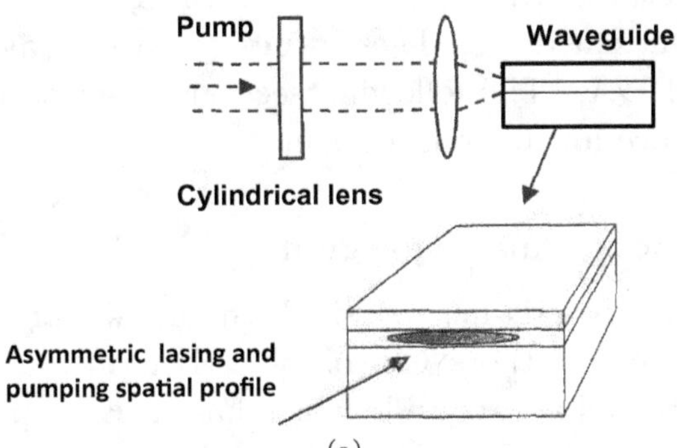

(a)

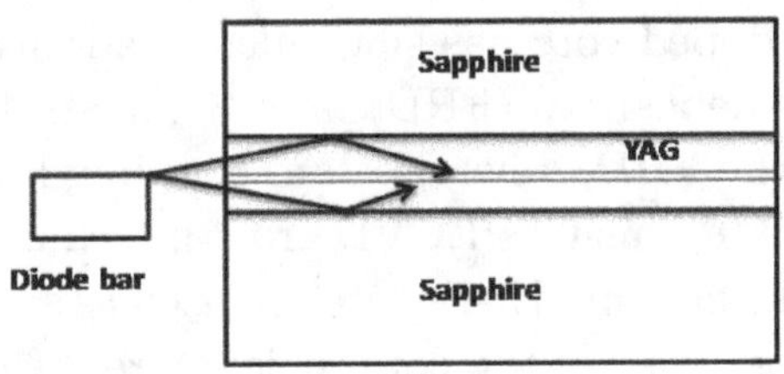

(b)

Figure 10.3: (a) Schematic of end pumping a planar waveguide using a diode laser. Also, shown is the pump laser mode in the waveguide. (b) Schematic of a double core waveguide pumping using a diode bar [8, 15].

waveguide with or without a lens and the light captured within the high numerical aperture of the double core waveguide gets coupled into the waveguide and serves as pump beam [Figure 10.3(b)].

As mentioned earlier, in order to achieve significant gains in ∼10 cm long waveguides, the concentration of rare earth dopant (e.g. Er) in planar waveguides needs to be two orders of magnitude greater than that for optical fibers. At higher concentration, the average spacing between the Er ions decrease. At close spacings, the Er ions lose the exited energy by a non-radiative process known as up-conversion. Consider two Er ions both in their upper energy levels, in the up-conversion process, one electron (in one Er ion) decays non-radiatively to the ground state, transferring its energy to the other Er ion, which gets excited to a higher energy level. Up-conversion is less pronounced in other glass types (e.g. soda lime glass, due to its cage-like structure) where the atoms in the glass surround each Er ion and prevent clustering. These types of glasses have been used in the study of PWAs. The following sections describe the fabrication and performance for different dopants.

10.2 Er-Doped planar waveguide

Er-doped silica-based nearly single-mode waveguide has been reported by several authors. One of the early reports is by Shuto *et al.* [19]. They reported a waveguide which has an Er-doped phosphosilicate glass core formed by plasma enhanced chemical vapor deposition (PECVD). The doped core has silica glass cladding layer formed by flame hydrolysis deposition (FHD) on a Si substrate. The core layer is deposited on the FHD grown bottom cladding layer at a substrate temperature of 400°C using PECVD growth technique using SiH and N_2O gases. P-alkoxide and Er-chelate are used as sources for P and Er doping. The lateral size of the core regions are etched using reactive ion etching (RIE) and then the core layer is embedded in the cladding region grown by FHD process. Thus, the Er forms a buried waveguide region. The refractive index of the core is ∼1.6% higher than that of the surrounding cladding regions. The Er concentration in the core is 0.48%. Prior to amplification studies, the propagation loss

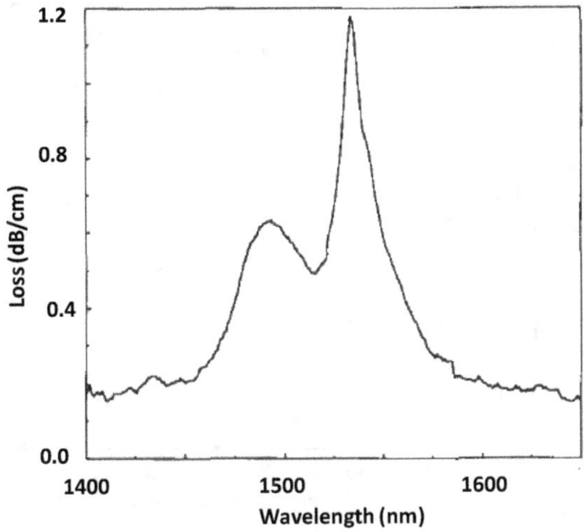

Figure 10.4: Absorption (optical loss) as a function of wavelength [19].

of the waveguide is measured. The absorption spectrum is shown in Figure 10.4. The optical loss at 1530 nm is ~1.1 dB/cm. The waveguide is 7.5 cm long and has a spot size of $8\times11\ \mu m^2$ at 1550 nm. Pump laser light at a wavelength of 980 nm is coupled into the waveguide. The output signal at 1530 nm is measured as a function of pump power. The optical gain is determined from the ratio of the output to input signal power. The data are shown in Figure 10.5. Using an S-shaped waveguide and a similar fabrication process as described earlier, a gain of ~10 dB has been reported at a pump power of 100 mW, and gain as high as 16 dB has been reported at a pump power of ~210 mW [20].

Soda lime glass has been used as a core for Er doping [21]. For high gain at low pump power, optimizing Er concentration is important in addition to waveguide dimensions and index difference between the core and the cladding regions.

Although high Er concentration provides high gain, it also increases the up-conversion rate [22–24]. For the experiment of Ref. [21], an Er concentration of $0.7\times10^{20}\text{cm}^{-3}$ is used. Er-doped films are deposited by RF sputtering from a soda lime glass target (doped with Er) on a silica lower cladding layer. This layer is 15 μm

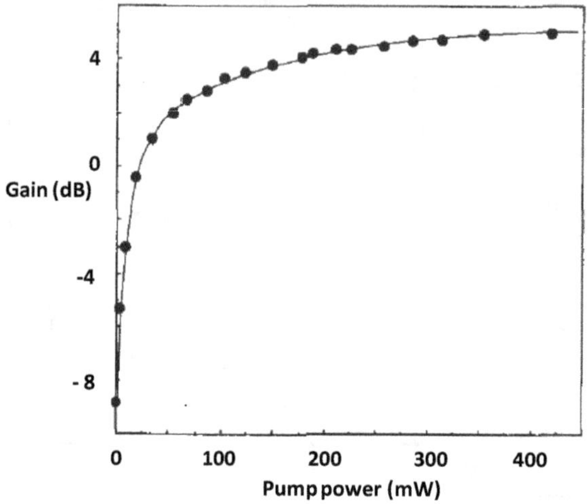

Figure 10.5: Net gain at 1530 nm as a function of input pump power at 980 nm [19].

thick and is deposited on silicon substrate by thermal oxidation. Ion milling is used to produce 3–9 μm wide waveguides on the deposited Er-doped film. The refractive index difference between the core and lower cladding layer is 3%. The optical gain of the amplifier is measured as a function of pump power. The observed low threshold (~10 mW) for net gain is due to low scattering loss, tightly confined mode, and low glass absorption loss [21].

10.2.1 *Optical gain and gain saturation*

Amplifier performance characteristics such as gain spectrum, gain saturation power, and noise figure discussed earlier (for fiber amplifiers) have been studied for Er-doped PWAs. As expected these amplifiers can be pumped with 1480 nm pump lasers in addition to 980 nm pump lasers [25–28].

The small signal gain spectrum of a PWA for various pump powers is shown in Figure 10.6 [28]. The PWA is 14 cm long and has alumino-silicate guide layers. The amplifier is pumped using a 1470 nm pump laser. The gain spectrum is measured using the input signal from a tunable (grating tuned) external cavity laser. The input

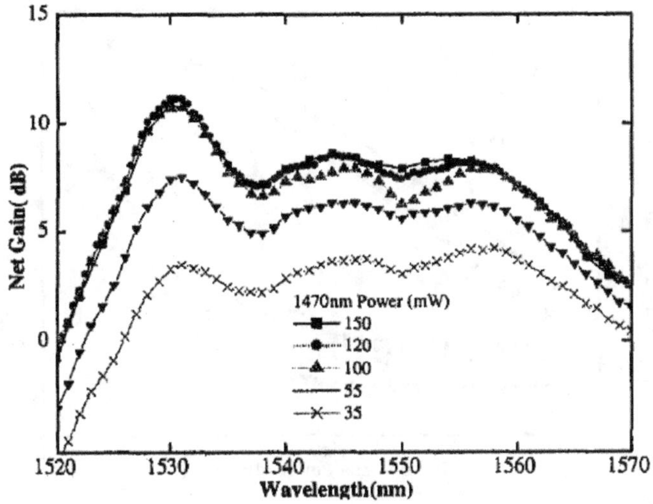

Figure 10.6: Gain spectrum for various pump powers for an Er-doped PWA. The pump laser wavelength is 1470 nm. The gain is reasonably flat at low pump powers [28].

laser wavelength can be tuned from 1520 to 1570 nm. The measured gain is reasonably flat for low pump powers suggesting the PWA is suitable for multiwavelength WDM transmission. Similar to fiber amplifiers, the PWA gain saturates with increasing signal power. The noise figure of the amplifier is about 8 dB in the unsaturated regime and $\sim$12 dB in the saturated regime.

The noise figure of the amplifier is measured for various input signal wavelengths and input signal powers. The data are shown in Figure 10.7. The noise figure is obtained using the amplified spontaneous emission (ASE) spectrum of the amplifier. Similar to fiber amplifiers, the PWA gain saturates with increasing signal power. The saturation of gain in amplifier is demonstrated by plotting the net gain as a function of output power. The data for an Er-doped PWA pumped using a 980 nm laser are shown in Figure 10.8.

10.3 Nd-doped planar waveguide

Single mode Nd-doped waveguide lasers fabricated on a Si substrate have been reported [38–41]. To fabricate the doped waveguide, $SiCl_4$

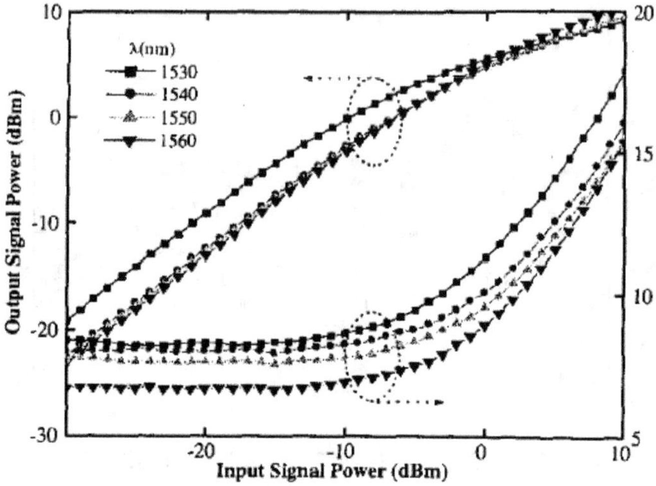

Figure 10.7: Output signal power and noise figure as a function of input signal power. The pump laser is at 1480 nm [28].

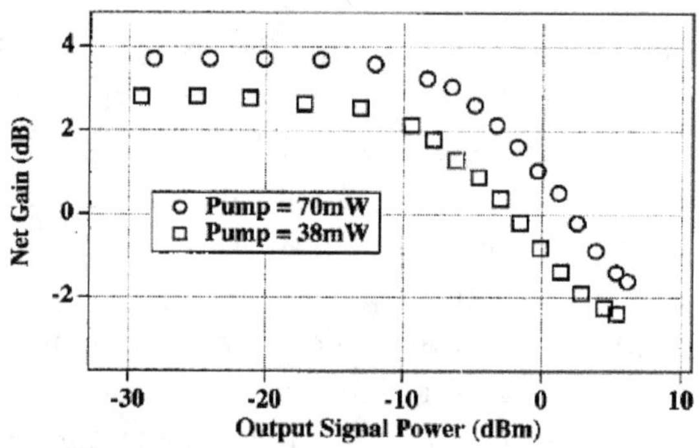

Figure 10.8: Gain is plotted as a function of output signal power of an Er-doped PWA. The pump laser is at 980 nm [21]. Gain decreases with increasing power due to gain saturation.

and PCl_3 are hydrolyzed (FHD) using an oxy-hydrogen burner to form a low-density soot on a Si substrate with a 10 μm thick SiO_2 layer (thermally grown). The soot is fused and immersed in an alcohol solution of NdCl. After drying, the porous structure is fused to form the Nd-doped SiO_2–P_2O_5 glass. Ridge waveguides are formed on the glass using standard photolithography and RIE. FHD is then used

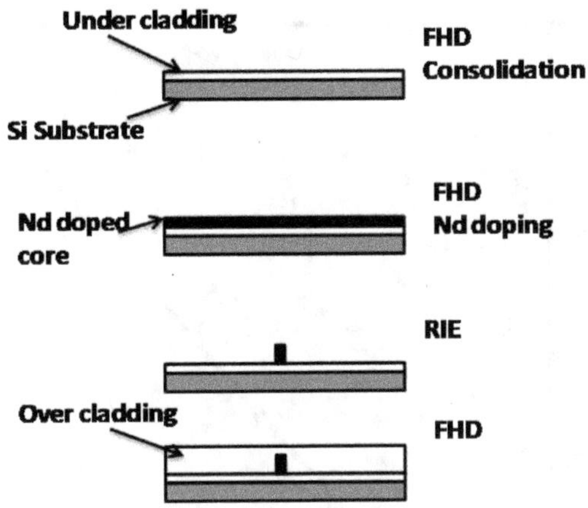

Figure 10.9: Schematic of the fabrication process [41]. The process starts at the top.

to form a cladding layer of SiO_2–P_2O_5-B_2O_3. The cladding layer is index matched to buffer layer and is ~20 μm thick. The schematic of the fabrication process is shown in Figure 10.9. In one experiment, the waveguide dimensions are 12 × 8 μm and the Nd-doping density is 0.38 wt%. The index difference between core and cladding is 0.87% and the waveguide device length is 5.7 cm. The laser cavity is formed by using dielectric coated mirror in close proximity to the waveguide ends using index matching liquids. The input mirror has a reflectivity of 99.9% at the laser wavelength (~1060 nm) and transmission of 90% at the pump wavelength (804 nm). The output mirror has a reflectivity of 95% at the laser wavelength and transmission of 95% at the pump wavelength. The laser output as a function of pump power is shown in Figure 10.10. The threshold power is ~20 mW.

The schematic of another waveguide laser design is shown in Figure 10.11. For this laser, the Nd-doped core has a Nd-doping level of 0.2 wt% and a thickness of 6 μm. The index difference between the core and cladding is 0.5%. The waveguide length is 4 cm and the widths varied in the 6–30 μm range. Dielectric mirrors are deposited on the ends of the waveguide cavity. The input mirror had high reflectivity ($R = 99.9\%$) at the lasing wavelength and high

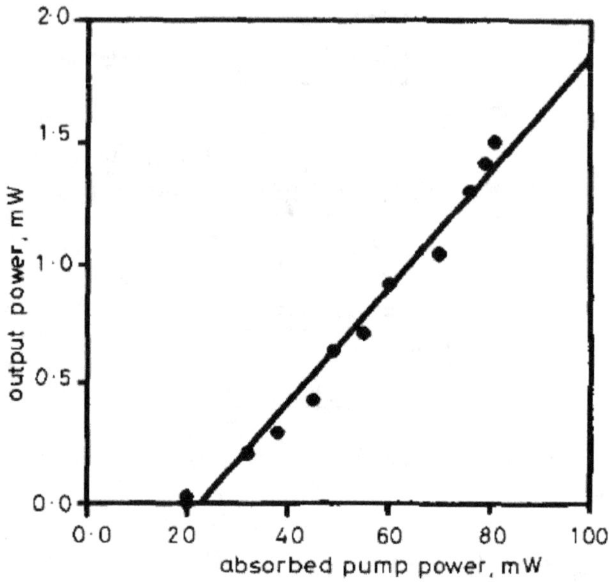

Figure 10.10: Laser output power as a function of absorbed pump power [42].

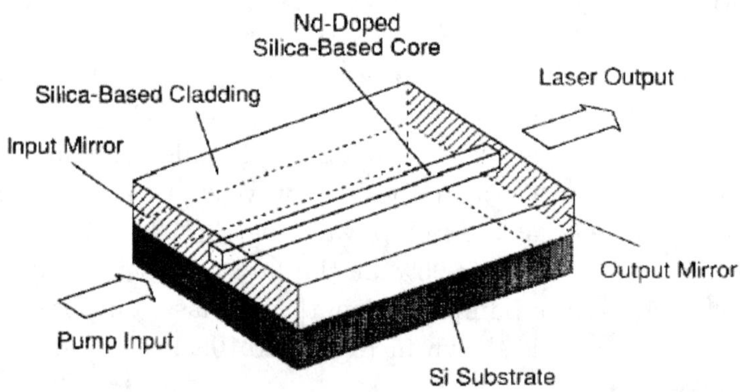

Figure 10.11: Schematic of the Nd-doped waveguide laser [41].

transmittance (>95%) in the pump wavelength range of 730–820 nm. The reflectivity of the output mirror is 95% at the lasing wavelength and the transmittance is >95% at the pump wavelengths. The lasing spectrum and the output power as a function of pump power for a Nd-doped waveguide laser of Figure 10.11 are shown in Figure 10.12.

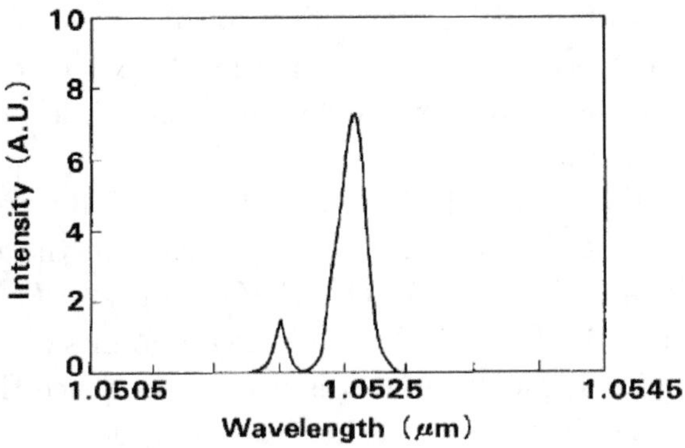

(a)

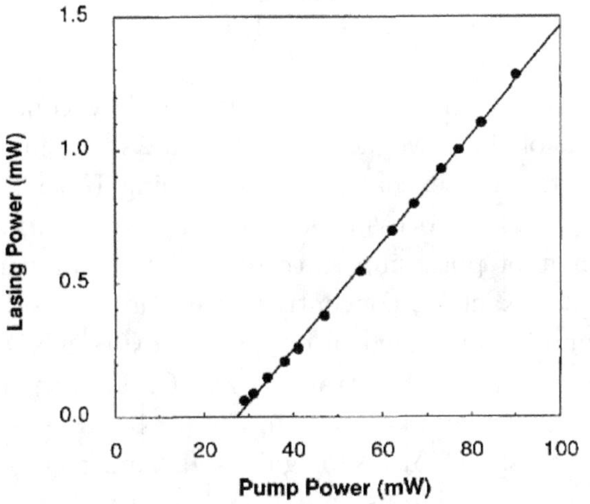

(b)

Figure 10.12: Lasing spectra (a) and laser output power as a function of pump power (b). The waveguide width is 8 μm [41]. Pump wavelength is 805 nm.

The threshold power is ~26 mW. The threshold pump power is found to vary with the pump wavelength. Lowest threshold power is obtained for 805 nm pump wavelength.

10.4 Er–Yb Co-doped planar waveguide

Similar to the case of Er-doped fiber amplifiers, Er-Yb co-doped waveguide amplifiers have been investigated [43–50]. This has the potential of reducing the effect of up-conversion which is a non-radiative process at high Er densities. In Er–Yb co-doped amplifiers, energy transfer from excited Yb to upper Er level results in gain for the Er transition.

The Er–Yb co-doped glasses containing ZnO are prepared by standard melting technique. The oxides used in the melt are SiO_2, Na_2O (14.2 mol%), ZnO, Al_2O_3, Er_2O_3, and Yb_2O_3. The samples prepared had 0.6 at% of Er, but differing amounts of Yb in the 0–1.2 at% range [47]. Typically Al_2O_3 is used to enhance the solubility of Er, ZnO is used in this case. The glass components are melted in a crucible at 1470°C. Mechanical stirring and quenching at 520°C are used to produce the bulk glass which is cut into 40 × 20 × 2 mm pieces for waveguide fabrication.

Waveguide boundaries are formed by standard contact lithography. The widths of the waveguides vary from 2 to 6 μm. Waveguides are formed by the ion exchange process using K to Na or Ag to Na ion exchange. The process of ion exchange with K ions was performed in a melt of potassium nitrate at 400°C for various times, depending on the exact Yb concentration in the glass substrate. The Ag ion exchange is performed in eutectic $NaNO_3$–KNO_3 melt with additional 3.6 wt% of silver nitrate at 280°C. The experimental set up used for measurements is shown in Figure 10.13. The pump laser (970 nm) and the signal (used for gain spectrum measurement) are coupled into the waveguide using a WDM multiplexer.

The intensity profiles of the guided modes and the waveguide scattering losses are experimentally determined. The mode field

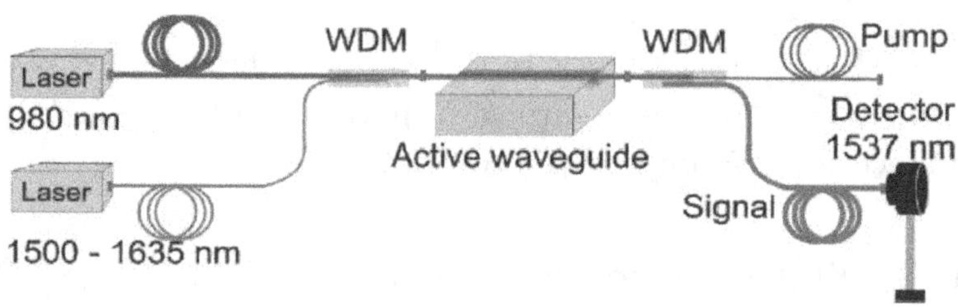

Figure 10.13: Schematic of the experimental set up used for Er–Yb co-doped ion-exchanged planar waveguide measurements [47].

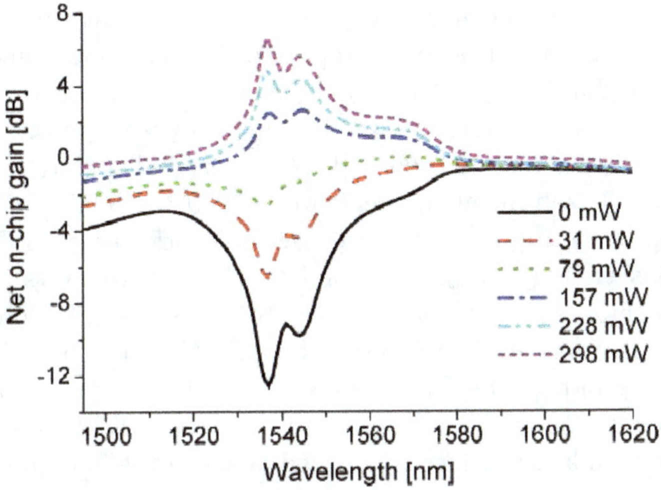

Figure 10.14: Spectral dependence of gain/absorption at various pump powers for the Er–Yb co-doped glass. Er, Yb concentrations at 0.6 and 1.2 at%, respectively [47].

profiles at the signal and pump wavelengths are obtained by near-field infrared imaging. In K and Ag ion-exchanged waveguides, the mode dimensions are found to be 11×14 μm and 3.5×5 μm at 1550 nm, respectively. In the Er emission band, the K and Ag ion-exchanged channel waveguide losses are estimated to be 0.18 and 0.8 dB/cm, respectively. The net on-chip gain as a function of pump power for K ion-exchanged waveguide is $\sim$5 dB at a pump power of 300 mW [47]. This Er–Yb co-doped glass has equal amounts (both 0.6 at%) of Er and Yb [47]. Optical gain at various input wavelengths is measured by using a tunable input laser. The spectral dependence of gain for several pump powers for a K ion-exchanged waveguide is shown in Figure 10.14. An up-conversion coefficient for Er of 1×10^{-24} m^{-3} s^{-1} is determined from the data on different Yb concentrations and from known model parameters.

10.5 Yb-doped planar waveguide

Yb-doped waveguide amplifiers have been investigated [51–55]. They generally have much higher gain due to high solubility and lack of non-radiative up-conversion process such as that present in Er. High optical gain has been reported in Yb-doped microstructured $KGd_xLu_yYb_{1-x-y}(WO_4)_2$ channel waveguides [51]. Yb-doped $KY(WO_4)_2$ layers were grown by liquid-phase epitaxy onto an undoped, (010)-oriented, polished, 1-cm^2-sized $KY(WO_4)_2$ substrate in a $K_2W_2O_7$ solvent at temperatures of 920–923°C [52]. Co-doping the layers with appropriate percentages of optically inert Gd^{3+} and Lu^{3+} ions enables the growth of lattice-matched layers [53]. The replacement of a fraction of Lu^{3+} by Yb^{3+} ions of similar ionic radius resulted in up to 47.5 at% Yb^{3+}-doped $KGd_xLu_yYb_{1-x-y}(WO_4)_2$ layers. The index difference between the Yb-doped layer and substrate is $\sim 2 \times 10^{-2}$. The samples are between 180–320 μm long and the Yb-doped layer surface is polished down to 2.2–3.5 μm thickness.

The ridge waveguides are fabricated on the wafer using Ar beam etching. This produced $\sim$1.4 μm deep ridges about 6 μm wide. Undoped $KY(WO_4)_2$ is then grown over the sample burying the ridges. The gain of the sample is measured using pump light from a Ti-sapphire laser emitting near 932 nm. The signal light is at 980 nm. The measured gain for different at% of Yb is shown in Figure 10.15.

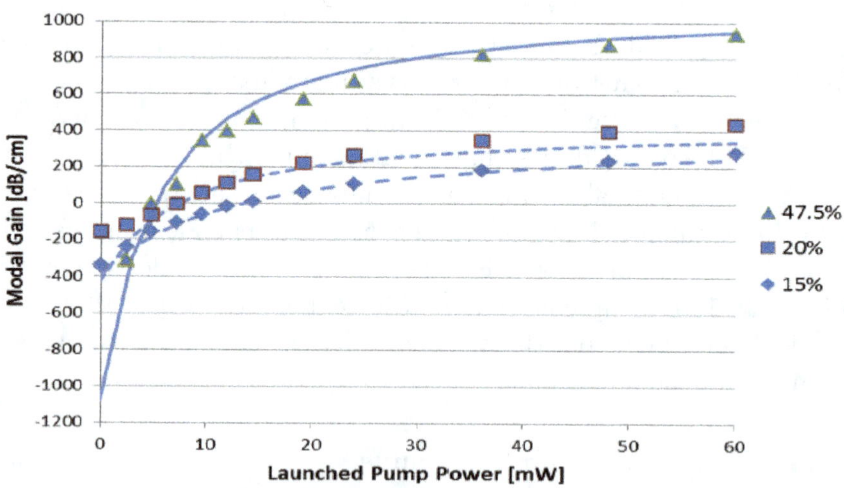

Figure 10.15: Measured small signal modal gain as a function of pump power for different at% of Yb. Signal wavelength is at 980 nm [51].

The measured maximum gain is ~900 dB/cm at 60 mW of pump power. The spectral dependence of gain is measured using a tunable laser for signal input. The optical gain exceeds 150 dB/cm over a 55 nm band. The above result illustrates that high gain can be obtained in Yb-doped planar waveguides. This suggests the possibility of high power laser emission using Yb doping with appropriate thermal management and pump input geometry.

10.6 High power planar waveguide lasers

For high power planar waveguide lasers, highest powers have been obtained using an array of diode bars in a side pumping geometry (Figure 10.2). This allows maximum coupling of pump light from diodes which have asymmetric beam profile. Nd or Yb or Tm are generally used as rare earth dopants for these high power lasers or amplifiers [56–61]. In addition to high pump power, another requirement is heat dissipation, i.e. a large fraction of the amount of pump power that is not converted to laser light gets absorbed and generates heat and this heat must be removed. Thus, mounting configurations with active cooling have produced high output powers. The planar waveguides for high power applications have a cladding pumped geometry [Figure 10.3(b)]. The calculated mode profile in the double clad Nd-doped yttrium aluminum garnet (YAG) structure is shown in Figure 10.16. The YAG is bounded by low-index sapphire on both sides. For end pumping, light from an 807 nm diode is coupled using suitable lenses to the Nd-doped waveguide. The Nd-doping level is 1 at%.

Sapphire has a lower index than YAG, thus, the light bounces at that interface and propagates along the waveguide much like a double clad fiber. The output mirror reflectivity is optimized for the particular Nd transition. It is ~78% for the 1064 nm transition. The measured results (including that of the most intense 1064 nm transition) are shown in Figure 10.17. For higher power operation side pumping and cooling is important [63–66]. The schematic diagram of a side-pumped cladding pumped waveguide laser in a water-cooled housing is shown in Figure 10.18. Two 20 W diode bars pump the

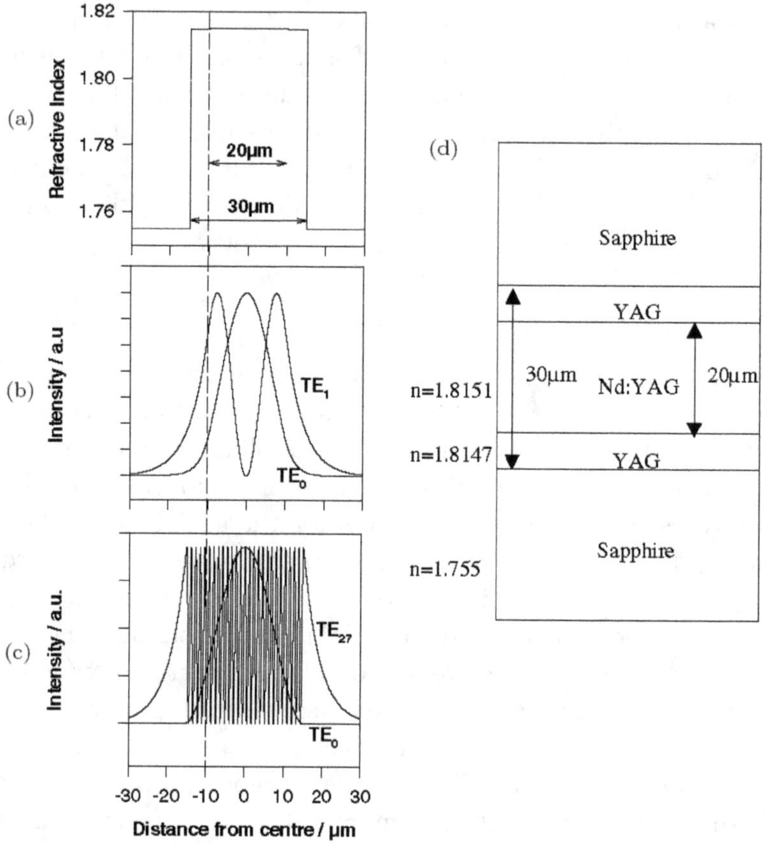

Figure 10.16: Double clad planar waveguide structure with Nd-doped YAG. The calculated mode profiles are also shown. Similar structures have been made using Yb- and Tm-doped YAG [56].

waveguide from each side. The waveguides are 1 cm long and 5 mm wide. The pump light propagates along the short dimension. The Nd-doping level is 1 at%. The schematic of the layers in the planar waveguide is shown in the inset. Sapphire has a lower index than YAG, thus the light propagates along the waveguide.

Double clad planar waveguide designs as shown in Figure 10.16 have been made using Nd, Yb, and Tm doping. The doping levels are 10 at% for Yb and Tm and 1 at% for Nd (as described previously). The waveguides have inner cladding widths of 18 and

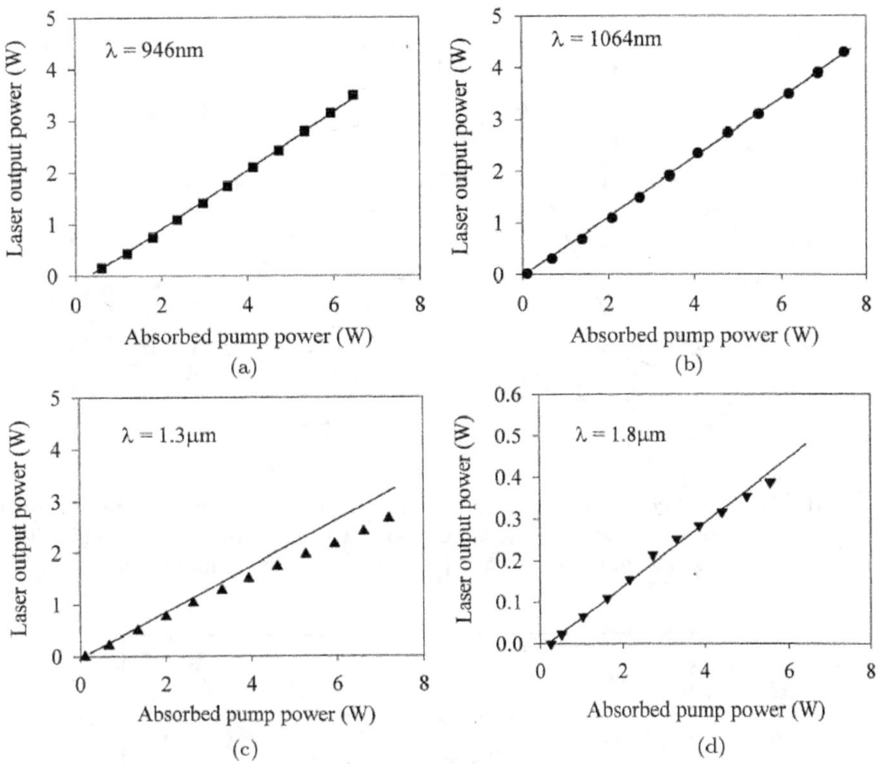

Figure 10.17: Output power for various laser transitions in Nd-doped YAG as a function of absorbed pump power [62]. Pump laser wavelength is 807 nm.

30 μm, and core widths of 8 and 20 μm, for Yb and Tm-doped structures, respectively. The reflectivity of output mirrors is 70 and 90%, respectively. The CW laser power as a function of pump power for Yb- and Tm-doped YAG is shown in Figure 10.19. The Yb-doped laser emits near 1030 nm and the Tm-doped laser emits near 2020 nm. Maximum output powers of 12.4 and 15.0 W are obtained for the Yb:YAG and Tm:YAG waveguide lasers. Higher powers have been reported. For example, ~1.08 kW of output power has been obtained using two composite Yb:YAG rods pumped by 3.93 kW of diode laser power [65].

Cooling is very important for high power operation [66]. Using a water cooling arrangement, 150 W of quasi CW operation has been

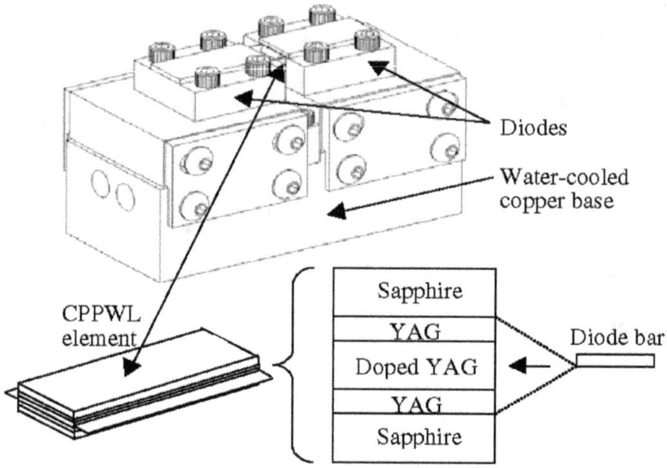

Figure 10.18: Schematic of a diode bar pumped high power planar waveguide laser. There are two diode bars located about 10 μm from the waveguide. The entire arrangement is mounted on a water-cooled Cu heat sink [56, 63].

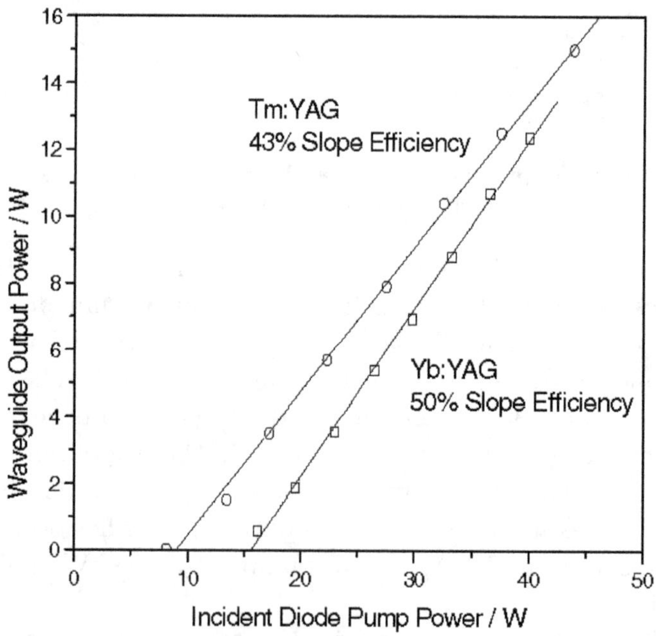

Figure 10.19: CW laser power as a function of pump power [56].

obtained using 430 W of pump power from 10 diode laser bars [64]. The active region of this laser has a planar Nd-doped YAG double-clad waveguide laser similar to that described previously. Supercontinuum generation (broadband source) due to nonlinear effects in planar waveguides at high input powers is discussed in Appendix 8.

Planar waveguides, as lasers, amplifiers, or as a broadband source, promise scalable geometries, low-cost fabrication, and the potential for integrated optical chip solutions.

References

1. J. Shmulovich, Y. H. Wong, G. Nykolak, P. C. Becker, R. Adar, A. J. Bruce, D. J. Muehlner, G. Adams, and M. Fishteyn, *OSA Optical Amplifier Meet.Opt. Soc. Amer.*, Paper PD18-1, 1993.
2. T. Kitagawa, K. Hattori, K. Shuto, M. Yasu, M. Kobayashi, and M. Horiguchi in *Optical Amplifiers and Their Applications,* Vol. 17 Tech. Dig. Ser. Opt. Soc. Amer., 1992, Paper PD-1.
3. T. Kitagawa, K. Hattori, M. Shimizu, Y. Ohmori, M. Kobayashi, and M. Horiguchi, *Electron. Lett.*, **28** (1992) 1818.
4. K. Shuto, K. Hattori, T. Kitagawa, and M. Horiguchiin, *Proc. Nineteenth Eur. Conf. Optical Commun.* (1993) 53.
5. T. Feuchter, E. K. Mwarania, J. Wang, L. Reekie, and J. **S.** Wilkinson, *IEEE Photon. Technol. Lett.*, **4**, (1992) 542.
6. P. Becker, R. Brinkmann, M. Dinand, W. Sohler, and H. Suche, *Appl. Phys. Lett.*, **61** (1992) 1257.
7. G. N. van den Hoven, E. Snoeks, **A.** Polman, C. van Dam, J. W. M. van Uffelen, and M. K. Smit, *Proc. Seventh Eur. Con$ Integrat. Opt.* (1995), p. 229.
8. D. P. Shepherd, Chapter B4.6, in C. Webb and J. Jones (eds.) *Handbook of Laser Technology and Applications,* Institute of Physics, 2004.
9. R. V. Ramaswamy and R. Srivastava, *IEEE J. Lightwave Tech.*, **6**, (1988) 984.
10. P. J. Chandler, L. Zhang, and P. D. Townsend, *Nucl. Instrument. Methods,* **B59** (1991) 1223.
11. R. V. Schmidt and I. P. Kaminow, *Appl. Phys. Lett.*, **25**, (1974) 458.
12. M. Svalgaard, C. V. Poulsen, A. Bjaerklev, and O. Paulsen, *Electron. Lett.*, **30** (1994) 1401.
13. C. T. A. Brown, C. L. Bonner, T. J. Warburton, D. P. Shepherd, A. C. Tropper, and D. C. Hanna, *Appl. Phys. Lett.*, **71** (1997) 1139.
14. M. Kawachi, *Opt. Quantum Electron.*, **22**, (1990) 391.
15. J. I. Mackenzie and D. P. Shepherd, *Opt. Lett.*, **27**, (2002) 2161.

16. T. Kitagawa, K. Hattori, K. Shuto, M. Yasu, M. Kobayashi, and M. Horiguchi, *Electron. Lett.*, **28** (1992) 1818.
17. J. Shmulovich, Y. H. Wong, G. Nykolak, P. C. Becker, R. Adar, and A. J. Bruce, *Electron. Lett.*, **28** (1992) 1181.
18. T. Feuchter, E. K. Mwarania, I. Wang, L. Reekie, and J. S. Wilkinson, *IEEE Photon. Technol. Lett.*, **4** (1992) 542.
19. K. Shuto, K. Hattori, T. Kitagawa, Y. Ohmori, and M. Horiguchi, *Electron Lett.*, **29** (1993) 139.
20. K. Hattori, T. Kitagawa, M. Oguma, M. Wada, J. Temmyo, and M. Horiguchi, *Electron Lett.*, **29**, (1993) 357.
21. R. N. Ghosh, J. Shmulovich, C. F. Kane, M. R. X. de Barros, G. Nykolak, A. J. Bruce, and P. C. Becker, *IEEE Photon. Tech. Lett.*, **8** (1996) 518.
22. G. Nykolak, P. C. Becker, J. Shmulovich, Y. H. Wong, D. J. DiGiovanni, and A. J. Bruce, *IEEE Photon. Technol. Lett.*, **5** (1993) 1014.
23. O. Lumholt, A. Bjarklev, T. Rasmussen, and C. Lester, *J. Lightwave Technol.*, **13** (1995) 275.
24. E. Snoeks, P. G. Kik, and A. Polman, *Opt. Muter.*, **5** (1996) 159.
25. J.-M. P. Delavaux, **S.** Granlund, O. Mizuhara, L. D. Tzeng, D. Barbier, M. Rattay, **F.** Saint Andre, and A. Kevorkian, *IEEE Photonic Tech. Lett.*, **9** (1997) 2.
26. J. Shmulovich., *Technical Digest*, OFC99, Paper PD42 (1999).
27. E. Desurvire, *Erbium-Doped Fiber Amplifiers, Principles and Applications* Wiley, New York, 1994.
28. J.-M. P. Delavaux, C. McIntosh, **J.** Shmulovich, A. J. Bruce, R. K. Pattnaik*, and B. R. Wirstiuk, *Technical Digest*, OFC 2000, Paper WA2-1 (2000).
29. M. Saruwatari and T. Izawa, *Appl. Phys. Lett.*, **24** (1974) 603.
30. Y. Hibino, T. Kitagawa, M. Shimizu, F. Hanawa, and A. Sugita, *IEEE Photon. Technol. Lett.*, **1** (1989) 349.
31. H. Aoki, O. Maruyama, and Y. Asahara, *IEEE Photon. Technol. Lett.*, **2** (1990) 459.
32. N. A. Sanford, K. J. Malone, and D. R. Larson, *Opt. Lett.*, **15** (1990) 366.
33. E. K. Mwarania, L. Reekie, J. Wang, and J. S. Wilkinson, *Electron. Lett.*, **26** (1990) 1317.
34. N. Takato, K. Jinguji, M. Yasu, H. Toba, and M. Kawachi, *IEEE J. Lightwave Technol.*, **6** (1988) 1003.
35. M. J. F. Digonnet and C. J. Gaeta, *Appl. Opt.*, **24** (1985) 333.
36. Y. Hibino, T. Kitagawa, M. Shimizu, A. Sugita, and Y. Ohmori, in *Proc. Integrated Photon. Res.*, Hilton Head, SC, 1990, Paper TuJ2, 1990.
37. H. Ioki, E. Ishikawa, and Y. Asahara, *Electron Lett.*, **27** (1991) 2352.
38. G. R. J. Robertson and P. E. Jessop, *Appl. Opt.*, **30** (1991) 276.
39. R. Tumnunelli, F. Halumi, and J. Haavisto, *Opt. Lett.*, **16** (1991) 1098.
40. K. Hattori and T. Kitagawa, *IEEE Photon. Technol. Lett.*, **4** (1992) 973.
41. T. Kitagawa, K. Hattori, Y. Hibino, and Y. Ohmori, *IEEE J. Lightwave Tech.*, **12** (1994) 436.

42. J. R. Bonar, J. A. Bebbington, J. S. Aitchison, G. D. Maxwell, and B. J. Ainslie, *Electron. Lett.*, **30** (1994) 229.
43. F. Ondracek, *Opt. Mater.*, **30** (2007) 457.
44. J. Jágerská, F. Ondráček, L. Salavcová, M. Míka, J. Špirková, and J. Ctyroký, *13th Eur. Conf. Integrated Optics ECIO 2007*, Paper ThG14 (2007).
45. F. Ondráček, M. Skalský, and J. Čtyroký, *Proc. SPIE*, **6180** (2006) 414.
46. M. P. Hehlen, N. J. Cockroft, and T. R. Gosnell, *Phys. Rev. B*, **56** (1997) 9302.
47. F. Ondráček, J. Jágerská, L. Salavcová, M. Míka, J. Špirková, and J. Čtyroký, *IEEE J. Quant. Electron.*, **QE-44** (2008) 536.
48. F. D. Patel, S. DiCarolis, P. l. Lum, S. Venkatesh, and J. N. Miller, *IEEE Photonic Tech. Lett.*, **16** (2004) 2607.
49. D. L. Veasey, D. S. Funk, P. M. Peters, N. A. Sanford, G. E. Obarski, N. Fontaine, M. Young, A. P. Peskin, W. C. Liu, S. N. Houde-Walter, and J. S. Hayden, *J. Non-Cryst. Solids*, **263&264** (2000) 369.
50. F. Di Pasquale and M. Federighi, *IEEE J. Quant. Electron.*, **30** (1994) 2127.
51. D. Geskus, S. Aravazhi, E. H. Bernhardi, L. Agazzi, S. M. García-Blanco, and M. Pollnau, *CLEO Technical Digest*, Paper CMA1.6 (2012).
52. Y. E. Romanyuk, C. N. Borca, M. Pollnau, S. Rivier, V. Petrov, and U. Griebner, *Opt. Lett.*, **31** (2006) 53.
53. F. Gardillou, Y. E. Romanyuk, C. N. Borca, R. P. Salathé, and M. Pollnau, *Opt. Lett.*, **32** (2007) 488.
54. D. Geskus, S. Aravazhi, C. Grivas, K. Wörhoff, and M. Pollnau, *Opt. Exp.*, **18** (2010) 8853.
55. F. Auzel, G. Baldacchini, L. Laversenne, and G. Boulon, *Opt. Mater.*, **24** (2003) 103.
56. D. P. Shepherd, S. J. Hettrick, C. Li, J. I. Mackenzie, R. J. Beach, S. C. Mitchell, and H. E. Meissner, *J. Phys. D: Appl. Phy.*, **34** (2001) 2424.
57. J. J. Chang, E. P. Dragon, C. A. Ebbers, L. I. Bass, and C. W. Cochran, in W. R. Bosenberg and M. M. Fejer (eds.), *OSA TOPS on Advanced Solid-State Lasers,* 19, Washington, DC: Optical Society of America, **19** (1998) 300.
58. D. P. Shepherd, C. L. Bonner, C. T. A. Brown, W. A. Clarkson, A. C. Tropper, D. C. Hanna, and H. E. Meissner, *Opt. Commun.*, **160** (1999) 47.
59. C. L. Bonner, T. Bhutta, D. P. Shepherd, and A. C. Tropper *IEEE J. Quantum Electron.*, **36** (2000) 236.
60. T. Tamanuki, T. Sasaki, and M. Kitamura, *Opt. Quantum Electron.*, **28** (1996) 513.
61. J. D. Minelly, L. A. Zenteno, M. J. Dejneka, W. J. Miller, D. V. Kuksenkov, M. K. Davis, S. G. Crigler, and M. E. Bardo, *Tech. Dig. Opt. Fiber Commun. Conf.*, Paper **PD2-1** (2000).
62. J. I. Mackenzie, C. Li, and D. P. Shepherd, *IEEE J. Quant. Electron.*, **QE-39** (2003) 493.
63. R. J. Beach, S. C. Mitchell, H. E. Meissner, O. R. Meissner, W. F. Krupke, J. M. McMahon, and W. J. Bennett, *Opt. Lett.*, **26** (2001) 881.

64. J. R. Lee, H. J. Baker, G. J. Friel, G. J. Hilton, and D. R. Hall, *Opt. Lett.*, **27** (2002) 524.
65. E. C. Honea, R. J. Beach, S. C. Mitchell, J. A. Skidmore, M. A. Emanuel, S. B. Sutton, and S. A. Payne, *Opt. Lett.*, **25** (2000) 805.
66. R. J. Beach, M. A. Emanuel, B. L. Freitas, J. A. Skidmore, N. W. Carlson, W. J. Bennett, and R. W. Solarz, *Proc. SPIE* **2383** (1995) 283.

Chapter 11

Optical Transmission System Application

11.1 Introduction

The semiconductor injection laser and photodetectors were originally developed in the 1970s and 1980s for optical fiber transmission system applications. Early systems used GaAs-based lasers emitting near 850 nm and later systems used InGaAsP-based lasers emitting near 1300 and 1550 nm. The 1500 nm systems use distributed feedback (DFB) single wavelength lasers. These systems evolved to wavelength division multiplexed (WDM) transmission which uses multiple DFB lasers for transmission. Current systems use WDM-based transmission where the output of 8 or 16 or more single frequency (DFB) lasers carrying data are multiplexed (combined) into a single fiber. Polarization multiplexing can be used to double the data rate per wavelength channel. The typical data rate per channel is in the 10–40 Gb/s range which aggregates to a transmission capacity of $(2 \times 16 \times 40)$ of $\sim$1 Tb/s. The WDM systems often use Er-doped amplifiers to compensate for losses in the multiplexer and in the demultiplexer prior to the receiver. The receiver consists of a detector (typically a photodiode) which converts the optical signal to an electrical signal. It is followed by an amplifier, clock recovery, and decision circuits. Er-doped fiber amplifier (EDFA) which operates in the telecommunication wavelength band (∼1530–1560 nm) is an important device for both terrestrial and submarine optical communication systems and next-generation all-optical networks [1–20]. The amplification factor or gain can be greater than 100 (>20 dB) in some of these applications. The amplifiers used are single mode amplifiers, i.e. the amplifier fiber has a single mode core doped with Er. For high power, an Er–Yb codoped double clad amplifier with a doped single mode core is also used. WDM systems are deployed in both terrestrial and undersea (or submarine) optical transmission systems. The Er-doped fiber amplifiers (EDFA) are deployed in terrestrial multichannel WDM networks where the typical transmission distance prior to regeneration is ∼40 km. In these systems, the amplifier could be used as a post amplifier (i.e. after transmitter) or as a preamplifier (i.e. prior to receiver). The first submarine application of optical amplification was for a link across the Atlantic ocean. Since then, many long-distance submarine links have been deployed. Instead of conventional regenerators, a chain of optical amplifiers spaced about 60 km for the entire link distance of ∼6000 km (for the Atlantic system) has been deployed. The link using optical amplifiers has the advantage that it can be upgraded to higher data rate simply by changing the transmitters and receivers which are located on the shores unlike a link using conventional regenerators where an entirely new system with new regenerators need to be deployed for higher capacity. Thus, increasing the capacity for amplifier links is much easier. Amplifiers have also been used for short distance submarine links, such would be the case for population centers (cities) located along the shores or from mainland to an island. The amplifier based systems, unlike regenerator-based systems, can be easily upgraded to a higher data rate simply by changing the transmitters and receivers located at the shore. That is why sometimes they are referred to as bit-rate flexible transmission systems.

11.2 Noise and bit-error-rate (BER)

Most optical transmission systems use two-level amplitude (or intensity) modulated signals for transmission. For digital transmission system, the "1's" and "0's" are determined by a high intensity (S_1) and low intensity (S_0) of the transmitted signal. Both the high- and low-intensity values have noise associated with them. A Gaussian distribution of both the high- and low-intensity of the signal is assumed here, the standard deviation of the Gaussian distribution being the noise. For example, the probability of a value S for either "1" or "0" is given by

$$P_1(s) = \frac{1}{\sigma_1\sqrt{2\pi}} \exp\left[-\frac{(s-s_1)^2}{2\sigma_1^2}\right] \tag{11.1}$$

$$P_0(s) = \frac{1}{\sigma_0\sqrt{2\pi}} \exp\left[-\frac{(s-s_0)^2}{2\sigma_0^2}\right] \tag{11.2}$$

where S_1 and S_0 are mean intensity of the "1" and "0" state and σ_1 and σ_0 are the noise in those states. The standard deviations of the Gaussian distribution are σ_1 and σ_0, respectively. Equations (11.1) and (11.2) for the probabilities are for bit "1" and bit "0", respectively. A decision intensity threshold (S_d) is set in the system such that values of S above S_d are considered "1" and values of S below S_d are considered "0". An error occurs when the value of S is lower than S_d for the high-intensity state and the value of S is higher than S_d for the low-intensity state. Since the probability of occurrence of "1's" and "0's" are on average 50% (i.e. 1/2), the probability of error of a single bit (i.e. bit-error-rate (BER)) is given by

$$\text{BER} = \frac{1}{2}\left[\int_{-\infty}^{S_d} P_1(S)dS + \int_{S_d}^{\infty} P_0(S)dS\right] \tag{11.3}$$

The above equation shows BER depends on the decision threshold (S_d). In a practical system, BER is minimized by varying S_d. For the above equation, BER can be minimized mathematically, and the minimum BER occurs when

$$(S_1 - S_d)/\sigma_1 = (S_d - S_0)/\sigma_0 = Q$$

where

$$S_d = \frac{\sigma_0 S_1 + \sigma_1 S_0}{\sigma_0 + \sigma_1} \tag{11.4}$$

For many situations, $\sigma_1 = \sigma_0$, i.e. the noise of the high- and low-intensity states are equal. When this happens $S_d = (S_1 + S_0)/2$, i.e. the decision threshold for minimum BER is at the average of the high- and low-intensities. Define a quantity Q given by:

$$Q = \frac{S_1 - S_0}{\sigma_1 + \sigma_0} \tag{11.5}$$

In this case, Eq. (11.3) for minimum BER can be simplified as

$$\text{BER} = \frac{1}{\sqrt{\pi}} \int_{Q/\sqrt{2}}^{\infty} \exp[-y^2] dy \simeq \frac{\exp(-Q^2/2)}{Q\sqrt{2\pi}} \tag{11.6}$$

BER decreases rapidly as Q increases. For $Q \sim 6$, BER is approximately equal to 10^{-9}. The noise primarily arises from the photodetectors. This is discussed in Chapter 6. The power P_1 in "1" bit is related to S_1 as $S_1 = RP_1$, where R is the responsivity (in unit of Amp/W) of the photodetector. The average received power is defined as $P_{\text{av}} = (P_1 + P_0)/2$. Using $P_0 \sim 0$, the quantity $S_1 = 2RP_{\text{av}}$ and $\sigma_1 + \sigma_0 = 2\sigma_t$, where σ_t is the thermal noise of the photodetector and it is assumed that noise for both "1" and "0" states are dominated by thermal noise. Thus, $Q \sim S_1/2\sigma_t = RP_{\text{av}}/\sigma_t$.

Using $\sigma_t = 100$ nA for a 0.5 GHz bandwidth (typically for a 1 Gb/s receiver), $R = 1$ A/W (typical value at 1550 nm for InGaAs photodiode), $P_{\text{av}} = 0.6\ \mu$W $= -32$ dBm for $Q = 6$. Thus, for a BER of 10^{-9}, an average power of -32 dBm is needed for digital signals at 1 Gb/s. In the thermal noise limit, the noise is proportional to the square root of bandwidth of the receiver ($\sigma_t \sim \Delta f^{1/2}$). The bandwidth, Δf, needed to receive a signal of bit rate B is given by $\Delta f \sim B/2$. Thus, the average received power needed for a BER of 10^{-9} in the thermal noise limit increases with bit rate as $B^{1/2}$. Another scheme currently being used for pulse amplitude modulation is a four-level scheme (PAM-4). The BER for this case is discussed in Section 11.9.3.

11.3 Long-distance transmission

Various transmission system experiments for long-distance transmission are described in this section. These are single wavelength links, multiwavelength WDM links, and very long-distance loop transmission experiments.

11.3.1 *Single wavelength link*

A typical optical transmitter and optical receiver schematic are shown in Figure 11.1. The transmitter has an InGaAsP distributed Bragg reflector (DBR) laser (operating near 1550 nm) the output of which is modulated using a $LiNbO_3$ modulator. The modulator is driven by a 5 Gb/s or 10 Gb/s data (PPG — pulse pattern generator). The output power of the modulator is amplified using an EDFA. The receiver has an EDFA preamplifier followed by an InGaAs photodiode and electronics.

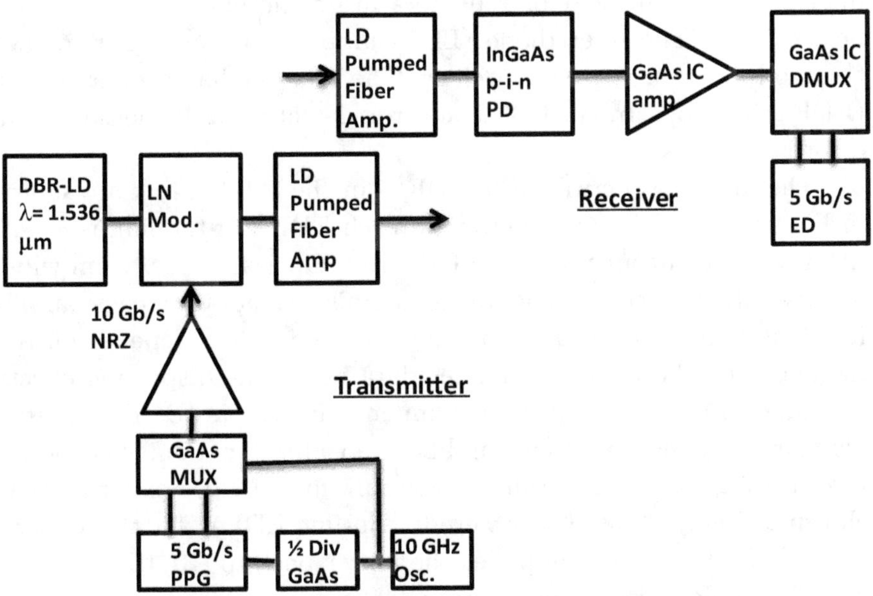

Figure 11.1: Transmitter and receiver configurations [19].

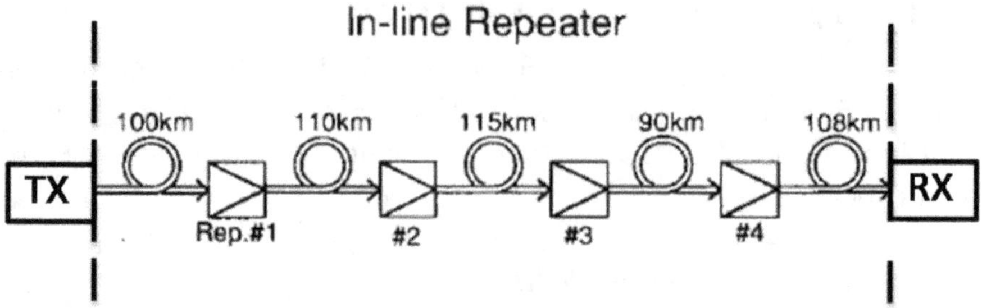

Figure 11.2: Link Schematic [19]. The devices marked Rep. #1, 2, 3, 4 are EDFAs. TX and RX are transmitters and receivers as shown in Figure 9.2. The transmitter is followed by an EDFA post amplifier and the receiver has an EDFA preamplifier prior to photodiode.

A 500 km long-bit-rate flexible link has been demonstrated using the above transmitter and receiver configuration. The link has four EDFAs as shown in Figure 11.2. The transmission link, viz. fiber and optical amplifiers (shown between dashed lines in Figure 11.2), can operate at different bit rates — as a result, systems using optical fiber amplifiers instead of repeaters are sometimes called bit-rate flexible link. The laser diode (LD) pumped fiber amplifier follows the modulator as a post amplifier. The receiver has a preamplifier (EDFA) pumped by an LD. The error detector (ED) measures the error rate.

The distances between the EDFAs in the 500 km link are shown in Figure 11.2. The input power for each EDFA in the link is ∼−15 dBm and the output power is ∼0–5 dBm. Optical filters, 1 nm wide, are used to reduce the spontaneous emission noise at each amplifier. The number of EDFAs in the transmission link depends on the distance and data rate. At higher data rates, the dispersion effects are larger. So, often dispersion compensating fibers (DCF) are used along with EDFA at each amplifier/repeater junction. The power level through a 505 km transmission link for 10 Gb/s experiment is shown in Figure 11.3. There were five in-line EDFAs for this experiment. Note the EDFA amplifies the power back to ∼0 dBm.

The BER performance for 10 Gb/s transmission is shown in Figure 11.4. The receiver sensitivity, i.e. the received power needed for 10^{-9} BER, increases with increasing transmission distance.

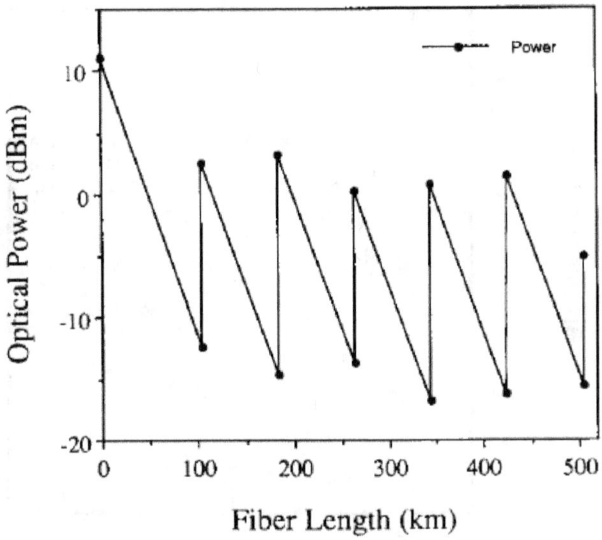

Figure 11.3: Power levels at the 505 km amplifier chain for 10 Gb/s operation [19].

11.3.2 *WDM transmission*

A WDM transmission experiment with 32 wavelength channels (100 GHz apart on the ITU grid) with each channel operating at 10 Gb/s over ~500 km transmission distance has been carried out [21]. The system uses four booster amplifiers. Each booster amplifier has two EDFAs with a DCF — fiber with high negative dispersion. The booster amplifier gains are ~30 dB. The schematic of the experiment is shown in Figure 11.5. The fiber loss is 0.19 dB/km.

The optical spectrum at the transmitter and at the receiver after ~500 km transmission shows all 32 wavelengths with some variation in power. Error-free transmission (BER $<10^{-9}$) for all wavelength channels is observed. It is found that adding an extra DCF reduces the BER sensitivity for longer wavelength channels.

11.3.3 *Loop transmission experiment*

To simulate very long-distance transmission (many thousands of km), a loop experiment is done. In a loop experiment, the optical data

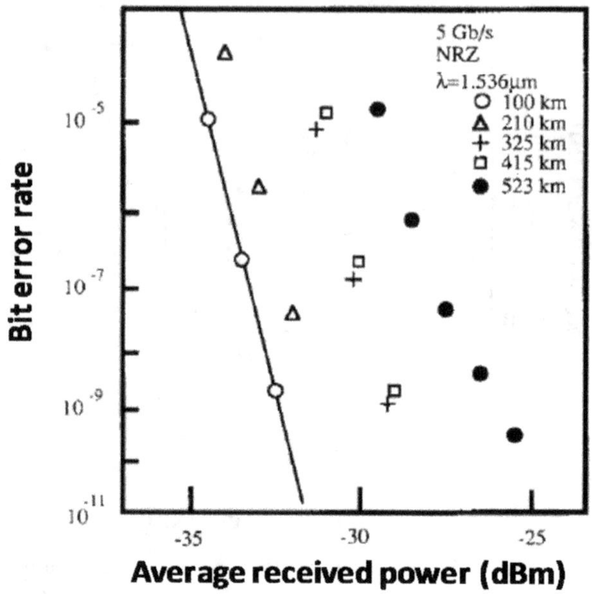

Figure 11.4: Bit-error-rate (BER) measurement as a function of received power for 10 Gb/s transmission for different distances [19].

circulates in a loop many times [22–27]. The loop (Figure 11.6) typically has all the elements of a transmission experiment such as fibers and chain of amplifiers.

Typical loop lengths are few tens of km to few hundreds of km. The transmitter, receiver, and the error rate test equipment are outside the loop. The experiment starts with the transmitter switch on to load the loop with data. The data is allowed to circulate in the loop for some specified number of revolutions. The number of revolutions corresponds to the total transmission distance. The delay per loop circulation is known from loop length (~4.9 ps/km of fiber). After a specific total transmission distance, the data is let out of the loop by a switch and the error rate is measured. Figure 11.7 shows the eye diagram and BER measurement after 9000 km of transmission at 5 Gb/s. The back-to-back (B2B) BER (i.e. ~0 km) is also shown in the figure. BER of less than 10^{-9} is obtained. Similar error-free performance was obtained after 20,000 km of transmission.

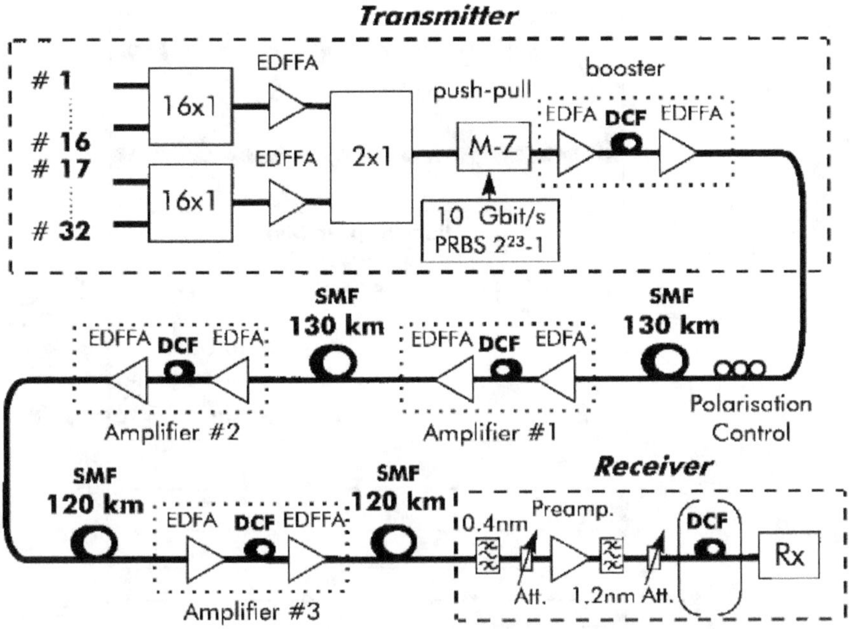

Figure 11.5: Schematic of the experiment [21]. The experiment uses two EDFAs with DCF in between as single amplifier blocks. DCFs are needed for dispersion compensation of high data rate signals. The dispersion causes pulse broadening.

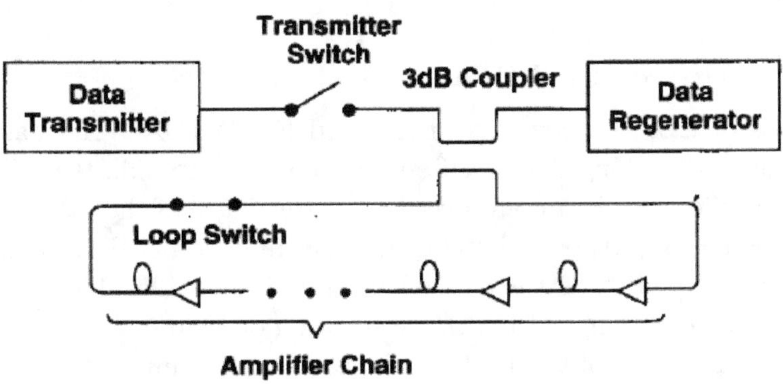

Figure 11.6: Schematic of a loop transmission experiment [27].

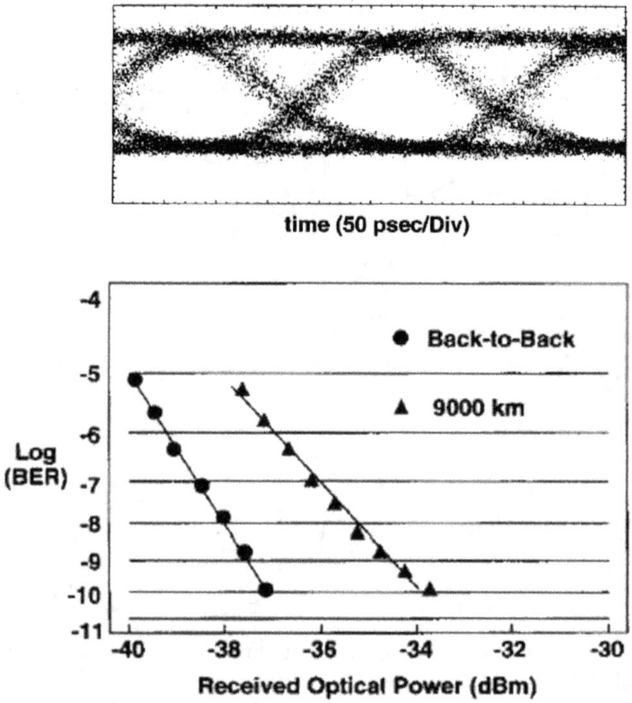

Figure 11.7: (a) Eye diagram after 9000 km of transmission at 5 Gb/s, (b) Bit-error-rate (BER) as a function of received power back to back (left) and after 9000 km (right) [27].

11.3.4 *Distributed EDFA*

A class of EDFA, known as distributed EDFA (DEDFA), has been investigated for long-distance fiber transmission [28–39]. For DEDFA, the amplifier gain is present over the entire length of the transmission fiber. For this purpose, the transmission fiber is lightly doped with Er and pumped along the entire length with multiple 1480 nm pump lasers. In the absence of Er, the absorption of 1480 nm pump is ~0.3 dB/km. The gain of the DEDFA depends on the number of 1480 nm pump sources used. Optical signal gain of ~10 dB has been reported using multiple pump sources. Initial experiments using DEDFA have been carried out for the demonstration of soliton propagation over long distances.

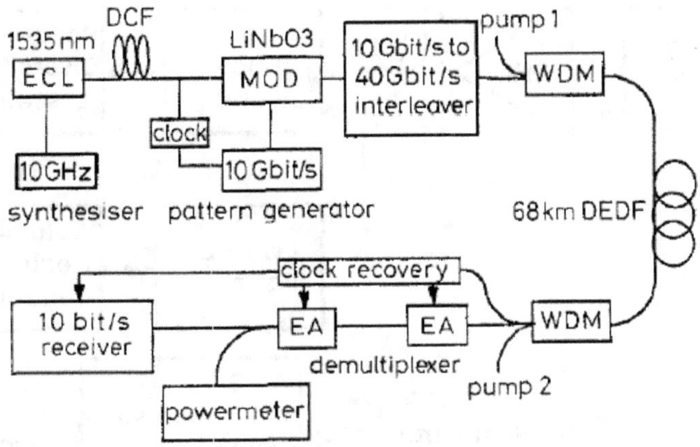

Figure 11.8: DEDFA transmission experiment at 40 Gb/s over 68 km of DEDFA [35].

An NRZ experiment using DEDFA is shown in Figure 11.8. The pulses are generated using a mode-locked external cavity laser. The pulse width is ~8 ps (repetition rate 10 GHz) with a time-bandwidth product of ~0.4. The NRZ data at 10 Gb/s is impressed using a $LiNbO_3$ modulator. For the 40 Gb/s experiment, the 10 Gb/s pulses are multiplexed to 40 Gb/s. The pulses are transmitted through 68 km of DEDFA pumped with 100 mW (total) of 1480 nm pump. The DEDFA net amplification is ~2 dB with −3 dBm input power. The DEDFA fiber is lightly doped with Er. The unpumped attenuation of the doped fiber is ~0.68 dB/km. At the receiving end, the signal is demultiplexed to 10 Gb/s and the error rate is measured using a BER detector. The demultiplexer is based on electroabsorption (EA) modulator. BER of less than 10^{-9} was observed indicating good performance.

11.4 Coherent transmission

Long-distance coherent transmission system experiments have been carried out [40–49]. Frequency shift keying (FSK) mode is used for the transmitter. The source is a DFB laser with the two sections (ME-DFB laser in Figure 11.9). One of the sections is biased CW

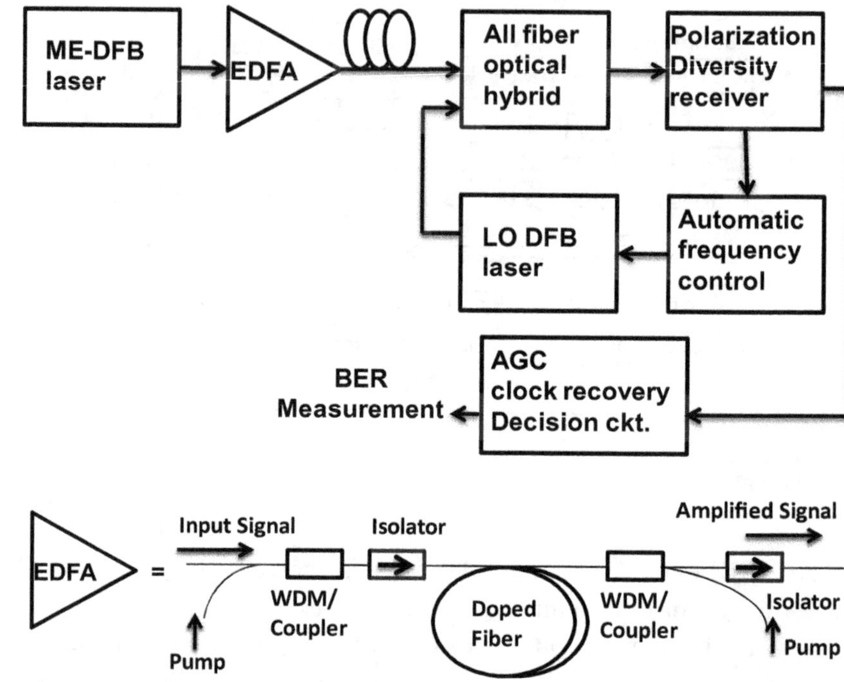

Figure 11.9: Schematic of a coherent FSK transmission system [44]. The EDFA is shown at the bottom.

and the current in the other section is modulated so that the emission frequency of the signal varies with the modulation current. This generates an FSK signal. The receiver has a DFB single wavelength laser as a local oscillator (LO DFB laser). A polarization diversity scheme is used to increase the receiver sensitivity. The EDFA is forward and backward pumped with LDs with output power of ~18 dBm each. The output of the EDFA is >15 dBm. Results of a transmission experiment at 1.7 Gb/s using the coherent FSK transmission format is shown in Figure 11.10. EDFAs are added to increase the transmission distance. A 253 km transmission span was achieved (in a system) without EDFA. The span length increases to 365 and 415 km using one and two EDFAs in the link, respectively.

A continuous phase FSK (CPFSK) transmission system using 10 EDFAs in a 1028 km transmission link has been reported. The system

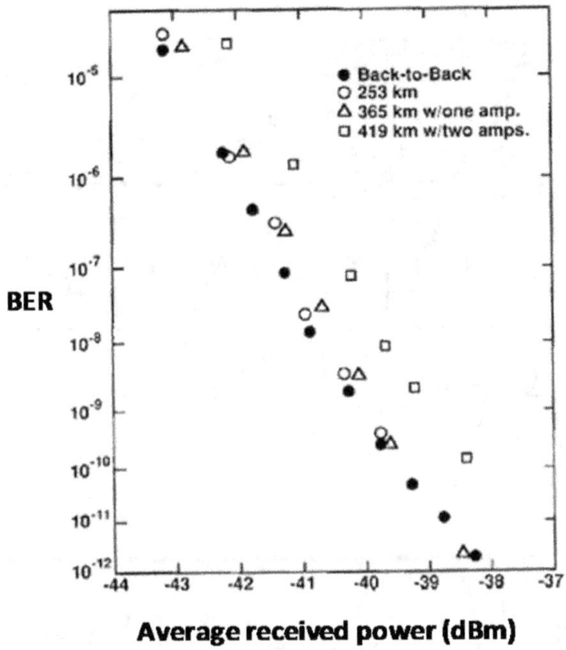

Figure 11.10: FSK coherent system transmission results using EDFAs. The data rate is 1.7 Gb/s [44].

operates at 565 Mb/s. It uses a polarization diversity receiver similar to that used in the previous experiment. The input to the receiver goes to a polarization beam splitter (PBS). The PBS module splits the data input into four beams. Two sets of these go to balanced receivers and another beam from the local oscillator module is added to the PBS module for mixing at the photodiodes.

The schematic of the transmission system experiment is shown in Figure 11.11. It consists of 10 EDFAs (noted as REP1 to 10 in the figure). The single-mode fiber (SMF) links between each EDFA is ∼100 km. The EDFA has a 15 m long Er-doped fiber and is pumped by an LD. The EDFAs amplify the power level to ∼0 dBm after transmission through the previous ∼100 km link (attenuation ∼20 dB). This is shown in Figure 11.11(b). The data signal from the link and a local oscillator (laser) signal are inputs to the polarization diversity receiver. The error-rate measurement through the link has

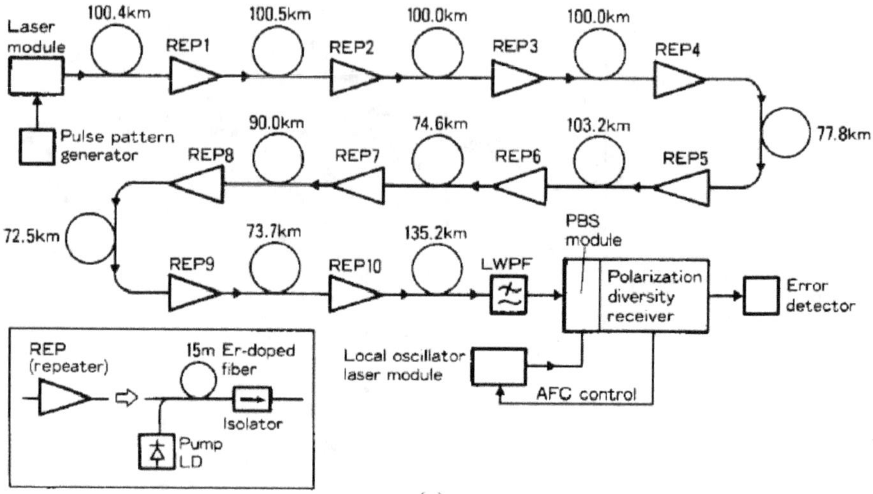

(a)

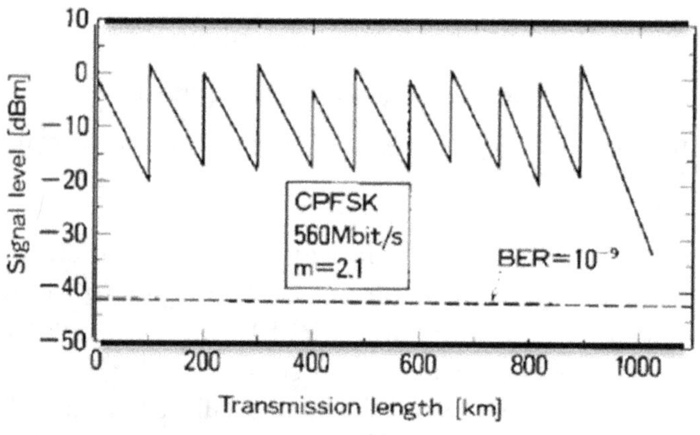

(b)

Figure 11.11: (a) Schematic of a 1028 km CPFSK transmission experiment, (b) Signal power levels through the link [47].

been carried out. The results show that error-free transmission (BER $<10^{-9}$) has been obtained.

11.5 Subscriber transmission

So far only long haul high data rate transmission has been discussed. Lasers and amplifiers (EDFAs) have also been studied for subscriber

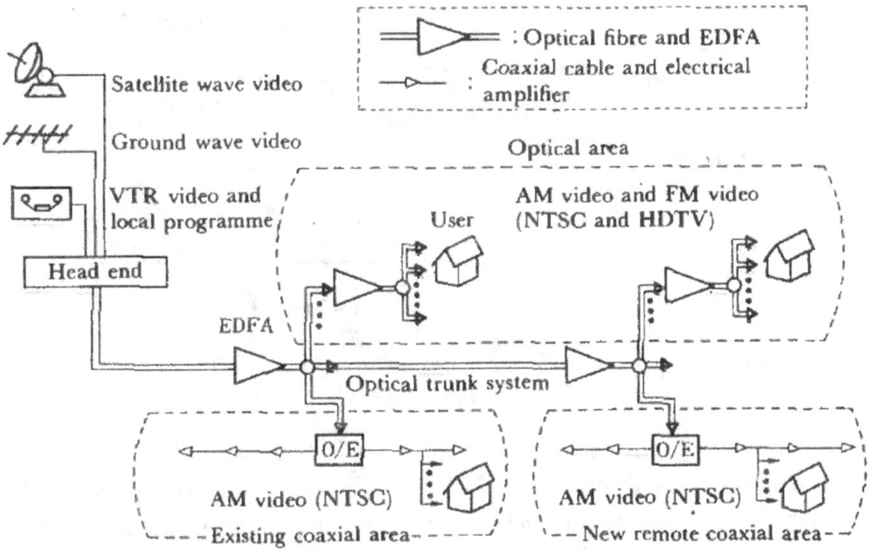

Figure 11.12: Subscriber distribution system using optical fibers, EDFAs, and coaxial cables [58].

transmission systems [50–60]. These systems transmit a wide range of information signals to the subscribers (home/office). The signals are low speed 64 kb/s voice signals, fax, data at few Mb/s to Gb/s, TV images, and broadband high-quality images (HDTV signals). The conventional TV signals are analog amplitude modulated (AM) signals. In the US, they are often called CATV systems; generally, there are many channels (~40–120 channels) in a multichannel AM CATV system. These channels are spaced 6 MHz apart. For example, US Channel 3 transmits at ~50 MHz. The frequencies and bandwidths of these AM transmission channels are standardized (NTSC standard). Most of the other subscriber feeds, such as voice, data, satellite-based transmission, and HDTV are digital signals. A schematic of a subscriber information distribution system using optical fibers, EDFAs, and coaxial cables are shown in Figure 11.12. All of the information types are combined at a unit called the "Head end". At the Head end, the electrical signal is converted to an optical signal using a laser and transmitted to the subscribers using optical fibers and EDFAs. The low-loss optical fibers and the EDFAs form an optical trunk system

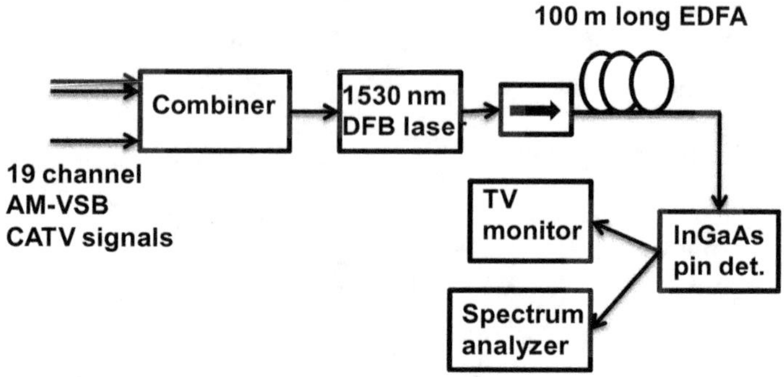

Figure 11.13: Schematic of an experimental setup for 19 television channel transmission [54]. The EFDA is pumped by a 1480 nm color center laser.

from which the signal is further split up for transmission to different areas using optical splitter. This signal is transmitted and amplified using EDFAs before conversion to electrical signal using an optical to electrical (O/E) converter. The latter is essentially an optical detector and amplifier and other electronic circuits for splitting the incoming data to different types of signal. After the O/E conversion, the signal is transmitted to home using electronic amplifiers and co-axial cables. This is the original format for transmission first used with the advent of EDFAs. More recently, for Gb/s broadband transmission, HDTV transmission, and CATV transmission this O/E converter is located at the subscriber site — this is often referred to as fiber to the home. Use of optical fibers and EDFAs allow much longer distance transmission of broadband signals without the need for amplification or regeneration. Of all the above types of signals, the multichannel analog AM CATV systems are most sensitive to any nonlinearities in the EDFA or in the incoming signal. The nonlinearity causes intermixing among frequencies through the generation of second-order and higher-order products (frequencies).

The first experiment on multichannel AM CATV channel transmission using EDFA was reported in Ref. [54]. A 19-channel system of AM television channels is combined and used to modulate a DFB laser (Figure 11.13). The laser output is coupled to an SMF and then sent through an EDFA pumped by a 1480 nm laser. Various optical isolators are used to reduce reflection effects. The EDFA operated under saturated gain of 11.5 dB with an output power of 8 dBm. The experimental system worked well. The effect of amplifier nonlinearity and its mitigation is discussed if Refs. [55–70].

11.6 Soliton transmission

The demand for the ultra large capacity transmission systems, i.e. with capacities approaching Tb/s continues to increase. For such systems, high-speed transmission with the bit rate of multiples of 10 Gb/s is quite attractive, since it can reduce the number of WDM channels. Time division multiplexing and transmission using optical solitons is important for these systems. Solitons allow error-free ultra-long-distance transmission in dispersion-managed systems [71–77].

A schematic of an optical soliton transmission system experiment at 40 Gb/s is shown in Figure 11.14. In this scheme, the cumulative chromatic dispersion is offset periodically by DCF with negative (normal) dispersion so that the total system dispersion is close to zero.

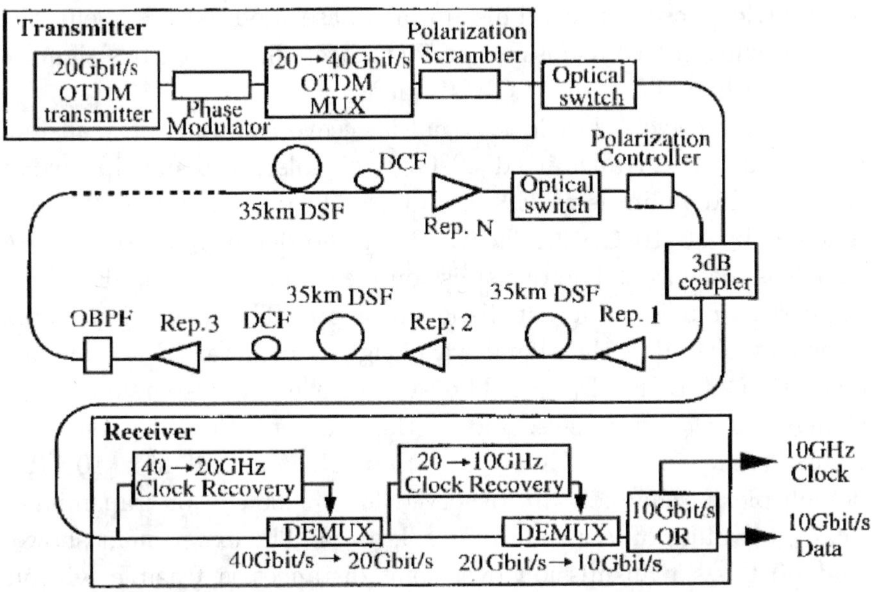

Figure 11.14: Soliton transmission experiment at 40 Gb/s [77].

The Gordon–Haus timing jitter is also reduced in this system with close to zero dispersion [78]. The experimental system uses a recirculating loop (described previously) for ultra-long-distance transmission. The recirculating loop consists of several units. Each unit has a dispersion shifted fiber (DSF), a 980 nm laser pumped erbium-doped amplifier (EDFA), and a dispersion compensating fiber (DCF). In addition, it has a polarization controller and an optical band pass filter (OBPF). The latter reduces the spontaneous emission noise.

For the transmitter, a 20 Gb/s optical soliton data stream is produced by optical time division multiplexing (OTDM) of 10 Gb/s RZ (return to zero) data pulses. These data pulses are generated by a DFB laser, sinusoidally-driven EA modulators, and two $LiNbO_3$ intensity modulators operated at 10 Gb/s with a pseudorandom binary sequence. The generated pulse width is about 9 ps. These pulses pass through a $LiNbO_3$ phase modulator driven by a 20 GHz clock. The phase condition is optimized by adjusting the delay between the optical data bits and the phase modulation signal using an electric phase shifter. This initial phase modulation is effective in improving the transmission performance. Then, 40 Gb/s signal is generated by OTDM of 20 Gb/s signals.

In the receiver, the clock signal is recovered from the transmitted signals. The transmitted 40 Gb/s signals are demultiplexed in the optical domain using EA modulators in two stages; 40–20 Gb/s followed by 20–10 Gb/s. The BER for the demultiplexed 10 Gb/s signals is measured. The transmission measurements (i.e. BER data) are carried out for different dispersion maps. The dispersion map is varied by changing the signal wavelength which varied the average dispersion of DSF (D_{local}). The system average dispersion (D_{ave}) changes as D_{local} changes, and as the compensating dispersion, by DCFs, are changed. BER of less than 10^{-9} for all four 10 Gb/s demultiplexed channels are observed for $\sim$9000 km of transmission for D_{local} values of less than 0.3 ps/km nm. The experiment showed that 40 Gb/s transmission over long distances is feasible without active dispersion compensation.

System performance impairment due to soliton–soliton interaction was reduced by the polarization division multiplexing. By using the polarization division multiplexing and optimum dispersion map, 40 Gb/s single-channel transmission over 10,200 km was successfully demonstrated without any active inline transmission control.

11.7 Optical transmission using raman amplifiers

Raman amplifiers have been used in optical transmission experiments [79–88]. Puc and coworkers [79] have demonstrated a cascaded Raman amplifier system using 11 Raman amplifiers. They demonstrated error-free transmission of 20 WDM channels (200 GHz apart) modulated at 10 Gb/s over 867 km of standard single-mode fiber. The transmitter wavelengths span from 1493 to 1521 nm. The typical spacing between the Raman amplifiers are ∼80–90 km and dispersion compensation is used after every amplification stage. The schematic of the experiment is shown in Figure 11.15.

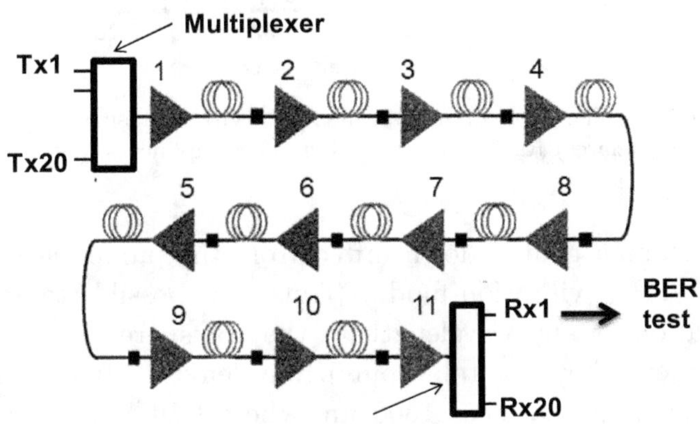

Figure 11.15: Schematic of a transmission experiment using Raman amplifiers [79]. There are 10 spans of single-mode fibers with a total length of 867 km. The average span loss including connectors is 21 dB. The output power is +14 dBm for all Raman amplifiers, i.e. +1 dBm per channel output. T×1, T×2, and T×20 are 20 transmitter modules each operating at a different wavelength. At the receiver, the signal is demultiplexed prior to photodiode followed by BER test.

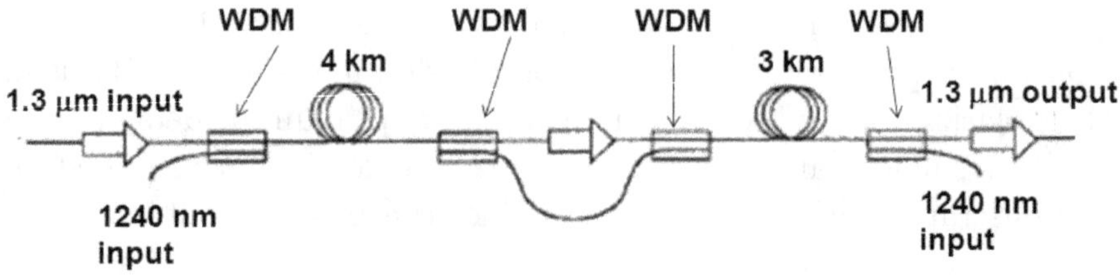

Figure 11.16: Schematic of Raman amplifier at 1300 nm [91]. The Raman gain occurs in the 4 and 3 km long fibers.

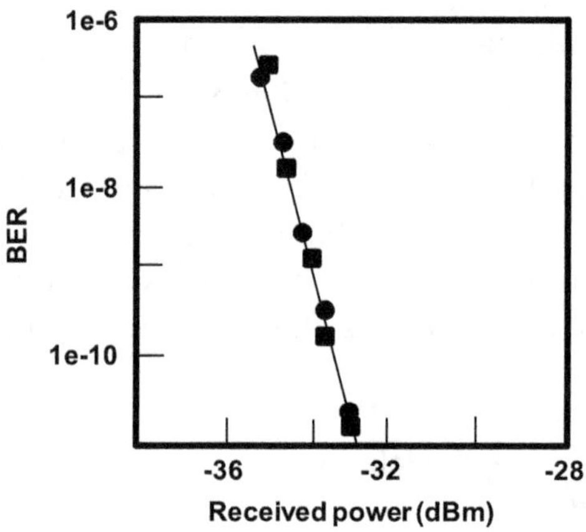

Figure 11.17: The bit-error-rate results for 2.5 Gb/s transmission with (circles) and without (squares) Raman amplification. The amplification is 29 dB [91].

The Raman gain is downshifted from the pump wavelength by about 13 THz. (vibration mode). Thus, it is possible to get Raman amplification at any wavelength in the transparency region of the optical fiber using a suitable pump wavelength. Many communication systems operate near 1300 nm where EDFAs do not operate. However, it was recognized early on in the development of Raman amplifiers that a Raman amplifier near 1300 nm is feasible [89–91]. A design of a 1300 nm Raman amplifier is shown in Figure 11.16 [91]. The figure shows a two-stage Raman amplifier with a counter-propagating pump at 1240 nm. The pump light is generated by a cascaded Raman laser (emitting at 1240 nm) which is pumped by a cladding pumped neodymium (Nd) fiber laser. The Nd laser operates at 1060 nm. The two-stage configuration reduces the noise for Raman amplification [92]. The 1300 nm Raman amplifier has a small signal gain of 29 dB. The output power could be as high as 16 dBm.

The absence of noise is verified using a transmission experiment at 2.5 Gb/s modulated data at 1300 nm. The result of the BER test is shown in Figure 11.17. The result shows no BER penalty with 29 dB of Raman gain.

11.8 Optical network using (SOAs)

Semiconductor optical amplifiers (SOAs) are important for several optical network applications such as wavelength conversion (i.e. converting data at one wavelength to data at different wavelength), optical demultiplexing, and optical logic circuits. Nonlinear effects in SOAs are discussed in Appendix 13.

11.8.1 *Wavelength conversion*

Four-wave mixing (FWM) and cross-phase/cross-gain modulation in SOAs has been proposed and demonstrated as a technique for performing wavelength conversion for WDM systems [93–97]. Wavelength conversion is defined as a process by which data at one wavelength can be converted to data at another wavelength. Such processes are important in all-optical networks where a subscriber node is designed to receive and transmit at one wavelength. The mechanism for wavelength conversion using SOA includes cross-gain modulation and cross-phase modulation of two simultaneously propagating signals. For cross-phase modulation, a Mach-Zehnder modulator (MZM) fabricated using two SOAs in each arm is often used. For wavelength conversion using MZI, a CW signal is injected into the central port. This signal splits and travels through the two SOAs. The two signals undergo a differential phase shift of π, as a result, there is no output. The modulated signal is injected into one of the SOAs. This changes the phase of the CW signal propagating through this SOA due to the phenomenon of cross-phase modulation, as a result,

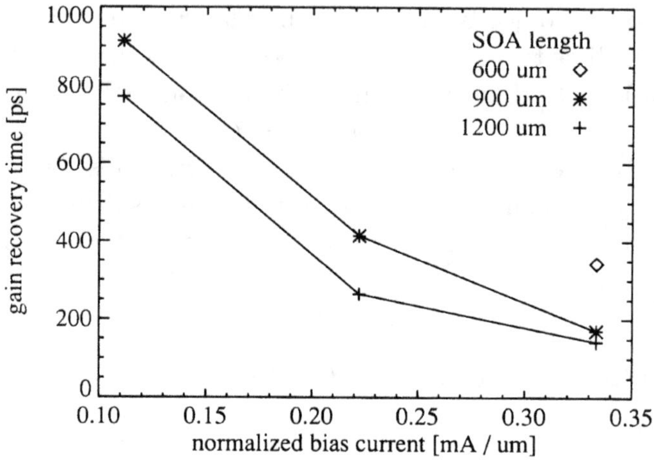

Figure 11.18: Gain recovery times for cross-gain modulation in 600, 900, and 1200 μm long SOAs, for bias currents up to 200, 300, and 400 mA, respectively. Bias currents are normalized with respect to the length of the devices in the x-axis of the above figure [99].

the phase difference between the CW signals traveling through the two SOAs is no longer π and an output is observed. The speed of wavelength conversion is determined by the phase recovery dynamics in the SOA. For long pulses, for all practical purposes, the phase recovery is limited by carrier recovery time. For short pulses (<2 ps), carrier heating, and spectral hole burning play a role and the response is faster. Careful adjustment of peak power of the modulated signal and SOA bias currents is necessary to get error-free performance at high data rates. The carrier lifetime in an SOA is an important parameter for high-speed operation of MZI devices. The measured lifetimes for various types of SOA devices are shown in Figure 11.18. The times plotted are for 10–90% gain recovery. The data shows that longer amplifiers are good for high data rates. Additional reduction in gain recovery times can be obtained using a holding beam [98, 99].

Wavelength conversion of high-speed signals has been demonstrated using an SOA-delayed interferometer (SOA-DI) combination [97, 100–102]. The DI can be formed using birefringent crystals (e.g. calcite), delay loop in waveguides, or a polarization maintaining

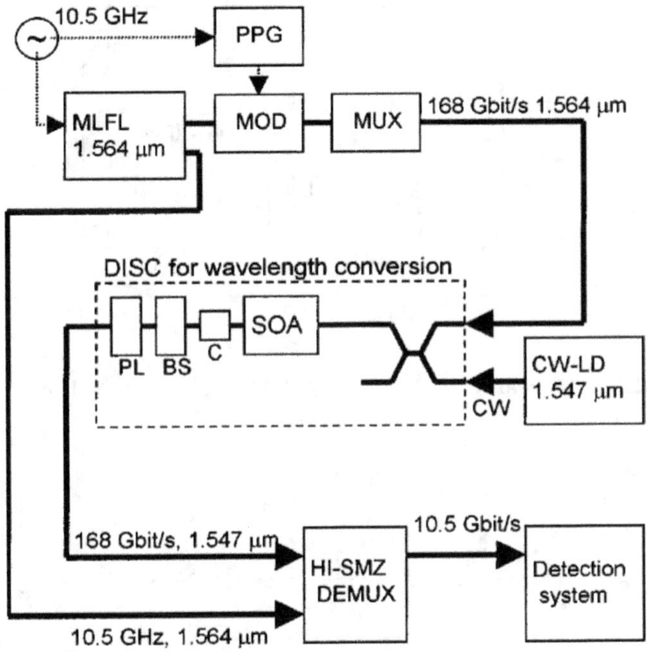

Figure 11.19: Schematic of a wavelength conversion experiment using SOA-delayed interferometer [97].

fiber. The experimental setup for a 168 Gb/s wavelength conversion experiment is shown in Figure 11.19. The modulated signal and the CW light (to which the data will be converted) are both injected into the SOA. In the SOA, the data pulses deplete the carriers which leads to cross-phase modulation of the CW light. The phase modulated CW light enters the delayed interferometer in which it splits into two paths (or two polarizations) and propagates in these two paths (or two polarizations) at different speeds. Thus, at the output, the two temporally displaced components of the phase modulated signal interfere producing an output pulse. The duration of the pulse is determined by the duration of the phase gate produced by the differential temporal delay between the two paths.

The 168 Gb/s signal is generated by multiplexing 16 times pulses at 10.5 GHz repetition rate from a mode-locked fiber laser. The pulse width is 1.0 ps. The pulsed signal wavelength is 1564 nm and the CW

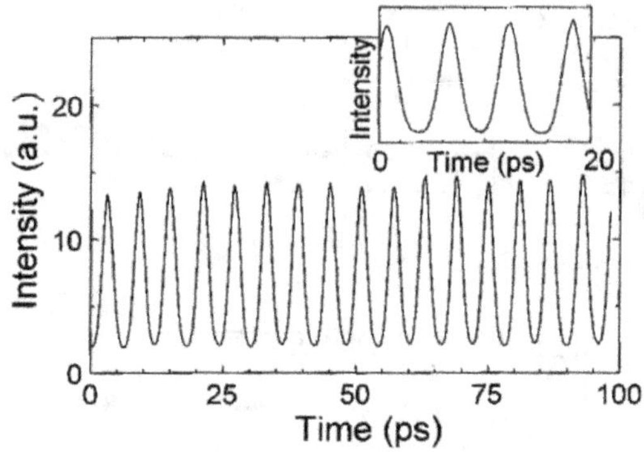

Figure 11.20: Streak camera trace of 1547 nm wavelength converted pulse train [100].

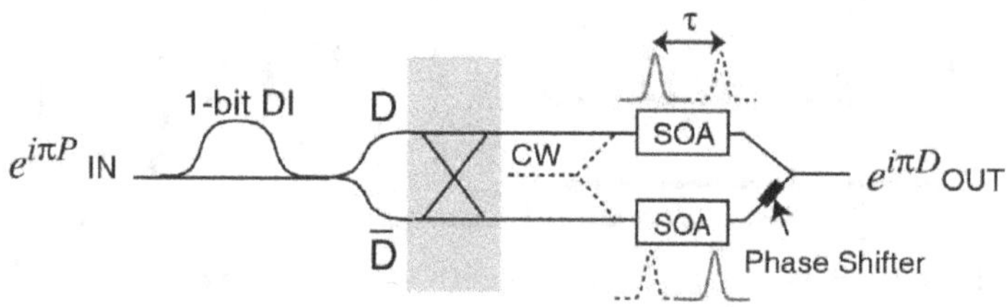

Figure 11.21: Schematic of a DPSK wavelength converter [101, 102].

signal wavelength is 1547 nm. The output 168 Gb/s pulses (viewed using a streak camera) is shown in Figure 11.20.

Wavelength conversion has been demonstrated using differential phase shift keyed (DPSK) signals. The schematic of the setup is shown in Figure 11.21 [101].

The wavelength converter has a DI built using silica waveguide at the front end. The front end converts the input DPSK signal into OOK binary data $D\psi$ and its complement $\bar{D}\psi$, followed by an SOA–MZI differentially driven by the $D\psi$ and $\bar{D}\psi$ in a "push–pull" mode. The SOA–MZI is a hybrid device fabricated by butt-coupling a bar of 2-mm long bulk InGaAsP/InP SOA arrays with an input and an output silica planar lightguide circuit (PLC) waveguides. The DPSK wavelength converter has an all-optical OOK-to-PSK data conversion system as shown in Figure 11.21. An EA modulator is used to

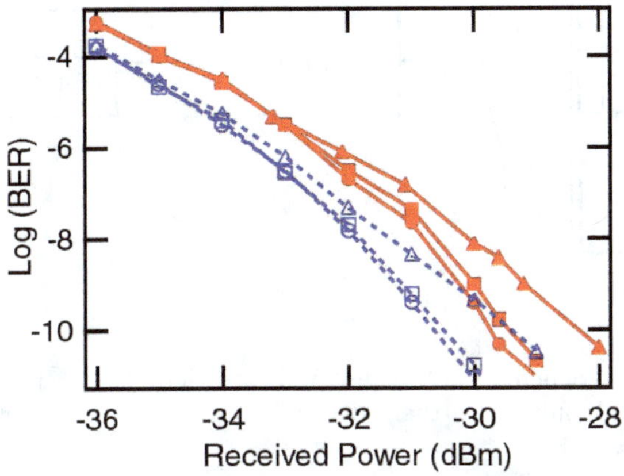

Figure 11.22: BER of back-to-back (hollow symbols with dashed curves) and wavelength-converted signals (solid symbols and curves) [102].

modulate the output of a CW laser ($\lambda = 1554$ nm) for generating ~6-ps optical pulses at 40-GHz repetition rate. These optical pulses are modulated by a lithium niobate MZM biased at the null position to encode the DPSK signals at 40 Gb/s. The wavelength converted signal at 1549 nm is separated from the input data using an optical filter with 1 nm bandwidth. The filtered wavelength converter output is demodulated using a one-bit delay DI prior to input to an erbium-doped fiber based preamplifier. The results of the BER measurement using a photoreceiver with 40 GHz bandwidth is shown in Figure 11.22.

11.8.2 *Optical demultiplexing*

Several reports have been published on optical demultiplexing as applications of FWM, and cross-gain/cross-phase modulation in SOAs [101–116]. FWM occurs both in SOAs and in optical fibers. FWM in fibers need long fiber lengths. It does not lead to compact devices and it will not be discussed here.

11.8.2.1 *Four wave mixing based scheme*

A schematic of optical demultiplexing using FWM is shown in Figure 11.23. This technique allows any regularly spaced set of bits

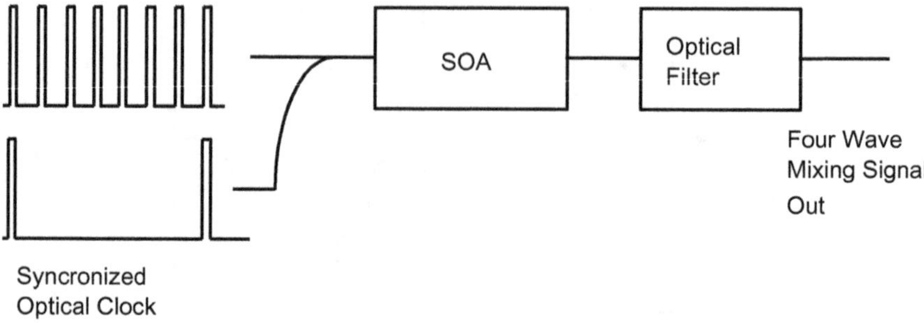

Figure 11.23: Schematic of an FWM based demultiplexer. The FWM takes place in the SOA only when there is a temporal overlap between the signal and the clock pulse.

to be simultaneously extracted from the incoming data stream using an SOA [101, 102].

Consider the high-speed data (at a data rate B) at wavelength λ_1 and an optical clock (at a clock rate B/N) at wavelength λ_2 injected into an optical amplifier. The output of the amplifier will include FWM signals at wavelengths $2\lambda_1 - \lambda_2$ and $2\lambda_2 - \lambda_1$. Typically, one of the FWM signals is filtered out using an optical filter at the exit port of the nonlinear element and represents the data signal before further processing. Thus, a signal representing the data every N^{th} bit time slot (where the mixing occurs with the clock signal) is generated by the FWM process (Figure 11.23). The clock signal can then be delayed by an additional bit each time to retrieve the original data in successive time slots as described below or a bit-delayed parallel system (similar to Figure 11.23) could be setup for simultaneous extraction of N data streams each at B/N data rate. A complete system would need N SOA elements in order to extract N data bits simultaneously. The FWM process can have modulation bandwidths as high as 1.5 Tb/s [115]. Thus, in principle demultiplexing from Tb/s optical data is feasible using the FWM process.

11.8.2.2 *Cross phase modulation based scheme*

Among the semiconductor devices suitable for high-speed-all-optical demultiplexing are (i) Mach–Zehnder interferometer (MZI), and (ii) Michelsen interferometer [101–103]. These devices operate on the

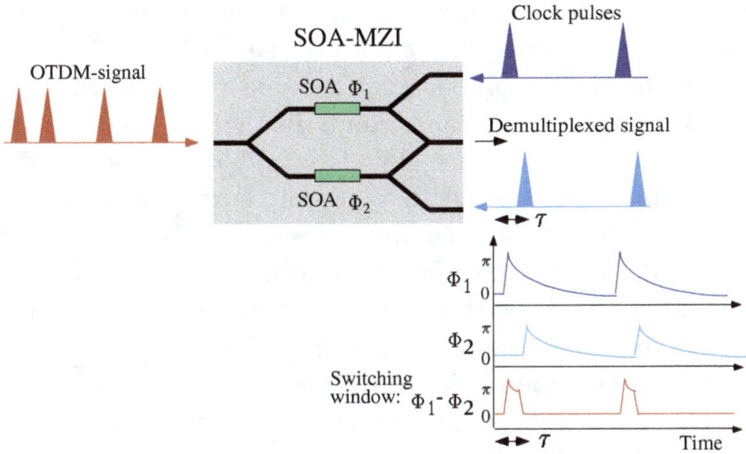

Figure 11.24: Schematic of an optical amplifier based Mach–Zehnder demultiplexer which uses a differential phase-shift scheme. The figure shows the principle of operation. ϕ_1 and ϕ_2 represent the phases produced by the two clock pulse sets, one of which is delayed with respect to the other in the two arms. The phase difference $(\phi_1 - \phi_2)$ between the two arms occurs for a time of duration τ. This increases the speed capability of the demultiplexer [101, 102].

principle of a phase change caused by an incident optical clock signal (cross-phase modulation). This phase change can be adjusted so that the interferometer only yields output when a data signal overlaps a clock signal for demultiplexing applications. The schematic of an MZI device operating as a demultiplexer is shown in Figure 11.24. The input signal beam is split into two beams by an input y-junction, which propagate through semiconductor amplifiers positioned in the upper and lower arms of the interferometer. These beams then merge and interfere at an output y-junction. One of the amplifiers has the clock pulse incident on it. The clock frequency is $1/N$ time the signal bit rate. The clock pulse is absorbed by the amplifier changing its carrier density, which induces an additional phase shift for the signal traveling through that amplifier arm. This changes the output of the device for every N^{th} signal bit. The demultiplexed signal appears as the output at the clock frequency.

The arms of the MZI can be configured with built-in phase shifters (or SOA current can be used as a phase shifter) so that the phase difference for the high-data rate signal propagating through the two arms is π (before the arrival of the clock pulse). Under this condition, the signals propagating through the two arms interfere destructively, resulting in no output. In the presence of the clock pulse, an additional π phase shift is introduced in one arm of the interferometer, which results in a non-zero output for every N^{th} bit. Similar to case of FWM, a complete system would need N MZI elements in order to extract N data bits simultaneously. The response speed, i.e. the signal speed which the MZI can demultiplex, is determined by the phase response time in the SOA.

The device can be made to operate considerably faster using a differential phase-shift scheme which involves introducing a second delayed clock signal in the other arm of the interferometer. The principle of this operation is illustrated in Figure 11.24. In the delayed pulse approach, the differential phase shift between the two arms is maintained over a short duration (1-bit period). As a result, a small gate is established for demultiplexing the high-speed OTDM signal. These devices have been used to demultiplex 80 and 160 Gb/s optical signals to 20 Gb/s signals.

Figure 11.25 shows the experimental results of demultiplexing an 80 Gb/s signal into eight channels at 10 Gb/s each [102]. This experiment utilized a Michelsen interferometer which operates on the same principle as Mach-Zehnder interferometer.

11.8.3 *OTDM system applications*

In an OTDM system, the data rates are high enough that the process of multiplexing (generation of high-data-rate signals from several low data rate signals) and demultiplexing (process of conversion of high-data-rate signals to several low-data-rate signals) are carried out optically. Several papers on various aspects of OTDM systems and all optical processing using optical amplifiers have been published [103–113]. For the demultiplexing process, an optical clock (a timing signal) needs to be extracted from the incoming high speed (>100 Gb/s) optical data. This clock recovery process has been carried out using cross-gain modulation, cross-phase modulation, and FWM [103, 114].

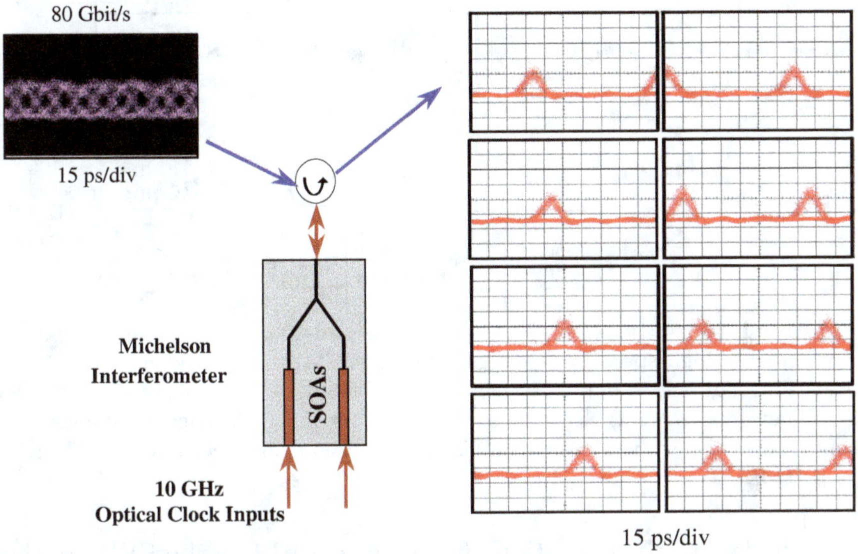

Figure 11.25: The figure shows the demultiplexing of an 80 Gb/s signal to eight 10 Gb/s channels using a Michelson demultiplexer. Scale = 15 ps/div in all traces [102].

11.8.3.1 *Clock recovery*

The schematic of an experimental setup used for clock recovery from high-speed data signals is shown in Figure 11.26. The input optical signal and the local oscillator optical signal produce an FWM signal at the output of the SOA.

In a transmission system, the timing signal (clock) extracted from the input optical data is used to demultiplex the data to lower data rate channels, the error rate of which is then measured using traditional electronic circuits. The clock recovery process could be carried out using an optical phase-locked loop (PLL) in conjunction with an electrical low-frequency PLL. This is shown schematically in Figure 11.26. In this process, the high-data rate optical signals are converted to a low-data rate electrical signal using an FWM process followed by a photodetector. The phase and frequency of this electrical signal is locked to that of a voltage controlled local oscillator using conventional electrical PLL techniques.

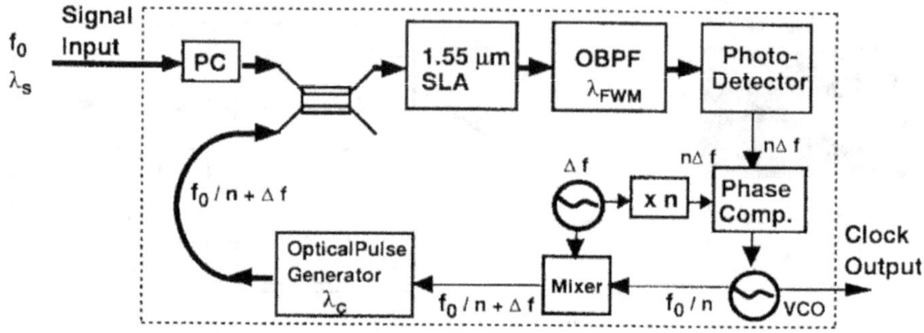

Figure 11.26: Schematic of a clock recovery diagram used for recovering the clock from a high-speed optical signal using FWM in SOA (labeled SLA in figure). The high-speed optical input (at >100 Gb/s) is down converted to a low frequency (at ~1 MHz) electrical signal using optical to electrical nonlinear down converter. This low-frequency signal is phase locked using a conventional electrical PLL [114].

The key to the operation of this optical PLL is the FWM optical to electrical converter. It is an SOA which has two optical signal inputs at wavelengths λ_1 and λ_2. The wavelength λ_1 is the input wavelength carrying data and the wavelength λ_2 is the optical clock generated by a laser at the demultiplexer. The wavelength λ_1 carries the data at B Gb/s and the wavelength λ_2 is the optical clock at a frequency $F \sim B/N$ GHz+ f kHz. The FWM process produces an optical signal at a wavelength of $2\lambda_1 - \lambda_2$. This optical signal has several electrical components including a low-frequency component close to Nf kHz produced by mixing of the N_{th} harmonic of the clock (NF $\sim B+$ Nf) and the incoming signal at B Gb/s. This low-frequency component in the optical signal is converted to an electrical signal using an avalanche photodiode, filtered, and amplified. This low-frequency electrical signal and the Nf kHz signal from a signal generator are used in a conventional PLL configuration to lock the frequency F to B/N. The conventional PLL has a phase detector (electrical mixer) and a voltage controlled oscillator. A stable optical clock is generated using a low-frequency electrical PLL which would provide the needed stability, and an optical front end (SOA) which would convert the high data rate signal to a low-frequency optical signal.

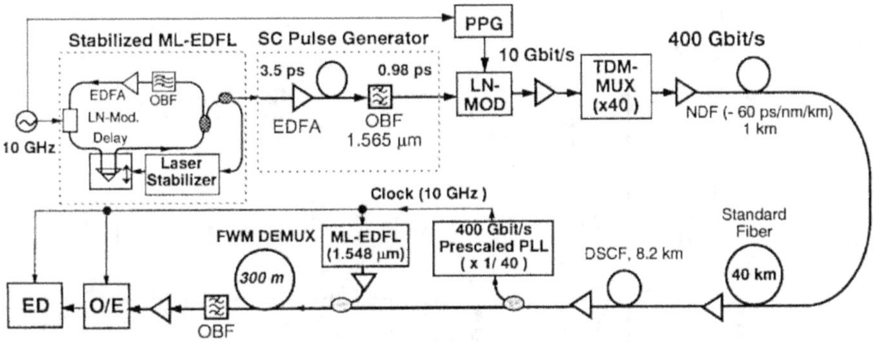

Figure 11.27: Schematic of a 400 Gb/s optical transmission experiment setup [114].

Although this conversion from a high-data rate optical signal to a low-frequency optical signal is described above using an FWM process, it can also be carried out using the gain saturation process in a semiconductor amplifier. In the latter process, the clock signal will be an input and the data signal will be the second input to the amplifier. The gain of the data signal is influenced by the high-power clock signal and this gain modulation has a component at the N_{th} harmonic (Nf) of the difference frequency. The phase variance of a PLL fabricated using the FWM and the gain modulation schemes have been analyzed [111–114]. The performance of the gain/phase modulation scheme is limited by the gain recovery time of the amplifier to ~100 Gb/s using a differential scheme. The FWM scheme can in principle operate at data rates of ~1 Tb/s [115].

11.8.3.2 *OTDM transmission*

The schematic of a OTDM transmission setup is shown in Figure 11.27. A stabilized mode-locked fiber laser generates short pulses at 10 GHz repetition rate. The 3.5 ps wide pulses are compressed to 0.98 ps pulses and are coded with data by transmission through a $LiNbO_3$ modulator. It then goes through a time division multiplexer (TDM) which converts the data stream to a 400 Gb/s data. The TDM consists of a suitably designed sequence of optical delay lines. The 400 Gb/s data is transmitted through various lengths

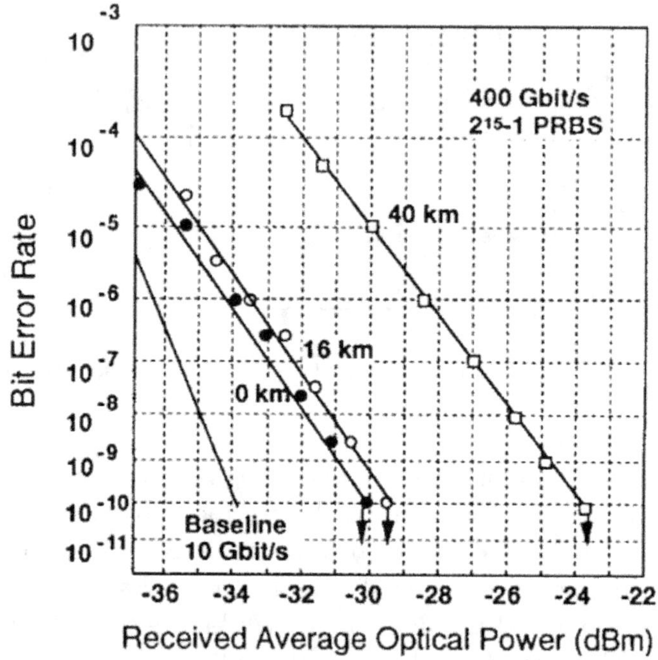

Figure 11.28: Measured bit-error-rate (BER results at 10 Gb/s after demultiplexing of the 400 Gb/s data into forty 10 Gb/s data channels. The baseline is the BER data after the modulator. The three other lines are the data after transmission through ~10 m (labeled 0 km in the figure), 16 and 40 km of fiber [114].

of fiber. The clock is extracted from the transmitted data using an FWM SOA-based clock recovery process. A mode-locked fiber laser operating at 10 GHz, the exact frequency of which is recovered by phase locking to the received 400 Gb/s data, provides the optical clock. The demultiplexing in this experiment is also carried out using an FWM process.

The error rate performance of the demultiplexed lower data rate channels are shown in Figure 11.28. BER of $<10^{-9}$ is obtained suggesting good performance.

Very high-speed OTDM transmission systems are now being developed in several research laboratories for future installment in commercial traffic. SOAs play a key role in the development of these systems. They can be used for both optical demultiplexing and optical clock recovery using either the FWM process or Mach–Zehnder type devices. A detailed review is given in Refs. [98, 116].

11.9 Advanced modulation formats

So far, optical transmission using two-level pulse amplitude modulation scheme has been discussed. Optical transmission systems using advanced modulation formats such as duobinary, DBPSK, differential quadrature phase shift keying (DQPSK), and four-level pulse amplitude modulation (PAM-4) have also been studied (see Section 5.5.2 of Chapter 5). Increasing data traffic requires a continuous expansion of network capacity, and advanced modulation formats play an important role in high-capacity transmission. This section describes experimental results using these formats. The advanced modulation formats provide higher capacity per optical carrier without increasing basic signaling rate or electronic speed. Long-distance communication requires high tolerance to both chromatic and polarization mode dispersion, and nonlinear effects. Generally, many (e.g. 32) WDM channels are transmitted over the same single-mode fiber. Complex digital signal processors (DSPs) are used to retrieve the information from DQPSK, 16 quadrature amplitude modulation (QAM) or coherent transmission systems.

11.9.1 *Optical duobinary transmission*

Duobinary transmission format has been studied for passive optical networks (PON) and ethernet networks [117–121]. The PON has both a transmitter and receiver. The duobinary signal is generated by low-pass filtering or delay-and-add filtering of a preencoded data sequence to obtain three-level pulses [120, Section 5.5]. The 3-dB bandwidth of the low-pass filter is generally 0.25 times of the signal-bit rate to realize an effective format conversion. For 28 Gb/s duobinary signal, this corresponds to a bandwidth of 7 GHz. A directly modulated laser with a bandwidth of 10 GHz and a photodiode with a bandwidth of 10 GHz can be used for duobinary optical data generation and

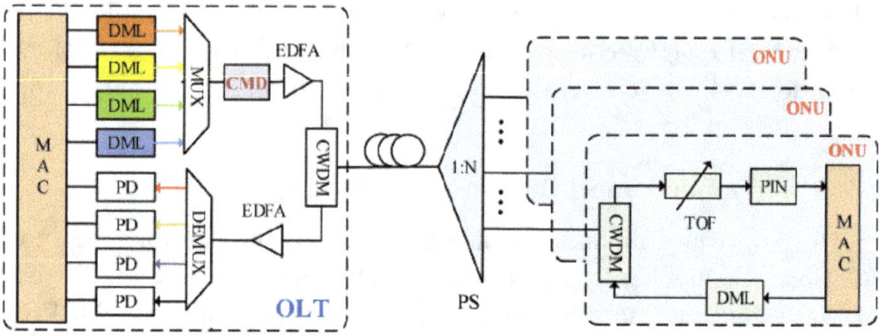

Figure 11.29: Schematic of a network architecture. *Notes*: CWDM, Coarse WDM; EDFA, erbium-doped fiber amplifier [120].

optical to electrical conversion at the receiver. Figure 11.29 shows the schematic of the network architecture of a WDM-PON system operating at 100 Gb/s. Each WDM channel operates at 28 Gb/s as opposed to 25 Gb/s to allow for overhead processing such as forward error correction. The outputs of lasers are combined by an optical multiplexer (MUX) followed by an EDFA which amplifies the signal power prior to transmission. After fiber transmission, the signal is distributed to all-optical network units (ONUs) by a power splitter. The tunable optical filter (TOF) in each ONU selects the specified wavelength and then launches the signal into the p-i-n for detection. Finally, the received electrical signal in duobinary format is demodulated back to binary signal by digital signal processor (DSP).

The laboratory experimental setup for testing the above architecture used one WDM channel. It is shown in Figure 11.30. The encoded data goes to a pulse pattern generator (PPG) operating at 28 Gb/s to get the OOK signal output. Then, the signal is modulated onto a laser emitting at 1543 nm. Before being launched into the fiber, the signal passes through a DI for chirp elimination, and an EDFA is used to boost the power. After 40-km transmission through SMF, the signal is attenuated by a variable optical attenuator (VOA), which emulates the function of power splitter in the downstream network of Figure 11.29 and varies the optical power for

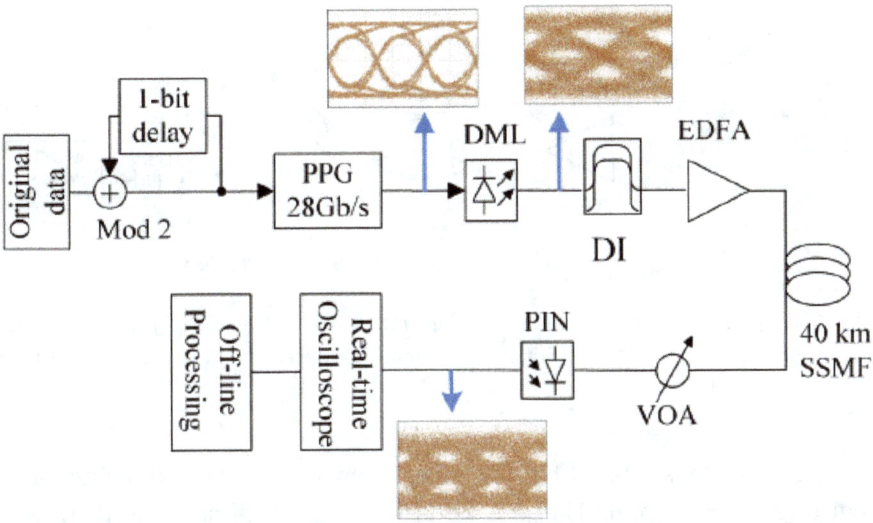

Figure 11.30: Experimental setup. *Notes*: PPG, pulse patern generator; DML, directly modulated laser; DI, delayed interferometer; EDFA, erbium-doped fiber amplifier; VOA, variable optical attenuator; PIN, p–i–n photodiode [120].

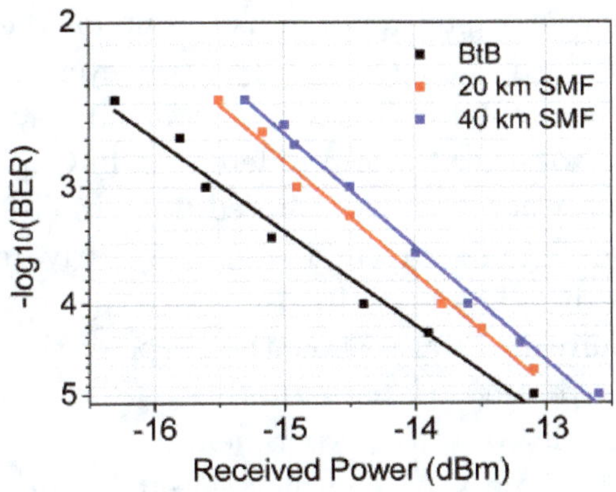

Figure 11.31: Results of BER measurement showing good performance [120].

BER testing. The measured BER as a function of received power is shown in Figure 11.31. The loss budget of the system of Figure 11.29 is found to be 31 dB which could potentially support 64 users within 40 km [120].

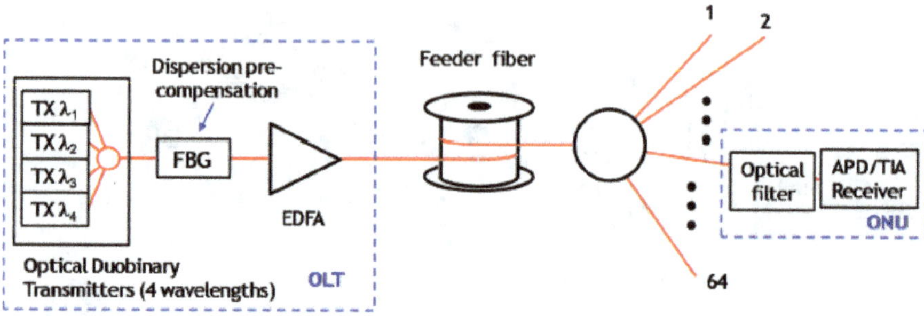

Figure 11.32: Concept of a 4 × 40 Gb/s duobinary signal transmission. The system has the capability of error-free operation over 42 km and 64 subscribers [117].

A 160 Gb/s (4 × 40 Gb/s) has been demonstrated using optical duobinary signals [117]. A time-wavelength division multiplexed (TWDM) concept is shown in Figure 11.32.

Figure 11.33 shows the detailed experimental setup of the optical duobinary transmitters. The four laser sources are DFB lasers operating on a 100-GHz grid from 193.5 to 193.2 THz (1549.32 to 1551.72 nm). Two transmitters are used, one for the two odd channels (first and third) and one for the two even channels (second and fourth). Four 10-Gbps pseudo-random bit sequences (PRBSs) are multiplexed together to form differential 40-Gbps PRBS signals. These signals are then low-pass filtered with 15-GHz filters to generate three-level signals which differentially drive a MZM with a 35 GHz bandwidth. The noise-free simulated eye diagrams included as insets in Figure 11.33 also show the shape of the electronic drive signals before and after low-pass filtering.

The schematic of the experimental receiver is shown in Figure 11.34. A single wavelength is selected using a fixed optical filter. The optical filter has insertion loss of 3 dB, and the cross-talk from adjacent channels is $\leq$35 dB. A commercial system would employ a tunable filter rather than a fixed filter at the ONU for wavelength selection. A 25-Gb/s APD-TIA receiver is used to achieve the required optical power budget typical of standardized PONs.

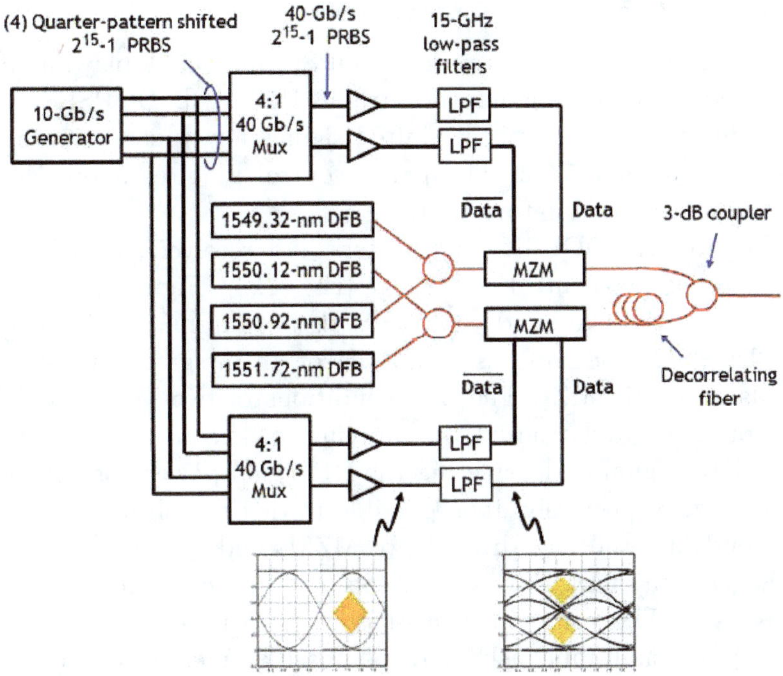

Figure 11.33: Schematic of a duobinary transmitter implementation for Figure 11.32. Insets show calculated (noise free) electrical drive signals before and after low-pass filtering [117].

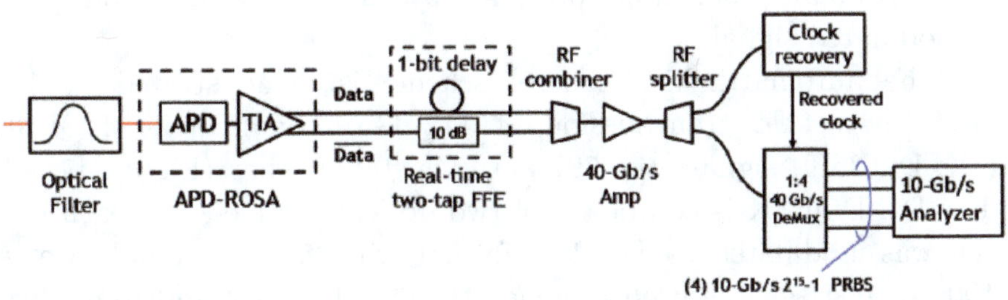

Figure 11.34: Schematic of a duobinary receiver for Figure 11.32 for one channel. The receiver is APD-based followed by a preamplifier (TIA). *Note*: APD-ROSA, APD based receiver subassembly (RSOA) developed for 100 G Ethernet standard [117].

11.9.2 *Optical DQPSK transmission*

Phase modulation formats such as DQPSK and QAM, play an important role in increasing network traffic [122–126]. DQPSK employ MZM for creating the phase shifted signals for transmission as discussed in Section 5.5.2. The phase control requires a precise bias voltage on the modulators.

In practical DQPSK transmitters and receivers, an automatic bias mechanism control is needed [123, 124]. Figure 11.35 shows DQPSK transmitter and receiver with an automatic bias control. The DQPSK modulator is composed of an integrated two-parallel MZMs nested in MZI. The bias conditions of two MZMs and MZI are optimized and tracked by applying low-frequency square-wave ($\sim$10 kHz) signal and detecting the RF signal power and RF power at 10 kHz, respectively. It is possible to distinguish and determine independently the bias drift of two MZMs and MZI. To distinguish I- and Q-data of DQPSK signal, common and differential tuning of DI is used. The common control is used to tune the phases of I- and Q-arm simultaneously, and the phase difference between I- and Q-arm is maintained at $\pi/2$ by differential control. A fraction (few percent) of the output signal of balanced detector is tapped and applied to voltage peak detector to produce bias control signal of DI. The proper bias points of optical DQPSK demodulator and tracking their optimum points were obtained by monitoring peak voltage of demodulated signal.

The automatic bias control scheme is demonstrated at 112 Gb/s in a field transmission of 112 Gb/s dual-carrier DQPSK (DC-DQPSK) signal over 797-km of installed fiber (Figure 11.36). The DC-DQPSK is composed of two optical carriers, and each carrier was modulated by DQPSK format with 28 Gb symbol rate. A 100 GE test set was used to generate 100 GE traffic and to analyze the transmission performance.

The fiber transmission consisted of 797 km of standard SMF and DSF with 9 ps/nm/km dispersion. The transmission fiber consisted of several span lengths, and EDFA are used to amplify the signal after every span. Span lengths varied from 40 to 106 km while span loss

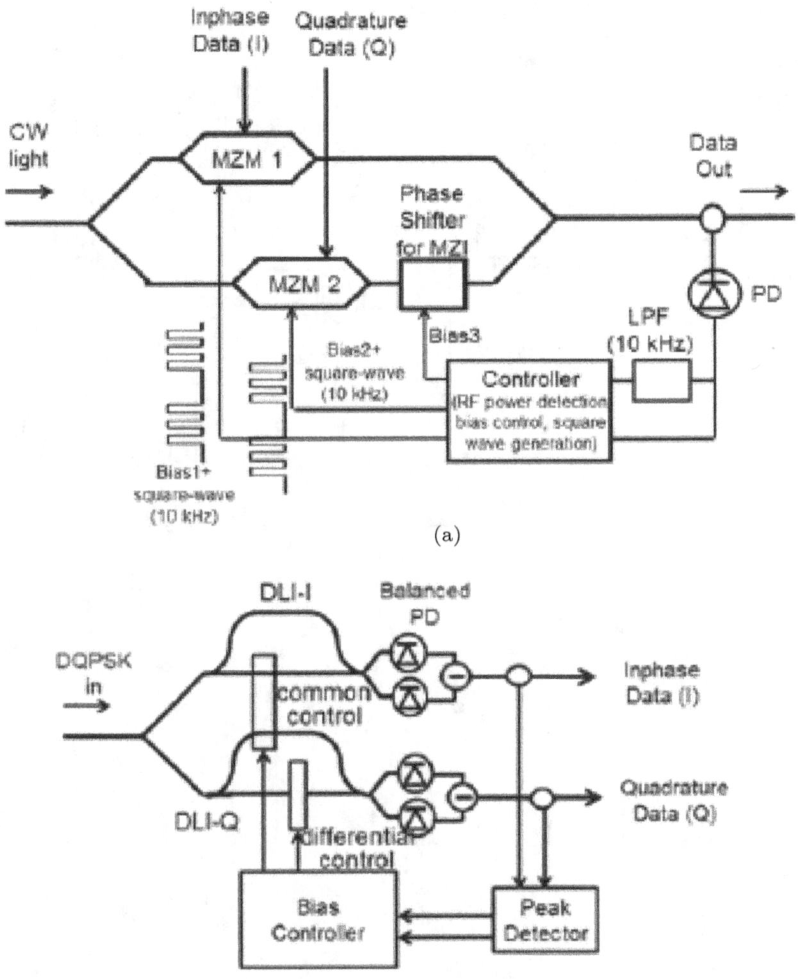

Figure 11.35: Schematic of DQPSK transmitter and receiver circuit with automatic bias control. (a) Bias control of optical IQ modulator is based on square wave signal input and RF power detection, (b) bias control of optical IQ demodulator based on peak voltage detection of demodulated signal [123].

varied in the range of 17–27 dB. The demultiplexed 112 Gb/s signal was connected to DI, and the demodulated outputs were converted to electrical signal via balanced photo-detector (BPD). Good BER results are obtained.

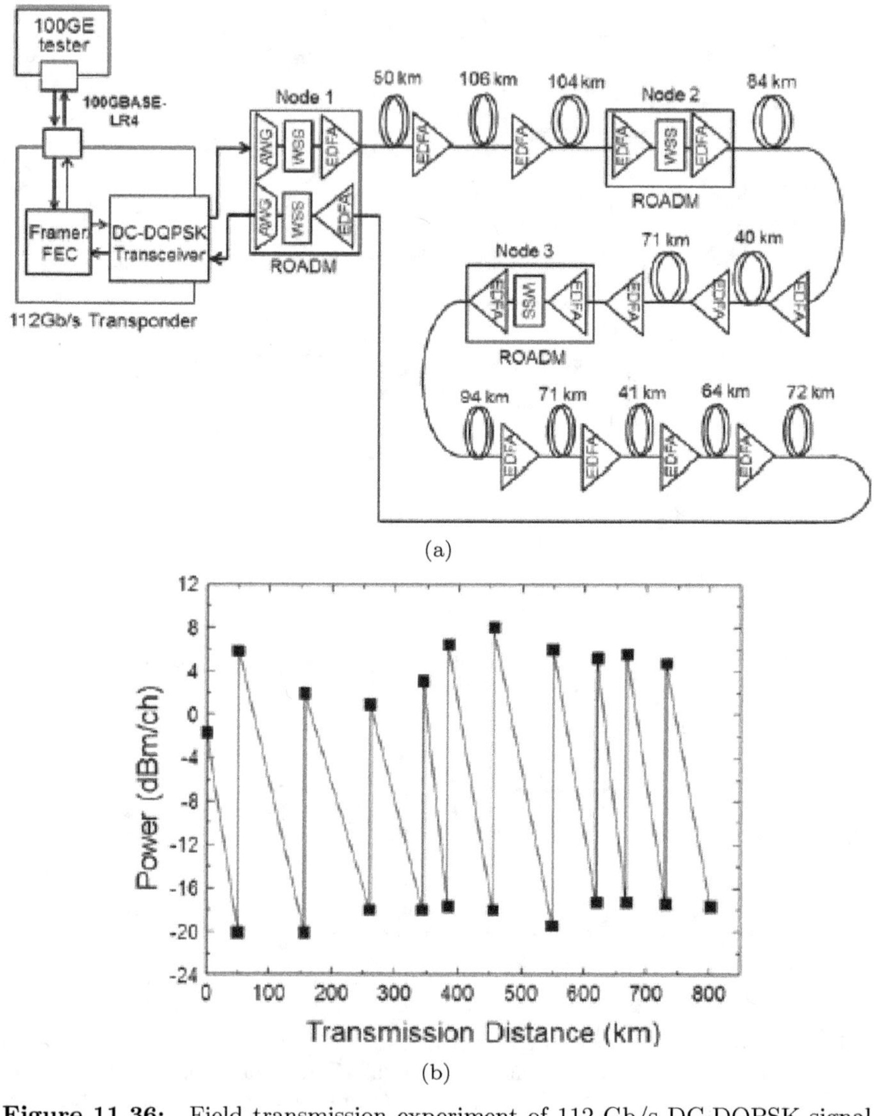

Figure 11.36: Field transmission experiment of 112 Gb/s DC-DQPSK signal with automatic bias control (a) experimental setup, (b) launching power and loss of each span [123].

11.9.3 *Optical transmission in data centers*

The needs of optical data communication links in data centers where a large number of computers or processors are interconnected are somewhat different [127–132]. A simpler low-cost modulation scheme along with ease of installation (such as use of multimode fibers) is more important. A four-level pulsed amplitude modulation scheme (PAM-4) is likely to be more suitable for this purpose. The BER for regular two-level PAM is discussed in Section 11.2 and that for PAM-4 is discussed later in this chapter.

An example of a 400 Gb/s aggregate data transmission using PAM-4 modulation scheme is shown in Figure 11.37. It uses 4 WDM channels (noted as optical transmitters in Figure 11.37) operating at wavelengths 1271, 1291, 1311, and 1331 nm, respectively. The wavelengths are far enough apart so that the dispersion effects are negligible over 2 km of SMF. Each wavelength channel operates at 100 Gb/s using PAM-4. PAM-4 offers the lowest implementation

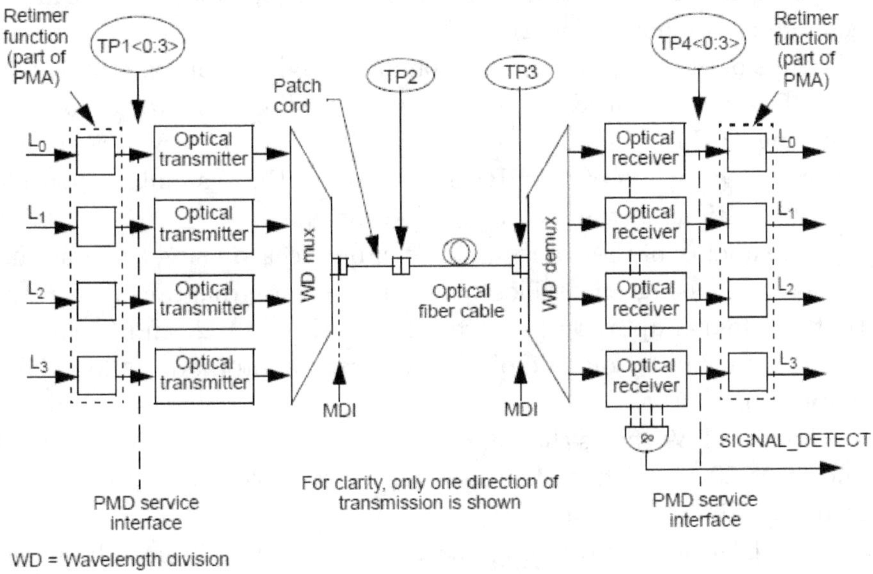

Figure 11.37: Schematic of a 4-channel WDM transmission system with an aggregate capacity of 400 Gb/s. Each channel operates at 100 Gb/s using PAM-4 modulation format [132].

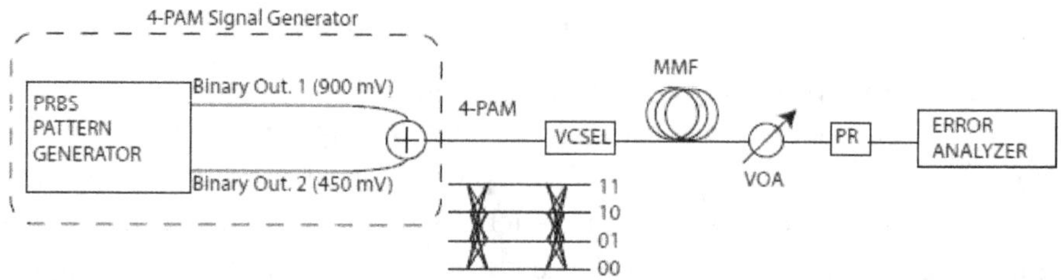

Figure 11.38: Schematic diagram of the experimental setup for PAM-4 transmission using VCSEL and multimode mode fiber (MMF). The PAM-4 signal generator (noted as 4-PAM) is shown in the dashed box [127].

complexity of all the modulation formats with spectral efficiency of 2 bits/s/Hz. Hardware for transmitters and receivers for PAM-4 have been developed [131]. Similar experiments have been carried out in the 850 nm wavelength region using both single mode and multimode fibers. Experiments using a multimode fiber and vertical cavity surface emitting lasers (VCSEL — Chapter 4) is now described [127]. VCSEL has the advantage that it can be easily aligned to a multimode fiber. In laboratory conditions, PAM-4 modulation can be generated as follows (Figure 11.38):

Consider two-pseudo random binary signals of approximately 900 and 450 mV amplitude and they are tuned slightly to adjust the resulting PAM-4 signal level spacing. The two binary signals are decorrelated by half of the PRBS pattern length, to simulate possible level transitions in the resulting PAM-4 signals.

The fidelity of the transmission can be checked using eye-diagram as observed using an oscilloscope. An open eye-diagram illustrates that the four amplitude (or intensity) levels (0, 1, 2, and 3) can be easily seen on the output of the receiver (photodiode followed by amplifier).

Figure 11.39 shows the eye-diagrams with very little fiber (~1 m) and after 200 m of fiber transmission. Also, shown is the measured error-rate data.

The BER for a two-level pulse amplitude modulation is discussed in Section 11.2. As discussed, a PAM-4 is important for increasing the total transmission rate particularly for short distance links. PAM-4 scheme can be easily implemented by adding two sets of data signals

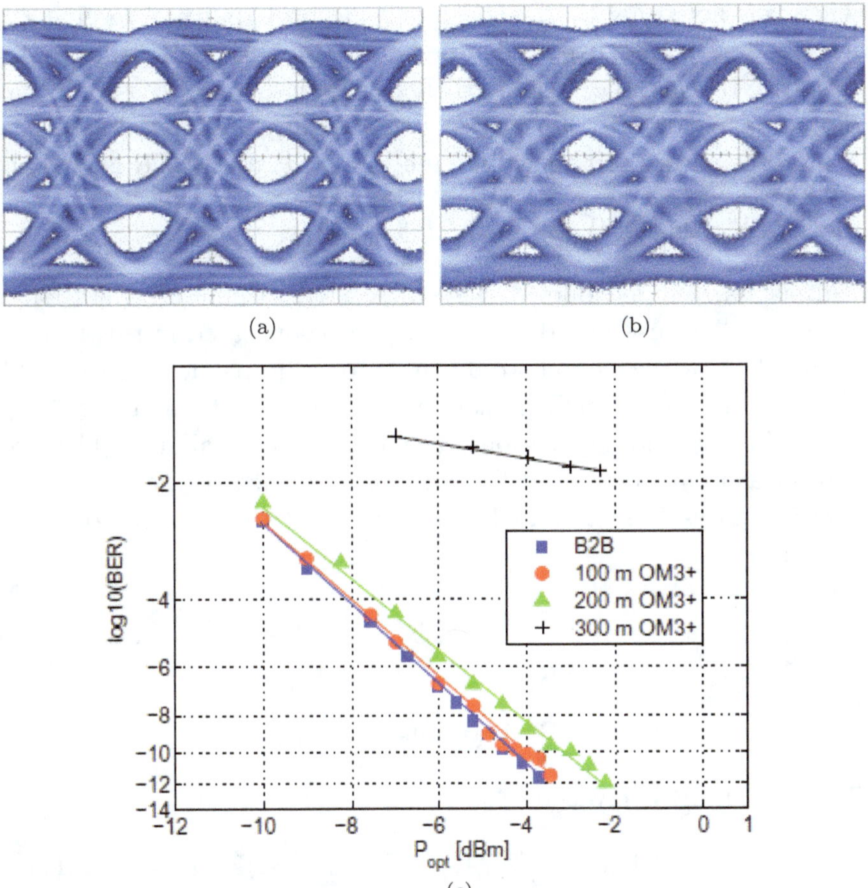

Figure 11.39: (a) Eye-diagram at 30 Gb/s after very little fiber — also known as back-to-back (B2B), (b) Eye-diagram at 30 Gb/s after 200 m of fiber. All four intensity levels (0, 1, 2, and 3) are visible. Received optical power is 0 dBm. The measured bit-error-rate (BER) at 30 Gb/s as a function of received power for various fiber lengths are shown. Note that the low BER ($<10^{-9}$) is feasible for fiber lengths less than 200 m [127].

as shown in Figure 11.38. Here, a simple discussion of BER for a PAM-4 is described. The discussion follows Ref. [127]. Consider that the four levels are S_0, S_1, S_2, and S_3. There are three decision levels between the adjacent symbols: L_1, L_2, and L$_3$. A symbol error is detected when a symbol S_i is transmitted and a symbol S_j is received, where i is not equal to j. The conditional probability of receiving Sj when S_i was transmitted is denoted with Pij. For high signal-to-noise ratio (SNR) case (i.e. for low BER), it is reasonable to assume that only errors between the nearest neighboring symbols happen, and given that natural bit to symbol mapping, the total BER is given by

$$\mathrm{BER} = (1/8)(P_{01} + P_{10} + 2P_{12} + 2P_{21} + P_{23} + P_{32}) \tag{11.7}$$

This is derived as follows: The high SNR region of operation is important because practical links would be operated in this fashion. Forward error correction would have to be implemented to enable operation in the low SNR region, which would probably be challenging in the targeted applications of short-range optical links. The error rate measurements ER1, ER2, and ER3 at each of the consecutive decision levels L1, L2, and L3 can be expressed in terms of Pij as follows:

$$\begin{aligned} \mathrm{ER}_1 &= (1/4)(P_{10} + P_{01}) \\ \mathrm{ER}_2 &= (1/4)(P_{12} + P_{21}) \\ \mathrm{ER}_3 &= (1/4)(P_{23} + P_{32}) \end{aligned} \tag{11.8}$$

The total BER is given by

$$\mathrm{BER} = (1/2)\mathrm{ER}_1 + \mathrm{ER}_2 + (1/2)\mathrm{ER}_3 \tag{11.9}$$

Experiments have been carried out using multicore fibers coupled to multiple VCSELs fabricated on the same chip. The photodiodes at the receivers are also fabricated on the same chip. A schematic of these devices are shown in Figure 11.40.

The BER data on a multicore fiber at 40 Gb/s (for one channel) is shown in Figure 11.41. The aggregate data transmission in this experiment is at $6 \times 40 = 240$ Gb/s.

WDM transmission at 40 Gb/s per wavelength channel using 4-VCSELs at 850, 880, 910, and 940 nm has been demonstrated. The data are shown in Figure 11.42 for PAM-4 transmission over 100 m of multimode fiber. A 32 nm spacing between each VCSEL

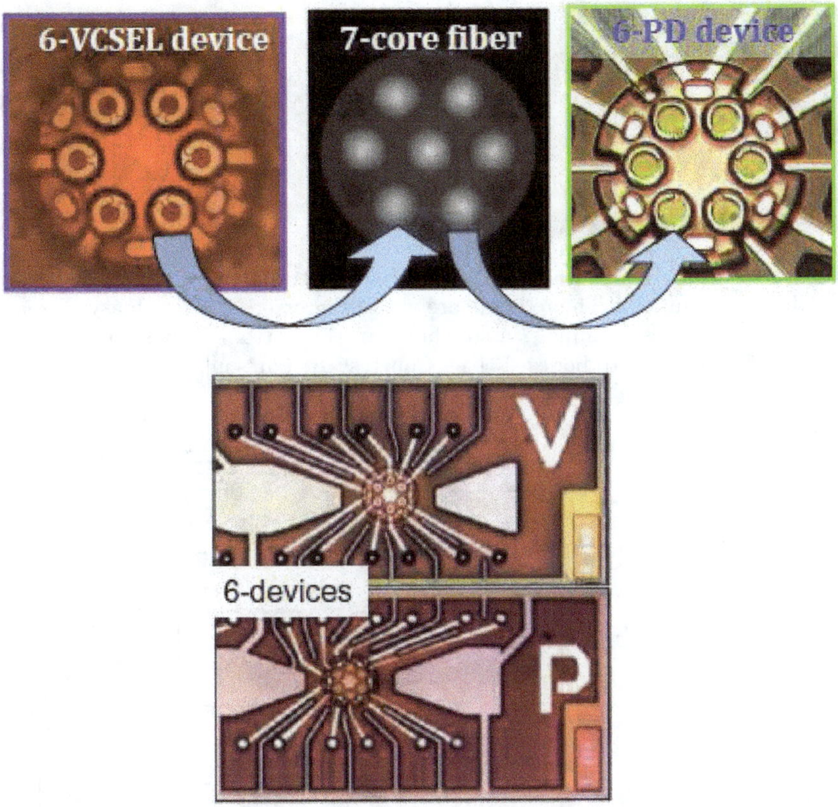

Figure 11.40: 6-VCSELs on a chip, 6-core multimode fiber, and a 6-photodiode device. The layout of electrical connections to a VCSEL and photodiode chip is shown at the bottom [131 — fabricated by Sumitomo Electric].

channel has been proposed as a standard for WDM interconnections in data centers.

Optical communication has been deployed for long-distance information transmission for nearly three decades now. Advanced modulation formats are being developed for increasing the transmission capacity of these systems. Optical communication for data center applications which need short distance (<1 km) very high throughput transmission are being developed. WDM transmission in addition to new modulation formats continue to be developed for data center applications.

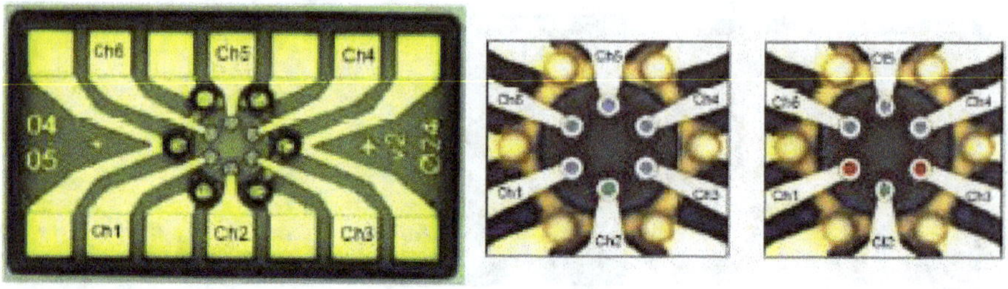

Figure 11.41: BER data (BERs a function of received power) at 25 Gb/s and 40 Gb/s for channel 2. The red dots are with "aggressors", i.e. with asynchronous data on two nearest channels. Also, shown in the bottom figures, are layouts of the VCSELs and photodiodes [131 — Courtesy A. Larsson].

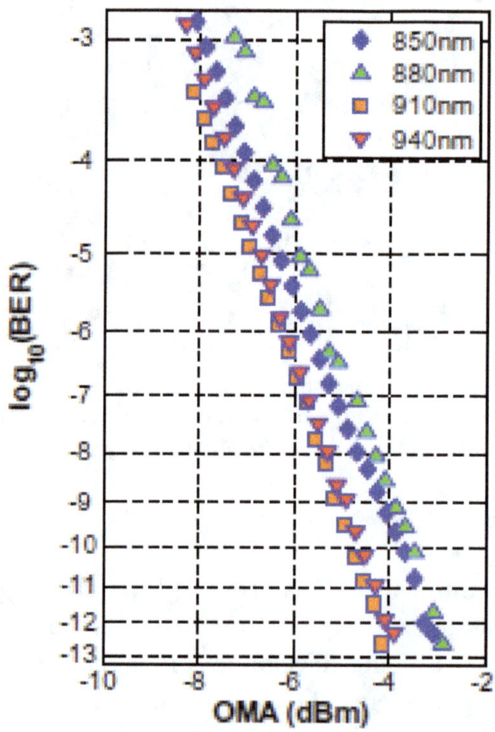

Figure 11.42: Plot of bit error rate (BER) as a function f received power for VCSEL-based WDM transmission. The data is on PAM-4 modulation format over 100 m of multimode fiber [131].

References

1. K. Hagimoto and K. Aida, *J. Lightwave Technol.*, **6** (1988) 1678.
2. A. H. Gnauck, J. E. Bowers, and J. C. Campbell, *Electron. Lett.*, **22** (1986) 600.
3. S. K. Korotky *et al.*, *J. Lightwave Technol.*, **LT-5** (1987) 1505.
4. S. Fujita *et al.*, *Electron. Lett.*, **25** (1989) 702.
5. Y. Yamamoto, *IEEE J.Quantum Elecrron.*, **QE-16** (1980) 1073.
6. K. Hagimoto *et al.*, presented at OFC'89, Paper pD-15, 1989.
7. R. I. Mears, L. Reekie, I. M. Jauncey, and D. N. Payne, *Electron. Lett.*, **23** (1987) 1026.
8. E. Desurvire, J. R. Simpson, and P. C. Becker, *Opt. Lett.*, **12** (1987) 888.
9. K. Hagimoto *et al.*, *Electron. Lett.*, **25** (1989) 662.
10. K. Hagimoto, Y. Miyagawa, A. Takada, K. Kawano, and Y. Tohmori, presented at ECOC'89, Paper TuA5-5, 1989.
11. K. Hagimoto *et al.*, presented at IOOC'89, Paper 20PDA-6, 1989; also Trans. *IEICE*, **E73** (1990) 27.
12. H. Nishimoto *et al.*, presented at IOOC'89, Paper 20PDA-8, 1989.
13. K. Aida, S. Nishi, Y. Sato, K. Hagimoto, and K. Nakagawa, presented at ECOC89, Paper PDA-7, 1989.
14. A. Righetti., presented at ECOC'89, Paper PDA -10, 1989.
15. D. Bayart, P. Bousselet, F. Chiquet, C. Le Sergent, B. Clesca, V. Havard, and J.-L. Beylat, presented at OFC'97, Dallas, TX, 1997, Paper WL1.
16. A. Takada. *Trans. IEICE*, **E72** (1989) 21.
17. S. Artigaud, M. W. Chbat, P. Nouchi, F. Chiquet, D. Bayart, L. Hamon, A. Pitel, F. Goudeseune, P. Bousselet, and J.-L. Beylat, presented at OFC'97, San Jose, CA, 1996, postdeadline Paper PD27.
18. A. K. Srivastava, J. B. Judkins, Y. Sun, L. Garrett, J. L. Zyskind, J. W. Sulhoff, C. Wolf, R. M. Derosier, A. H. Gnauck, R. W. Tkach, J. Zhou, R. P. Espindola, A. M. Vengsarkar, and A. R. Chraplyvy, presented at OFC'97, Dallas, TX, 1997, postdeadline Paper PD18.
19. K. Hagimoto, S. Nishi, and K. Nakagawa, *J. Lightwave Tech.*, **8**, (1990) 1387.
20. N. Edagawa *et al.*, presented at ECOC'89, Paper PDA-8, 1989.
21. S. Bigo, A. Bertaina, M. W. Chbat, S. Gurib, J. Da Loura, J.-C. Jacquinot, J. Hervo, P. Bousselet, S. Borne, D. Bayart, L. Gasca, and J.-L. Beylat, *IEEE Photonics Tech. Letts.*, **10** (1998) 1045.
22. T. Tanifuji and M. Ikeda, *Opt. Lett.*, **16** (1977) 2175.
23. P. R. Trischitta, P. Sannuti, and C. Chamzas, *IEEE Trans. Commun.*, **36** (1988) 675.
24. L. F. Mollenauer and K. Smith, *Opt. Lett.*, **13** (1988) 675.
25. D. Malyon, T. Widdowson, E. G. Bryant, S. F. Carter, J. V. Wright, and W. A. Stallard, *Electron. Lett.*, **27** (1991) 120.

26. N. S. Bergano, J. Aspell, C. R. Davidson, P. R. Trischitta, B. M. Nvman, and F. W. Kerfoot, *Electron. Lett.*, **27** (1991) 879.
27. N. S. Bergano and C. R. Davidson, *J. Lightwave Tech.*, **13** (1995) 879.
28. M. Nakazawa, *IEEE, J. Quantum Electron*, **26** (1990) 2103.
29. D. M. Spirit, *Electron. Lett.*, **27** (1991) 222.
30. K. Kurukawa and M. Nakazawa, *IEEE J. Quantum Electron*, **28** (1992) 1922.
31. C. Lester, *Electron. Lett.*, 1995, **31** (1995) 219.
32. D. M. Spirit, G. E. Wickens, T. Widdowson, G. R. Walker, D. L. Williams, and L. C. Blank, *Electron. Lett.*, **28** (1992) 1218.
33. M. Nakazawa, Y. Kimura, and K. Suzuki, *IEEE J. Quant. Electron.*, **26** (1990) 2103.
34. F. Chiaraluce, E. Gambi, and P. Pierleoni, *Opt. Commun.*, **105** (1994) 47.
35. A. Altuncu, L. Noel, W. A. Pender, A. S. Siddiqui, T. Widdowson, A. D. Ellis, M. A. Newhouse, A. J. Antos, G. Kar, and P. W. Chu, *Electron. Lett.*, **32** (1996) 233.
36. C. Lester, K. Bertilsson, K. Rottwitt, P. A. Andrekson, M. A. Newhouse, and A. J. Antos, *Electron. Lett.*, **31** (1995) 219.
37. A. Altuncu, A. S. Siddiqui, and A. D. Ellis, *Electron. Lett.,* **33** (1997) 1558.
38. S. Wen, *Opt. Lett.*, **19** (1994) 22.
39. H. Kawakami, T. Kataoka, Y. Miyamoto, K. Hagimoto, and H. Toba, *J. Lightwave Technol.*, **19** (2001) 1887.
40. G. R. Walker, N. G. Walker, R. C. Steele, M. J. Creaner, and M. C. Brain, *J. Lightwave Technol.*, **9** (1991) 182.
41. R. E. Wagner, *Proc. ECOC'89*, Vol. 2, (1989), Paper ThA21-1 .
42. S. Yamazaki, T. Ono, H. Shimizu, M. Kitamura, and K. Emura, in *Proc. OFC'9* , 1990, Paper PD12. (1990).
43. G. R. Walker, R. C. Steele, and N. G. Walker, *J. Lightwave Technol.*, **8** (1990) 1409.
44. Y. K. Park, J-M., P. Delavaux, T. W. Cline, and C. F. Flores, *Proc. ICC* (1991), p. 1205.
45. M. J. Creaner, *Electron. Lett.*, **24** (1988) 1354.
46. M. J. Creaner, *Electron. Lett.*, **26** (1990) 442.
47. S. Ryu, S. Yamamoto, H. Taga, N. Edagawa, Y. Yoshida, and H. Wakabayashi, *J. Lightwave Technol.*, **9** (1991) 2551.
48. S. Ryu, N. Edagawa, Y. Yoshida, and H. Wakabayashi, *IEEE Photon. Technol. Lett.*, **2** (1990) 428.
49. R. Welter, R. I. Laming, W. B. Sessa, R. S. Vodhanel, M. W. Maeda, R. E. Wagner, *Electron. Lett.*, **25** (1989) 1335.
50. H. Shinohara, *FTTH Exp. Japan — JON,* **6** (2007) 616.
51. K-I. Suzuki, Y. Fukada, D, Nesset, and R. Davey, *JON* **6** (2007) 422.
52. Y. Fukada, K-I. Suzuki, H. Nakamura, N. Yoshimoto, and M. Tsubokawa, ECOC, Paper We.2.F.4. (2008).

53. K-I. Suzuki, Y. Fukada, N. Yoshimoto, K. Kumozaki, and M. Tsubokawa, PON, OFC Paper OTuH1.(2009).
54. W. I. Way, M. M. Choy, A. Yi-Yan, M. Andrejco, M. Saifi, and C. Lin, *IEEE Photon. Technol. Lett.*, **1** (1989) 343.
55. C. R. Giles, E. Desurvire, and J. R. Simpson, *Opt. Lett.*, **14** (1989) 880.
56. A. A. Saleh, T. E. Darcie, and R. Jopson, *Electron. Lett.*, **25** (1989) 79.
57. S. Y. Huang, T. W. Cline, L. C. Upadhyayula, R. E. Tench, and J. Lipson, *Top. Meet. Broadband Analog Optoelectron*, Devices and Systems, Tech. Dig., Monterey, CA, (1990), Paper BAM7.
58. K. Kikushima, Chapter 10, in S. Shimada and H. Ishio (eds.), *Optical Amplifiers and Their Applications*,, John Wiley, 1994.
59. W. I. Way, *IEEE J. Lightwave Technol.* **7** (1989) 1806.
60. K. Kikushima and E. Yoneda, *Trans. IEICE Jpn.* **E74** (1991) 2042.
61. S. L. Hansen, P. Thorsen, K. Dybdal, and S. B. Andreasen, *IEEE Photon. Technol Lett.*, **4** (1993) 409.
62. B. Clesca, P. Bousselet, and L. Hamon, *IEEE Photon. Technol. Lett.*, **5** (1993) 9.
63. K. Kikushima, *J. Lightwave Technol.*, **12** (1994) 3.
64. F. Lai, C. Liu, and J. Jou, *IEEE Photon. Technol. Lett.*, **11** (1999) 545.
65. C. R. Giles, E. Desurvire, and J. R. Simpson, *Opt. Lett.*, **14** (1989) 880.
66. K. Kikushima and H. Yoshinaga, *IEEE Photon. Technol. Lett.*, **3** (1991) 945.
67. P. C. Becker, N. A. Olsson, and J. R. Simpson, Chapter 7, in *Erbium Doped Fiber Amplifiers-Fundamentals and Technology*, Academic Press, USA 1999.
68. P. Dua, K. Lu, N. K. Dutta, and J. Jaques, Chapter 11, in B. P. Pal (ed.), *Guided Wave Optical Components and Devices*, Elsevier Academic Press, USA, 2006.
69. C. Y. Kuo and E. E. Bergmann, *IEEE J. Lightwave Technol.*, **10** (1992) 1751.
70. C. Y. Kuo and E. E. Bergmann, *IEEE Photon. Technol. Lett.*, **3** (1991) 829.
71. M. Suzuki, N. Edagawa, H. Taga, H. Tanaka, S. Yamamoto, and S. Akiba, *Electron. Lett.*, **30** (1994) 1083.
72. F. Favre and D. LeGuen, *Electron. Lett.*, **31** (1995) 991.
73. G. Aubin, E. Jeanney, T. Montalant, J. Moulu, F. Pirio, J.-B. Thomine, and F. Devaux, *Electron. Lett.*, **31** (1995) 1079.
74. J. King, I. Hardcastle, J. Harvey, P. Greene, B. Shaw, M. Jones, D. Forbes, and M. Wright, *Electron. Lett.*, **31** (1995) 1090.
75. M. Suzuki, I. Morita, N. Edagawa, S. Yamamoto, H. Taga, and S. Akiba, *Electron. Lett.*, **31** (1995) 2027.
76. G. Aubin, T. Montalant, J. Moulu, F. Pirio, J.-B. Thomine, and F. Devaux, *Electron. Lett.*, **32** (1996) 2188.
77. I. Morita, K. Tanaka, M. Suzuki, and N. Edagawa, *IEEE J. Lightwave Technol.*, **17** (1999) 2506.

78. J. Gordon and H. Haus, *Opt. Lett.*, **11** (1986) 665.
79. A. B. Puc, M. W. Chbat, J. D. Henrie, N. A. Weaver, H. Kim, A. Kaminski, A. Rahman, and H. A. Fevrier, *Optical Fiber Communication Conf.*, OSA Tech. Dig., Washington, DC, 2001.
80. V. Dominic, E. Mao, J. Zhang, B. Fidric, S. Sanders, and D. Mehuys, *Optical Amplifiers and Their Applications*, OSA Tech. Dig., Washington, DC, 2001.
81. A. J. Stentz, T. Nielsen, S. G. Grubb, T. A. Strasser, and J. R. Pedrazzani, *Optical Fiber Communication Conf.*, OSA Tech. Dig., Washington, DC, 1996.
82. C. R. S. Fludger and V. Henderek, *Optical Fiber Communication Conf.*, OSA Tech. Dig., Washington, DC, 2001.
83. F. Galeener, J. Millelsen, R. Geils, and W. Mosby, *Appl. Phys. Lett.*, **32** (1978) 34.
84. M. Nissov, C. R. Davidson, K. Rottwitt, R. Menges, P. C. Corbett, D. Innis, and N. S. Bergano, *European Conference Optical Communications*, 1997, PD Paper, p. 9.
85. M. N. Islam, *IEEE J. Sel. Top. of Quant. Electron.* **8** (2002) 248.
86. S. Namiki and Y. Emori, *IEEE J. Sel. Top. Quant. Electron.* **7** (2001) 3.
87. C. R. S. Fludger, V. Handerek, and R. J. Mears, *Optical Fiber Communication Conf.*, Tech. Dig., 2002, Paper TuJ3, pp. 60–62.
88. T. Naito, T. Tanaka, K. Torii, N. Shimojoh, H. Nakamoto, and M. Suyama, in *Optical Fiber Communication Conf.*, Tech. Dig., 2002, Paper TuR1, pp. 116–117.
89. S. G. Grubb *et al.*, in *Optical Amplifiers and Their Applications*, 1994, PD1.
90. Y. Aoki, *J. Lightwave Technol.* **6** (1988) 1225.
91. A. J. Stentz, S. G. Crubb, C. E. Headley, J. R. Simpson, T. Strasser, and N. Park, *Optical Fiber Communication Conf.*, Tech. Dig., 1996, Paper TuD3, pp. 16–17.
92. J. L. Gimlett and N. K. Cheung, *J. Lightwave Technol.*, **7** (1989) 888.
93. S. L. Danielsen, P. Hansen, and K. E. Stubkjaer, *IEEE, J. Quant. Electron.*, **16** (1998) 2095.
94. W. Shieh, E. Park, and A. E. Willner, *IEEE. Photon. Technol. Lett.*, **8** (1996) 524.
95. K. Obermann, S. Kindt, D. Breuer, K. petermann, C. Schmidt, S. Diez, and H. G. Weber, *IEEE Photon. Technol. Lett.*, **9** (1997) 312.
96. S. Kawanishi and M. Saruwatari, *Electron. Lett.*, **24** (1998) 1452.
97. S. Nakamura, Y. Ueno, and K. Tajima, *IEEE Photon. Technol. Lett.*, **13** (2001) 1091.
98. For a review see N. K. Dutta, *Semiconductor Optical Amplifiers*, Chapter 8, World Scientific, 2012.
99. J. Leuthold, C. H. Joyner, B. Mikkelsen, G. Raybon, J. L. Pleumeekers, B. I. Miller, K. Dreyer, and C. A. Burrus, *Electron. Lett.*, **36** (2000) 1129.
100. Y. Ueno, S. Nakamura, K. Tajima, and S. Kitamura, *IEEE Photon. Technol. Lett.*, **10** (1998) 346.

101. A. Piccirilli, H. Presby, and N. K. Dutta, *MillCom 99*, Atlantic City, NJ, 1999.
102. I. Kang *et al.*, *IEEE J. Selected Areas Quant. Electron.*, **14** (2008) 258.
103. S. Kawanishi, and M. Saruwatari, *IEEE J. Lightwave Technol*, **11** (1993) 2123.
104. R. S. Tucker, G. Eisenstein and S. K. Korotky, *J. Lightwave Technol.*, **6** (1988) 1737.
105. S. Nakamura, Y. Ueno, and K. Tajima, *IEEE Photon. Technol. Lett.*, **10** (1998) 1575.
106. K. L. Hall and K. A. Rauschenbach, *Opt. Lett.*, **23**, (1996) 1271.
107. S. Kawanishi, *2nd Optoelectronics & Communication Conference (OECC '97)*, Tech. Digest, 12, Seoul, Korea, 1997.
108. K. E. Stubkjaer, *et al. 2nd Optoelectronics & Communication Conference (OECC '97)*, Tech. Dig., 178, Seoul, Korea, 1997.
109. K. L. Hall and K. A. Rauschenbach, *Electron. Lett.*, **32** (1996) 1214.
110. N. S. Patel, K. L. Hall, and K. A. Rauschenbach, *Appl. Opt.*, **37** (1998) 2831.
111. O. Kamatani and S. Kawanishi, *J. Lightwave Technol.*, **14** (1996) 1757.
112. S. Kawanishi, T. Morioka, O. Kamatani, H. Takara, J.M. Jacob, and M. Saruwatari, *Electron. Lett.*, **30** (1994) 981.
113. O. Kamatani, S. Kawanishi, M. Saruwatari, *Electron. Lett.*, **30** (1994) 807.
114. S. Kawanishi, *IEEE J. Quantum Electron*, **QE-34** (1998) 2064.
115. C. Wu and N. K. Dutta, *J. Appl. Phys.*, **87** (2000) 2076.
116. H. Sotobayashi and T. Ozeki, Chapter 2, in A. K. Dutta, N. K. Dutta, and M. Fujiwara (eds.), *WDM Technologies- Optical Networks*, Elsevier Science, USA, 2004.
117. A. Gnauck, P. Iannone, D. van Veen, and V. Houtsma. *Opt. Exp.*, **23** (2015) 24113.
118. D. van Veen, V. Houtsma, A. Gnauck, and P. Iannone, *J. Lightwave Technol.*, **33** (2015) 1675.
119. Y. Ban, T. D. Keulenaer, G. Torfs, J. H. Sinsky, B. Kozicki, and J. Bauwelinck, *Electron. Lett.*, **51** (2015) 617.
120. Z. Li, L. Yi, X. Wang, and Y. Hu, *Opt. Exp.,* **23** (2015) 20249.
121. K. Yonenaga, S. Kuwano, S. Norimatsu, and N. Shibata, *Electron. Lett.*, **3** (1995) 302.
122. Y. Ji, Y. Li, J. Wu, B. Luo, Y. Dai, K. Xu, W. Li, X. Hong, and J. Lin, *IEEE Photon. Technol. Lett.*, **24** (2012) 772.
123. H. S. Chung, S. H. Chang, J. H. Lee, and K. Kim, *Opt. Exp.*, **21** (2013) 24962.
124. H. Kawakami, E. Yoshida, and Y. Miyamoto, *J. Lightwave Technol.*, **30** (2012) 962.
125. H. Yoon, D. Lee, and N. Park, *Opt. Exp.*, **13** (2005) 371.
126. P. S. Cho and M. Nazarathy, *IEEE Photon. Technol. Lett.*, **22** (2010) 1030.
127. K. Szczerba,1, P. Westbergh, J. Karout, J. Gustavsson, A. Haglund, M. Karlsson, P. Andrekson, E. Agrell, and A. Larsson, *Opt. Exp.*, **19** (2011) B204.

128. P. Westbergh, J. Gustavsson, A. Haglund, A. Larsson, F. Hopfer, G. Fiol, D. Bimberg, and A. Joel, *Electron. Lett.*, **45** (2009) 366.
129. D. Watanabe, A. Ono, and T. Okayasu, *IEEE International Solid-State Circuits Conference* 106, 2009.
130. P. Westbergh, J. Gustavsson, B. Kögel, A. Haglund, A. Larsson, A. Mutig, A. Nadtochiy, D. Bimberg, and A. Joel, *Electron. Lett.,* **46** (2010) 1014.
131. D. M. Kuchta, *Optical Fiber Communication Conference* (2017).
132. M. J. Wale, Ocaloro, *Corp. 2nd Symposium on Optical Interconnects in Data Centres ECOC*, Cannes, 23rd September 2014.

Appendix 1

Waveguide Modes in Planar Waveguide

First, the concept of numerical aperture is discussed. Numerical aperture (NA) is a measure of the cone of the incident light that can be coupled into the waveguide.

Consider Figure A.1.1, a ray of light incident at an angle θ follows the path shown in the figure. The maximum value of θ_2 is the critical angle (θ_c) at the active layer–cladding layer interface. It is given by $\text{Sin}(\theta_c) = n_2/n_1$, where n_2 and n_1 are the refractive indices of the active and cladding layer, respectively. It follows that there is a maximum incident angle (θ_m) for which the ray would be reflected at the active layer–cladding layer interface. This angle is given by

$$\text{Sin}(\theta_m) = n_1\text{Sin}(\theta_1) = n_1\text{Cos}(\theta_c) = (n_1^2 - n_2^2)^{1/2} = (2n_1\Delta n)^{1/2} \tag{A.1.1}$$

where it is assumed $n_1 \sim n_2$ and $\Delta n = n_1 - n_2$. NA is defined by,

$$\text{NA} = \text{Sin}(\theta_m) = (2n_1\Delta n)^{1/2} \tag{A.1.2}$$

The dielectric waveguide is now discussed. The treatment in this Appendix follows that of Refs. [1–2]. The mathematical description of the laser modes is based on the time-independent wave equation

$$\nabla^2 E + \varepsilon(x, y)k_0^2 E = 0$$

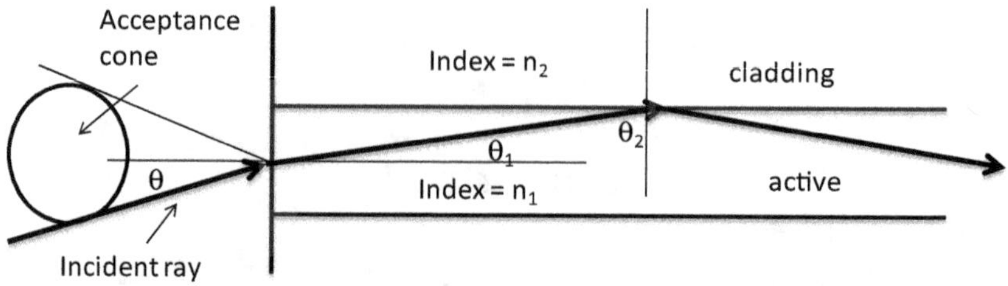

Figure A.1.1: Schematic of a ray propagating through the waveguide.

where the x-axis is parallel and the y-axis is perpendicular to the heterojunction. The spatially varying dielectric constant is generally of the following form, i.e. ε varies slowly along the x-direction, where j represents various layers of the heterostructure

$$\varepsilon(x, y) = \varepsilon_j(x)$$

A linear polarization is considered. The electric field in the wave equation is approximated by

$$E \cong \hat{e}\phi(y; x)\psi(x)\exp(i\beta z) \tag{A.1.3}$$

Substitute (A.1.3) to (A.1.1), we obtain

$$\frac{1}{\psi}\frac{\partial^2\psi}{\partial x^2} + \frac{1}{\phi}\frac{\partial^2\phi}{\partial y^2} + [\varepsilon(x, y)k_0^2 - \beta^2] = 0 \tag{A.1.4}$$

The transverse field distribution $\phi(y;\ x)$ is obtained first by solving

$$\frac{\partial^2\phi}{\partial y^2} + [\varepsilon(x, y)k_0^2 - \beta_{\mathit{eff}}^2(x)]\phi = 0 \tag{A.1.5}$$

The lateral field distribution $\psi(x)$ is then obtained by solving

$$\frac{\partial^2\psi}{\partial x^2} + [\beta_{\mathit{eff}}^2 - \beta^2]\psi = 0 \tag{A.1.6}$$

The effective index of refraction of the waveguide is defined as

$$n_{\mathit{eff}}(x) = \beta_{\mathit{eff}}(x)/k_0 \tag{A.1.7}$$

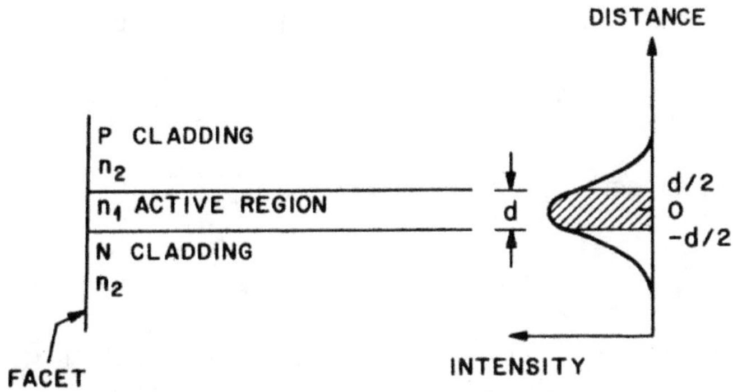

Figure A.1.2: The dielectric waveguide of the semiconductor laser. n_2 is the refractive index of the cladding layers and n_1 that of the active region. $n_1 > n_2$. The cladding layers are of a higher band gap material than the active region. The intensity distribution of the fundamental mode is shown. The cross-hatched region represents the fraction of the mode (Γ) within the active region.

The dielectric constant $\varepsilon(x, y)$ is of the form

$$\varepsilon(x, y) = n_b^2(y) + \Delta\varepsilon(x, y) \tag{A.1.8}$$

where n_b is the real refractive index of each of the layers and $\Delta\varepsilon(x, y)$ represents changes to this due to current and temperature ($\Delta\varepsilon(x, y) \ll n_b^2$). Using first-order perturbation theory, the eigenvalue given by n_{eff} becomes

$$n_{\mathit{eff}}(x) = n_e(x) + \Delta n_e(x) \tag{A.1.9}$$

where $n_e(x)$ is obtained by solving the unperturbed eigenvalue equation

$$\frac{d^2\phi}{dy^2} + k_0^2[n_b^2(y) - n_e^2(x)]\phi = 0 \tag{A.1.10}$$

The perturbation Δn_e is obtained using

$$2n_e\Delta n_e = \frac{\int_{-\infty}^{\infty} \Delta\varepsilon(x, y)\phi^2(y; x)dy}{\int_{-\infty}^{\infty} \phi^2(y; x)dy} \tag{A.1.11}$$

Figure A.1.2 shows the dielectric waveguide of a semiconductor laser normal to the junction. The laser has an active region bounded by cladding regions of same index.

In general, Eq. (A.1.11) can be simplified to become

$$\Delta n_e(x) = \frac{1}{2n_e}\sum_j \Gamma_j(x)\Delta\varepsilon_j(x) \tag{A.1.12}$$

$\Delta\varepsilon_j$ is the dielectric perturbation of jth layer, and

$$\Gamma_j(x) = \frac{\int_{d_j}\phi^2(y;x)dy}{\phi^2(y;x)dy} \tag{A.1.13}$$

is the fraction of the mode intensity contained in that layer. Schematic of a three-layer waveguide for a typical laser is shown in Figure A.1.2. Consider the even transverse electric (TE) modes. For these modes, a general solution (which is symmetric in y: even modes) of Eq. (A.1.10) for the waveguide shown in Figure A.1.1 is of the form

$$\phi(y) = \begin{cases} A_e\cos(\kappa y) & |y| \le d/2 \\ B_e\exp[-\gamma(|y| - d/2)] & |y| \ge d/2 \end{cases} \tag{A.1.14}$$

where

$$\kappa = k_0(n_2^2 - n_e^2)^{1/2} \tag{A.1.15}$$

$$\gamma = k_0(n_e^2 - n_1^2)^{1/2} \tag{A.1.16}$$

and n_2 and n_1 are the material refractive indices for the active and cladding layers, respectively, with $n_2 > n_1$. The continuity of ϕ and $d\phi/dy$ at $|y| = d/2$ requires that

$$B_e = A_e\cos(\kappa d/2) \tag{A.1.17}$$

$$\gamma B_e = \kappa A_e\sin(\kappa d/2) \tag{A.1.18}$$

Dividing Eq. (A.1.18) by Eq. (A.1.17), the following eigenvalue equation is obtained

$$\gamma = \kappa\tan(\kappa d/2) \tag{A.1.19}$$

whose solution yields the effective mode index n_e. For the odd TE modes, the eigenvalue equation is

$$\gamma = -\kappa\cot(\kappa d/2) \tag{A.1.20}$$

whose solutions yield n_e for the odd TE modes. The transverse magnetic (TM) modes are obtained using the same procedure. The even and odd TM modes eigenvalue equations are

$$n_2^2\gamma = n_1^2\kappa\tan(\kappa d/2) \tag{A.1.21}$$

$$n_2^2\gamma = -n_1^2\kappa\cot(\kappa d/2) \tag{A.1.22}$$

In heterostructure semiconductor lasers, the TE modes are generally favored over the TM modes since the facet reflectivity is higher for TE modes [3]. In the following discussion we therefore consider only TE modes. The TE-mode eigenvalues are obtained using Eq. (A.1.19) together with the relation

$$\kappa^2 + \gamma^2 = k_0^2(n_2^2 - n_1^2) \tag{A.1.23}$$

obtained by squaring and adding Eqs. (A.1.15) and (A.1.16). The cut-off condition is determined by $\gamma = 0$ and occurs when

$$\kappa d = p\pi \tag{A.1.24}$$

where p is an integer whose even and odd values correspond to even and odd TE modes. Substitute κ from Eq. (A.1.23) after using $\gamma = 0$, we obtain

$$D = p\pi \tag{A.1.25}$$

where

$$D = k_0(n_2^2 - n_1^2)^{1/2}d \tag{A.1.26}$$

is the normalized waveguide thickness. For $D < \pi$, the waveguide can support only fundamental TE mode. A heterostructure semiconductor laser can therefore be made to emit light in a single transverse mode if the active layer thickness d is chosen to satisfy the condition $D < \pi$ or from Eq. (A.1.26) after using $k_0 = 2\pi/\lambda$,

$$d < \frac{\lambda}{2}(n_2^2 - n_1^2)^{-1/2} \tag{A.1.27}$$

For an InGaAsP laser with InP ($n_2 = 3.51$ for $\lambda \sim 1300$ nm InGaAsP, $n_1 \quad = 3.22$ for InP) cladding layers,

$$\lambda(n_2^2 - n_1^2)^{-1/2} \cong 0.95\mu\text{m} \tag{A.1.28}$$

Condition (A.1.27) then simply becomes d <0.48 μm for fundamental transverse mode operation of InGaAsP/InP laser emitting at 1300 nm. Similarly, using Eqs. (A.1.26) and (A.1.28), one finds that

$$D \cong 6.6d \tag{A.1.29}$$

where d is the active-layer thickness in micrometers.

Transverse confinement factor Γ_T plays an important role for heterostructure semiconductor laser. Using $\phi(y)$ from Eq. (A.1.14) in

$$\Gamma_T = \frac{\int_{-d/2}^{d/2} \phi^2(y)dy}{\int_{-\infty}^{\infty} \phi^2(y)dy} \tag{A.1.30}$$

and carrying out the integrations, the following equation is obtained:

$$\Gamma_T = \frac{1 + 2\gamma d/D^2}{1 + 2/\gamma d} \tag{A.1.31}$$

For the fundamental transverse mode, a remarkably simple expression

$$\Gamma_T \cong D^2/(2 + D^2) \tag{A.1.32}$$

is found to be accurate to within 1.5% [4] and can be used to obtain Γ_T for any slab waveguide with the help of Eq. (A.1.26). For an InGaAsP laser with a 0.15 μm thick active layer, $D \approx 1$ from Eq. (A.1.29) and $\Gamma_T \approx 1/3$. The effective index n_e of the fundamental TE mode can be approximated by

$$n_e^2 \cong n_1^2 + \Gamma_T(n_2^2 - n_1^2) \tag{A.1.33}$$

The lateral modes are obtained by solving Eq. (A.1.6) which after using Eqs. (A.1.7) and (A.1.9) becomes

$$\frac{\partial^2 \psi}{\partial x^2} + \{k_0^2[n_e(x) + \Delta n_e(x)]^2 - \beta^2\}\psi = 0 \tag{A.1.34}$$

In a gain-guided device, $n_e(x)$ is a constant given by Eq. (A.1.33). In an index-guided (buried heterostructure) device (see Chapter 3), the slab-waveguide problem is solved separately in the two regions, and

$$n_e(x) = \begin{cases} n_e^{in} & |x| \leq w/2 \\ n_e^{out} & \text{otherwise} \end{cases} \tag{A.1.35}$$

where n_e^{in} and n_e^{out} are the effective indices corresponding to the two regions. The lateral index step

$$\Delta n_L = n_e^{\text{in}} - n_e^{\text{out}} \tag{A.1.36}$$

determines the extent of index guiding. In strongly index-guided semiconductor lasers, the mode-propagation constant is given by

$$\beta = k_0 \bar{n} + i\bar{\alpha}/2 \tag{A.1.37}$$

where $\bar{n}$ and $\bar{\alpha}$ are the refractive index and the absorption coefficient of the mode supported by the rectangular waveguide of width w and thickness d. The lateral modes are obtained by solving the three-layer slab-waveguide problem using the following equation:

$$\frac{\partial^2 \psi}{\partial x^2} + k_0^2 [n_e^2(x) - \bar{n}^2] \psi = 0 \tag{A.1.38}$$

The mode-absorption coefficient $\bar{\alpha}$ is obtained using the first-order perturbation theory and is given by

$$\bar{\alpha} = \frac{k_0}{\bar{n}} \text{Im} \left(\frac{\int 2n_e(x) \Delta n_e(x) \psi^2(x) dx}{\int \psi^2(x) dx} \right) \tag{A.1.39}$$

where $\Delta n_e(x)$ is given by Eq. (A.1.12). For the symmetric three-layer waveguide considered previously, $\Gamma_2 = \Gamma_T$, and $\Gamma_1 = (1 - \Gamma_T)/2$,

$$\Delta n_e = \frac{1}{2n_e} [\Gamma_T n_2 (2\Delta n_p - ig/k_0) + i(1 - \Gamma_T) n_1 \alpha_c / k_0] \tag{A.1.40}$$

where α_c is the cladding-layer absorption coefficient, Δn_p is the carrier-induced index change, and g is the active-layer gain. Substituting Eq. (A.1.40) in Eq. (A.1.39) and carrying out the integrations, the following simple expression is obtained:

$$\bar{\alpha} \cong -\Gamma_L\Gamma_T g + (1 - \Gamma_L\Gamma_T)\alpha_c \tag{A.1.41}$$

where the approximation $\bar{n} \cong n_1 \cong n_2$ has been used and

$$\Gamma_L = \frac{\int_{-w/2}^{w/2} |\psi(x)|^2 dx}{\int_{-\infty}^{\infty} |\psi(x)|^2 dx} \tag{A.1.42}$$

is the lateral confinement factor.

In analogy to Eqs. (A.1.25) and (A.1.26), one can define the normalized waveguide width W as

$$W = k_0 w[(n_e^{in})^2 - (n_e^{out})^2]^{1/2} \tag{A.1.43}$$

and the cut-off condition becomes

$$W = q\pi \tag{A.1.44}$$

where q is an integer whose even and odd values correspond to even and odd lateral modes, respectively. In particular, only the lowest-order ($q = 0$) lateral mode is supported by the waveguide if $W \leq \pi$ or equivalently, the active-layer width

$$w \leq \lambda/(8n_e\Delta n_L)^{1/2} \tag{A.1.45}$$

where Δn_L is the lateral index step given by Eq. (A.1.35) and n_e is the average effective index. The lateral confinement factor Γ_L and the mode refractive index $\bar{\mu}$ for the fundamental lateral mode are given by

$$\Gamma_L \cong W^2/(2 + W^2) \tag{A.1.46}$$

and

$$\bar{n}^2 \cong (n_2^{out})^2 + \Gamma_L[(n_e^{in})^2 - (n_e^{out})^2]. \tag{A.1.47}$$

A semiconductor laser emits light in the form of a narrow spot of elliptical cross-section, the spatial-intensity distribution of the emitted light near the laser facet is known as the near field. The angular intensity distribution far from the laser facet is known as the far field. The active-layer dimensions are often chosen such that only the lowest-order transverse and lateral modes are supported by the waveguide. The near field just outside the laser facet can be written as

$$E = E_0\psi(x)\phi(y)\exp(ik_0 z) \tag{A.1.48}$$

where E_0 is a constant related to the output power ($P_{\text{out}} \propto |E_0|^2$). $\Psi(x)$ and $\phi(y)$ are the lateral and transverse mode profiles discussed previously.

For a three-layer waveguide model, the near-field $\phi(y)$ is given by Eq. (A.1.14). By fitting $\phi(y)$ with a Gaussian distribution, the width can be approximated by [1]

$$w_\perp \cong d(2\ln 2)^{1/2}(0.321 + 2.1D^{-3/2} + 4D^{-6}) \tag{A.1.49}$$

where the normalized thickness D is given by Eq. (A.1.26). The expression is reasonably accurate for $1.8 < D < 6$.

The far field is obtained by taking a two-dimensional Fourier transform of the near field given by Eq. (A.1.48) [2]. It can be expressed as a product of the two one-dimensional Fourier transforms corresponding to the lateral and transverse directions x and y. The lateral far-field intensity distribution is given by

$$|\tilde{\phi}_{FF}(\theta)|^2 = \cos^2(\theta)|\int_{-\infty}^{\infty} \phi(y)\exp(ik_0 y\sin\theta)dy| \tag{A.1.50}$$

with a similar expression for the transverse far field. The full width at half maximum (FWHM) $\theta_\perp$ (perpendicular to the junction) is obtained from the above. When $\phi(y)$ is a Gaussian function

$$\phi(y) = \exp\left[-\frac{1}{4}\left(\frac{y}{\bar{y}}\right)^2\right] \tag{A.1.51}$$

$$|\phi(\theta)|^2 = \cos^2\theta\exp\left[-\frac{1}{2}(\bar{y}k_0\sin\theta)^2\right] \tag{A.1.52}$$

$\bar{y}$ is the "spot size" in the y-direction (normal to junction) at the laser facet. If the effect of $\cos\theta$ is neglected,

$$|\phi(\theta)|^2 \sim \exp\left[-\frac{1}{2}(\bar{y}k_0\sin\theta)^2\right] \tag{A.1.53}$$

The FWHM (full width at half maximum) $\theta_\perp$ is then given by

$$\theta_\perp = 2\sin^{-1}\{(2\ln 2)^{1/2}/\bar{y}k_0\} \tag{A.1.54}$$

An expression for FWHM along the junction can be obtained using similar approximations.

A.1.1 Laser output power for unequal facet reflectivity

The intracavity power is not uniform along the cavity length when the facet reflectivities are different. The average intracavity power is obtained by using

$$P_{in} = (\alpha_m + \alpha_{int})\int_0^L |A\exp(-i\beta z) + B\exp(i\beta z))|^2 dz \tag{A.1.55}$$

where A and B are the amplitudes of the forward and backward propagating waves normalized such that $|A|^2$ and $|B|^2$ represent the corresponding optical power at $z = 0$. The complex propagation constant β is given by

$$\beta = n_R k_0 + i(\Gamma g - \alpha)/2 \tag{A.1.56}$$

The boundary conditions at the laser facets relate A and B such that

$$\begin{aligned} A &= \sqrt{R_1}B, \\ B &= \sqrt{R_2}A\exp(-2i\beta L) \end{aligned} \tag{A.1.57}$$

where R_1 and R_2 are the reflectivities of the facets located at $z = 0$ and $z = L$, respectively. The integration in Eq. (A.1.55) is easily performed. The interference term can be neglected because of the rapidly oscillating nature. Byusing Eqs. (A.1.55) and (A.1.56) together with $\alpha_m = \Gamma g - \alpha_{\rm int}$, the result is given by

$$P_{in} = \frac{\alpha_m + \alpha_{int}}{\alpha_m}\frac{(\sqrt{R_1} + \sqrt{R_2})(1 - \sqrt{R_1 R_2})}{R_1\sqrt{R_2}}|A|^2 \tag{A.1.58}$$

The average intracavity power $P_{\rm in}$ is related to the photon density $N_{\rm ph}$

$$P_{in} = h\nu V N_{\rm ph}/\tau_p \tag{A.1.59}$$

where $h\nu N_{\rm ph}$ represents the total energy and τ_p is photon lifetime in the cavity. Equations (A.1.58) and (A.1.59) can be used to obtain $|A|^2$ in terms of $N_{\rm ph}$. The power emitted from the facet located at $z = 0(P_2)$ and $z = L(P_1)$ are related to $|B|^2$ and $|A|^2$ as

$$P_1 = (1 - R_1)|B|^2,$$
$$P_2 = (1 - R_2)|A\exp(-i\beta L)|^2 \quad \text{(A.1.60)}$$

For unity internal efficiency ($\eta_{\rm i} = 1$), $1/\tau_{\rm p} = v_g(\alpha_m + \alpha)$, where v_g is the group velocity. By using Eqs. (A.1.58)–(A.1.60), the output powers are related to $N_{\rm ph}$ as

$$P_1 = \frac{(1 - R_1)\sqrt{R_2}}{(\sqrt{R_1} + \sqrt{R_2})(1 - \sqrt{R_1 R_2})} h\nu v_g \alpha_m V N_{\rm ph} \quad \text{(A.1.61)}$$

$$P_2 = \frac{(1 - R_2)\sqrt{R_1}}{(\sqrt{R_1} + \sqrt{R_2})(1 - \sqrt{R_1 R_2})} h\nu v_g \alpha_m V N_{\rm ph} \quad \text{(A.1.62)}$$

When the facet reflectivities are equal ($R_1 = R_2$) then $P_1 = P_2 = P_{\rm out}$ of Eq. (3.12). Note that $P_1 + P_2 = 2P_{\rm out}$ for all values of R_1 and R_2, since a change in facet reflectivities only partitions the power between the two facets.

References

1. G. P. Agrawal and N. K. Dutta, *Semiconductor Lasers*, Chapter 2, van Nostrand Reinhold, NY, 1992.
2. H. Kressel and J. K. Butler, *Semiconductor Lasers and Heterojunction LEDs*, Academic Press, NY, 1977.
3. T. Ikegami, *IEEE J. Quant. Electron.*, **QE-8** (1972) 470.
4. D. Botez, *IEEE J. Quant. Electron.*, **QE-17** (1981) 178.

Appendix 2

Waveguide Mode in Optical Fibers

An optical fiber consists of a central core surrounded by a cladding region. The core region has a slightly higher index (~0.3%) than the cladding region, and the two regions form a dielectric waveguide which guides the light in the core region. Conceptually, stimulated emission in a fiber amplifier arises from the atomic transitions in the doped core region, and the light generated is confined and guided by a circular waveguide consisting of the core and cladding regions (Figure A.2.1). Figure A.2.1 shows schematically the cross-section and refractive index profile of a step-index fiber.

A graded index fiber has a graded index region between the core and the cladding and a step-index fiber has an index step ($\Delta n = n_1 - n_2$) between the core and the cladding region. A fraction of the optical mode is confined in the core region. The parameters that characterize the fiber are the core radius and refractive index difference Δ given by

$$\Delta = (n_1 - n_2)/n_1 \tag{A.2.1}$$

The core radius (a) is typically in the 3–5 μm range for a standard single-mode fiber. The cladding layer radius (b) is ~50 μm. The cladding radius (as long as it is large enough to confine the mode, i.e. >20 μm) is not an important parameter for the waveguide in a typical single-mode transmission fiber or a typical single-mode fiber amplifier. Index raising dopants for silica are germanium, phosphorus, and aluminum and index lowering dopants are boron and

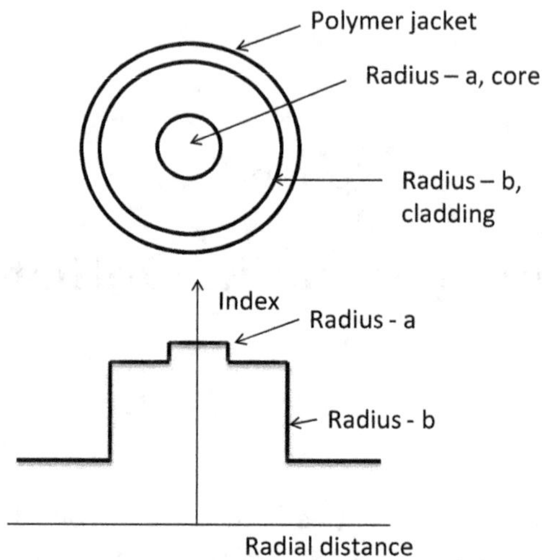

Figure A.2.1: The circular waveguide of a fiber amplifier. n_2 is the refractive index of the cladding region and n_1 that of the core region. $n_1 > n_2$. A jacket (polymer) is also shown.

fluorine. A typical fiber has a Ge-doped SiO_2 core and the cladding is SiO_2. The quantity Δ is $\sim 3 \times 10^{-3}$. The numerical aperture (NA) of a fiber waveguide is given by Eq. (A.1.2), i.e. NA $= (2n_1\Delta n)^{1/2}$. Another parameter, V, is important for characterizing waveguide properties, it is given by [1]

$$V = k_0 a(n_1^2 - n_2^2)^{1/2} \tag{A.2.2}$$

The quantity $k_0 = 2\pi/\lambda$. The step-index fiber (where there is a single index step as shown in Figure A.2.1 as opposed to a graded index fiber) supports a single fiber mode for $V < 2.405$. The optical wave propagation equation in the fiber is derived from Maxwell's equation. The Fourier transform of the electric field is defined as

$$\vec{\mathbf{E}}(\vec{\mathbf{r}}, t) = \frac{1}{2\pi} \int_{-\infty}^{\infty} \vec{\mathbf{E}}(\vec{\mathbf{r}}, \omega) \exp(-i\omega t) d\omega \tag{A.2.3}$$

In cylindrical coordinates (ρ, ϕ, z — which is representative of wave propagation in fibers) the propagation equation for $\mathbf{E}$ ($\mathbf{r}$, ω) is (z is along fiber length)

$$\frac{\partial^2 \vec{\mathbf{E}}}{\partial \rho^2} + \frac{1}{\rho}\frac{\partial \vec{\mathbf{E}}}{\partial \rho} + \frac{1}{\rho^2}\frac{\partial^2 \vec{\mathbf{E}}}{\partial \phi^2} + \frac{\partial^2 \vec{\mathbf{E}}}{\partial z^2} + n^2 k_0^2 \vec{\mathbf{E}} = 0 \tag{A.2.4}$$

where $k_0 = \omega/c = 2\pi/\lambda$. The components of **E** and **H** (electric and magnetic fields) are related by Maxwell's equations, only two components out of the six are independent. It is convenient to choose E_z and H_z as independent variables. The wave propagation for E_z is solved by assuming (based on symmetry)

$$E_z = E_0 F(\rho) \exp(\text{i}m\phi) \exp(i\beta z) \tag{A.2.5}$$

where E_0 is a constant, β is the propagation constant, and m is an integer. By substituting Eq. (A.2.5) in Eq. (A.2.4), the following equation for F (ρ) is obtained:

$$\frac{d^2 F}{d\rho^2} + \frac{1}{\rho}\frac{dF}{d\rho} + \left(k^2 - \frac{m^2}{\rho^2}\right) F = 0 \tag{A.2.6}$$

with

$$k^2 = n^2 k_0^2 - \beta^2$$

The refractive index (n) for a step-index fiber of core radius a (where ρ is the radial coordinate) is given by

$$\begin{aligned} n &= n_1 \quad \text{for } \rho < a \\ &= n_2 \quad \text{for } \rho > a \end{aligned} \tag{A.2.7}$$

Equation (A.2.6) is a well-known equation whose solutions are Bessel functions.

$$\begin{aligned} F(\rho) &= A_1 J_m(\kappa\rho) \quad \text{for } \rho < a \\ &= A_2 K_m(\gamma\rho) \quad \text{for } \rho > a \end{aligned} \tag{A.2.8}$$

with $\kappa = n_1^2 k_0^2 - \beta^2$, $\gamma = \beta^2 - n_2^2 k_0^2$. J_m and K_m are Bessel functions of mth order. The function K_m $(\gamma\rho)$ decays exponentially for large ρ. The quantities A_1, A_2, and β are constants. The ratio between A_1 and A_2, and β are determined by the boundary condition that the electric field and its first derivative is continuous at $\rho = a$. For light polarized along the propagation axis (z-axis) in the fundamental mode ($m = 0$),

$$F(\rho) = J_0(\kappa\rho) \quad \text{for } \rho < a;$$
$$F(\rho) = (a/\rho)^{1/2} J_0(\kappa a) \exp[-\gamma(\rho - a)] \quad \text{for } \rho > a. \quad \text{(A.2.9)}$$

Since the exact solutions are somewhat complex, it is often convenient to use a Gaussian approximation

$$F(\rho) = \exp[-\rho^2/\omega^2] \quad \text{(A.2.10)}$$

where $\rho^2 = x^2 + y^2$, ρ is the radial coordinate. The mode width parameter ω in the Gaussian approximation is obtained by fitting to the exact solution. Figure A.2.2 shows the mode width parameter as a function of V. The same figure also shows that the Gaussian approximation is quite good.

The fit is good for V around 2. The spot size (ω) in the Gaussian approximation can be determined from fits similar to that in Figure A.2.2. For a good fit, a good approximation for ω (accurate to within 1%) for 1.2 < V < 2.4 is given by (a is the radius of core)

$$\omega/a = 0.65 + 1.619\text{V}^{-3/2} + 2.879\text{V}^{-6} \quad \text{(A.2.11)}$$

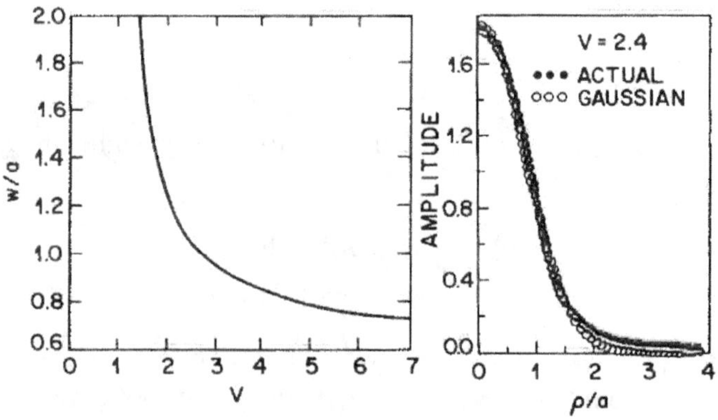

Figure A.2.2: Variation of the mode width parameter ω with the fiber parameter V. Also, shown is the actual field distribution (calculated using Bessel functions) and a Gaussian fit for $V = 2.4$ [2].

The mode index (n_m) is approximately given by

$$n_m = n_2 + (1 + b\Delta) \quad \text{where } \Delta = (n_1 - n_2)/n_1 \tag{A.2.12}$$

The quantity b is a function of V and is given by

$$b(V) = (1.1428 - 0.9960/V)^2 \tag{A.2.13}$$

The optical gain of a signal in the fundamental mode traveling through the amplifier is given by

$$g_m = \Gamma_g \tag{A.2.14}$$

The mode gain (g_m) is proportional to the mode confinement factor Γ. Since the optical amplifier is circularly symmetric, it generally exhibits a polarization independent gain (except for stress-induced polarization dependence or for birefringent fibers). If P_s is the saturation power of the gain medium, the observed saturation power is given by

$$P_s(\text{observed}) = P_s/\Gamma \tag{A.2.15}$$

Thus, amplifiers with small mode confinement factor (small Γ), for example, microstructure fiber gain medium, exhibit high saturation power.

References

1. G. P. Agrawal, *Nonlinear Fiber Optics*, Chapter 1, Academic Press, USA, 1989.
2. G. P. Agrawal, *Nonlinear Fiber Optics*, Chapter 2, Academic Press, USA, 1989.

Appendix 3

Optical Pulse Propagation Equation

The treatment in this appendix follows that of Ref. [1]. The equations derived are used for the calculations (supercontinuum generation) in Appendices 7 and 8. The propagation of the optical fields in fibers is governed by Maxwell's equations, which in the International System of Units take the form

$$\nabla \times \mathbf{E} = -\frac{\partial \mathbf{B}}{\partial t} \tag{A.3.1}$$

$$\nabla \times \mathbf{H} = \mathbf{J} + \frac{\partial \mathbf{D}}{\partial t} \tag{A.3.2}$$

$$\nabla \cdot \mathbf{D} = \rho \tag{A.3.3}$$

$$\nabla \cdot \mathbf{B} = 0 \tag{A.3.4}$$

The flux densities $\mathbf{D}$ and $\mathbf{B}$ arise in response to the electric and magnetic fields $\mathbf{E}$ and $\mathbf{H}$ propagating inside the medium and are related to them through the constitutive relations given by

$$\mathbf{D} = \varepsilon_0 \mathbf{E} + \mathbf{P} \tag{A.3.5}$$

$$\mathbf{B} = \mu_0 \mathbf{H} + \mathbf{M} \tag{A.3.6}$$

By taking the curl of Eq. (A.3.1) and using Eqs. (A.3.2), (A.3.5), and (A.3.6), one can obtain the wave equation that describes light propagation in optical fibers

$$\nabla \times \nabla \times \mathbf{E} = -\frac{1}{c^2}\frac{\partial^2 \mathbf{E}}{\partial t^2} - \mu_0 \frac{\partial^2 \mathbf{P}}{\partial t^2} \tag{A.3.7}$$

where c is the speed of light in vacuum. For the study of nonlinear optical effects in optical fibers, in the wavelength range of 0.5–2 μm, the induced polarization consists of two parts

$$\mathbf{P}(\mathbf{r}, t) = \mathbf{P}_L(\mathbf{r}, t) + \mathbf{P}_{\text{NL}}(\mathbf{r}, t) \tag{A.3.8}$$

where the linear part $\mathbf{P}_L$ and the nonlinear part $\mathbf{P}_{\text{NL}}$ are related to the electric field. If we include only the third-order nonlinear effects governed by $\chi^{(3)}$ (second order is zero in crystals due to inversion symmetry)

$$\mathbf{P}_L(\mathbf{r}, t) = \varepsilon_0 \int_{-\infty}^{t} \chi^{(1)}(t - t') \cdot \mathbf{E}(\mathbf{r}, t') dt' \tag{A.3.9}$$

$$\mathbf{P}_{\text{NL}}(\mathbf{r}, t) = \varepsilon_0 \int_{-\infty}^{t} dt_1 \int_{-\infty}^{t} dt_2 \int_{-\infty}^{t} dt_3$$

$$\times \chi^{(3)}(t - t_1, t - t_2, t - t_3) \vdots \mathbf{E}(\mathbf{r}, t_1)\mathbf{E}(\mathbf{r}, t_2)\mathbf{E}(\mathbf{r}, t_3) \tag{A.3.10}$$

These relations are valid in the electric-dipole approximation and assume that the medium response is local. By using Eq. (A.3.8) and $\nabla \times \nabla \times \mathbf{E} = -\nabla^2 \mathbf{E}$, Eq. (A.3.7) can be written in the form

$$\nabla^2 \mathbf{E} - \frac{1}{c^2}\frac{\partial^2 \mathbf{E}}{\partial t^2} = \mu_0 \frac{\partial^2 \mathbf{P}_L}{\partial t^2} + \mu_0 \frac{\partial^2 \mathbf{P}_{NL}}{\partial t^2} \tag{A.3.11}$$

To solve Eq. (A.3.11), it is necessary to make several simplifying assumptions. $\mathbf{P}_{\text{NL}}$ is treated as a small perturbation to $\mathbf{P}_{\text{L}}$ and the optical field is assumed to be quasi-monochromatic. In the slowly varying envelope approximation, it is useful to separate the rapidly varying part of the electric field by writing it in the form

$$\mathbf{E}(\mathbf{r}, t) = \frac{1}{2}\hat{x}[E(\mathbf{r}, t)\exp(-i\omega_0 t) + c.c.] \tag{A.3.12}$$

where $\hat{x}$ is the polarization unit vector, and $\mathbf{E}(\mathbf{r}, t)$ is a slowly varying function of time. The polarization components $\mathbf{P}_L$ and $\mathbf{P}_{\rm NL}$ can also be expressed in a similar way by writing

$$\mathbf{P}_L(\mathbf{r},t) = \frac{1}{2}\hat{x}\lceil P_L(\mathbf{r},t)\exp(-i\omega_0 t) + c.c.\rceil \tag{A.3.13}$$

$$\mathbf{P}_{\rm NL}(\mathbf{r},t) = \frac{1}{2}\hat{x}\lceil P_{\rm NL}(\mathbf{r},t)\exp(-i\omega_0 t) + c.c.\rceil \tag{A.3.14}$$

The linear component $\mathbf{P}_L$ is obtained by substituting Eq. (A.3.13) in Eq. (A.3.9)

$$\begin{aligned} P_L(\mathbf{r},t) &= \varepsilon_0 \int_{-\infty}^{\infty} \chi^{(1)}(t-t')E(\mathbf{r},t')\exp[i\omega_0(t-t')]dt' \\ &= \frac{\varepsilon_0}{2\pi}\int_{-\infty}^{\infty} \tilde{\chi}^{(1)}(\omega)\tilde{E}(\mathbf{r},\omega-\omega_0)\exp[-i(\omega-\omega_0)t]d\omega \end{aligned} \tag{A.3.15}$$

where $\tilde{E}(\mathbf{r},\omega)$ is the Fourier transform of $E(\mathbf{r},t)$. Substituting Eq. (A.3.14) in Eq. (A.3.10) and assuming the nonlinear response to be instantaneous, then nonlinear component $\mathrm{P_{NL}}$ is given by

$$\mathbf{P}_{\rm NL}(\mathbf{r},t) = \varepsilon_0\chi^{(3)} \vdots \mathbf{E}(\mathbf{r},t)\mathbf{E}(\mathbf{r},t)\mathbf{E}(\mathbf{r},t) \tag{A.3.16}$$

By making use of Eq. (A.3.14), $\mathrm{P_{NL}}(\mathbf{r}, \mathrm{t})$ is given by

$$P_{\rm NL}(\mathbf{r},t) \approx \varepsilon_0\varepsilon_{\rm NL}E(\mathbf{r},t) \tag{A.3.17}$$

where the nonlinear contribution to the dielectric constant is defined as

$$\varepsilon_{\rm NL} = \frac{3}{4}\chi^{(3)}|E(\mathbf{r},t)|^2 \tag{A.3.18}$$

Substituting Eqs. (A.3.12) through (A.3.14) in (A.3.11), the Fourier transform defined as

$$\tilde{E}(\mathbf{r},\omega-\omega_0) = \int_{-\infty}^{\infty} E(\mathbf{r},t)\exp[i(\omega-\omega_0)t]dt \tag{A.3.19}$$

is found to satisfy the Helmholtz equation

$$\nabla^2\tilde{E} + \varepsilon(\omega)k_0^2\tilde{E} = 0 \tag{A.3.20}$$

where $k_0 = \omega/c$ and

$$\varepsilon(\omega) = 1 + \tilde{\chi}^{(1)} + \varepsilon_{\text{NL}} \tag{A.3.21}$$

is the dielectric constant, which can be used to define the refractive index $\tilde{n}$ and the absorption coefficient $\tilde{\alpha}$. It is customary to introduce

$$\tilde{n} = n + n_2|E|^2 \tag{A.3.22}$$

$$\tilde{\alpha} = \alpha + \alpha_2|E|^2 \tag{A.3.23}$$

Using $\varepsilon = (\tilde{n} + i\tilde{\alpha}/2k_0)^2$ and Eqs. (A.3.18) and (A.3.21), the nonlinear-index coefficient n_2 and the two-photon absorption coefficient α_2 are given by

$$n_2 = \frac{3}{8n}\text{Re}(\chi^{(3)}) \tag{A.3.24}$$

$$\alpha_2 = \frac{3\omega_0}{4nc}\text{Im}(\chi^{(3)}) \tag{A.3.25}$$

Equation (A.3.20) can be solved by using the method of separation of variables. Assume a solution of the form

$$\tilde{E}(r, \omega - \omega_0) = F(x, y)\tilde{A}(z, \omega - \omega_0)\exp(i\beta_0 z) \tag{A.3.26}$$

Equation (A.3.20) leads to the following two equations for $F(x, y)$ and $\tilde{A}(z, \omega)$:

$$\frac{\partial^2 F}{\partial x^2} + \frac{\partial^2 F}{\partial y^2} + [\varepsilon(\omega)k_0^2 - \tilde{\beta}^2]F = 0 \tag{A.3.27}$$

$$2i\beta_0\frac{\partial \tilde{A}}{\partial z} + (\tilde{\beta}^2 - \beta_0^2)\tilde{A} = 0 \tag{A.3.28}$$

The wavenumber $\tilde{\beta}$ is determined by solving the eigenvalue equation (A.3.27). The dielectric constant $\varepsilon(\omega)$ in Eq. (A.3.27) can be approximated by

$$\varepsilon = (n + \Delta n)^2 \approx n^2 + 2n\Delta n \tag{A.3.29}$$

where Δn is a small perturbation given by

$$\Delta n = n_2|E|^2 + \frac{i\tilde{\alpha}}{2k_0} \tag{A.3.30}$$

Equation (A.3.27) can be solved using first-order perturbation theory. We first replace ε with n^2 and obtain the modal distribution $F(x, y)$ and the corresponding wave number $\beta(\omega)$. Then include the effect of Δn, and the eigenvalue becomes

$$\tilde{\beta}(\omega) = \beta(\omega) + \Delta\beta(\omega) \tag{A.3.31}$$

where

$$\Delta\beta(\omega) = \frac{\omega^2 n(\omega)}{c^2\beta(\omega)} \frac{\int\int_{-\infty}^{\infty} \Delta n(\omega)|F(x,y)|^2 dxdy}{\int\int_{-\infty}^{\infty} |F(x,y)|^2 dxdy} \tag{A.3.32}$$

Using Eqs. (A.3.12) and (A.3.26), the electric field $\mathbf{E}(\mathbf{r}, t)$ can be written as

$$\mathbf{E}(\mathbf{r}, t) = \frac{1}{2}\hat{x}\{F(x,y)A(z,t)\exp[i(\beta_0 z - \omega_0 t)] + c.c.\} \tag{A.3.33}$$

$A(z, t)$ satisfies Eq. (A.3.28), and the Fourier transform $\tilde{A}(z, \omega - \omega_0)$ of $A(z, t)$ satisfies the following equation:

$$\frac{\partial \tilde{A}}{\partial z} = i[\beta(\omega) + \Delta\beta(\omega) - \beta_0]\tilde{A} \tag{A.3.34}$$

where we used Eq. (A.3.31) and approximated $\tilde{\beta}^2 - \beta_0^2$ by $2\beta_0(\tilde{\beta} - \beta_0)$. It is useful to expand $\beta(\omega)$ in a Taylor series around the carrier frequency ω_0 as

$$\begin{aligned}\beta(\omega) &= \beta_0 + (\omega - \omega_0)\beta_1 + \frac{1}{2}(\omega - \omega_0)^2\beta_2 \\ &\quad + \frac{1}{6}(\omega - \omega_0)^3\beta_3 + \cdots\end{aligned} \tag{A.3.35}$$

where $\beta_0 \equiv \beta(\omega_0)$ and other parameters are defined as

$$\beta_m = \left(\frac{d^m\beta}{d\omega^m}\right)_{\omega=\omega_0} \tag{A.3.36}$$

Similarly,

$$\Delta\beta(\omega) = \Delta\beta_0 + (\omega - \omega_0)\Delta\beta_1 + \frac{1}{2}(\omega - \omega_0)^2\Delta\beta_2 + \cdots \tag{A.3.37}$$

The cubic and higher-order terms in the expansion (A.3.35) are negligible if the spectral width of the pulse satisfies the condition $\Delta\omega \ll \omega_0$. If $\beta_2 = 0$ for same cases, it may be necessary to include the β_3 term. After these simplifications, we take the inverse Fourier transform using

$$A(z,t) = \frac{1}{2\pi}\int_{-\infty}^{\infty} \tilde{A}(z, \omega - \omega_0)\exp[-i(\omega - \omega_0)t]d\omega \tag{A.3.38}$$

Replace $\omega - \omega_0$ by the differential operator in the Fourier transform, the equation for $A(z,t)$ becomes

$$\frac{\partial A}{\partial z} + \beta_1\frac{\partial A}{\partial t} + \frac{i\beta_2}{2}\frac{\partial^2 A}{\partial t^2} = i\Delta\beta_0 A \tag{A.3.39}$$

$\Delta\beta_0$ term includes the effects of fiber loss and nonlinearity. Using $\beta(\omega) \approx n(\omega)\,\omega/c$ and assuming $F(x,y)$ in Eq. (A.3.32) does not vary significantly over the pulse bandwidth, Eq. (A.3.39) takes the form

$$\frac{\partial A}{\partial z} + \beta_1\frac{\partial A}{\partial t} + \frac{i\beta_2}{2}\frac{\partial^2 A}{\partial t^2} + \frac{\alpha}{2}A = i\gamma(\omega_0)|A|^2 A \tag{A.3.40}$$

where the nonlinear parameter γ is defined as

$$\gamma(\omega_0) = \frac{n_2(\omega_0)\omega_0}{cA_{\text{eff}}} \tag{A.3.41}$$

The parameter A_{eff} is the effective mode area and is defined as

$$A_{\text{eff}} = \frac{\left(\int\int_{-\infty}^{\infty}|F(x,y)|^2 dxdy\right)^2}{\int\int_{-\infty}^{\infty}|F(x,y)|^4 dxdy} \tag{A.3.42}$$

To include the high-order nonlinear effects, we need to go back to Eq. (A.3.10). The intensity-dependent nonlinear effects can be included by assuming the following form for the third-order susceptibility

$$\chi^{(3)}(t-t_1, t-t_2, t-t_3) = \chi^{(3)}R(t-t_1)\delta(t_1-t_2)\delta(t-t_3) \tag{A.3.43}$$

where $R(t)$ is the normalized nonlinear response function. Substitute Eq. (A.3.42) in Eq. (A.3.10) and introduce a slowly varying optical field through Eq. (A.3.12), the scalar form $\mathbf{P}_{\mathrm{NL}}$ is given by

$$P_{NL}(\mathbf{r},t) = \frac{3\varepsilon_0}{4}\chi^{(3)}E(\mathbf{r},t)\int_{-\infty}^{t} R(t-t_1)E^*(\mathbf{r},t_1)E(\mathbf{r},t_1)dt_1 \tag{A.3.44}$$

The response function $\mathrm{R}(\mathrm{t}-\mathrm{t}_1)$ must be zero for $\mathrm{t}_1 > \mathrm{t}$. Following the same procedure and using Eqs. (A.3.12)–(A.3.14), $\tilde{E}$ is found to satisfy

$$\begin{aligned}
&\nabla^2\tilde{E} + n^2(\omega)k_0^2\tilde{E} \\
&\quad = -ik_0\alpha - \chi^{(3)}k_0^2 \int\int_{-\infty}^{\infty} \tilde{R}(\omega_1-\omega_2) \\
&\qquad \times \tilde{E}(\omega_1,z)\tilde{E}^*(\omega_2,z)\tilde{E}(\omega-\omega_1+\omega_2,z)d\omega_1 d\omega_2
\end{aligned} \tag{A.3.45}$$

where $\tilde{R}(\omega)$ is the Fourier transform of $R(t)$. Consider the frequency dependence of $\Delta\beta$ as indicated in Eq. (A.3.37) and expand the parameters γ and α as

$$\gamma(\omega) = \gamma(\omega_0) + \gamma_1(\omega-\omega_0) + \frac{1}{2}\gamma_2(\omega-\omega_0)^2 + \cdots \tag{A.3.46}$$

$$\alpha(\omega) = \alpha(\omega_0) + \alpha_1(\omega-\omega_0) + \frac{1}{2}\alpha_2(\omega-\omega_0)^2 + \cdots \tag{A.3.47}$$

where $\gamma_m = (d^m\gamma/d\omega^m)_{\omega=\omega_0}$. In most cases of practical interest, it is sufficient to retain the first two terms in this expansion. Then after some algebra, we obtain the following equation for pulse evolution inside a single-mode fiber

$$\frac{\partial A}{\partial z} + \frac{1}{2}\left(\alpha(\omega_0) + i\alpha_1\frac{\partial}{\partial t}\right)A + \beta_1\frac{\partial A}{\partial t} + \frac{i\beta_2}{2}\frac{\partial^2 A}{\partial t^2} - \frac{\beta_3}{6}\frac{\partial^3 A}{\partial t^3}$$
$$= i\left(\gamma(\omega_0) + i\gamma_1\frac{\partial}{\partial t}\right)\left(A(z,t)\int_0^\infty R(t')|A(z,t-t')|^2 dt'\right) \tag{A.3.48}$$

Adding high-order dispersion terms, this equation is generalized to

$$\frac{\partial A}{\partial z} + \frac{1}{2}\left(\alpha + i\alpha_1\frac{\partial}{\partial t}\right)A + \sum_{m=2}^{M} i^{m-1}\frac{\beta_m}{m!}\frac{\partial^m A}{\partial t^m}$$
$$= i\left(\gamma + i\gamma_1\frac{\partial}{\partial t}\right)\left(A(z,t)\int_0^\infty R(t')\left|A(z,t-t')\right|^2 dt'\right) \tag{A.3.49}$$

where m represents the order up to which dispersive effects are included. The nonlinear response function R(t) including both the electronic and nuclear contributions can be written as

$$R(t) = (1 - f_R)\delta(t - t_e) + f_R h_R(t) \tag{A.3.50}$$

where t_e is short delay in electronic response ($t_e < 1$ fs) and f_R represents the fractional contribution of the delayed Raman response to nonlinear polarization P_{NL}. The temporal form of $h_R(t)$ can be deduced from experimentally measured spectrum of the Raman gain. A useful approximate form of Raman response function $h_R(t)$ is given by

$$h_R(t) = \frac{\tau_1^2 + \tau_2^2}{\tau_1\tau_2^2}\exp(-t/\tau_2)\sin(t/\tau_1) \tag{A.3.51}$$

The parameters τ_1 and τ_2 are two adjustable parameters chosen to provide a good fit to the actual Raman-gain spectrum.

Reference

1. G. P. Agrawal, *Nonlinear Fiber Optics* 4th Edn., Academic Press, NY, 2007.

Appendix 4

Three-Level and Four-Level Laser Equations

Rare earth ions have been studied extensively for optical and magnetic applications. These ions are generally doped in crystalline or glassy materials. Glasses provide broad emission and absorption spectrum, as a result glass hosts are used in many optical applications. Typical glass hosts are oxide and fluoride glasses. Fiber amplifiers used in optical transmission generally use silica (SiO_2) as the host material. Table A.4.1 shows a simplified table of the stimulated emission wavelengths for various dopants in silica.

For rare earth-doped lasers and amplifiers, population inversion is important, i.e. there need to be more ions in the high energy excited state compared to that in the ground state.

The energy stored in the excited state ions are released in the form of stimulated emission when an incident signal is present and this signal undergoes amplification. The energy levels in rare doped ions are generally classified as three- or four-level laser systems (Figure A.4.1). A simplified form of a laser is a two-level system discussed in Section 9.1.1. Erbium (Er) is often considered a three-level system, whereas Nd and Pr are considered four-level systems. In reality, the lower laser energy level of Er in glass consists of a group of levels broadened by the crystal field in glass, i.e. it is not an ideal three-level system.

Table A.4.1: Stimulated emission wavelength for various rare earths. Silica-based fiber host is desirable due to ease of fabrication [1].

Rare Earth Dopant (Host)	Principal Gain Wavelength Region (nm)
Nd (Silica)	1060, 940
Yb (Silica)	1100
Er (Silica)	1550
Pr (Flouride)	1300
Pr (Silica)	1050
Tm (Silica)	1650, 2000
Tm (Flouride)	2300
Ho (Silica)	2040
Ho (Flouride)	2900

For three-level systems, the low-energy level of the stimulated emission transition is the ground state of the ion and for four-level systems, the ground state of the ion is at a lower energy level than the low-energy level of the amplifying transition. For a four-level system, the low-energy level is naturally depopulated by transition to the ground state, hence it is easier to achieve population inversion in such systems.

For a three-level laser at least half of the ions have to be excited, otherwise the system will have optical loss and not optical gain. This results in an optimum length for Er amplifiers (three-level system) pumped at one end for a given pump power.

If the length is too large, the ions at the far end may not achieve population inversion and that region will contribute to optical loss of the transmitted signal. The four-level system remains transparent. A schematic of the energy level diagrams of three-level and four-level lasers are shown in Figure A.4.1. An example of a three-level laser is the Cr- doped Al_2O_3 (ruby) laser operating on the $^2E \rightarrow ^4A_2$ transition at 690 nm. An example of a four-level laser is the Nd-doped Yttrium–Aluminum–Garnet (YAG) laser operating on the $^4F_{3/2} \rightarrow ^4I_{11/2}$ transition at 1064 nm.

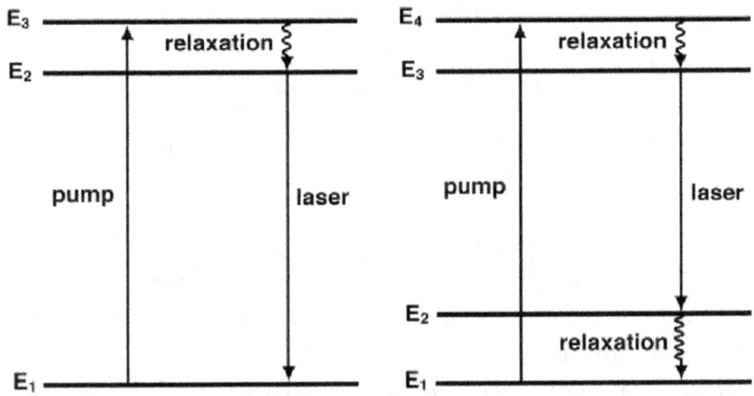

(a) **Three level laser** (b) **Four level laser**

Figure A.4.1: Schematic diagram of the energy levels in (a) three-level lasers, (b) four-level lasers [2, 3]. The principal energy levels for laser transition are shown.

A.4.1.1 *Three-level laser*

For a three-level laser, (Figure A.4.1), the ground state (1) has energy (E_1), the upper laser level (2) has energy E_2, and there is a higher energy level (3), of energy E_3. The pump excites the atom from ground state to excited state (3), the atom makes a fast transition from 3 to 2, and the laser transition is from 2 to 1, i.e. the ground state of the atom or ion is the lower energy level of the laser transition. The rate equation for a three-level laser is described below [2].

The population density at levels 1, 2, 3 are labeled N_1, N_2, N_3, respectively. The pump rate is denoted by P, so that PN_1 is the number of atoms excited per cm^{-3}. The total number of atoms per unit volume is

$$N = N_1 + N_2 + N_3$$

The rate equations due to pump are: (assuming fast decay from level 3 to 2)

$$\left(\frac{dN_1}{dt}\right)_{\text{pump}} = -PN_1 = \left(\frac{dN_3}{dt}\right)_{\text{pump}} = \left(\frac{dN_2}{dt}\right)_{\text{pump}} \qquad \text{(A.4.1)}$$

Thus, the rate equations for N_1, N_2 are

$$\begin{aligned}\frac{dN_1}{dt} &= -PN_1 + \Gamma N_2 + \sigma(\upsilon)\phi_\upsilon(N_2 - N_1)\\ \frac{dN_2}{dt} &= PN_1 - \Gamma N_2 - \sigma(\upsilon)\phi_\upsilon(N_2 - N_1)\end{aligned} \tag{A.4.2}$$

where ϕ_ν is the optical photon density at frequency ν, $\sigma(\nu)$ is the stimulated emission cross-section at frequency ν, and Γ is the spontaneous decay rate from level 2 to level 1. Γ includes both spontaneous radiation rate and non-radiative rates. Γ is the inverse of the lifetime of level 2. If the decay rate from level 3 to level 2 is fast, $N_3 \sim 0$, thus the total atom density (N) is given by

$$N_1 + N_2 = N \tag{A.4.3}$$

If the intensity is small (i.e. $\phi_\nu = 0$), then in steady state

$$\bar{N_1} = \frac{\Gamma}{P + \Gamma}N \quad \bar{N_2} = \frac{P}{P + \Gamma}N \tag{A.4.4}$$

where $\bar{N_1}$ and $\bar{N_2}$ are steady-state values of N_1 and N_2, respectively. It follows from the above,

$$\bar{N_2} - \bar{N_1} = \frac{P - \Gamma}{P + \Gamma}N \tag{A.4.5}$$

Thus, the minimum pump rate needed for population inversion ($\bar{N_2} > \bar{N_1}$) is $P > \Gamma$, i.e. $P_{\text{min}} = \Gamma$.

Since N is the number of pump atoms per unit vol. and $h\upsilon_{31}$ is the energy needed per pump photon, it follows that the minimum pump power (in W/cm^3) for population inversion is

$$P_{\text{min}} = \frac{1}{2}\Gamma h\upsilon_{31}N \tag{A.4.6}$$

Thus pump power larger than the minimum value given by the above equation is needed for producing population inversion or gain. Equation (A.4.5) shows that more than half of the atoms are in the excited state for producing gain. This phenomenon is known as population inversion.

For a ruby ($Cr:Al_2O_3$) laser, $\Gamma \sim 500\ s^{-1}$. The density of atoms $N \sim 1.6 \times 10^{19}\ cm^{-3}$, and assuming pump photon energy $h\nu_{31} \sim 2$ eV, the minimum pump power per unit volume is

$$P \sim (1/2)\Gamma N h\nu_{31} \sim 1\text{kWcm}^{-3} \tag{A.4.7}$$

For a 5-cm long ruby rod with a stimulated emission area radius of 2 mm, the required pump threshold power is ~600 W.

Another three-level laser system is Nd-doped glass operating at 920 nm. For this system, $\Gamma \sim 100\ s^{-1}$, $h\nu_{31} \sim 1$ eV, and $N \sim 10^{19}$ cm^{-3}. Using the above formula, the threshold pump power is ~70 Wcm^{-3}.

So far, the rate equations have been solved in the absence of stimulated emission (i.e. $\phi_\nu = 0$). For a non-zero ϕ_ν in steady state from Eqs. (A.4.2)

$$\begin{aligned} \bar{N}_2 &= \frac{PN}{P + \Gamma + 2\sigma(\upsilon)\phi_\upsilon} \\ \bar{N}_2 - \bar{N}_1 &= \frac{(P - \Gamma)N}{P + \Gamma + 2\sigma(\upsilon)\phi_\upsilon} \end{aligned} \tag{A.4.8}$$

and the gain, $g(\nu)$, is given by $\sigma(\nu)$ (the stimulated emission cross-section at frequency ν)

$$g(\upsilon) = \frac{\sigma(\upsilon)(P - \Gamma)N}{P + \Gamma + 2\sigma(\upsilon)\phi_\upsilon} \tag{A.4.9}$$

Thus, the gain is smaller in the presence of stimulated emission, i.e. the gain saturates. The gain saturation is rewritten as

$$g(\upsilon) = \frac{g_0(\upsilon)}{1 + \phi_\upsilon/\phi_{\upsilon s}} \tag{A.4.10}$$

The quantity $g_0(\nu)$ is the unsaturated gain and $\phi_{\nu s}$ is the saturation intensity. They are given by

$$\begin{aligned} g_0(\upsilon) &= \frac{\sigma(\upsilon)(P - \Gamma)N}{P + \Gamma} \\ \phi_{\upsilon s} &= \frac{P + \Gamma}{2\sigma(\upsilon)} \end{aligned} \tag{A.4.11}$$

The saturation power is $I_{\nu s} = h\nu\phi_{\nu s}$. For a laser, gain = loss, i.e. $g = \alpha$, where α is the loss per unit length. It follows from Eq. (A.4.10), removing the subscript ν,

$$\phi = \left(\frac{g_0}{\alpha} - 1\right)\phi_s \tag{A.4.12}$$

Since the intensity is proportional to the photon density, the laser intensity I is given by

$$I = \left(\frac{g_0}{\alpha} - 1\right)I_s \tag{A.4.13}$$

where I_s is the saturation intensity. Using (A.4.11), I_s is given by

$$I_s = h\upsilon v_g\left(\frac{P+\Gamma}{2\sigma}\right) \tag{A.4.14}$$

where v_g is the group velocity. If the number of pump photons is N_p, the pump rate is given by

$$P = N_p v_g \sigma_a \tag{A.4.15}$$

where σ_a is the pump absorption cross-section and v_g is the group velocity. If $h\nu_p$ is the pump photon energy, the pump power I_p is given by

$$I_p = h\upsilon_p N_p v_g \tag{A.4.16}$$

From Eqs. (A.4.13) to (A.4.16)

$$I_{\text{laser}} = \left(\frac{g_0}{\alpha} - 1\right)\frac{h\upsilon v_g}{2\sigma}\left[\left(\frac{\sigma_a}{h\upsilon_p}\right)I_p + \Gamma\right] \tag{A.4.17}$$

Thus, the laser power is linearly proportional to the pump power for pump power greater than threshold power ($P > P_{\text{min}}$), i.e. when $g_o > \alpha$. For the specific case of Er-doped amplifier, the upper- and lower energy levels are broadened by Stark splitting. The pump laser wavelengths are at 980 and 1480 nm, respectively. For 980 nm pump, the laser can be represented by a three-level system and for the 1480 nm pump, the laser is similar to a two-level laser. The population distribution in the band (stark split levels) is determined by usual Boltzmann distribution. Nearly complete inversion of the Er levels can be achieved using 980 nm pumping whereas for 1480 nm pumping the inversion is generally lower. The conversion efficiency is higher for 1480 nm pump due to closer matching with the emission wavelength (1520–1570 nm). The noise figure (Chapter 9) with 980 nm pump is generally lower than that for 1480 nm pump due to more complete inversion.

A.4.1.2 *Four-level laser*

An example of a four-level laser is the Nd:YAG laser operating at 1064 nm on the $^4F_{3/2} \rightarrow {}^4I_{11/2}$ transition. In these lasers, the lower laser level is emptied very rapidly to the ground state (Figure A.4.1). The quantities N_1, N_2, N_3, and N_4 are the populations of the four levels shown in Figure 2.2.1. While deriving the rate equations, it is assumed, as before, that the transition from level 4 where pump absorption takes place to level 3 (high energy laser level) is very fast, i.e.

$$\left(\frac{dN_4}{dt}\right)_{\text{pump}} = \left(\frac{dN_3}{dt}\right)_{\text{pump}} \tag{A.4.18}$$

The rate equations are

$$\begin{aligned}
\frac{dN_1}{dt} &= -PN_1 + \Gamma_{21}N_2 \\
\frac{dN_2}{dt} &= -\Gamma_{21}N_2 + \Gamma_{32}N_3 + \sigma(\upsilon)(N_3 - N_2)\phi_\upsilon \\
\frac{dN_3}{dt} &= PN_1 - \Gamma_{32}N_3 - \sigma(\upsilon)(N_3 - N_2)\phi_\upsilon
\end{aligned} \tag{A.4.19}$$

where P is the pump rate to level N_4, but all level 4 atoms relax to level 3 very rapidly. Hence, P is also pumping rate for level 3. The quantities Γ_{32} and Γ_{21} are decay rates from levels 3 to 2 and levels 2 to 1 (ground state), respectively. The last term is stimulated emission. The total number of atoms per unit volume in all levels is N, a constant, i.e.

$$N = N_1 + N_2 + N_3 \tag{A.4.20}$$

The steady state population inversion of the above in the absence of stimulated emission (i.e. $\phi_\nu = 0$) is

$$\bar{N}_3 - \bar{N}_2 = \frac{P(\Gamma_{21} - \Gamma_{32})N}{\Gamma_{21}\Gamma_{32} + \Gamma_{21}P + \Gamma_{32}P} \tag{A.4.21}$$

Thus, the condition for population inversion is $\Gamma_{21} > \Gamma_{32}$ for a non-zero P, i.e. any amount of pump would cause population inversion. This shows there is no threshold (i.e. no minimum pump power requirement) for population inversion for 4-level case compared to such a requirement for a 3-level case. This condition, $\Gamma_{21} > \Gamma_{32}$, implies that the lower laser level decays faster than the upper laser level. When, $\Gamma_{21} \gg \Gamma_{32}$ and P, Eq. (A.4.21) becomes

$$\Delta N = \bar{N}_3 - \bar{N}_2 \approx \bar{N}_3 \approx \frac{P}{P + \Gamma_{32}} N \tag{A.4.22}$$

The power per unit vol. (P_w) needed (using photons of energy $h\nu_{41}$) for a population inversion per unit volume ΔN is given by

$$P_w = hv_{41}\Delta N\Gamma_{32} \tag{A.4.23}$$

The gain of the laser is given by

$$g(v) = \sigma(v)\Delta N \tag{A.4.24}$$

The threshold for a four-level laser depends on the condition gain = loss for a laser. This is estimated as follows: For a 5-cm long Nd rod with 96% and 100% reflectivity at the ends with a round-trip scattering loss of 3% ($0.03 = e^{-2\alpha l}$, $\alpha = 0.003$ cm^{-1} for $l = 5$ cm), the gain at threshold (g_t) is calculated using the formula which assumes gain = loss at threshold.

$$g_t = \frac{1}{2l} \ln\left(\frac{1}{r_1 r_2}\right) + \alpha \tag{A.4.25}$$

The calculated threshold gain from above is $g_t = 7 \times 10^{-3}$ cm^{-1}. For the 1060 nm transition in Nd, $\Gamma_{32} = 1800$ s^{-1}, and $\sigma \sim 9 \times 10^{-19}$ cm^2. Using the above equations, the threshold population inversion for this gain is $\Delta N = 8 \times 10^{15}$ cm^{-3}, and the threshold power per unit volume for a pump photon energy of 1.5 eV is ~3.4 W cm^{-3}.

This is about a factor of 20 less than that for a Nd three-level laser. For a 5-cm long rod of radius 2 mm (radius of stimulated emission area), the threshold power is ~2 W.

The optical gain for a fiber amplifier depends on the pump power and the energy level of the dopants. As stated earlier, the dopants are generally modeled as a three-level or four-level laser. An efficient design for the conversion of pump photons to signal photons (amplification) uses a configuration where both the pump and signal beams are confined in the core of the optical fiber. For this design, the launched power needed to provide gain (threshold pump power — P_{th}) is a reasonable figure of merit. For high efficiency, low P_{th} is desirable. From Eqs. (A.4.23) and (A.4.24), this quantity is proportional to (for a fixed gain g — gain at threshold is determined by loss)

$$P_{\text{th}} \sim h\nu_p g A_{\text{eff}}/(\sigma(\nu)\varepsilon\tau) \tag{A.4.26}$$

where A_{eff} is the effective area of the core (~100 μm^2), $h\nu_p$ is the pump photon energy (1.5 eV), σ is the stimulated emission cross-section (9×10^{-19} cm^2), ε is the fraction of absorbed pump power (~0.2), and τ ($=1/\Gamma_{32}$) is the fluorescence lifetime of the excited energy level ($\Gamma_{32} = 1800$ s^{-1}). For a gain (g) of 1 cm^{-1}, the calculated pump power needed using the above values and Eq. (A.4.26) is ~3 mW. In a fiber amplifier, the above result gets modified due to the broadening of energy levels, pump depletion, and gain saturation. The efficiency of the device can be increased by decreasing the core area, by increasing the stimulated emission cross section, and by increasing the pump absorption cross-section. The cross-section $\sigma(\nu)$ has a line shape. For homogeneous broadening, the line shape function is a Lorentzian:

$$\sigma(v) = \frac{\sigma(v_0)}{1 + (v - v_0)^2/\Delta v^2} \tag{A.4.27}$$

where $\Delta\nu$ is the spectral width. Since $\sigma(\nu)$ varies with frequency, the gain also varies with frequency and power as given by Eqs. (A.4.22)–(A.4.27). For a rare earth-doped amplifier, the gain spectrum is more complicated due to the presence of multiple energy levels and homogeneous and inhomogeneous broadening.

References

1. N. K. Dutta, *Fiber Amplifies and Fiber Lasers*, World Scientific, Singapore, 2015.
2. B. M. Walsh, N. P. Barnes, and B. Di Bartolo, *J. Lumin.*, **75** (1997) 89; B. M. Walsh, *Nasa Langley Res.* Center Report.
3. P. W. Milloni and J. H. Eberly, *Lasers*, John Wiley, NY, 1988.

Appendix 5

Spontaneous Emission and Absorption for Parabolic Bands

The absorption (or gain) and spontaneous emission rate for parabolic conduction and valence bands are calculated here. The transition probability or the emission rate W is given by (Fermi's Golden rule)

$$W = \frac{2\pi}{\hbar}|H'_{if}|^2\rho(E_f)\delta(E_i - E_f) \tag{A.5.1}$$

$E = h\nu = E_i - E_f$. E_i, E_f are the energies of the initial and final states.

$$H'_{if} = \langle i|H'|f\rangle \tag{A.5.2}$$

$$\begin{aligned} H' &= -\frac{e}{m}\mathbf{A}\cdot\mathbf{P} \\ &= \frac{ie\hbar}{m}\mathbf{A}\cdot\boldsymbol{\nabla} \end{aligned} \tag{A.5.3}$$

$\vec{\mathbf{A}}$ is vector potential in MKS units, m is the mass of electron, and e is charge of electron. We need to determine $\vec{\mathbf{A}}$ per photon. Electrical field

$$\varepsilon = -\frac{\partial\vec{\mathbf{A}}}{\partial t} \tag{A.5.4}$$

$$\vec{\mathbf{A}} = A_0\hat{\epsilon}\sin(\omega t - \mathbf{k}\cdot\mathbf{r}) \tag{A.5.5}$$

$\hat{\epsilon}$ is the polarization vector and $\mathbf{k} \times \hat{\epsilon} = 0$. The magnitude of the Poynting vector that represents the energy flux is

$$S = |\langle \varepsilon \times \vec{\mathbf{H}} \rangle| \tag{A.5.6}$$

The angle brackets denote time averaging and $\vec{\mathbf{H}}$ is the magnetic field. Using

$$\vec{\mathbf{A}} = \vec{\mathbf{A}}_0 \sin(\omega t - \mathbf{k} \cdot \mathbf{r}) \tag{A.5.7}$$

$$\varepsilon = -\frac{\partial \vec{\mathbf{A}}}{\partial t} = -\omega \vec{\mathbf{A}}_\mathbf{0} \cos(\omega t - \mathbf{k} \cdot \mathbf{r}) \tag{A.5.8}$$

$$\vec{\mathbf{B}} = \boldsymbol{\nabla} \times \vec{\mathbf{A}} = -\mathbf{k} \times \vec{\mathbf{A}}_\mathbf{0} \cos(\omega t - \mathbf{k} \cdot \mathbf{r}) \tag{A.5.9}$$

$$S = |\langle \varepsilon \times \vec{\mathbf{H}} \rangle| = \left| \left\langle \varepsilon \times \frac{\vec{\mathbf{B}}}{\mu_0} \right\rangle \right| = \frac{1}{2} \frac{k\omega}{\mu_0} \vec{\mathbf{A}}_0^2 \tag{A.5.10}$$

The energy flux per photon is photon energy × group velocity, which is $\hbar\omega \times c/\bar{\mu}$, $\bar{\mu}$ is the refractive index.

$$\frac{1}{2} \frac{k\omega}{\mu_0} A_0^2 = \frac{c\hbar\omega}{\bar{\mu}} \tag{A.5.11}$$

$$\begin{aligned} A_0 &= \left(\frac{2\hbar c \mu_0}{k \bar{\mu}} \right)^{1/2} \\ &= \left(\frac{2\hbar}{\epsilon_0 \bar{\mu}^2 \omega} \right)^{1/2} \end{aligned} \tag{A.5.12}$$

$$\vec{\mathbf{A}} = \hat{\epsilon} \left(\frac{2\hbar}{\epsilon_0 \bar{\mu}^2 \omega} \right)^{1/2} \sin(\omega t - \mathbf{k} \cdot \mathbf{r}) \tag{A.5.13}$$

Above is the vector potential per photon.

$$H' = \frac{e}{m} \left(\frac{2\hbar}{\epsilon_0 \bar{\mu}^2 \omega} \right)^{1/2} (\hat{\epsilon} : \cdot \mathbf{P}) \sin(\omega t - \mathbf{k} \cdot \mathbf{r}) \tag{A.5.14}$$

$$|H'_{if}|^2 = \frac{e^2}{m^2} \frac{2\hbar}{\epsilon_0 \bar{\mu}^2 \omega} \frac{1}{4} |\langle i | \hat{\epsilon} \cdot \mathbf{P} | f \rangle|^2 \tag{A.5.15}$$

$$\sin(\omega t - \mathbf{k} \cdot \mathbf{r}) = \frac{1}{2i} \{ e^{i(\omega t - \mathbf{k} \cdot \mathbf{r})} - c.c. \} \tag{A.5.16}$$

We have made the usual assumption $\mathbf{k} \cdot \mathbf{r} \ll 1$, i.e. electron wave function is small compared to the wavelength of light. For absorption or stimulated emission, $\rho(E_f) = 1$, hence the transition rate

$$W = \frac{\pi e^2}{m^2 \epsilon_0 \bar{\mu}^2 \omega} |M_{if}|^2 \delta(E_i - E_f - \hbar\omega) \tag{A.5.17}$$

where $|M_{if}| = \langle i|\hat{\epsilon} \cdot \mathbf{P}|f\rangle$; and E $= \hbar\omega$.

For absorption, the number of photons absorbed per second is W; the photons travel a distance of $c/\bar{\mu}$ in a second. Hence, the absorption coefficient (for a photon of energy E) is

$$\alpha = \frac{\bar{\mu} W}{c} = \frac{\pi e^2}{m^2 \epsilon_0 c \bar{\mu} \omega} |M_{if}|^2 \delta(E_i - E_f - E) \tag{A.5.18}$$

$$\alpha(E) = \frac{h e^2}{2\epsilon_0 m_0^2 c \bar{\mu} E} |M_{if}|^2 \delta(E_i - E_f - E) \tag{A.5.19}$$

For spontaneous emission, the photons are emitted in all directions, hence,

$$\rho(E_f) = \frac{2}{(2\pi)^3} \frac{d^3k}{dE} = 2 \frac{4\pi k^2}{(2\pi)^3} \frac{dk}{dE} \tag{A.5.20}$$

The parameter 2 arises from two polarization states. Since

$$k = \frac{\bar{\mu}\omega}{c} \tag{A.5.21}$$

$$\frac{dk}{dE} = \frac{\bar{\mu}}{c\hbar} \tag{A.5.22}$$

$$\rho(E_f) = \frac{\bar{\mu}^3 \omega^2}{\pi^2 c^3 \hbar} \tag{A.5.23}$$

$$\begin{aligned} r_{sp}(E) &= \frac{2\pi}{\hbar} \frac{e^2}{m_0^2} \frac{2\hbar}{\epsilon_0 \bar{\mu}^2 \omega} \frac{1}{4} \frac{\bar{\mu}^3 \omega^2}{\pi^2 c^3 \hbar} |M_{if}|^2 \delta(E_i - E_f - \hbar\omega) \\ &= \frac{4\pi e^2 \bar{\mu} E}{m_0^2 \epsilon_0 c^3 h^2} |M_{if}|^2 \delta(E_i - E_f - \hbar\omega) \end{aligned} \tag{A.5.24}$$

We now calculate the spontaneous emission rate and the absorption coefficient for direct-gap semiconductors. These quantities are obtained by integrating Eqs. (A.5.19) and (A.5.24) over the occupied electron and hole states in a semiconductor. A realistic band model for a III–V direct gap semiconductor is the four-band model shown in Figure 2.9 in Chapter 2. In this model, the valence band is represented by three subbands: the heavy-hole band, the light-hole band, and the spin-split-off-hole band. The light- and heavy-hole bands are degenerate at $\mathbf{k} = 0$. Usually, the split-off energy Δ is large compared with the thermal energy k_BT; hence the split-off band is full of electrons, i.e. no holes are present. The quasi-Fermi levels for the conduction and valence bands can be obtained from the known effective masses and the density of states in each band. For the conduction band, E_{fc} is obtained using

$$n = \int \frac{\rho_c(E)dE}{1 + \exp[(E - E_{fc})/k_BT]} = \int \rho_c(E) f_c(E) dE \qquad \text{(A.5.25)}$$

where $\rho_c(E)$ is the conduction-band density of states; $f_c(E)$ is the Fermi factor; E_{fc} is the quasi-Fermi level in the conduction band, k_B is the Boltzmann constant, T is the temperature, and n is the number density of electrons in the conduction band. The density of states for electrons of energy E can be obtained using Eq. (A.5.20) with the dispersion relation ($E = \hbar^2k^2/2m_c$) becoming

$$\rho_c(E) = (2)\frac{4\pi k^2}{(2\pi)^3}\frac{dk}{dE} = 4\pi\left(\frac{2m_c}{\hbar^2}\right)^{3/2} E^{1/2} \qquad \text{(A.5.26)}$$

where m_c is the conduction band effective mass. The factor 2 in Eq. (A.5.26) arises from the two electronic spin states. Equation (A.5.25) can be written as

$$n = N_c \frac{2}{\pi^{1/2}} \int \frac{\epsilon^{1/2} d\epsilon}{1 + \exp(\epsilon - \epsilon_{fc})} \qquad \text{(A.5.27)}$$

where $N_c = 2(2\pi m_c k_B T/\hbar^2)^{3/2}$ and $\epsilon_{fc} = E_{fc}/k_BT$. This equation can be used to calculate ϵ_{fc} for a given electron concentration. A useful approximation for ϵ_{fc} is as follows:

$$\epsilon_{fc} = \ln\left(\frac{n}{N_c}\right) + \sum_{i=1}^{\infty} A_i \left(\frac{n}{N_c}\right)^i \qquad \text{(A.5.28)}$$

where the first few coefficients are $A_1 = 3.53553 \times 1^{-1}$, $A_2 = 4.95009 \times 10^{-3}$, $A_3 = 1.48386 \times 10^{-4}$, and $A_4 = 4.42563 \times 10^{-6}$.

For a non-degenerate electron gas ($n \ll N_{\text{c}}$), all terms except the first in Eq. (A.5.28) can be neglected and the Fermi energy is given by

$$E_{fc} = k_B T \ln\left(\frac{n}{N_c}\right) \tag{A.5.29}$$

Then the occupation probability becomes

$$f_c(E) \cong \frac{n}{N_c} \exp\left(\frac{-E}{k_B T}\right) \tag{A.5.30}$$

This is often referred to as the Boltzmann approximation. The use of Eq. (A.5.30) simplifies considerably the calculation of recombination rates in semiconductors. The non-degenerate approximation $n \ll N_c$, however, does not always hold for semiconductor lasers. The quasi-Fermi level for the holes in the valence band can be similarly calculated. The hole density is given by

$$\begin{aligned} p &= \sum_{i=l,h} \int \frac{\rho_{vi}(E)dE}{1+\exp[(E-E_{fv})/k_B T]} \\ &= \sum_{i=l,h} \int \rho_{vi}(E) f_{vi}(E) dE \end{aligned} \tag{A.5.31}$$

where p is the density of holes and the summation is over the light and heavy-hole bands. Using an expression similar to Eq. (A.5.26) for the density

$$p = N_v \frac{2}{\pi^{1/2}} \int \frac{\epsilon^{1/2} d\epsilon}{1+\exp(\epsilon - \epsilon_{fv})} \tag{A.5.32}$$

where $N_v = 2(2\pi k_B T/\hbar^2)^{3/2}(m_{\text{lh}}^{3/2} + m_{\text{hh}}^{3/2})$ and m_{lh} and m_{hh} are the effective masses of light and heavy holes, respectively. For the non-degenerate case where $p \ll N_{\text{v}}$, Eq. (A.5.32) can be simplified to obtain the following expression for the quasi-Fermi energy and the hole occupation probability:

$$E_{fv} = k_B T \ln\left(\frac{p}{N_v}\right) \tag{A.5.33}$$

$$f_v(E) = \frac{p}{N_v} \exp\left(\frac{-E}{k_B T}\right) \tag{A.5.34}$$

If transitions follow k-selection rule, $|M_{if}|^2 = 0$ unless $\mathbf{k}_c = \mathbf{k}_v$. Since photon has negligible momentum

$$|M_{if}|^2 = |M_b|^2 \frac{(2\pi)^3}{V} \delta(\mathbf{k}_c - \mathbf{k}_v) \tag{A.5.35}$$

where V is the volume of the semiconductor,

$$|M_b|^2 = |\langle i|\vec{\mathbf{P}}|f\rangle|^2 \tag{A.5.36}$$

where $|i\rangle$ and $|f\rangle$ are initial and final states in the conduction and valence bands.

Using the Kane model ($\mathbf{k} \cdot \mathbf{p}$ method) of semiconductors

$$|M_b|^2 = \frac{m_0^2 E_g (E_g + \Delta)}{12 m_c (E_g + \frac{2}{3}\Delta)} = \gamma m_0 E_g \tag{A.5.37}$$

For GaAs, $E_g = 1.424$ eV, $m_c = 0.067\ m_0$, and $\Delta = 0.33$ eV, the calculated $\gamma = 1.3$. Valence band states are $\frac{1}{\sqrt{2}}|x + iy\rangle$, $\frac{1}{\sqrt{2}}|x - iy\rangle$, $|z\rangle$. Each state has a degeneracy of 2 due to two spin states. Thus using k-selection rule, the spontaneous emission rate for parabolic bands is (at energy E)

$$r_{sp}(E) = \frac{4\pi e^2 \bar{\mu} E}{m_0^2 \epsilon h^2 c^3} |M_b|^2 \frac{(2\pi)^3}{V} 2 \left[\frac{V}{(2\pi)^3}\right] \frac{1}{V} \sum \int \cdots$$
$$\times \int f_c(E_c) f_v(E_v) d^3k_c d^3k_v \delta(\mathbf{k}_c - \mathbf{k}_v) \delta(E_i - E_f - E) \tag{A.5.38}$$

The Σ stands for summation over valence band states.

$$r_{sp}(E) = \frac{2\bar{\mu} e^2 E |M_b|^2}{\pi m_0^2 \epsilon_0 h^2 c^3} \left(\frac{2m_\Gamma}{\hbar}\right)^{3/2} (E - E_g)^{1/2} f_c(E_c) f_v(E_v) \tag{A.5.39}$$

where

$$E_c = \frac{m_r}{m_c}(E - E_g) \tag{A.5.40}$$

$$E_v = \frac{m_r}{m_{hh}}(E - E_g) \tag{A.5.41}$$

$$m_r = m_c m_{hh}/(m_c + m_{hh}) \tag{A.5.42}$$

$$\begin{aligned} R &= \int_{E_g}^{\infty} r_{sp}(E)dE \\ &= \frac{2\bar{\mu}e^2 E|M_b|^2}{\pi m_0^2 \epsilon_0 h^2 c^3}\left(\frac{2m_r}{\hbar^2}\right)^{3/2} \int_{E_g}^{\infty} (E - E_g)^{1/2} f_c(E_c) f_v(E_v) dE \end{aligned} \tag{A.5.43}$$

$$\begin{aligned} \alpha(E) &= \frac{e^2 h}{2e_0 m_0^2 c\bar{\mu}E}|M_b|^2 \frac{(2\pi)^3}{V} 2\left[\frac{V}{(2\pi)^3}\right]^2 \cdot \frac{1}{V} \\ &\quad \times \int \cdots \int (1 - f_c - f_v) d^3\bar{k}_c d^3\bar{k}_v \delta(\bar{k}_c - \bar{k}_v)\delta(E_i - E_f - E) \\ &= \frac{e^2 h|M_b|^2}{4\pi^2 \epsilon_0 m_0^2 c\bar{\mu}E}\left(\frac{2m_r}{\hbar^2}\right)^{3/2} (E - E_g)^{1/2}(1 - f_c - f_v) \end{aligned} \tag{A.5.44}$$

In the absence of parabolic bands, the absorption coefficient $\alpha(E)$ is given by (Chapter 3)

$$\begin{aligned} \alpha(E) = \frac{e^2 h}{2\epsilon_0 m_0^2 c\bar{\mu}E} &\int_{-\infty}^{\infty} \rho_c(E')\rho_v(E'')|M_{if}|^2 \\ &\times [1 - f_v(E'') - f_c(E')]dE' \end{aligned} \tag{A.5.45}$$

Going back to the previous case of parabolic bands and **k**-selection

$$\begin{aligned} \alpha(E) &= \frac{e^2 h|M_b|^2}{4\pi^2 \epsilon_0 m_0^2 c\bar{\mu}E}\left(\frac{2m_r}{\hbar^2}\right)^{3/2} (E - E_g)^{1/2}(1 - f_c - f_v) \\ &= C\frac{(E - E_g)^{1/2}}{E}(1 - f_c - f_v) \\ &= C'(1 - E_g/E)^{1/2}(1 - f_c - f_v) \end{aligned} \tag{A.5.46}$$

For absorption far from the band edge, $f_c \approx 0$, $f_v \approx 0$,

$$\alpha(E) = C'(1 - E_g/E)^{1/2} \tag{A.5.47}$$

C' is a constant whose value is defined by the terms above. For Boltzmann approximation

$$f_c = \frac{1}{1 + \exp(E - E_{fc})/k_B T} \simeq e^{-(E - E_{fc})/k_B T} \tag{A.5.48}$$

$$\frac{E_{fc}}{kT} = \ln\left(\frac{n}{N_c}\right) + A_1\left(\frac{n}{N_c}\right) + A_2\left(\frac{n}{N_c}\right)^2 + \cdots \tag{A.5.49}$$

$E_{fc} = kT\ln(n/N_c)$ in Boltzmann approximation.

$$f_c \simeq \frac{n}{N_c} e^{-E/k_B T} \tag{A.5.50}$$

$$N_c = 2 \cdot \left(\frac{m_c k_B T}{2\pi\hbar^2}\right)^{3/2}$$

$$= 25 \times 10^{18} \left(\frac{m_c}{m_0}\right)^{3/2} \left(\frac{T}{300}\right)^{3/2} \text{cm}^{-3} \tag{A.5.51}$$

$$r_{\text{sp}}(E) = C'(E - E_g)^{1/2} f_c(E_c) f_v(E_v) \tag{A.5.52}$$

where

$$E_c = \frac{m_r}{m_c}(E - E_g)$$

$$E_v = \frac{m_r}{m_{hh}}(E - E_g)$$

$$m_r = m_c m_{hh}/(m_c + m_{hh}) \tag{A.5.53}$$

and C' is a constant shown in Eq. (A.5.32). The total spontaneous emission rate is given by Eq. (A.5.46), which in the Boltzmann approximation becomes,

$$
\begin{aligned}
R &= \int r_{sp}(E)dE \\
&= C' \int_{E_g}^{\infty} (E - E_g)^{1/2} \frac{np}{N_C N_V} \exp(-E_c - E_v) \\
&= C' \frac{np}{N_C N_V} \int_{E_g}^{\infty} (E - E_g)^{1/2} e^{-(E-E_g)/k_B T} dE \\
&= C' \frac{np}{N_C N_V} \Gamma\left(\frac{3}{2}\right) (k_B T)^{3/2} = Bnp
\end{aligned}
\tag{A.5.54}
$$

One mechanism of carrier loss is spontaneous emission, i.e. $dn/dt = -R = -\text{Bnp}$. Carrier lifetime is defined by

$$\frac{dn}{dt} = -\frac{n}{\tau_r} = -Bnp \tag{A.5.55}$$

$$\tau_r = (Bp)^{-1} \tag{A.5.56}$$

For undoped semiconductors, $n = p$, $R = \text{Bn}^2$, and $\tau_r = (\text{Bn})^{-1}$. Due to the presence of donor and acceptors states, parabolic bands are an approximation, but it represents an important case for which closed-form solutions can be obtained. Numerical solutions including the donor and acceptor states (also known as band-tail states) are discussed in Chapter 3.

A.5.1.1 Modulation Bandwidth of LEDs

The current injection in LEDs results in light output due to radiative recombination of injected carriers. Assume that the LED has a low background doping, in this case, the radiative recombination rate $R = Bn^2$ and the radiative lifetime $\tau_r = (Bn)^{-1}$, where n is the injected carrier density per unit volume. The light output $L = fBn^2$, where f is the fraction of spontaneous emission from LED active region which is captured.

In addition to radiative lifetime, there is a non-radiative lifetime τ_n, the value of which depends on the defects and quality of interfaces.

Thus, the total lifetime of carriers τ is given by

$$1/\tau = 1/\tau_r + 1/\tau_n \tag{A.5.57}$$

Generally, for good epitaxial material (e.g. GaAs) at high-injected carrier density $\tau_\mathrm{r} \ll \tau_n$. The carrier density equation in the LED is given by

$$dn/dt = J/(\mathrm{ed}) - n/\tau \tag{A.5.58}$$

where J is the current density (A/cm^2), d is the thickness of the active region, and e is the charge of the electron. The steady state value of n from above is $n_s = J\tau/(\mathrm{ed})$. Now consider a small modulation of current $J = J_0 + J_m \exp(i\omega t)$, with $J_m \ll J_0$. Using $n = n_0 + n_m \exp(i\omega t)$ with $n_0 = J_0\tau/(\mathrm{ed})$ in Eq. (A.5.58), the following equation for n_m is obtained

$$n_m = J_m\tau/(\mathrm{ed}(1 + i\omega\tau)) \tag{A.5.59}$$

The magnitude of n_m is $J_m\tau/(\mathrm{ed}(1+\omega^2\tau^2)^{1/2})$. The light output L is written as $L = L_0+L(\omega)\exp(i\omega\tau)$ with $L(\omega) \ll L_0$. Using $L = fBn^2$ state earlier, $L(\omega) = 2fBn_0n_m$ and $L_0 = fBn_0^2$. Thus, the magnitude of $L(\omega)$ is

$$L(\omega) = 2fBn_0J_m\tau/(\mathrm{ed}(1 + \omega^2\tau^2)^{1/2}) \tag{A.5.60}$$

The above may be rewritten as

$$L(\omega) = L(0)/(1 + \omega^2\tau^2)^{1/2} \tag{A.5.61}$$

where $L(0)$ is the value of $L(\omega)$ at $\omega = 0$. The modulation bandwidth Δf is defined as the frequency at which the detected electrical power ($P_\mathrm{elc} \sim L(\omega)^2$) is one half of the value at the zero modulation frequency, i.e.

$$\Delta f = \Delta\omega/2\pi = 1/(2\pi\tau) \tag{A.5.62}$$

When radiative recombination is the dominant recombination mechanism, i.e. $\tau_r \ll \tau_n$, $\tau = (Bn_0)^{-1}$. Using $n_0 = J_0\tau/(\mathrm{ed})$ derived earlier,

$$\Delta f = (1/2\pi)(BJ_0/\mathrm{ed})^{1/2} \tag{A.5.63}$$

Thus, the modulation bandwidth of a double heterostructure LED is proportional to square root of the current for the approximations used here.

Appendix 6

Quantum Cascade Laser — Gain Calculation

In order to estimate the gain of the quantum cascade laser (QCL), a rate equation method is used to determine the population levels. Below threshold, the stimulated emission is negligible, therefore, the rate equation for levels 3 and 2, from Section 4.3 in Chapter 4 (with the Section 4.3 equation numbers are):

$$\frac{dN_3}{dt} = \eta_3 I - \frac{N_3}{\tau_{32}} - \frac{N_3}{\tau_{31}} \tag{4.2}$$

$$\frac{dN_2}{dt} = \eta_2 I + \frac{N_3}{\tau_{32}} - \frac{N_2}{\tau_2} \tag{4.3}$$

where N_2 and N_3 are the population of levels 2 and 3, respectively. η_2 and η_3 describe the fraction of the total current injected to levels 2 and 3, respectively [1]. The quantities τ_{32} and τ_{31} describe the lifetimes associated with LO phonon induced transitions between levels 3 and 2 and between levels 3 and 1. τ_2 is the lifetime of level 2 dominated by resonant depopulation. Due to the existence of mini-stopband, the lifetime of level 3 tunneling is not considered here.

The lifetime of the LO phonon transition is evaluated using Fermi's golden rule. The transition rate is given by [2]

$$w = \frac{2\pi}{\hbar}|c|^2(n_q+1)\sum_{k_t}\sum_{q_t q_z}\delta_{k_t,k_0-q_t}\frac{1}{q^2}f(q_z)\delta \times\left(\frac{\hbar^2}{2m_3}k_0^2+E_g-\frac{\hbar^2}{2m_2}k_t^2-\hbar\omega_{LO}\right) \quad \text{(A.6.1)}$$

with

$$|c|^2 = \frac{e^2\hbar\omega_{LO}}{2\varepsilon_0 V}\left(\frac{1}{\varepsilon_\infty}-\frac{1}{\varepsilon_s}\right) \quad \text{(A.6.2)}$$

$$f(q_z) = \left|\int_{-\infty}^{+\infty} e^{-iq_z z}\varphi_2(z)\varphi_1(z)dz\right|^2 \quad \text{(A.6.3)}$$

where w is the transition rate, n_q is the LO phonon number per mode, m_3 and m_2 are the effective mass of the conduction subbands, and E_g is the energy separation of levels 3 and 2. $\hbar\omega_{LO}$ is the energy of LO phonon.

Define $\frac{\hbar^2}{2m_3}k_0^2+E_g-\hbar\omega=\frac{\hbar^2}{2m_2}k_{t0}^2$, then,

$$\begin{aligned}\delta\left(\frac{\hbar^2}{2m_3}k_0^2+E_g-\frac{\hbar^2}{2m_2}k_t^2-\hbar\omega_{LO}\right) &= \delta\left(\frac{\hbar^2}{2m_2}k_{t0}^2-\frac{\hbar^2}{2m_2}k_t^2\right)\\ &= \frac{2m_2}{\hbar^2}\delta[(k_{t0}+k_t)(k_{t0}-k_t)]\\ &= \frac{2m_2}{\hbar^2}\frac{1}{2k_{t0}}\delta(k_{t0}-k_t)\cdot H(k_{t0}^2)\end{aligned} \quad \text{(A.6.4)}$$

where $H(x)$ is the Heavyside unit step function.

From Eqs. (A.6.3) and (A.6.4), it follows

$$w = \frac{2\pi}{\hbar}|c|^2(n_q+1)\sum_{k_t}\sum_{q_t q_z}\delta_{k_t,k_0-q_t}\frac{1}{q^2}f(q_z)\delta \times\left(\frac{\hbar^2}{2m_3}k_0^2+E_g-\frac{\hbar^2}{2m_2}k_t^2-\hbar\omega\right)$$

$$= \frac{2\pi}{\hbar}|c|^2(n_q+1)\sum_{k_t}\sum_{q_t q_z}\delta_{k_t,k_0-q_t}\frac{1}{q^2}f(q_z)\frac{2m_2}{\hbar^2}\frac{1}{2k_{t0}}\delta(k_{t0}-k_t)$$

$$\cdot H(k_{t0}^2)$$

$$= \frac{2\pi}{\hbar}|c|^2(n_q+1)\sum_{k_t}\sum_{q_z}\frac{f(q_z)}{q_z^2+(k_0-k_t)^2}\frac{m_2}{\hbar^2}\frac{1}{k_{t0}}\delta(k_{t0}-k_t)$$

$$\cdot H(k_{t0}^2)$$

$$= \frac{2\pi}{\hbar}|c|^2(n_q+1)\frac{V}{8\pi^3}\iint k_t dk_t d\theta$$

$$\times \int dq_z \frac{f(q_z)}{q_z^2+k_0^2-2k_0k_t\cos\theta+k_t^2}\frac{m_2}{\hbar^2}\frac{1}{k_{t0}}\delta(k_{t0}-k_t)\cdot H(k_{t0}^2)$$

$$= \frac{m_2}{2\pi\hbar^3}\frac{e^2\hbar\omega_{LO}}{2\varepsilon_0}\left(\frac{1}{\varepsilon_\infty}-\frac{1}{\varepsilon_s}\right)(n_q+1)$$

$$\times \int dq_z \frac{f(q_z)}{\sqrt{(q_z^2+k_0^2+k_{t0}^2)^2-4k_0^2k_{t0}^2}}H(k_{t0}^2) \tag{A.6.5}$$

From (A.6.5), the lifetime (τ) is calculated as $\tau = 1/w$. Using equation (A.6.5), the calculated lifetime is shown in Figure A.6.1. This calculation is done for the transition between $n = 1$ state and $n = 2$ state of a quantum well (with infinite well depth) of the width 20 nm. The energy of the subband gap is varied while keeping all the other parameters unchanged. The figure shows that if the energy separation between two states is chosen to be the same as the LO phonon energy, the lifetime is significantly reduced. That is the main reason why level 1 is introduced to resonantly depopulate level 2 (Figure 4.7). In order to evaluate the gain of the QCL, a density matrix theory is used [3]. The wave function of the electron is assumed to be of the following form:

$$\psi_a = u_c(r)\frac{e^{ik_t\rho}}{\sqrt{A}}\varphi_2(z) \tag{A.6.6}$$

$$\psi_b = u_c(r)\frac{e^{ik_t\rho}}{\sqrt{A}}\varphi_3(z) \tag{A.6.7}$$

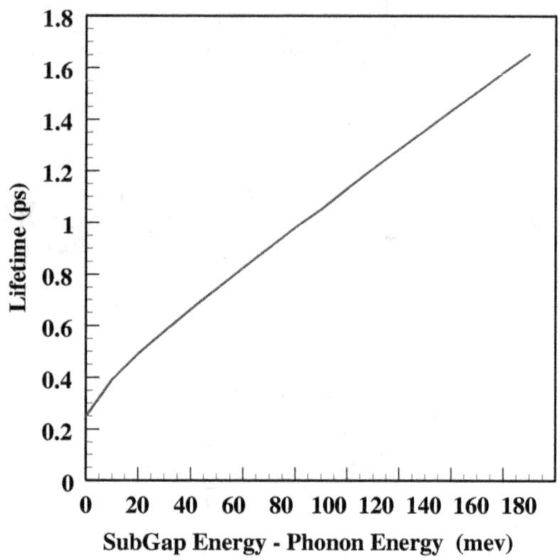

Figure A.6.1: Calculated lifetime as a function of the separation of transition energy and LO phonon energy.

where $u_c(r)$ is the unit cell function, $\varphi_2(z)$ and $\varphi_3(z)$ are the quasi-localized functions in the z-direction associated with levels 2 and 3, respectively. A is the area in the quantum well plane. The density matrix for the off-diagonal term is given by [3, 4]

$$\hbar\omega\rho_{12} - [H,\rho]_{12} + i\hbar\gamma\rho_{12} = eF[z,\rho]_{12} \tag{A.6.8}$$

$$\hbar\omega\rho_{21} - [H,\rho]_{21} + i\hbar\gamma\rho_{21} = eF[z,\rho]_{21} \tag{A.6.9}$$

where $\hbar\omega$ is the energy of the phonon, F is electric field of the phonon, and γ is a dephasing constant, usually of the order 100 fs. By solving (A.6.8, A.6.9) in the rotating wave approximation [3, 4], the following expression for absorption coefficient is obtained:

$$\alpha(\hbar\omega) = \frac{\omega}{n_r c\varepsilon_0}\frac{2}{V}\sum_{k_t}\frac{|\mu_{21}|^2\gamma/2}{(E_b - E_a - \hbar\omega)^2 + (\gamma/2)^2}(f_b - f_a) \tag{A.6.10}$$

where

$$E_b = E_3 + \frac{\hbar^2 k^2}{2m_3}, \quad E_a = E_2 + \frac{\hbar^2 k^2}{2m_2}.$$

The mass of the conduction subbands are significantly modified from that of the mass of free electron. The effective mass for each subband is evaluated using [4]

$$m_n = m_e\left(1 + \frac{2E_n(0)}{E_G}\right) \tag{A.6.11}$$

By introducing a new variable $\varepsilon = \frac{\hbar^2 k^2}{2m_0}$, where m_0 is the free-electron mass, Eq. (A.6.10) is rewritten as:

$$\begin{aligned}\alpha(\hbar\omega) &= \frac{\omega}{n_r c\varepsilon_0}\frac{2}{V}\sum_{k_t}\frac{|\mu_{21}|^2\gamma/2}{(E_b - E_a - \hbar\omega)^2 + (\gamma/2)^2}(f_b - f_a)\\ &= \frac{\omega}{n_r c\varepsilon_0}\frac{2}{L_z A}\frac{A}{4\pi^2}\iint dk^2\frac{|\mu_{21}|^2\gamma/2}{(E_b - E_a - \hbar\omega)^2 + (\gamma/2)^2}(f_b - f_a)\end{aligned}$$

Taking into account the mode confinement factor and the number of stages,

$$\begin{aligned}\alpha(\hbar\omega) = \Gamma\cdot N\frac{m_0\omega}{\hbar^2 n_r c\varepsilon_0}\frac{|\mu_{21}|^2\gamma/2}{L_z\pi}\\ \int\frac{d\varepsilon}{(\frac{m_0}{m_3}\varepsilon - \frac{m_0}{m_2}\varepsilon + E_g - \hbar\omega)^2 + (\gamma/2)^2}(f_b - f_a)\end{aligned} \tag{A.6.12}$$

where Γ is the optical confinement factor, and N is the total number of stages.

The Fermi factors f_a and f_b are given by:

$$f_b = \frac{1}{Exp\left[\left(\frac{m_0}{m_3}\varepsilon - E_{f3}\right)/k_B T\right] + 1} \tag{A.6.13}$$

$$f_a = \frac{1}{Exp\left[\left(\frac{m_0}{m_2}\varepsilon - E_{f2}\right)/k_B T\right] + 1} \tag{A.6.14}$$

The Fermi levels E_{f2} and E_{f3} are evaluated using the following expression (in Boltzmann approximation) once the population is

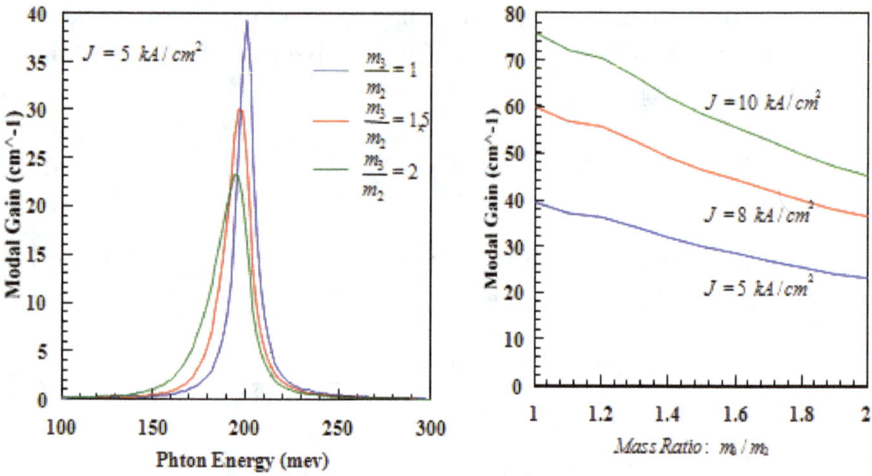

Figure A.6.2: Gain spectrum of the QCL for different mass ratios between the conduction subbands.

determined from Eqs. (4.2) and (4.3) for a given current density.

$$E_{f3} = k_B T \cdot \mathrm{Log}\left[\frac{n_3 \pi \hbar^2}{m_3 k_B T}\right] \tag{A.6.15}$$

$$E_{f2} = k_B T \cdot \mathrm{Log}\left[\frac{n_2 \pi \hbar^2}{m_2 k_B T}\right] \tag{A.6.16}$$

Using the method developed above, the gain of the QCL has been evaluated. The calculated results are shown by the Figures A.6.2–A.6.4. In Figure A.6.2, we show the gain spectrum of the QCL as a function of mass ratio between levels 3 and 2. It is clear that the non-parabolicity of effective mass makes the gain shape deviate from Lorentzian and lowers the gain.

Further, in Figure A.6.3, the temperature dependence of the gain spectrum is shown. The gain decreases with increasing temperature for the following reasons. First, the increasing of the temperature excites more LO phonons and therefore decreases the lifetime of each state. Hence, for the same injection current, the population of the higher energy level is lowered. Second, the phase damping constant

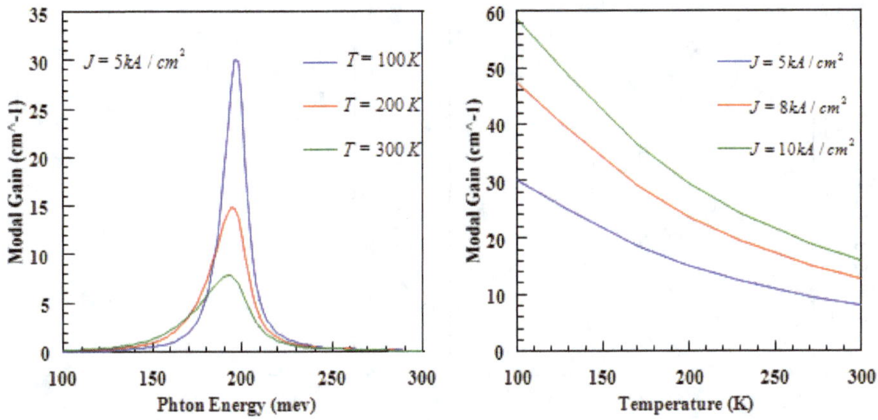

Figure A.6.3: Gain spectrum of the QCL for different temperatures.

also changes as a function of the temperature, and at high temperature, the gain spectrum gets broadened and therefore lower the peak value.

The analysis shows that the lifetime of the lower transition level should be made as short as possible in order to increase the gain. In order to achieve wide tunability of the QCL, the first requirement is to achieve a broad gain spectrum. Gmachl *et al.* have demonstrated a QCL with a broad emission spectrum [5]. This device had several QCL active regions which emit at different wavelengths (different well thicknesses of the light-emitting layer) grown on top of each other. We analyze here the gain spectrum of such a QCL which is approximately determined by the sum of the gain spectrum of each active region well. Starting from the previously derived equations the following is obtained:

$$\alpha(\hbar\omega) = \sum_{i=1}^{N} \Gamma(i) \frac{m_0\omega}{\hbar^2 n_r c\varepsilon_0} \frac{|\mu_{21}(i)|^2\gamma/2}{L_z\pi} \times \int \frac{d\varepsilon}{\left[\frac{m_0}{m_3}\varepsilon - \frac{m_0}{m_2}\varepsilon + E_g(i) - \hbar\omega\right]^2 + (\gamma/2)^2}(f_b - f_a) \tag{A.6.17}$$

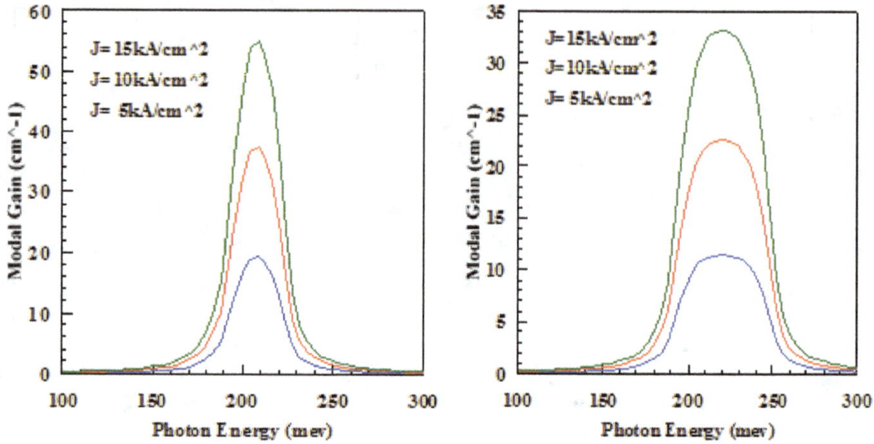

Figure A.6.4: The calculated modal gain is plotted as a function of photon energy for different injection current densities. Note the broad gain spectrum. For the left figure, the energy difference between two nearby active regions is 1 Mev, and for the right figure, this value is 2 MeV.

where the summation is over all the active regions. The quantities $\Gamma(i)$, $\mu_{21}(i)$, and $E_g(i)$ are the values for ith active region, and could be calculated for each of the layers. In the following numerical calculation (Figure A.6.4), $\Gamma(i)$ and $\mu_{21}(i)$ are assumed to be equal for all active regions. The calculated results are shown in Figure A.6.4.

References

1. J. Faist *et al.*, *IEEE J. Quant. Electron.*, **34** (1998) 336; J. Faist *et al.*, *Appl. Phys. Lett.*, **68** (1996) 3680.
2. D. Ahn *et al.*, *Phy. Rev. B*, **37** (1988) 2529.
3. A. Yariv, *Quantum Electronics*, John & Wiley, USA, 1989.
4. B. Gelmont, *App. Phy. Lett.*, **68** (1996) 2171.
5. C. Gmachl *et al.*, *Nature*, **415** (2002) 883.

Appendix 7

Supercontinuum Generation in Optical Fibers

Broadband supercontinuum generation (SCG) schemes have gained significant interests due to their applications in optical coherence tomography, spectroscopy, and dense wavelength division multiplexing (DWDM) sources for high-capacity optical communication systems. In SCG, extreme broadening of pulse spectrum can be realized for sufficiently intense narrow-band incident pulses. The primary mechanism of spectral broadening is considered as a combined action of self-phase modulation (SPM), cross-phase modulation (XPM), stimulated Raman scattering (SRS), parametric four-wave mixing (FWM), and the dispersive properties of the fiber. The processes of the SCG are sensitive to the dispersive and nonlinear properties of the optical fiber. However, for broad spectrum, the control of the fiber dispersion is more important than achieving the maximum possible nonlinearity per unit length. Therefore, there has been considerable research focusing on dispersion tailoring to maximize the bandwidth of the generated continuum. Many schemes have been proposed by using dispersion-decreasing fiber, dispersion-flattened fiber, dispersion-shifted fiber, tapered fiber, and photonic crystal fiber [1–5]. The mechanism of SCG has been analyzed numerically [6, 7]. The results showed that a dispersion-decreasing fiber with a convex dispersion profile can generate a flat broadened spectrum. The progress in the design of photonic crystal fiber (PCF) enables the fabrication of fibers with two zero dispersion wavelengths (ZDWs) and SCG in these fibers has been studied extensively [8–13]. In Refs. [12–14], the authors investigated the possibility of obtaining the efficient mid-IR supercontinuum sources based on non-silica microstructured optical fibers (MOFs) technology. The results show that small core MOFs with two ZDWs can have large efficiency for mid-IR continuum generation using convenient pump wavelengths by phase-matching across the upper (longer wavelength) ZDW. They found that a seed wavelength approximately midway between the ZDWs can produce the broadest spectrum. In Ref. [15], a ZBLAN fiber taper with exponentially-decreasing core size was numerically modeled and the coherent continuum could be generated. The authors showed that the spectrum generated across the upper ZDW can be progressively shifted with this non-uniform fiber taper design.

Although it is possible to enhance the SC broadening in optical fibers with two ZDWs, the bandwidth of the spectrum is still limited in fibers with two ZDW fibers if the longer wavelength ZDW is fixed in wavelength. The reason is the power will transfer from the soliton to the other wavelengths at frequencies determined by the phase-matching condition (PMC) [15], so the spectrum cannot be further extended to longer wavelength where the PMC is not satisfied. Also Raman self-frequency shift of the soliton will be canceled by the spectral recoil effect [16], which will stop the red shift of the soliton near the second ZDW. However, if the fiber taper has an increasing fiber core radius, the second ZDW can be shifted to longer wavelength and the PMC will be continuously modified as well. The generated wavelengths can be guided towards longer wavelength eventually to generate a flat and broad spectrum.

A.7.1 Theoretical model

SF57 (lead-silicate glass) is a material of choice for the fiber taper since the lead silicate glasses are promising materials

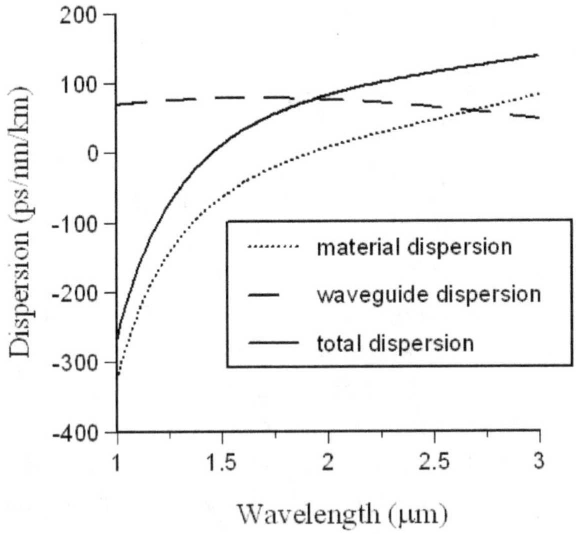

Figure A.7.1: Wavelength-dependent dispersion profile of SF57 fiber with 1.5-μm core radius [15].

for SCG. Although their material nonlinearity is lower than that for chalcogenide [16] and heavy metal oxide glass, they offer higher thermal and crystallization stability and less steep viscosity-temperature-curves, while exhibiting low softening temperatures [17–19]. With microstructure fabrication technique, SF57 fibers with tight mode confinement and tailored dispersion have been demonstrated. Here for simplification, SF57 dispersion profile is simulated by using step-index waveguide model, the total dispersion in the fiber depends on both material and waveguide contributions. The wavelength dependent dispersion parameter is calculated by numerically solving the eigenvalue equation for the fundamental HE11 mode of step-index fiber. Figure A.7.1 shows the calculated wavelength-dependent dispersion profile of SF57 fiber with 1.5-μm core radius. The similar dispersion profile can also be achieved in photonics crystal fiber.

The SCG can be numerically described by the generalized nonlinear Schrödinger equation (GNLS) under the slowly varying envelope approximation (Appendix 3),

$$\frac{\partial A}{\partial z} + \frac{\alpha}{2}A + \sum_{m\geq 2} \frac{i^{m-1}}{m!}\beta_m \frac{\partial^m A}{\partial T^m}$$
$$= i\gamma\left(1 + \frac{i}{\omega_0}\frac{\partial}{\partial T}\right)\left(\mathrm{A}\int_{-\infty}^{+\infty} R(T')|A(z,T-T')|^2 dT'\right) \tag{A.7.1}$$

where A(z,T) is the complex slowly varying amplitude of the light field in the reference frame moving with the group velocity of the input pulse and α is the fiber loss, γ is the effective nonlinearity parameter of the fiber that will be calculated by $\gamma = \frac{2\pi n_2}{\lambda A_{\text{eff}}}$, where n_2 is the nonlinear coefficient of the fiber and λ is the center wavelength of the pulse. The effective mode-field area of a guided mode for SF57 fiber is given by

$$A_{\text{eff}} = \frac{\left(\int_{-\infty}^{+\infty}\int_{-\infty}^{+\infty}|E^2|dxdy\right)^2}{\int_{-\infty}^{+\infty}\int_{-\infty}^{+\infty}|E|^4 dxdy}.$$

β_m are the coefficients of Taylor series expansion of the propagation constant $\beta(\omega)$ about the carrier frequency ω_0, where $\beta_m = n(\omega)\omega/c$, $n(\omega)$ is the frequency dependent refractive index, and c is the speed of light. Though $A(z,T)$ is a slowly varying amplitude, reasonably accurate simulation results can be obtained if sufficient number of higher order dispersive terms are included. The dispersion operator is $\beta(\omega)-\beta_0-\omega\beta_1$ in the calculation so that all orders of dispersion are included [14]. The response function $R(T) = (1-f_R)\delta(T)+f_R h_R(T)$ includes both the instantaneous electronic and non-instantaneous Raman responses, where the delta function represents the instantaneous electronic response, and $h_R(T)$ represents the delayed Raman response of the ions. In the simulations, $h_R(T)$ of the SF57 glass is described by means of an analytic form based on experimental measurements and $f_R = 0.1$ [20]. A hyperbolic-secant pulse (at 1550 nm wavelength) with 100 fs full-width-at-half-maximum (FWHM) and 6 kW peak power is the input pulse (energy is ~0.6 nJ per pulse). The pulse is launched into the SF57 fiber or fiber taper. The designed fiber taper has an increasing core radius r_{core} along the propagation length z with the relationship $r_{\text{core}}(z) = r_0(1 + z/z_0)^3$, where initial radius $r_0 = 0.7$ μm and $z_0 = 27.6$ cm. This fiber taper has non-uniform structure which is similar to the one used in Ref. [14]. However, in the design, the 2nd ZDW is red shifted as taper radius increases and input pulse has much higher soliton number so that the soliton fission and compression can take place during propagation.

A.7.2 Results and discussion

Figure A.7.2 shows the wavelength dependent dispersion profile of SF57 fiber with different core radii. The second ZDW is formed for fibers with core radius smaller than ~0.9 μm. With decreasing core radius, the second ZDW shifts to shorter wavelengths. The spectrum evolution in the fiber with 1.2-μm core radius which does not have the second ZDW is studied first. The seed wavelength is at 1550 nm, which is close to the first ZDW at ~1.38 μm. Now consider a dispersion profile with two ZDWs and a seed wavelength (1550 nm) near the middle of two ZDWs and a little closer to the first ZDW.

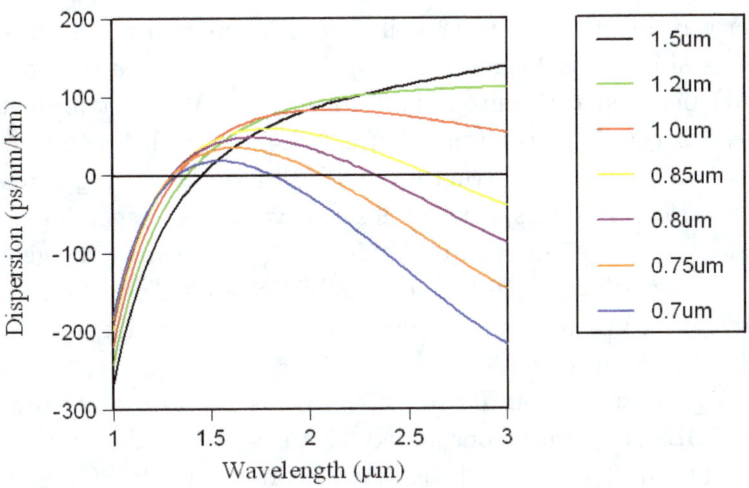

Figure A.7.2: Wavelength dependent dispersion profile of SF57 fiber with different core radii. Note the fibers with small core have two ZDWs [15].

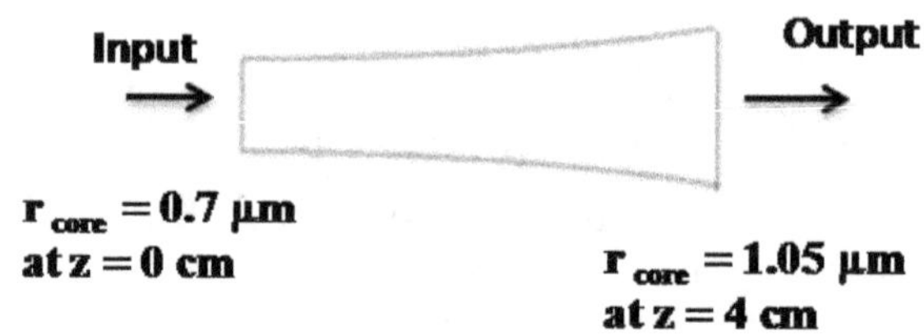

Figure A.7.3: A schematic of the SF57 fiber taper with nonlinear coefficient $n_2 = 4.1 * 10^{-19}$ m^{2}W^{-1} and A$_{\text{eff}} \sim 3.5\ \mu$m^2 at 1550 nm input wavelength used in Ref. [15].

The schematic of the SF57 fiber taper that is proposed earlier is shown in Figure A.7.3.

The results of the calculation for different z_0 values are shown in Figure A.7.4. The SC spectrum at the output of 4 cm SF57 fiber taper extends to ~4.8 μm (for $z_0 = 27.6$ cm). The shifting of second ZDW along the fiber makes possible the generation of new dispersive waves (DWs) towards longer wavelength with continuously modified PMC. Note that the spectrum looks less dense at wavelengths larger than ~2 μm. This is because the energy of the solitons transfer to DWs and spreads out over a larger spectrum region, resulting in a broad and relatively uniform SC spectrum. The solitons also red shift to longer wavelength since the recoil effect is decreased when the second ZDW is shifting to longer wavelength. The cross-phase-modulation (XPM) between the solitons and generated DWs contribute to the flatness of the SC spectrum. Figure A.7.4 shows the output spectra for various fiber taper structures with different tapering rates. The generated continuum spectrum is sensitive to the dispersion properties along the SF57 fiber taper, which is determined by the geometry of the taper. This suggests the possibility of carefully tailoring the dispersion profile to achieve a desired output spectrum based on the application requirements.

SC generation from 0.8 to 4.5 μm has been obtained by using 4.5 m ZBLAN fluoride fiber joined with a ~1 m of silica single-mode fiber [19]. In Ref. [20], a broad bandwidth, mid-IR SC generation using an 8 mm length of highly nonlinear telluride microstructured PCF has been demonstrated. It is pumped with 100 fs pulses of

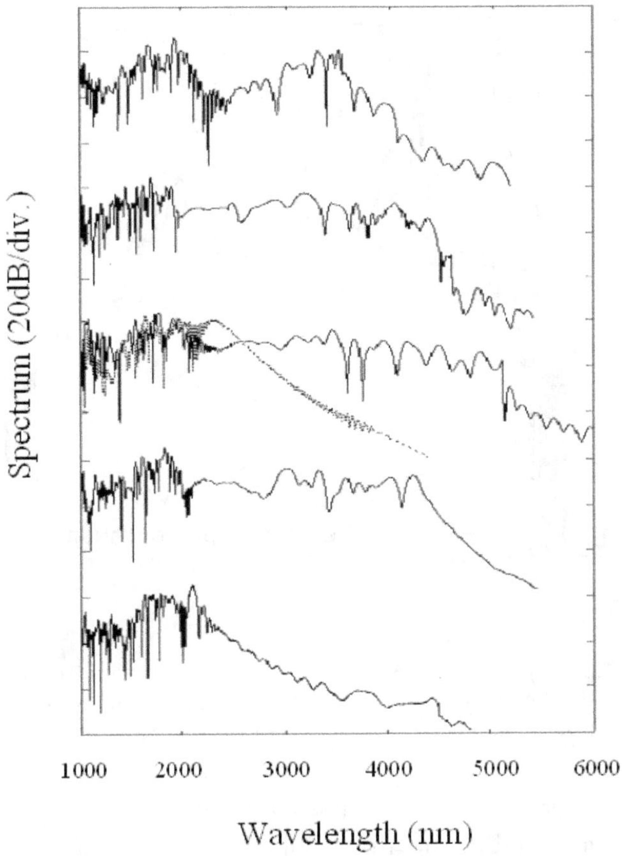

Figure A.7.4: Output spectra for SF57 fiber taper with various tapering rates with (a) $z_0 = 80$ cm, (b) $z_0 = 29$ cm, (c) $z_0 = 27.6$ cm, (d) $z_0 = 21$ cm, and (e) $z_0 = 12$ cm. (a)–(e) are labeled top to bottom. The output spectrum, a uniform SF57 fiber (least broadening), is shown with the dotted line near the spectrum labeled (c) [15].

energy 1.9 nJ at 1550 nm. The resulting bandwidth is from 789 to 4870 nm which covers over 4000 nm spectrum range. The experimental data are shown in Figure A.7.5. For the tellurite PCF and pump pulses ($\beta_2 = 80$ ps^{2}km^{-1}, $P_0 = 7954$ W, $\gamma = 596$ km^{-1}W^{-1}, and $T_{FWHM} = 110$ fs) the nonlinear length is calculated to be 0.21 mm and the dispersion length is 38 mm [20]. The soliton number found in the tellurite fiber is $N = 14$, indicating that higher order soliton

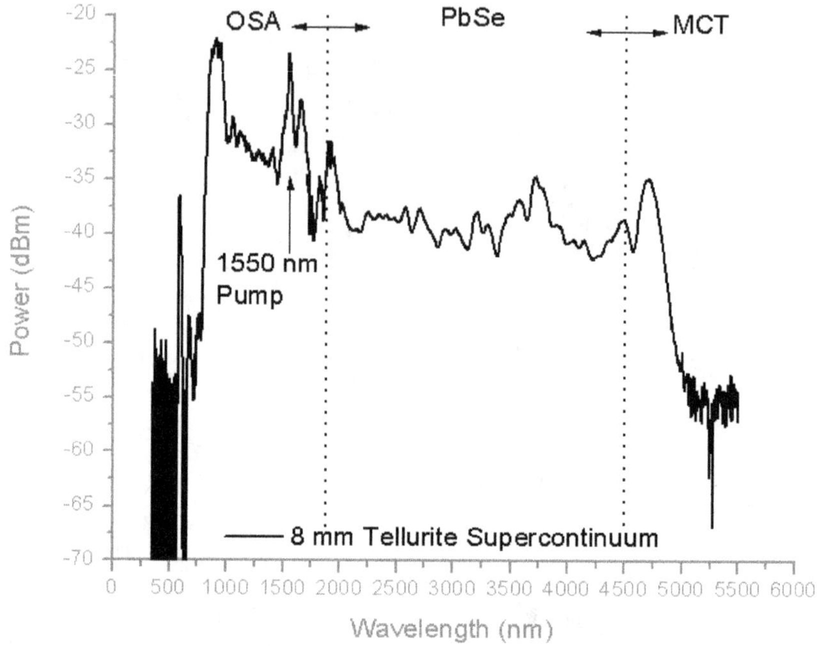

Figure A.7.5: Spectrum of SCG in an 8 mm long tellurite microstructured fiber [20]. The different detector types are noted on top. Notes: OSA: optical spectrum analyzer, PbSe detector, MCT — mercury cadmium telluride.

dynamics affect the spectral characteristics of the supercontinuum radiation generated in this fiber [20]. For the tapered fiber described above a spectrum of width ~4000 nm (shown in Figure A.8.4(c)) is predicted with ~1/3 input pulse energy (~0.6 nJ) than that for tellurite fiber. For some applications, it is important to understand the coherence properties of the generated SC. To study the coherence properties of the generated SC, the complex degree of first-order coherence function introduced by Dudley and Coen [21, 22] is used:

$$g_{12}(\lambda) = \frac{\langle E_1^*(\lambda)E_2(\lambda)\rangle}{[\langle|E_1(\lambda)|^2\rangle\langle|E_2(\lambda)|^2\rangle]^{1/2}} \tag{A.7.2}$$

where the angle brackets denote averaging over the independently generated SC pairs $[E_1(\lambda), E_2(\lambda)]$. The spectral coherence degree of the SC pulses is quantitatively measured by the modulus of g_{12}, which

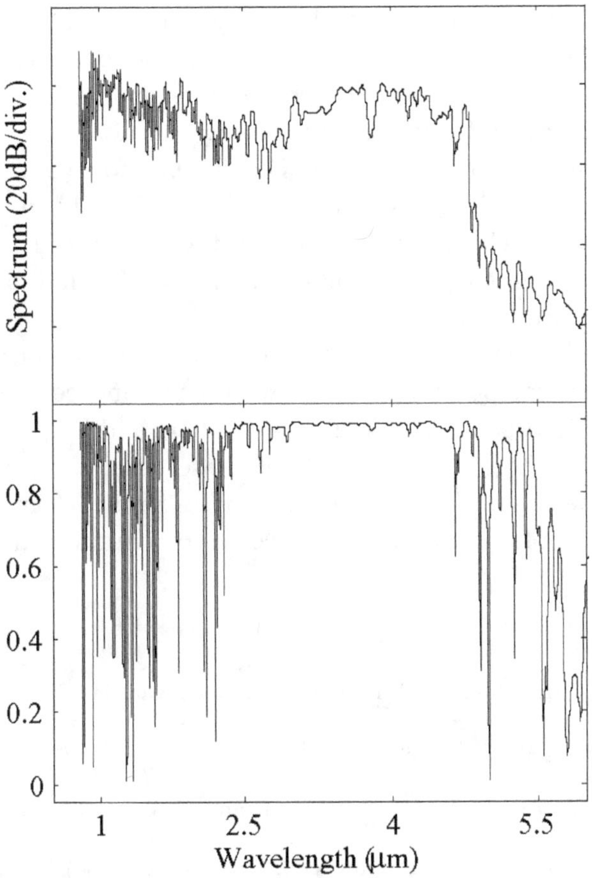

Figure A.7.6: Calculated spectrum (top) and degree of coherence (bottom) for 200 fs pulses at 6 kW peak power [23].

can be approximated as [21, 22]:

$$|g_{12}(\lambda)| = \frac{1}{N^2 - N} \frac{N^2 |\langle E(\lambda)\rangle|^2 - N\langle |E(\lambda)|^2\rangle}{\langle |E(\lambda)|^2\rangle} \qquad \text{(A.7.3)}$$

where N is the total number of SC realizations. It is useful to introduce the overall degree of coherence across the SC spectrum as [21, 22]

$$\langle |g_{12}|\rangle = \frac{\int |g_{12}(\lambda)||E(\lambda)|^2 d\lambda}{\int |E(\lambda)|^2 d\lambda} \qquad \text{(A.7.4)}$$

Both $|g_{12}(\lambda)|$ and $<|g_{12}|>$ lie in the interval [0, 1], a value close to 1 indicates more coherence. The degree of coherence of SC generated in the fiber is simulated by adding quantum noise seeds in the input field, and numerically solving the nonlinear Schrodinger equation applying the split-step Fourier method (SSFM) [15, 23]. The random noise in every temporal discrete bin of the input field is modeled as a stochastic variation with the standard deviation equal to the square root of the number of photons in that bin. The calculated spectrum and degree of coherence $|g_{12}(\lambda)|$ as a function of wavelength is shown in Figure A.7.6. A high degree of coherence is obtained for peak powers and pulse widths within a certain range [23].

Nonlinear processes in single mode fibers and fiber amplifiers are important for the generation of Raman lasers and coherent supercontinuum generation. These types of sources provide new laser wavelengths for applications in spectroscopy, tomography, and sensors.

References

1. G. Genty, M. Lehtonen, H. Ludvigsen, and M. Kaivola, *Opt. Exp.*, **12** (2004) 3471.
2. T. A. Birks, W. J. Wadsworth, and P. St. J. Russell, *Opt. Lett.*, **25** (2000) 1415.
3. T. Hori, J. Takayanagi, N. Nishizawa, and T. Goto, *Opt. Exp.*, **12** (2004) 3317.
4. C. Kakkar and K. Thyagarajan, *Opt. Express*, **14** (2006) 10292.
5. T. Okuno, M. Onishi, T. Kashiwada, S. Ishikawa, and M. Nishimura, *IEEE J. Sel. Top. Quant. Electron.*, **5** (1999) 1385.
6. K. Mori, H. Takara, S. Kawanishi, M. Saruwatari and T. Morioka, *Electron. Lett.*, **33** (1997) 1806.
7. K. Mori, H. Takara, and S. Kawanishi, *J. Opt. Soc. Am. B*, **18** (2001) 1780.
8. T. Schreiber, T. V. Andersen, D. Schimpf, J. Limpert, and A. Tunnermann, *Opt. Exp.*, **13** (2005) 9556.
9. M. H. Frosz, P. Falk, and O. Bang, *Opt. Exp.*, **13** (2005) 6181.
10. A. Mussot, M. Beaugeois, M. Bouazaoui, and T. Sylvestre, *Opt. Exp.*, **15** (2007) 11553.
11. J. H. V. Price, T. M. Monro, H. Ebendorff-Heidepriem, F. Poletti, P. Horak, V. Finazzi, J. Y. Y. Leong, P. Petropoulos, J. C. Flanagan, G. Brambilla, X. Feng, and D. J. Richardson, *IEEE J. Quant. Electron.*, **13** (2007) 738.
12. Zhigang Chen, Antoinette J. Taylor, and Anatoly Efimov, *Opt. Exp.*, **17** (2009) 5852.

13. V. Skryabin, F. Luan, J. C. Knight, and P. St. J. Russell, *Science*, **301** (2003) 1705.
14. P. Petropoulos, T. M. Monro, H. Ebendorff-Heidepriem, K. Frampton, R. C. Moore, and D. J. Richardson, *Opt. Exp.*, **11** (2003) 3568.
15. Z. Chen, S. Ma, and N. K. Dutta, *Opt. Commun.*, **283** (2010) 3076.
16. S. Fujino, H. Ijiri, F. Shimizu, and K. Morinaga, *J. Jpn. Inst. Met.*, **62** (1998) 106.
17. J. Y. Y. Leong, P. Petropoulos, J. H. V. Price, H. Ebendorff Heidepriem, S. Asimakis, R. C. Moore, K. E. Frampton, V. Finazzi, X. Feng, T. M. Monro, and D. J. Richardson, *J. Lightwave. Technol.*, **24** (2006) 183.
18. V. L. Kalashnikov, E. Sorokin, I. T. Sorokina, *Appl. Phys. B.*, **87** (2007) 37.
19. C. Xia, M. Kumar, O. P. Kulkarni, M. N. Islam, F. L. Terry Jr., M. J. Freeman, M. Poulain, and G. Mazé, *Opt. Lett.*, **31** (2006) 2553.
20. P. Domachuk, N. A. Wolchover, M. Cronin-Golomb, A. Wang, A. K. George, C. M. B. Cordeiro, J. C. Knight, and F. G. Omenetto, *Opt. Exp.*, **16** (2008) 7161.
21. J. M. Dudley, and S. Coen, *Opt. Lett.*, **27** (2002) 1180.
22. J. M. Dudley, and S. Coen, *IEEE J. Sel. Top. Quant. Electron.*, **8** (2002) 651.
23. H. Hu, W. Li, and N. K. Dutta, *Fiber and Integrated Opt.*, **32** (2019) 209.

Appendix 8

Supercontinuum Generation in Planar Waveguides

Supercontinuum generation (SCG) is the spectral broadening of narrow-band incident pulses by the propagation through nonlinear media. The physical mechanism of supercontinuum (SC) is demonstrated to be a combination of linear and nonlinear processes including self-phase modulation (SPM), cross-phase modulation (XPM), stimulated Raman scattering (SRS), four-wave mixing (FWM), soliton fission, and dispersive wave (DW) (Cherenkov) radiation [1–8]. Various types of glasses such as lead silicate, fluoride [9], and chalcogenide glass [10–14] have been introduced as the core material, which benefit from higher nonlinearity and transparency than conventional silica fiber [6–17]. For example, photonic crystal fibers (PCFs) using lead silicate SF57 glass have been fabricated and a nonlinear parameter of $\gamma = 1860\ \mathrm{W}^{-1}\mathrm{km}^{-1}$ was obtained, approximately 2000 times that of a standard single-mode fiber [6, 7].

Supercontinuum generation in planar waveguides has gained much interest recently [17–19]. In Ref. [19], dispersion-engineered highly nonlinear chalcogenide planar waveguides were experimentally demonstrated to have SC with a bandwidth of 750 nm. The schematic of the waveguide is shown in Figure A.8.1. A 0.87 μm layer of As_2S_3 is deposited by thermal evaporation onto a thermally oxidized silicon substrate. The 2 μm wide waveguides are defined

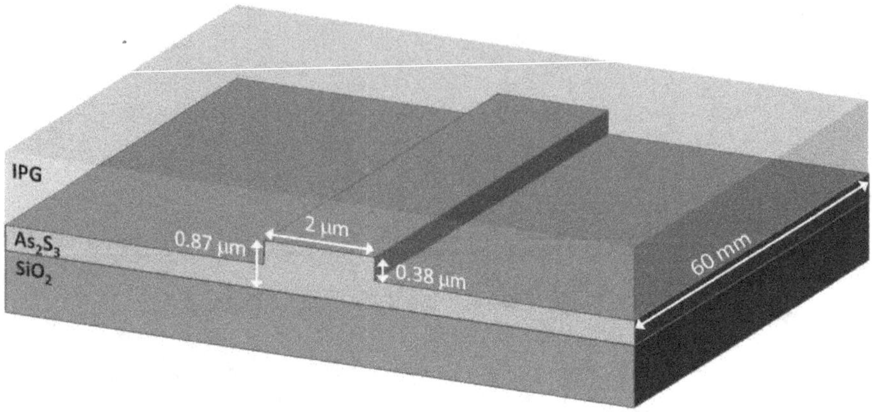

Figure A.8.1: Schematic of the planar waveguide used for supercontinuum generation. The waveguide cross-section is 2 μm × 0.87 μm [19].

using photolithography and created using reactive ion etching with CHF_3 gas. The waveguide chip is then coated in a protective coating layer of inorganic polymer glass (IPG). The waveguides are 6 cm long and the optical loss (α) of the waveguide is ~0.6 dB/cm. The non-linear parameter of the material is 10 $W^{-1}m^{-1}$, and the dispersion is 29 ps/km nm. A mode-locked fiber laser operating at a repetition rate of 10 MHz with ~610 fs wide pulses at 1550 nm is used as the input pulsed source. The maximum peak power is ~136 kW. The input pulses could be coupled into transverse electric (TE) and transverse magnetic (TM) mode of the waveguide using a polarization controller. The measured SC spectrum extends from 1100 to 2000 nm. TE coupled powers produced considerably less broadened spectrum.

A study of tapered planar rib waveguide which allows dispersion management for SCG is now discussed. The Air-SF57-SiO_2 rib waveguide continuously increases its etch depth along the propagation distance, which shifts the second zero dispersion wavelength (ZDW) similar to that described in Appendix 7. By launching hyperbolic secant pulses at 1550 nm into this waveguide, supercontinuum with 30 dB bandwidth of ~1 to ~4.6 μm is obtained, limited by the multiphonon absorption edge of the material.

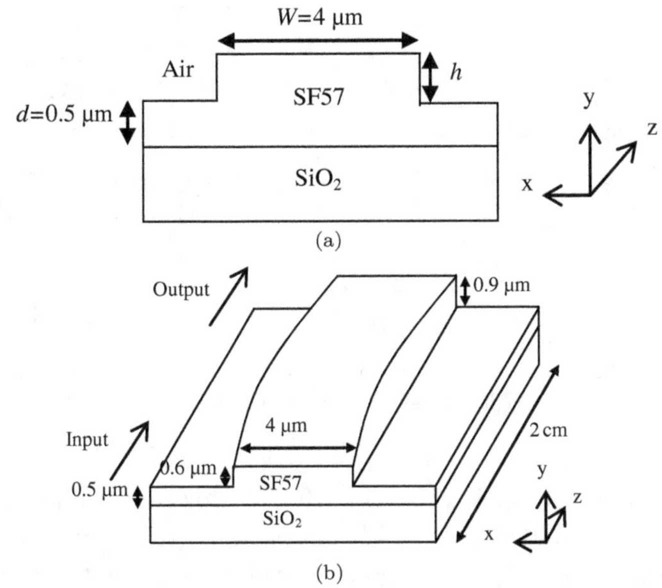

Figure A.8.2: (a) Cross-section of the Air-SF57-SiO_2 planar rib waveguide with the rib width W, outer slab thickness d, and etch depth h, respectively, (b) schematic of a tapered waveguide.

A numerical study of SCG in an Air-SF57-SiO_2 planar rib waveguide is now described. The waveguide schematic (with a rib width W, outer slab thickness d, and etch depth h) is shown in Figure A.8.2. Scott SF57 is a commercially available lead-silicate glass mainly composed of SiO_2 and PbO, with a Pb cationic content of 40–50% [20]. The linear refractive index of SF57 at 1550 nm is $n = 1.8$ [21], which results in an index contrast of 17.6% to the SiO_2 substrate, allowing for tight modal confinement. The nonlinear refractive index of SF57 measured at 1060 nm is $n_2 = 4.1 \times 10^{-19}$ m^2/W, which is much higher than that of pure silica glass (2.7×10^{-20} m^2/W at 1.06 μm) [22]. The material loss of SF57 glass is 1.6 dB/m at 1500 nm [22]. The transmission window of SF57 glass is limited by the multiphonon absorption edge to $\sim$5 μm.

The wavelength-dependent dispersion profiles of the rib waveguide for various h values (etch depth) are shown in Figure A.8.3. W and d are set to be 4 and 0.5 μm, respectively.

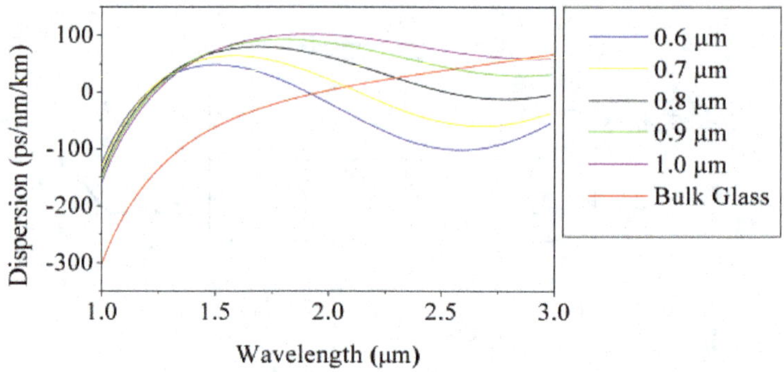

Figure A.8.3: Dispersion profile of SF57 rib waveguide with varying etch depths h. The rib width W and outer slab thickness d are 4 and 0.5 μm, respectively.

The computation is done by a full-vector finite difference mode solver for the fundamental TM mode. Since the waveguide dispersion depends strongly on the geometrical design, the total dispersion characteristic, which is the combined effect of material and waveguide dispersion, can be controlled by varying the etch depth, as shown in Figure A.8.3. Note that for small h (<0.83 μm), a ZDW exists, and it shifts to shorter wavelengths as h decreases. For comparison, the material dispersion of SF57 bulk glass is also plotted, which has one ZDW at 1970 nm.

Based on the dispersion property, a tapered waveguide illustrated in Figure A.8.2 (b) has been designed [28]. The waveguide is 2 cm long, with $W = 4$ μm and $d = 0.5$ μm. The etch depth $h$continuously increases from 0.6 to 0.9 μm along the propagation distance z by the following relationship:

$$h = h_0 \left[1 + \left(\frac{z}{z_0}\right)^{\frac{1}{3}}\right] \tag{A.8.1}$$

where $h_0 = 0.6$ μm and $z_0 = 16$ cm. Hyperbolic secant pulses centered at 1550 nm are launched into the designed waveguide for SCG. The optical field of the input pulse takes the form $P^{1/2}\text{sech}\,(T/T_0)$, where P denotes the peak power and the full width at half maximum (FWHM) is $1.763T_0$. The pulse evolution is numerically simulated by the generalized nonlinear Schrödinger equation (GNLSE) with slowly-varying envelope approximation [Appendices 3 and 7, 23]:

$$\frac{\partial A}{\partial z} + \frac{\alpha}{2}A + \sum_{k\geq 2} \frac{i^{k-1}}{k!}\beta_k \frac{\partial^k A}{\partial T^k}$$
$$= i\gamma\left(1 + \frac{i}{\omega_0}\frac{\partial}{\partial T}\right)\left(A\int_{-\infty}^{+\infty} R(T')|A(z, T - T')|^2 dT'\right) \tag{A.8.2}$$

where $A(z,t)$ is the pulse envelope and $T = t - z/v_g$ is the retarded time frame moving at the pulse group velocity v_g. α is the linear propagation loss and β_k denote the Taylor-series expansion coefficients of the propagation constant $\beta(\omega)$ around the center frequency ω_0. γ is the nonlinear parameter defined as $\gamma(\omega_0) = n_2(\omega_0)\omega_0/(cA_{eff})$, where $A_{eff} = (\iint |E|^2 dxdy)^2/(\iint |E|^4 dxdy)$ is the effective mode area. The nonlinear response function $R(T) = (1 - f_R)\delta(T) + f_R h_R(T)$ includes both the instantaneous electronic response and the delayed Raman contribution, indicated by the delta function and $h_R(T)$, respectively. For SF57, f_R is 0.1 and the Raman response function can be expressed as [24–28]:

$$h_R(t) = \frac{\tau_1^2 + \tau_2^2}{\tau_1\tau_2^2}\exp\left(-\frac{t}{\tau_2}\right)\sin\left(\frac{t}{\tau_1}\right) \tag{A.8.3}$$

in which $\tau_1 = 5.5$ fs and $\tau_2 = 32$ fs. The left side of equation (A.8.2) models the linear propagation effects, while the right side models the nonlinear effects. Equation (A.8.2) is numerically solved by the split-step Fourier method (SSFM) [26, 27].

A.8.1.1 Results and discussions

First, consider uniform waveguide structure with constant etch depth. For this case, the output SC spans from 0.9 μm out to 2.5 μm. In this situation, the spectral components on the long-wavelength side are mainly formed by soliton-self-frequency shift (SSFS), while the emission of dispersive waves (DWs) is responsible for the short-wavelength components.

Figure A.8.4(a) shows the spectrum evolution of the incident pulse (P = 5 kW, FWHM = 100 fs) traveling in the tapered

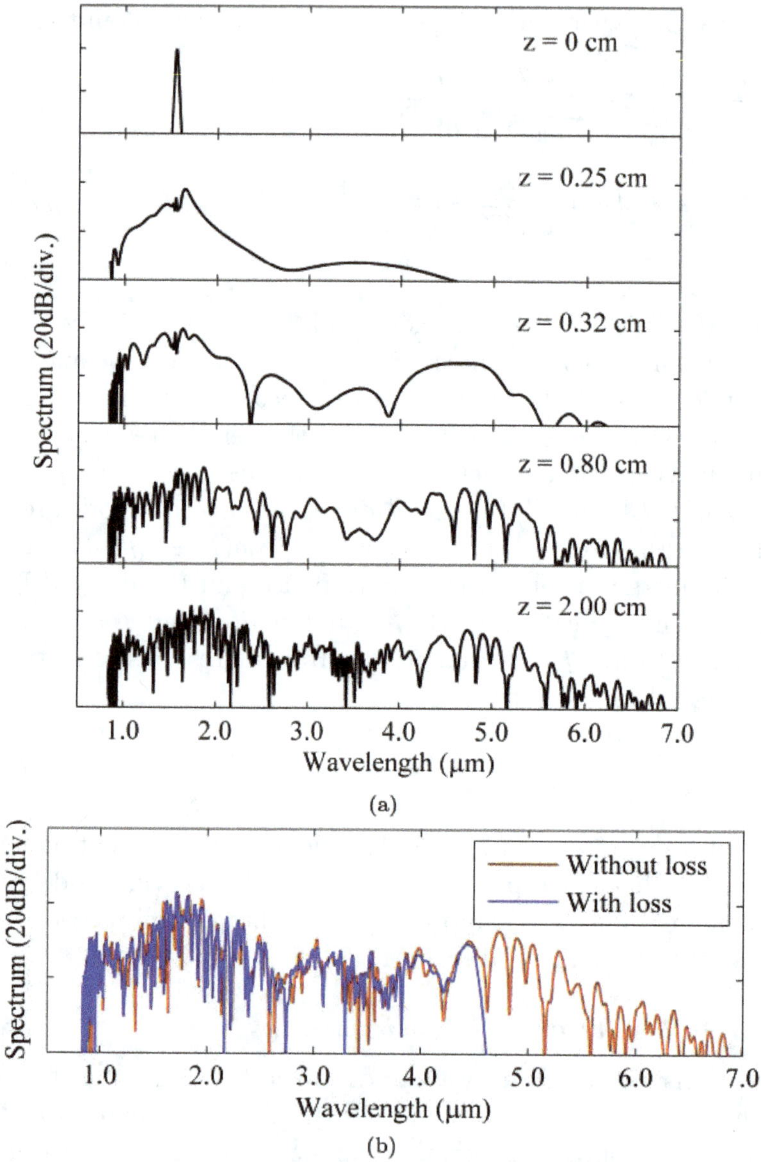

Figure A.8.4: (a) Spectrum evolution along the propagation distance for the tapered SF57 rib waveguide without loss. (b) Effect of loss on the output continuum [28].

waveguide (with etch depth given by A.8.1) if no optical loss is included in the simulation. The effective mode area A_{eff} at the input is equal to 3.2 μm^2, yielding a nonlinear parameter γ of 0.52 /W/m at 1550 nm, about 500 times that of a conventional single-mode fiber. The generated spectrum has a 30 dB bandwidth from ~1 to ~6 μm. Since the continuum extends over the multiphonon edge, the loss should be considered. The spectrum including loss is shown in Figure A.8.4(b). It can be seen that the output spectrum reaches ~4.6 μm, which is considerably larger than that of uniform waveguides [28].

The waveguide is carefully designed so the center wavelength of the incident pulse is within the anomalous dispersion regime. As a result, the input pulses undergo soliton fission and DW emission across the upper ZDW, as described before. For the tapered waveguide with an increasing value of h, the 2nd ZDW shifts along the propagation distance, leading to a continuous modification of the phase-matched wavelength for DW emission. This allows the DW to be produced towards longer wavelengths and this significantly broadens the supercontinuum at the output. Clearly, different tapered structures other than that described by Eq. (A.8.1) may be used, but the tapering rate should be optimized to ensure efficient energy transfer from solitons to DWs.

Planar waveguides, as lasers, amplifiers, or as a broadband sources, promise scalable geometries, low-cost fabrication, and the potential for integrated optical chip solutions.

References

1. J. M. Dudley, G. Genty, and S. Coen, *Rev. Mod. Phys.*, **78** (2006) 1135.
2. J. M. Dudley and J. R. Talor, *Supercontinuum Generation in Optical Fibers*, Cambridge University Press, New York, 2010.
3. S. Coen, A. H. L. Chau, R. Leonhardt, J. D. Harvey, J. C. Knight, W. J. Wadsworth, and P. S .J. Russell, *J. Opt. Soc. Am.*, B **19** (2002) 753.
4. W. J. Wadsworth, A. Ortigosa-Blanch, J. C. Knight, T. A. Birks, T. P. M. Man, and P. S. J. Russell, *J. Opt. Soc. Am.*, B **19** (2002) 2148.
5. T. Yamamoto, H. Kubota, S. Kawanishi, M. Tanaka, and S. Yamaguchi, *Opt. Exp.* **11** (2003) 1537.

6. J. Y. Y. Leong, P. Petropoulos, J. H. V. Price, H. Ebendorff-Heidepriem, S. Asimakis, R. C. Moore, K. E. Frampton, V. Finazzi, X. Feng, T. M. Monro, and D. J. Richardson, *J. Lightwave Technol.*, **24** (2006) 183.
7. Z. Chen, S. Ma, and N. K. Dutta, *Opt. Commun.* **283** (2010) 3076.
8. J .H .V. Price, X. Feng, A. M. Heidt, G. Brambilla, P. Horak, F. Poletti, G. Ponzo, P. Petropoulos, M. Petrovich, J. Shi, M. Ibsen, W. H. Loh, H. N. Rutt, and D. J. Richardson, *Opt. Fiber Technol.* **18** (2012) 327.
9. Z. Chen, A. J. Taylor, and A. Efimov, *Opt. Exp.*, **17** (2009) 5852.
10. D. I. Yeom, E. C. Mägi, M. R. Lamont, M. A. Roelens, L. Fu, and B. J. Eggleton, *Opt. Lett.*, **33**, (2008) 660.
11. J. Hu, C. R. Menyuk, L. B. Shaw, J. S. Sanghera, and I. D. Aggarwal, *Opt. Exp.*, **18** (2010) 6722.
12. N. Granzow, S. P. Stark, M. A. Schmidt, A .S. Tverjanovich, L. Wondraczek, and P. S. J. Russell, *Opt. Exp.*, **19** (2011) 21003.
13. R. R. Gattass, L. Brandon Shaw, V. Q. Nguyen, P. C. Pureza, I. D. Aggarwal, and J. S. Sanghera, *Opt. Fiber Technol.*, **18** (2012) 345.
14. A. Marandi, C.W. Rudy, V. G. Plotnichenko, E. M. Dianov, K. L. Vodopyanov, and R. L. Byer, *Opt. Exp.*, **20** (2012) 24218.
15. A.V. Husakou and J. Herrmann, *Appl. Phys. B*, **77** (2003) 227.
16. K. M. Hilligsøe, T. Andersen, H. Paulsen, C. Nielsen, K. Mølmer, S. Keiding, R. Kristiansen, K. Hansen, and J. Larsen, *Opt. Exp.*, **12** (2004) 1045.
17. O. Fedotova, A. Husakou, and J. Herrmann, *Opt. Exp.*, **14** (2006) 1512.
18. L. Yin, Q. Lin, and G .P. Agrawal, *Opt. Lett.*, **32**, (2007) 391.
19. M. R. Lamont, B. Luther-Davies, D.Y. Choi, S. Madden, and B. J. Eggleton, *Opt. Exp.*, **16** (2008) 14938.
20. G. Genty, M. Surakka, J. Turunen, and A. T. Friberg, *Opt. Lett.*, **35,** (2010) 3057.
21. V. Diez-Blanco and J. Siegel, *J. Solis, Appl. Surf. Sci.*, **252**, (2006) 4523.
22. Schott Glass Catalogue, 2003.
23. G. P. Agrawal, *Nonlinear Fiber Optics*, 4th Edn., Academic Press, Burlington, MA, London, 2007.
24. W. Bolaños, J. J. Carvajal, X. Mateos, G. S. Murugan, A. Subramanian, J. S. Wilkinson, E. Cantelar, G. Lifante, M. Aguiló, and F. Díaz, *Opt. Mater.*, **34** (2011) 475.
25. V. L. Kalashnikov, E. Sorokin, and I. T. Sorokina, *Appl. Phys. B*, **87** (2007) 37.
26. J. M. Dudley and S. Coen, *Opt. Lett.*, **27**, (2002) 1180.
27. J. M. Dudley and S. Coen, *IEEE J. Sel. Top. Quant. Electron.* **8** (2002) 651.
28. H. Hu, W. Li, and N. K. Dutta, *Appz. Opt.*, **52** (2013) 7336.

Appendix 9

Theory of Harmonic and Rational Harmonic Mode-Locking

In Chapter 4, a simple explanation of mode locking is given. This Appendix develops the theory of harmonic and rational harmonic mode-locking of fiber lasers discussed in Chapter 9 (see also Refs. [1–11]). The analysis below is similar to that of Refs. [10–11].

A.9.1 Harmonic mode-locking

The envelope of mode-locked pulse $a(t)$ consists of its Fourier components a_q at frequency ω_q. There are various sources of losses in the fiber lasers, they are lumped into one single parameter l here, so that after one round trip, the qth mode amplitude (a_q) is multiplied by $1 - l$ as it circulates once through the fiber laser. Similarly it is assumed that due to gain, the qth mode amplitude (a_q) is multiplied by $G(q) = 1 + g(q)$ as it circulates once through the fiber laser where $g(q)$ is the gain for the qth Fourier component at frequency ω_q. Assuming a Lorentzian spectral profile for the gain, and $\omega_q = \omega_0 + q\omega_c$, where ω_0 is the center frequency,

$$G(q) = 1 + \frac{g}{1 + \frac{(q\omega_c)^2}{\Omega_g^2}} = 1 + g\left[1 - \frac{(q\omega_c)^2}{\Omega_g^2}\right] \tag{A.9.1}$$

where Ω_g is the gain bandwidth and g is the gain at the center frequency. The modulator is assumed to pass the light at periodically timed sequences so that the Fourier component $a_q \exp(i(\omega_0 + q\omega_c)t)$ becomes after one pass (using $\omega_m = n\omega_c$)

$$a_q \exp(i(\omega_0 + q\omega_c)t)M(1 - \cos(\omega_m t))$$
$$= -\left(\frac{M}{2}\right) a_q \exp(i\omega_0 t)[e^{i(q-n)\omega_c t} - 2e^{iqn\omega_c t} + e^{i(q+n)\omega_c t}] \tag{A.9.2}$$

In the above, the modulation factor is given by $M(1 - \cos(\omega_m t))$ at each pass of the components, where M is modulation depth. So each Fourier component acquires sidebands that are injected into the nth adjacent modes.

Considering the above three elements, a recursion equation is obtained for the Fourier component $a_q^{(k+1)}(\omega)$ traveling around the cavity the $(k + 1)$th time in terms of its value $a_q^{(k)}$ upon the kth time

$$a_q^{(k+1)} = a_q^{(k)} + g\left[1 - \frac{(q\omega_c)^2}{\Omega_g^2}\right] a_q^{(k)} - la_q^{(k)}$$
$$+ \frac{M}{2}(a_{q+n}^{(k)} - 2a_q^{(k)} + a_{q-n}^{(k)}) \tag{A.9.3}$$

The equation (A.9.3) is the basis of discussing harmonic mode-locking. If q′ is defined $q' = q/n$, then Eq. (A.10.3) becomes

$$a_{q'}^{(k+1)} = a_{q'}^{(k)} + g\left[1 - \frac{(nq'\omega_c)^2}{\Omega_g^2}\right] a_{q'}^{(k)} - la_{q'}^{(k)}$$
$$+ \frac{M}{2}(a_{q'+1}^{(k)} - 2a_{q'}^{(k)} + a_{q^1-1}^{(k)}) \tag{A.9.4}$$

Equation (A.9.4) shows that there are n pulses traveling around the cavity. In steady state, $a_{q'}^{(k+1)} = a_{q'}^{(k)}$, the superscript (k) can be dropped, and Eqs. (A.9.3) and (A.9.4) reduce to a second-order difference equation. Moreover, when the number of oscillating modes is large, it is useful to replace the second order difference by a second order derivative, so Eq. (A.9.4) is written as

$$\left[g\left(1-\frac{\omega^2}{\Omega_g^2}\right)-\ell+\frac{M\omega_m^2}{2}\frac{d^2}{d\omega^2}\right]a(\omega)=0 \tag{A.9.5}$$

where a_q has been replaced by the function a(ω) of the continuous frequency variable $\omega = q'\omega_c$. The solution of Eq. (A.9.5) is the Hermite Gaussian function and is given by

$$a(\omega)=H_v(\omega\tau)\exp\left(\frac{-\omega^2\tau^2}{2}\right) \tag{A.9.6}$$

where

$$\tau=\left(\frac{2g}{M}\right)^{1/4}\left(\frac{1}{\omega_m\Omega_g}\right)^{1/2} \quad \text{and} \quad g-l=M\omega_m^2\tau^2\left(\nu+\frac{1}{2}\right) \tag{A.9.7}$$

In general, the 0th$^-$ order ($\nu = 0$) solution dominates. The following expressions for $a(\omega)$ and $a(t)$ follows:

$$a(\omega)=A\exp\left(\frac{-\omega^2\tau^2}{2}\right)$$

$$a(t)=\frac{\sqrt{2\pi}}{\tau}A\exp\left(\frac{-t^2}{2\tau^2}\right) \tag{A.9.8}$$

So for harmonic mode-locking case $\omega_m = n\omega_c$, the solution is a Gaussian pulse, where the pulse width is proportional to τ, and hence to $\omega_m^{-1/2}$. This is in agreement with the measurement (Chapter 9). Furthermore, from Eq. (A.9.7), lower pulse width is obtained by increasing the modulation frequency and modulation power while reducing the pump power (which reduces gain) for the cavity of the fiber laser.

A.9.2 Rational harmonic mode-locking

The technique of rational harmonic mode locking has attracted considerable interest, where an optical pulse train with a repetition rate of $(np+1)f_c$ can be produced when the modulation frequency is set at $(n+1/p)f_c$, where f_c is the longitudinal mode spacing or fundamental cavity frequency of the fiber laser ring, and n, p are integers. The theoretical analysis of the rational harmonic mode locking follows the analysis of harmonic mode locking. For the rational harmonic mode-locking case, the modulation frequency is $\omega_m = (n + 1/p)\omega_c$. So, the modulation factor is given by $M\{1 - \cos[(n+1/p)t]\}$ and the recursion equation (A.9.3) becomes

$$a_q^{(k+1)} = a_q^{(k)} + g\left[1 - \frac{(q\omega_c)^2}{\Omega_g^2}\right] a_q^{(k)} - la_q^{(k)} + \frac{M}{2}\left(a^{(k)}_{q+n+\frac{1}{p}} - 2a_q^{(k)} + a^{(k)}_{q-n-\frac{1}{p}}\right) \quad \text{(A.9.9)}$$

For this case as $(n+ 1/p)$ is not an integer, the modes in the cavity do not couple. But if a cavity of equivalent length pL is considered, then the modes can couple. This means that the pulse has to propagate p times through the cavity before it can get enough gain. Following the steps in the previous section, the above equations can be rewritten as

$$a_q^{(k+1)} = a_q^{(k)} + g\left[1 - \frac{\left(\frac{q}{p}\omega_c\right)^2}{\Omega_g^2}\right] a_q^{(k)} - la_q^{(k)} + \frac{M}{2}(a^{(k)}_{qp+np+1} - 2a_q^{(k)} + a^{(k)}_{qp-np-1}) \quad \text{(A.9.10)}$$

where $\omega_c = 2\pi f_c$. The quantity q' is redefined as

$$q' = \frac{q}{np+1}$$

Then, Eq. (A.9.10) is written as

$$a_{q'}^{(k+1)} = a_{q'}^{(k)} + g\left[1 - \frac{\left[\left(n+\frac{1}{p}\right)q'\omega_c\right]^2}{\Omega_g^2}\right] a_{q'}^{(k)} - la_{q'}^{(k)} + \frac{M}{2}(a^{(k)}_{q'+1} - 2a_{q'}^{(k)} + a^{(k)}_{q'-1}) \quad \text{(A.9.11)}$$

Equation (A.9.11) is similar to Eq. (A.9.4). Following the same process as in the previous section, it is possible to conclude that the solution for rational harmonic mode locking is also a Gaussian pulse given by Eq. (A.9.8) and the pulse width τ is expressed as

$$\tau = \left(\frac{2g}{M}\right)^{1/4}\left(\frac{1}{\left(n+\frac{1}{p}\right)\omega_c\Omega_g}\right)^{1/2} \quad \text{(A.9.12)}$$

The pulse width parameter τ is proportional to $(n + 1/p)^{-1/2}$. Since n is much larger than 1 for the modulation frequencies used here, it is reasonable to conclude that the pulse width for rational harmonic mode-locking is almost same as that for harmonic mode locking.

References

1. Z. Ahmed and N. Onodera, *Electron. Lett.*, **32** (1996) 455.
2. M. Y. Jeon, H. K. Lee, J. T. Ahn, D. S. Lim, H. Y. Kim, K. H. Kim, and E.-H. Lee, *Electron. Lett.,* **34** (1998) 182.
3. S. Li, C. Lou, and K. T. Chan, *Electron. Lett.*, **34** (1998) 375.
4. P. Das, W. Kaechele, J. P. Theimer, and A. R. Pirich, *SPIE*, **3075** (1997) 21.
5. C. Wu and N. K. Dutta, *IEEE J. Quant. Electron.*, **36** (2000) 145.
6. T. Pfeiffer and G. Veith, *,Electron. Lett.*, **29** (1993) 1849.
7. M. Nakazawa, E. Yoshida, and Y. Kimura, *Electron. Lett.*, **30** (1994) 1603.
8. E. Yoshida and M. Nakazawa, *Electron. Lett.*, 1996, **32** (1996) 1370.
10. H. A. Haus, Short pulse generation in I. N. Duling (ed.), *Compact Sources of Ultrashort Pulses* 3rd Edn., Cambridge, 1995.
11. N. K. Dutta, *Fiber Amplifies and Fiber Lasers,* World Scientific, Singapore, 2015.

Appendix 10

Reflectivity of Periodic Structure

A medium whose properties are constant throughout each plane perpendicular to a fixed direction is called a stratified medium. A stratified periodic medium with period h is characterized by a dielectric constant ε and magnetic permeability μ which are functions of z (this fixed direction) only and are such that

$$\begin{aligned} \varepsilon(z + jh) &= \varepsilon(z) \\ \mu(z + jh) &= \mu(z) \end{aligned} \tag{A.10.1}$$

where j is any integer in some fixed range $1 \leq j \leq N$.

We take the plane of incidence to be the yz-plane and z being the direction of stratification. For a TE wave, $E_x = E_z = 0$ and Maxwell's equations are reduced to the following six scalar equations (time dependence $\exp(-i\omega t)$ being assumed):

$$\frac{\partial H_z}{\partial y} - \frac{\partial H_y}{\partial z} + \frac{i\varepsilon\omega}{c} E_x = 0$$

$$\frac{\partial H_x}{\partial z} - \frac{\partial H_z}{\partial x} = 0$$

$$\frac{\partial H_y}{\partial x} - \frac{\partial H_x}{\partial y} = 0$$

$$\frac{i\omega\mu}{c} H_x = 0$$

$$\frac{\partial E_x}{\partial z} - \frac{i\omega\mu}{c} H_y = 0$$

$$\frac{\partial E_x}{\partial y} + \frac{i\omega\mu}{c} H_z = 0 \tag{A.10.2}$$

These equations show that H_y, H_z, and E_x are functions of y and z only. Eliminating H_y and H_z, it follows that

$$\frac{\partial^2 E_x}{\partial y^2} + \frac{\partial^2 E_x}{\partial z^2} + n^2 k_0^2 E_x = \frac{d(\log \mu)}{dz} \frac{\partial E_x}{\partial z} \tag{A.10.3}$$

where $n^2 = \varepsilon\mu$, $k_0 = \omega/c = 2\pi/\lambda_0$. To solve Eq. (A.10.3), we take as a trial solution, a product of two functions, one involving y only and the other involving z only

$$E_x(y, z) = Y(y)U(z) \tag{A.10.4}$$

Equation (A.10.3) becomes

$$\frac{1}{Y}\frac{d^2Y}{dy^2} = -\frac{1}{U}\frac{d^2U}{dz^2} - n^2k_0^2 + \frac{d(\log \mu)}{dz}\frac{1}{U}\frac{dU}{dz} \tag{A.10.5}$$

Equation (A.10.5) can only hold if each side is equal to a constant ($-K^2$ say):

$$\frac{1}{Y}\frac{d^2Y}{dy^2} = -K^2 \tag{A.10.6}$$

$$\frac{d^2U}{dz^2} - \frac{d(\log \mu)}{dz}\frac{dU}{dz} + n^2k_0^2 U = K^2 U \tag{A.10.7}$$

It will be convenient to set

$$K^2 = k_0^2 \alpha^2 \tag{A.10.8}$$

Then, Eq. (A.10.6) gives

$$Y = const.\ e^{ik_0\alpha y} \tag{A.10.9}$$

and consequently, E_x is of the form

$$E_x = U(z)e^{i(k_0\alpha y - \omega t)} \tag{A.10.10}$$

From Eq. (A.10.2), we see that H_y and H_z are given by expressions of the same form

$$H_y = V(z)e^{i(k_0\alpha y - \omega t)} \tag{A.10.11}$$

$$H_z = W(z)e^{i(k_0\alpha y - \omega t)} \tag{A.10.12}$$

According to Eq. (A.10.2), the amplitude functions U, V, and W are related by

$$V' = ik_0[\alpha W + \varepsilon U] \tag{A.10.13}$$

$$U' = ik_0\mu V \tag{A.10.14}$$

$$\alpha U + \mu W = 0 \tag{A.10.15}$$

The prime denotes differentiation with respect to z. Eliminating W from the three equations above, we get a pair of simultaneous first-order differential equations for U and V

$$\begin{aligned} U' &= ik_0\mu V \\ V' &= ik_0\left(\varepsilon - \frac{\alpha^2}{\mu}\right)U \end{aligned} \tag{A.10.16}$$

Elimination between these equations finally gives the following second-order linear differential equations for U and V

$$\frac{d^2U}{dz^2} - \frac{d(\log\mu)}{dz}\frac{dU}{dz} + k_0^2(n^2 - \alpha^2)U = 0 \tag{A.10.17}$$

$$\frac{d^2V}{dz^2} - \frac{d\left[\log(\varepsilon - \frac{\alpha^2}{\mu})\right]}{dz}\frac{dV}{dz} + k_0^2(n^2 - \alpha^2)V = 0 \tag{A.10.18}$$

Since the functions of U and V each satisfy a second-order linear differential equation, it follows that U and V may each be expressed as a linear combination of two particular solutions, say U_1, U_2 and V_1, V_2. These particular solutions cannot be arbitrary; they must be coupled by the first-order differential equation (A.10.16)

$$\begin{aligned}
U_1' &= ik_0\mu V_1 \\
V_1' &= ik_0\left(\varepsilon - \frac{\alpha^2}{\mu}\right)U_1 \\
U_2' &= ik_0\mu V_2 \\
V_2' &= ik_0\left(\varepsilon - \frac{\alpha^2}{\mu}\right)U_2
\end{aligned} \tag{A.10.19}$$

From these equations, it follows that

$$\begin{aligned}
V_1U_2' - U_1'V_2 &= 0 \\
U_1V_2' - V_1'U_2 &= 0
\end{aligned} \tag{A.10.20}$$

So that

$$\frac{d}{dz}(U_1V_2 - U_2V_1) = 0 \tag{A.10.21}$$

This relation implies that the determinant

$$D = \begin{vmatrix} U_1 & V_1 \\ U_2 & V_2 \end{vmatrix} \tag{A.10.22}$$

associated with any two arbitrary solutions of Eq. (A.10.16) is a constant. For our purpose the most convenient choice of the particular solution is

$$\begin{aligned}
U_1 &= f(z) \\
V_1 &= g(z) \\
U_2 &= F(z) \\
V_2 &= G(z)
\end{aligned} \tag{A.10.23}$$

such that $f(0) = G(0) = 0$ and $F(0) = g(0) = 1$. Then, the solutions with $U(0) = U_0$, $V(0) = V_0$ may be expressed in the matrix notation

$$\begin{bmatrix} U(z) \\ V(z) \end{bmatrix} = \begin{bmatrix} F(z) & f(z) \\ G(z) & g(z) \end{bmatrix} \begin{bmatrix} U_0 \\ V_0 \end{bmatrix} \tag{A.10.24}$$

so that $Fg - fG = 1$. It is usually more convenient to express U_0 and V_0 as functions of $\underline{U}(z)$ and $V(z)$. Solving for U_0 and V_0 we obtain the inverse matrix

$$\mathbf{M} = \begin{bmatrix} g(z) & -f(z) \\ -G(z) & F(z) \end{bmatrix} \tag{A.10.25}$$

This matrix is also unimodular, $|\mathbf{M}| = 1$. This matrix $\mathbf{M}$ relates the x and y components of the electric (or magnetic) vectors in the plane $z = 0$, to the components in an arbitrary plane $z =$ constant. Now we saw that the knowledge of U and V is sufficient for the complete specification of the field. Hence, for the purposes of determining the propagation of a plane monochromatic wave through a stratified medium, the medium only need be specified by an appropriate two by two unimodular matrix $\mathbf{M}$. For this reason, we shall call $\mathbf{M}$ the characteristic matrix of the stratified medium.

In a homogeneous dielectric film, ε, μ, and n are constants. If θ denotes the angle which the normal to the wave makes with the z-axis, we have $\alpha = n \sin\theta$. For a TE wave, according to Eqs. (A.10.7) and (A.10.8)

$$\frac{d^2U}{dz^2} + (k_0^2 n^2 \cos^2\theta)U = 0$$
$$\frac{d^2V}{dz^2} + (k_0^2 n^2 \cos^2\theta)V = 0 \tag{A.10.26}$$

The solutions of these equations, subject to the relations (A.10.16), are easily seen to be

$$U(z) = A\cos(k_0 nz\cos\theta) + B\sin(k_0 nz\cos\theta)$$
$$V(z) = \frac{1}{i}\sqrt{\frac{\varepsilon}{\mu}}\cos\theta\{B\cos(k_0 nz\cos\theta) - A\sin(k_0 nz\cos\theta)\} \tag{A.10.27}$$

Hence, the particular solutions (A.10.23) which satisfy the boundary conditions are

$$
\begin{aligned}
U_1 &= f(z) = \frac{i}{\cos\theta}\sqrt{\frac{\mu}{\varepsilon}}\sin(k_0 nz\cos\theta) \\
V_1 &= g(z) = \cos(k_0 nz\cos\theta) \\
U_2 &= F(z) = \cos(k_0 nz\cos\theta) \\
V_2 &= G(z) = i\sqrt{\frac{\mu}{\varepsilon}}\cos\theta\sin(k_0 nz\cos\theta)
\end{aligned} \tag{A.10.28}
$$

If we set

$$
\begin{aligned}
\beta &= k_0 nz\cos\theta \\
p &= \sqrt{\frac{\varepsilon}{\mu}}\cos\theta
\end{aligned} \tag{A.10.29}
$$

the characteristic matrix is seen to be

$$
\mathbf{M} = \begin{bmatrix} \cos\beta & -\dfrac{i}{p}\sin\beta \\ -ip\sin\beta & \cos\beta \end{bmatrix} \tag{A.10.30}
$$

Let $\mathbf{M}(h)$ be the characteristic matrix corresponding to one period

$$
\mathbf{M}(h) = \begin{bmatrix} m_{11} & m_{12} \\ m_{21} & m_{22} \end{bmatrix} \tag{A.10.31}
$$

Then,

$$
M(Nh) = M(h)\cdot M(h)\cdots M(h) = (M(h))^N \tag{A.10.32}
$$

A multilayer usually consists of a succession of homogeneous layers of alternately low and high refractive indices n_2 and n_3 of thickness h_2 and h_3, respectively, placed between two homogeneous media of refractive indices n_1 and n_l. To simplify, consider the incidence is normal ($\theta_2 = \theta_3 = \pi/2$, Figure A.10.1). Of particular interest is the case when the two basic layers are of the same optical thickness ($n_2h_2 = n_3h_3 = \lambda_0/4$). Assume the media to be non-magnetic

($\mu = 1$), then,

$$\beta_2 = \frac{2\pi}{\lambda} n_2 h_2 \cos\theta_2 = \frac{\pi}{2}$$
$$\beta_3 = \frac{2\pi}{\lambda} n_3 h_3 \cos\theta_3 = \frac{\pi}{2}$$
$$p_2 = n_2 \cos\theta_2 = n_2$$
$$p_3 = n_3 \cos\theta_3 = n_3 \tag{A.10.33}$$

Then, the characteristic matrix $M_2(h)$ of one period (two layers) is given by

$$M_2(h) = \begin{bmatrix} \cos\beta_2 & -\frac{i}{p_2}\sin\beta_2 \\ -ip_2 \sin\beta_2 & \cos\beta_2 \end{bmatrix} \begin{bmatrix} \cos\beta_3 & -\frac{i}{p_3}\sin\beta_3 \\ -ip_3 \sin\beta_3 & \cos\beta_3 \end{bmatrix} \tag{A.10.34}$$

$$M_2(h) = \begin{bmatrix} 0 & -\frac{i}{n_2} \\ -in_2 & 0 \end{bmatrix} \begin{bmatrix} 0 & -\frac{i}{n_3} \\ -in_3 & 0 \end{bmatrix} = \begin{bmatrix} -\frac{n_3}{n_2} & 0 \\ 0 & -\frac{n_2}{n_3} \end{bmatrix} \tag{A.10.35}$$

The characteristic matrix of the multilayer whose basic period in such a double layer is, as can be directly verified by multiplying Eq. (A.10.35) N time by itself,

$$M_{2N}(Nh) = \begin{bmatrix} \left(-\frac{n_3}{n_2}\right)^N & 0 \\ 0 & \left(-\frac{n_2}{n_3}\right)^N \end{bmatrix} \tag{A.10.36}$$

Let, A, R, and T denote the amplitudes (possibly complex) of the electric vectors of the incident, reflected, and transmitted waves. Further, let n_1, n_l be the refractive index of the first and the last medium; consider normal incidence, so θ_1 and θ_l are 0. The boundary conditions demand that the tangential components of **E** and **H**

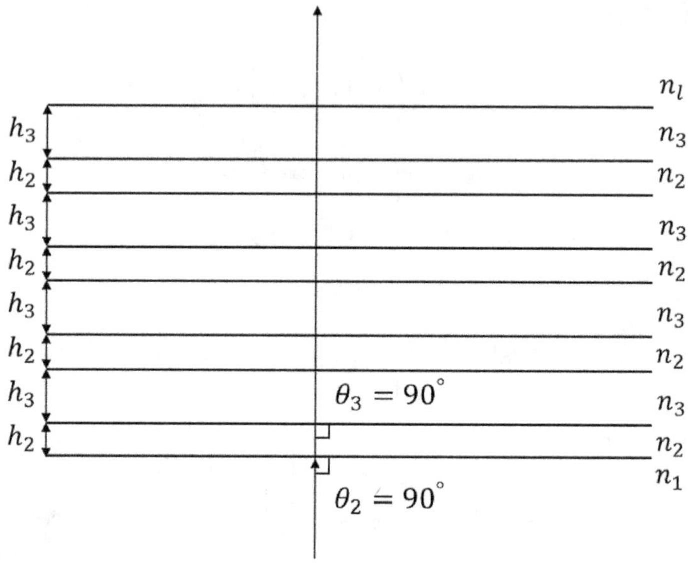

Figure A.10.1: A periodic multilayer structure. The light is incident normal to the layers.

shall be continuous across each of the two boundaries of the stratified medium. Thus, the following relations for a TE wave

$$\begin{aligned} U_0 &= A + R \\ U(z_1) &= T \\ V_0 &= n_1(A - R) \\ V(z_1) &= n_l T \end{aligned} \tag{A.10.37}$$

The four quantities U_0, V_0, U, and V given by Eq. (A.10.37) are given by the basic relation Eq. (A.10.24); hence,

$$\begin{aligned} A + R &= (m_{11} + m_{22} n_l)T \\ n_1(A - R) &= (m_{21} + m_{22} n_l)T \end{aligned} \tag{A.10.38}$$

m_{ij} being the elements of the characteristic matrix of the medium. The reflection coefficient of the film is given by

$$r = \frac{R}{A} \tag{A.10.39}$$

In term of r, the reflectivity is $|r|^2$.

Substituting the characteristic matrix Eq. (A.10.36) into Eq. (A.10.38), and solve for the reflectivity. Then the reflectivity for a fixed number N of the double layers (or layer pairs) is

$$R_{2N} = \left(\frac{1 - \frac{n_l}{n_1}\left(\frac{n_2}{n_3}\right)^{2N}}{1 + \frac{n_l}{n_1}\left(\frac{n_2}{n_3}\right)^{2N}} \right)^2 \tag{A.10.40}$$

Reference

1. M. Born and E. Wolf, *Principles of Optics: Electromagnetic Theory of Propagation, Interference and Diffraction of Light*, Elsevier, Amsterdam, 2013.

Appendix 11

Semiconductor Laser Rate Equations

The rate equations involving the phase of the laser is examined in order to obtain information on chirp, linewidth, phase noise, intensity noise, etc. The wave equation is

$$\nabla^2\mathbf{E} = \frac{1}{c^2}\frac{\partial^2}{\partial t^2}(\epsilon\mathbf{E}) \tag{A.11.1}$$

For the light propagating along $\hat{z}$ (laser cavity direction), polarized along $\hat{x}$ (along the junction),

$$\bar{E}(x, y, z, t) = \frac{1}{2}\hat{x}\varphi(x)\phi(y)\sum_j \sin(k_j z)E_j(t)e^{-i\omega t} + c.c. \tag{A.11.2}$$

where $\varphi(x)$ and $\phi(y)$ are lateral and transverse mode profiles and c.c. is complex conjugate.

$$k_j = \frac{2\pi}{\lambda_j} = \frac{\pi m_j}{\mu L} \tag{A.11.3}$$

where λ_j is the jth longitudinal mode that satisfies $2\mu L = m_j\lambda_j$, $m_j = 0, 1, 2, \ldots$; k_j are discrete values of wavenumber, and μ is the refractive index. For simplicity, we consider one longitudinal mode and drop the subscript j,

$$E(x, y, z, t) = \frac{1}{2}\hat{x}\varphi(x)\phi(y)\sin(kz)E(t)e^{-i\omega t} + c.c. \tag{A.11.4}$$

Substitute Eq. (A.11.4) in (A.11.1) and integrating over x, y we get

$$\frac{2i\omega}{c^2}\left(\langle\epsilon\rangle + \frac{\omega}{2}\frac{\partial\langle\epsilon\rangle}{\partial\omega}\right)\frac{dE}{dt} + \left(\frac{\omega^2}{c^2}\langle\epsilon\rangle - k^2\right)E = 0 \qquad (A.11.5)$$

where $\langle\epsilon\rangle = \int_{-\infty}^{+\infty}\int_{-\infty}^{+\infty}\epsilon(x,y)\varphi^2(x)\phi^2(y)dxdy$ is the average dielectric constant,

$$\begin{aligned}\langle\epsilon\rangle &= (\bar{\mu} + \Gamma\Delta\mu)^2\\ &= \bar{\mu}^2 + 2\Gamma\bar{\mu}\Delta\mu \qquad (A.11.6)\\ \Delta\mu &= \Delta\mu_r + i\Delta\mu_i \qquad (A.11.7)\end{aligned}$$

$\Delta\mu$ is the change in index (both real and imaginary) due to carrier injection and $\bar{\mu}$ is the mode index.

$$\Gamma\Delta\mu_i = \bar{\alpha}/2k_0 \qquad (A.11.8)$$

where $\bar{\alpha}$ is the mode absorption coefficient,

$$\bar{\alpha} = -\Gamma g + \alpha_{int} + \alpha_m \qquad (A.11.9)$$

α_{int} and α_m are scattering and mirror losses. Thus,

$$\langle\epsilon\rangle = \bar{\mu}^2 + 2\Gamma\bar{\mu}\Delta\mu_r + i\bar{\mu}\bar{\alpha}/k_0 \qquad (A.11.10)$$

In Eq. (A.11.5) now we use $\bar{k} = \bar{\mu}\Omega/c$, Ω is specific longitudinal mode frequency, $(\omega^2 - \Omega^2) \approx 2\omega(\omega - \Omega)$ and $\langle\epsilon\rangle \simeq \bar{\mu}^2$ we get

$$\frac{dE}{dt} = \frac{i\bar{\mu}}{\mu_g}(\omega - \Omega)E + \frac{i\omega}{\mu_g}\left(\Gamma\Delta\mu_r + \frac{i\bar{\alpha}}{2k_0}\right) \qquad (A.11.11)$$

$$\langle\epsilon\rangle + \frac{\omega}{2}\frac{\partial\langle\epsilon\rangle}{\partial\omega} = \bar{\mu}\mu_g \qquad (A.11.12)$$

where μ_g is the group index corresponding to mode index $\bar{\mu}$. The electric field E can be represented by

$$E = A\exp(-i\phi) \qquad (A.11.13)$$

A is amplitude and ϕ is phase. Separate into real and imaginary part

$$\frac{dA}{dt} = \frac{1}{2}v_g[\Gamma g - (\alpha_{\text{int}} + \alpha_m)]A \tag{A.11.14}$$

$$\frac{d\phi}{dt} = -\frac{\bar{\mu}}{\mu_g}(\omega - \Omega) - \frac{\omega}{\mu_g}\Gamma\Delta\mu_r \tag{A.11.15}$$

where $v_g = c/\mu_g$ is the group velocity. The amplitude equation above can be written in terms of photon number P defined as

$$P = \frac{\epsilon_0\bar{\mu}\mu_g}{2\hbar\omega}\int |E|^2 dV \tag{A.11.16}$$

Thus, $P \propto A^2$, we get

$$\begin{aligned} \frac{dP}{dt} &= (G - \gamma)P + \beta R_{\text{sp}} \\ G &= \Gamma v_g g \\ \gamma &= v_g(\alpha_m + \alpha_{\text{int}}) = 1/\tau_p \end{aligned} \tag{A.11.17}$$

where τ_p is defined as the photon lifetime, βR_{sp} is the spontaneous emission in the stimulated photon mode, and β is the spontaneous emission factor. The α factor is defined as

$$\begin{aligned} \alpha &= \frac{dn_R/dN}{dn_I/dN} \\ &= -2k_0\left(\frac{\partial n_R/\partial N}{\partial g/\partial N}\right) \end{aligned} \tag{A.11.18}$$

$n = n_R + in_I$, n_R, and n_I are the real and imaginary parts of the index, g is the gain. Thus,

$$\Delta\mu_r = \Delta n_R = -\left(\frac{\alpha}{2k_0}\right)\Delta g \tag{A.11.19}$$

If $\Delta G = \Gamma v_g \Delta g$ is replaced by $G - \gamma$, the following is obtained:

$$\Delta\mu_r = -\left(\frac{\alpha}{2k_0}\right)\frac{(G-\gamma)}{\Gamma v_g} \tag{A.11.20}$$

$$\frac{d\phi}{dt} = -\frac{\bar{\mu}}{\mu_g}(\omega - \Omega) + \frac{1}{2}\alpha(G-\gamma) \tag{A.11.21}$$

Steady-state solution of the above is

$$\omega \simeq \Omega + \frac{\mu_g}{\bar{\mu}}\frac{1}{2}\alpha(G-\gamma) \tag{A.11.22}$$

Appendix 12

Nonlinear Effects in Fibers

The third-order nonlinearity in silica can result in various observables such as four-wave mixing, self-focusing, stimulated Raman scattering (SRS), stimulated Brillouin scattering, and supercontinuum generation. For transmission fibers, SRS and stimulated Brillouin scattering are also important nonlinear effects. Nonlinear effects in fibers (primarily transmission fibers) have been discussed in detail in several articles and books [1–17]. In this appendix, many of these nonlinear effects are described.

A.12.1 Four-wave mixing

Four-wave mixing (FWM) is an important nonlinear optical process in single-mode silica fibers. In wavelength division multiplexed (WDM) transmission systems, where data at multiple equally spaced wavelengths are transmitted, FWM process can result in intermixing of data through creation of new wavelengths, which coincides with an existing WDM wavelength channel. The transmission fiber provides very long interaction lengths with low dispersion near 1550 nm which allows creation of high intensity for the FWM signal [1–4]. Although the interaction lengths are smaller for Er-doped fibers, the FWM efficiency could be significant due to smaller cross-section and high signal gain.

When two signals propagate in the fiber with frequencies f_1 and f_2, a new set of optical waves at frequencies $2f_1 - f_2$ and $2f_2 - f_1$

are produced. These new set of frequencies are sometimes known as sidebands in analogy with electronics. These new frequencies copropagate with the original signals and continue to grow at their expense. A set of three propagating waves (at frequencies f_i, f_j, f_k) also generate new optical frequencies (nine in this case) at frequencies $f_{ijk} = f_i + f_j - f_k$, where i, j, k are 1, 2, 3, respectively. These are generally known as FWM products and are schematically shown in Figure A.12.1 [17]. The appearance of these new waves would degrade the performance of a multichannel transmission system through cross-talk.

In a multichannel WDM transmission system, the channel spacing is fixed, i.e. the frequency of the nth channel is given by $f_m = f_0 + m\Delta f$, where m is an integer. The quantity Δf is the channel spacing, its typical value is 50, 100, or 200 GHz and f_0 is $\sim 1.93 \times 10^5$ GHz (i.e. $\sim$1550 nm wavelength). In addition to the value of the nonlinear coefficient, an important factor that determines the magnitude of the FWM signal is the phase matching factor. For the case where an FWM signal at frequency $f_3 = 2f_2 - f_1$ is generated from original signal frequencies f_1 and f_2, (neglecting optical loss) the phase matching factor (F) after traveling through a fiber of length

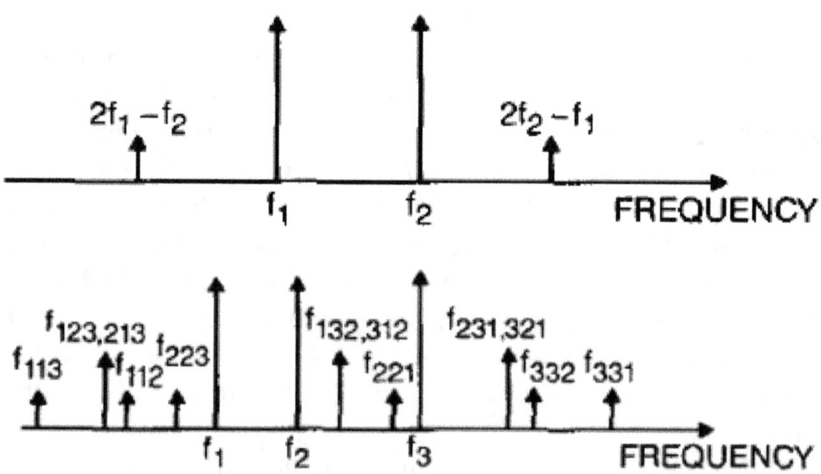

Figure A.12.1: Schematic diagram of four wave mixing (FWM) processes [4].

L is (in a dimensionless form)

$$F = \sin^2 \frac{\left(\frac{\Delta k L}{2}\right)}{\left(\frac{\Delta k L}{2}\right)^2} \tag{A.12.1}$$

where $\Delta k = 2k_2 - k_1 - k_3 = 2\pi(2n_2 f_2 - n_1 f_1 - n_3 f_3)/c$, k_1, k_2, k_3 are the wave vectors of waves at frequencies f_1, f_2, f_3 and n_1, n_2, n_3 are the refractive indices at frequencies f_1, f_2, f_3, respectively. Using $f_3 = 2f_2 - f_1$, the quantity $\Delta k = 2\pi(2(n_2 - n_3)f_2 - (n_1 - n_3)f_1)/c$, and for the case of a WDM system ($f_1 = f_0 + m_1\Delta f$, etc.) this reduces to

$$\Delta k = (2n_2 - n_1 - n_3)f_0 + \{2(n_2 - n_3)m_2 - (n_1 - n_3)m_1\}\Delta f \tag{A.12.2}$$

Thus, the phase mismatch Δk depends on the dispersion and channel spacing (Δf). If the dispersion is ~0 ($n_1 \sim n_2 \sim n_3$), then Δk is small and the quantity F will be near unity. For a fiber, the dispersion is non-zero and, hence, $F < 1$, but it depends on channel spacing and length of fiber traversed, L. For the same dispersion, for larger channel spacing, the phase matching factor is smaller. Figure A.12.2 shows the FWM efficiency for two dispersion parameters [17]. For a conventional fiber ($D = 16$ ps/km nm), the FWM process is efficient for channel spacing <20 GHz whereas for dispersion shifted fibers ($D = 1$ ps/km nm), the FWM process is efficient for channel spacing to 50 GHz.

The effect of low dispersion is illustrated in Figure A.12.3. Four CW optical signals each with a power of 3 dBm is injected into 25 km dispersion shifted fibers with zero dispersion around the injected wavelengths. There are $(N^3 - N^2)/2 = 24$ FWM products generated after traveling through the fiber. The spectrum of the output light with all the FWM products is shown in Figure A.12.3(a). Similar experiment was carried out with a fiber with an average dispersion of 2 ps/km nm near the wavelengths of the injected signal. The fiber length is 50 km. The output spectrum is shown in Figure A.12.3(b). Results show that a small amount of dispersion reduces FWM significantly. In the presence of FWM, a cross-talk between different

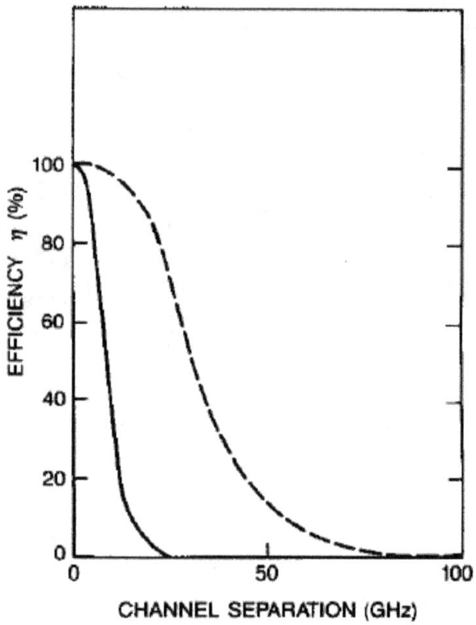

Figure A.12.2: FWM efficiency as a function of channel separation for two dispersion values. Both input wavelengths are near 1550 nm. The solid curve represents standard single-mode fiber with a dispersion of 16 ps/nm km. The dashed curve is for dispersion shifted fiber with dispersion of 1 ps/nm km [17].

channels in a WDM system takes place. This cross-talk results in errors in the transmission system.

A.12.2 Stimulated Raman scattering

SRS is a nonlinear process which takes place in optical fibers [12, 19–21]. The basis of the process is Raman scattering where a monochromatic wave propagating in the fiber generates a wave of different frequency. Typically, the energy of the photons ($h\nu_\mathrm{p}$) at the incident wavelength is larger than the energy of Raman process generated photon ($h\nu_\mathrm{s}$) by an amount equal to the separation between two energy levels (δE) in the material, i.e. $h\nu_\mathrm{p} - h\nu_\mathrm{s} = \delta E$ (see Section 9.7). The photons at the lower energy are generated by both spontaneous and stimulated emission. In many materials, such as silica, δE represents the energy of the vibrational modes of Si and

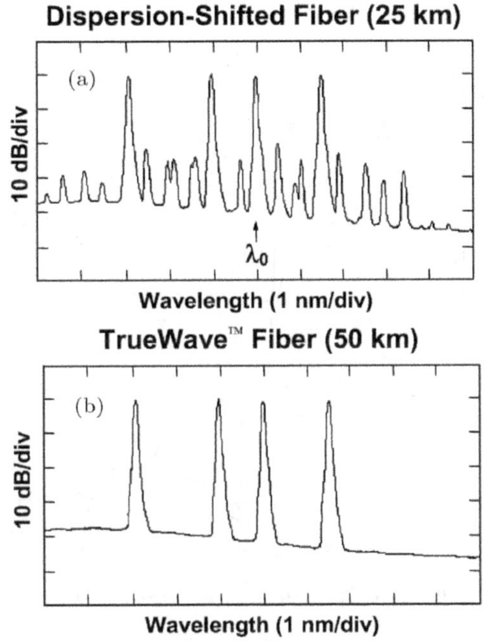

Figure A.12.3: (a) Output spectrum using zero-dispersion fiber. The input power of each channel is 3 dBm. (b) Output spectrum using TrueWave fiber with average dispersion of 2 ps/km nm for the same input powers [18].

O atoms. Typically, a small fraction of the energy of the incident wave (also called pump wave with photon energy $h\nu_{\mathrm{p}}$) gets converted to the lower photon energy ($h\nu_{\mathrm{s}}$) wave. As in the case of a laser, the spontaneously generated photons cause stimulated emission of photons with same frequency and wave vector. This leads to amplification of spontaneously generated photons (by Raman scattering) and can cause laser action (Raman laser). This is discussed in detail in Section 9.7. Typically the Raman scattered photons in silica fibers are generated in a wide frequency range ($\sim$40 THz). So, an injected signal within this Raman scattered wavelength range could be amplified by SRS if a sufficiently large amount of incident light (pump light) is present. If the generated wave is at a lower frequency than the pump, it is called the Stokes wave and if it at a higher frequency than the pump, it is called an antiStokes wave. If I_{s} is the intensity of the Stokes signal

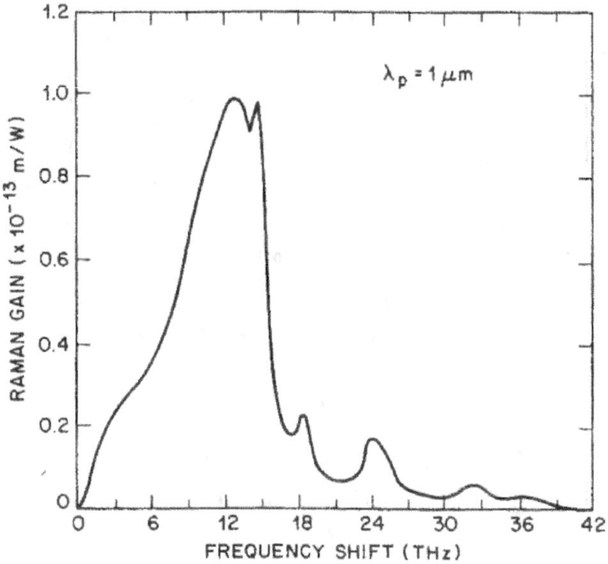

Figure A.12.4: Measured Raman gain for fused silica at a pump wavelength of 1 μm [20].

and $I_{\rm p}$ is the intensity of the pump, the growth of the Stokes signal due to SRS is given by the equation,

$$\frac{{\rm d}I_{\rm s}}{{\rm d}z} = g_{\rm R} I_{\rm p} I_{\rm s} \qquad \text{(A.12.3)}$$

where $g_{\rm R}$ is the Raman gain coefficient. The Raman spectrum in fused silica has been measured [20]. The data for a pump wavelength of 1 μm is shown in Figure A.12.4. The Raman gain is $\sim 10^{-13}$ m/W for a frequency shift of $\sim$13 THz. Thus, for an input signal power of 1 W into a single-mode fiber (cross-section $\sim$50 μm^2, $I_{\rm p} = 2 \times 10^{10}$ W/m^2), the Raman gain is $\sim 2 \times 10^{-3}$ m^{-1}, i.e. for a 100-m long fiber, the single pass gain is $\sim$0.2 which is large enough for Stokes wave (Raman) laser action. In the case of pure silica, the Stokes wave is downshifted from the pump by $\sim$13.2 THz (440 cm^{-1}).

While propagating through the fiber, the intensity of both the pump and the Stokes wave decrease due to losses. Thus, the propagation equations for the stokes ($I_{\rm s}$) and pump ($I_{\rm p}$) intensity are written as

$$\frac{\mathrm{d}I_\mathrm{s}}{\mathrm{d}z} = g_\mathrm{R} I_\mathrm{p} I_\mathrm{s} - \alpha_\mathrm{s} I_\mathrm{s} \tag{A.12.4}$$

$$\frac{\mathrm{d}I_\mathrm{p}}{\mathrm{d}z} = g_\mathrm{R} I_\mathrm{p} I_\mathrm{s} - \alpha_\mathrm{p} I_\mathrm{p} \tag{A.12.5}$$

where α_s and α_p are absorption coefficient of the Stokes and pump wave, respectively, ω_s and ω_p are the angular frequencies of the stokes and pump wave, respectively, and g_R is the Raman gain coefficient. Equations (A.12.4) and (A.12.5) show that the Stokes intensity increases and pump intensity decreases due to SRS as the light propagates in the fiber. For the purpose of estimating the pump threshold (which is defined as threshold pump power needed at input so that pump power equals Stokes power at the output), the depletion of the pump due to SRS is neglected. Equation (A.12.5) is then easily solved which is then substituted in Eq. (A.12.4) to yield the following equation:

$$\frac{\mathrm{d}I_\mathrm{s}}{\mathrm{d}z} = g_\mathrm{R} I_0 \exp(-\alpha_\mathrm{p} z) I_\mathrm{s} - \alpha_\mathrm{s} I_\mathrm{s} \tag{A.12.6}$$

where I_0 is the incident pump input intensity at $z = 0$. Solving Eq. (A.12.6), it follows

$$I_\mathrm{s}(L) = I_\mathrm{s}(0) \exp(g_\mathrm{R} I_0 L_\mathrm{eff} - \alpha_\mathrm{s} L) \tag{A.12.7}$$

$$L_\mathrm{eff} = \frac{1}{\alpha_\mathrm{p}}[1 - \exp(-\alpha_\mathrm{p} L)] \tag{A.12.8}$$

L_eff is the effective interaction length which is less than L due to pump absorption. Equation (A.12.7) requires a Stokes power at fiber input ($z = 0$). In practice, the Stokes intensity builds up from spontaneous Raman scattering throughout the length of the fiber. It has been shown [12], this is equivalent to injecting one photon per mode at the input. The Stokes power is then calculated using $I_s(0) = h\omega$ at the input and integrating over all frequency components. Thus,

$$P_\mathrm{s}(L) = \int_{-\infty}^{\infty} \hbar\omega \exp[g_\mathrm{R}(\omega) I_0 L_\mathrm{eff} - \alpha_\mathrm{s} L]\mathrm{d}\omega \tag{A.12.9}$$

It is assumed that the fiber supports a single mode at each frequency component. $g_R(\omega)$ is the frequency response of the gain spectrum shown in Figure A.12.4. Since the main contribution to the integral comes from a narrow region near the gain peak ($\omega = \omega_s$), Eq. (A.12.9) can be approximated (using method of steepest descent) as

$$P_s(L) = P_{s0} \exp[g_R(\omega_s) I_0 L_{\text{eff}} - \alpha_s L] \qquad \text{(A.12.10)}$$

The effective input power at $z = 0$, in the above is

$$P_{s0} = \hbar\omega_s B_{\text{eff}} \qquad \text{(A.12.11)}$$

with

$$B_{\text{eff}} = \left[\frac{2\pi}{g_R''(\omega_s) I_0 L_{\text{eff}}}\right]^{1/2} \quad \text{and} \quad g_R''(\omega_s) = \left[\frac{\partial^2 g_R}{\partial \omega^2}\right]_{\omega=\omega_s} \qquad \text{(A.12.12)}$$

B_{eff} can be interpreted as the effective bandwidth of the Stokes radiation although it depends on the pump intensity and fiber length. The Raman threshold is defined as the input power at which the Stokes power becomes equal to the pump power at the fiber input, i.e.

$$P_s(L) = P_p(L) = P_0 \exp(-\alpha_p L) \qquad \text{(A.12.13)}$$

where $P_0 = I_0 A_{\text{eff}}$ where P_0 is the input pump power and A_{eff} is the effective area of propagating pump mode. Using $\alpha_s = \alpha_p$, the threshold condition becomes,

$$P_{s0} \exp\left(\frac{g_R P_0 L_{\text{eff}}}{A_{\text{eff}}}\right) = P_0 \qquad \text{(A.12.14)}$$

Note that from Eq. (A.12.12) P_{s0} also depends on P_0. Thus, the above needs to be solved numerically for different gain spectrum. For a Lorentzian gain spectrum, the threshold power to a good approximation is given by

$$\frac{g_{\mathrm{R}} P_{\mathrm{oth}} L_{\mathrm{eff}}}{A_{\mathrm{eff}}} \sim 16 \tag{A.12.15}$$

Raman threshold power is now estimated using Eq. (A.12.15). For long fibers ($\alpha_{\mathrm{p}} L \gg 1$), $L_{\mathrm{eff}} \sim 1/\alpha_{\mathrm{p}}$. For wavelengths near 1550 nm, loss = 0.2 dB/km, hence, $L_{\mathrm{eff}} \sim 20$ km. Since g_{R} is inversely proportional to the pump wavelength, assuming $g_{\mathrm{R}} = 7 \times 10^{-14}$ m W^{-1} near 1,550 nm, the pump threshold power using Eq. (A.12.15) is ~0.6 W. Thus, SRS is not important for typical single-channel optical fiber transmission systems with input powers of ~3 mW. A numerical model of SRS taking into account pump depletion has been carried out [19]. The result for threshold power is similar to that stated above.

For high pump powers, the generated Stokes signal can become large enough to act as a pump and create a second order Stokes signal. In a single pass Raman generation experiment, pump pulses (~150 ns long) from Nd–YAG lasers are used to generate Stokes pulses. The first Stokes is observed at a pump power of 70 W. Higher order stokes signals appear at higher pump powers. At ~1 kW pump power, five Stokes lines are observed [21]. This multiple SRS scattering process is much more easily feasible in an optical cavity and has been used to demonstrate cascaded Raman lasers (see Chapter 9).

A.12.3 Stimulated Brillouin scattering

Stimulated Brillouin scattering (SBS) is a nonlinear process similar to SRS except it involves acoustic waves [22–26]. The pump frequency is downshifted to Stokes frequency by an acoustic wave frequency, i.e.

$$\omega_{\mathrm{s}} = \omega_{\mathrm{p}} - \omega_{\mathrm{A}} \tag{A.12.16}$$

$$\mathbf{k}_{\mathrm{s}} = \mathbf{k}_{\mathrm{p}} - \mathbf{k}_{\mathrm{A}} \tag{A.12.17}$$

where ω_p and ω_s are the frequencies of the pump and Stokes signal and ω_{A} is the frequency of the acoustic wave. $\mathbf{k}_A$, $\mathbf{k}_p$, and $\mathbf{k}_s$ are the wave vectors of the acoustic wave and pump and stokes wave, respectively. Since the magnitudes of $\mathbf{k}_p$ and $\mathbf{k}_s$ are nearly equal, it follows from Eq. (A.12.17)

$$\omega_{\rm A} = |k_{\rm A}|v_{\rm A} = 2v_{\rm A}|k_{\rm P}|\sin\frac{\theta}{2} \tag{A.12.18}$$

where $v_{\rm A}$ is the velocity of the acoustic wave and θ is the angle between $\mathbf{k}_p$ and $\mathbf{k}_s$. Thus, the frequency shift depends on the scattering angle. For forward scattering ($\theta = 0$), there is no frequency shift and for backward scattering ($\theta = \pi$), the frequency shift is maximum. The frequency shift in the backward direction is given by

$$v_0 = \frac{\omega_{\rm A}}{2\pi} = \frac{2nV_{\rm A}}{\lambda_{\rm p}} \tag{A.12.19}$$

where $\mathbf{k}_{\rm p} = 2\pi/\lambda_p$, λ_p is the pump wavelength and n is the refractive index. For optical fibers, only forward and backward propagation directions are useful. For forward direction, small Brillouin shifts ($\sim$10 to 1000 MHz) with low intensity have been observed due to guided nature of the optical and acoustic waves. The Brillouin scattering is primarily in the backward direction. Using $v_A = 6$ km/s and $n = 1.45$ for silica optical fibers in Eq. (A.12.19), the calculated $\nu = 11$ GHz at $\lambda_p \sim 1550$ nm.

Similar to the case of SRS, the SBS can be represented by a gain function. The width of the gain spectrum is small ($\sim$10 MHz), thus the gain can be assumed to be a Lorentzian with the spectral profile given by

$$g(v) = \frac{(\Delta v/2)^2}{(v - v_0)^2 + (\Delta v/2)^2} g(v_0) \tag{A.12.20}$$

with

$$g(v_0) = \frac{2\pi n^7 p^2}{c\lambda_p^2 \rho_0 v_A \Delta v} \tag{A.12.21}$$

where p is the longitudinal elasto-optic coefficient and ρ_0 is the density of the material. Using the parameter values of fused silica, the SBS gain at 1550 nm is $g_b \sim 5 \times 10^{-11}$ m/W. This is about 500 times larger than the peak SRS gain coefficient shown in Figure A.12.4.

Thus, the SBS threshold in a single-mode fiber is expected to be significantly smaller than the SRS threshold. Brillouin shift in single-mode fibers have been measured using a heterodyne technique and an external cavity pump laser at 1525 nm with a narrow linewidth. The measured shift of 11.25 GHz is in good agreement with the result from Eq. (A.12.19).

Similar to the SRS case, the pump and Stokes propagation equation for SBS is given by (note that Stokes power grows in the backward direction, i.e. $-z$)

$$\frac{\mathrm{d}I_\mathrm{s}}{\mathrm{d}z} = -g_b I_\mathrm{p} I_\mathrm{s} + \alpha I_\mathrm{s} \tag{A.12.22}$$

$$\frac{\mathrm{d}I_\mathrm{p}}{\mathrm{d}z} = -g_b I_\mathrm{p} I_\mathrm{s} - \alpha I_\mathrm{p} \tag{A.12.23}$$

where it is assumed $\omega_p \sim \omega_s$ and $\alpha_p \sim \alpha_s$. As before, for the purpose of estimating threshold, the pump depletion due to SBS is neglected. The Stokes intensity then grows exponentially in the backward direction as follows:

$$I_\mathrm{s}(0) = I_\mathrm{s}(L)\exp(g_b P_0 L_\mathrm{eff}/A_\mathrm{eff} - \alpha L) \tag{A.12.24}$$

with

$$L_\mathrm{eff} = \frac{1}{\alpha}[1 - \exp(-\alpha L)] \tag{A.12.25}$$

Using a procedure as described earlier, the SBS threshold (P_0) is given by the relation [12]:

$$g_b P_{0\mathrm{th}} L_\mathrm{eff}/A_\mathrm{eff} \sim 21 \tag{A.12.26}$$

where g_b is the peak SBS gain. Using typical values for single-mode fibers, $A_\mathrm{eff} \sim 50\ \mu\mathrm{m}^2$, $L_\mathrm{eff} \sim 20$ km, and $g_b \sim 5 \times 10^{-11}$ m/W, Eq. (A.12.26) yields a threshold power of $\sim$1 mW. Due to its low threshold, SBS is the dominant nonlinear process in optical fibers. Measured threshold are in the 1–10 mW range, and it depends on the linewidth of the laser.

A.12.4 Nonlinear refraction

Many nonlinear effects in optical fibers originate from nonlinear refraction, i.e. an intensity dependence of the refractive index of the fiber. This intensity dependence arises from the third-order nonlinear susceptibility and is frequency dependent. The index is given by

$$n(\omega) = n_0(\omega) + n_2 I \quad \text{(A.12.27)}$$

where I is the optical intensity, $n_0(\omega)$ is the refractive index arising from linear susceptibility, and $n_2(\omega)$ is the nonlinear index coefficient. For silica optical fibers, the typical value of $n_2 = 3.2 \times 10^{-16}$ cm^2/W [1]. This value is smaller by at least two orders of magnitude from that of typical nonlinear crystals. However, optical fibers provide long-interaction lengths which can enhance nonlinear effects.

References

1. G. P. Agrawal, *Nonlinear Fiber Optics* 4th Edn., Academic Press, NY, 2007.
2. N. K. Dutta *Fiber Amplifiers and Fiber Lasers*, World Scientific, Singapore, NJ, 2015.
3. M. W. Maeda, W. B. Sessa, W. I. Way, A. Yi-yan, L. Curtis, R. Spicer, and R. I. Laming, *J. Lightwave Technol.*, **8** (1990) 1402.
4. N. Shibata, R. Braun, and R. Waarts, *IEEE J. Quant. Electron.*, **23** (1987) 1205.
5. F. Forgnieri, W. T. Robert, A. R. Chraplyvy, and D. Marcuse, Tech. Deg. IOOC-OFC'93, Paper FC4, 1993.
5. P. A. Andrekson, N. A. Olsson, J. R. Simpson, T. Tanban-EK, R. A. Logan, and M. Haner, *Electron. Lett.*, **27** (1991) 922.
6. A. H. Gnauck, R. M. Jopson, and R. M. Derosier, *IEEE Photon. Technol. Lett.*, **5** (1993) 663.
7. K. O. Hill, D. C. Johnson, B. S. Kawasaki, and I. R. Macdonald, *J. Appl. Phys.*, **49** (1978) 5098.
8. K. Inoue and H. Toba, *IEEE Photon. Technol. Lett.*, **4** (1992) 69.
9. R. Hui and A. Mecozzi, *Appl. Phys. Lett.*, **60** (1992) 2454.
10. B. P. Pal (ed.), *Guided Wave Optical Components and Devices*, Elsevier Academic Press, USA, 2006.
11. S. Shimoda and H. Ishio (eds.), *Optical Amplifiers and Their Applications*, John Wiley and Sons, USA, 1994.
12. R. G. Smith, *Appl. Opt.*, **11** (1972) 2489.
13. R. H. Stolen, Chapter, 10, in S. E. Miller and A. G. Chynoweth (eds.), *Optical Fiber Telecommunications*, Academic Press, New York, 1979.

14. R. H. Stolen and J. E. Bjorkholm, *IEEE J. Quant. Electron.*, **QE-18** (1982) 1062.
15. N. Bloembergen, *Am. J. Phys.*, **35** (1967) 989.
16. W. Kaiser and M. Maier, Stimulated Rayleigh, Brillouin and Raman spectroscopy, in F. T. Arecchi and E. O. Schulz-Dubois (eds.), *Laser Handbook*, North-Holland, Amsterdam, p. 1077.
17. A. R. Chraplyvy, *IEEE J. Lightwave Tech.*, **8** (1990) 1548.
18. A. H. Gnauck, R. W. Tkach, A. R. Chraplyvy, and T. Li, *IEEE J. Lightwave Technol.*, **26** (2008) 1032.
19. J. Auyeung and A. Yariv, *IEEE J. Quant. Electron.*, **QE-14** (1978) 347.
20. R. H. Stolen, *Proc. IEEE*, **68** (1980) 1232.
21. L. G. Cohen and C. Lin, *IEEE J. Quant. Electron.*, **QE-14** (1978) 855.
22. E. P. Ippen and R. H. Stolen, *Appl. Phys. Lett.*, 21 (1972) 539.
23. D. Heiman, D. S. Hamilton, and R. W. Hellwarth, *Phys. Rev.*, **B19** (1979) 6583.
24. E. Lichtman and A. A. Friesem, *Opt. Comm.*, **64** (1987) 544.
25. E. Lichtman, R. G. Waarts, and A. A. Friesem, *J. Lightwave Technol.*, **7** (1989) 171.
26. R. W. Tkach, A. R. Chraplyvy, and R. M. Derosier, *Electron. Lett.*, **22** (1986) 1011.

Appendix 13

Nonlinear Effects in Semiconductor Optical Amplifier

In this appendix, the four-wave mixing (FWM) and cross-gain modulation in semiconductor optical amplifiers are discussed. These phenomena are important when SOA is used in optical networking applications [1].

A.13.1 Four-wave mixing

FWM is a process by which optical signals at different (but closely spaced) wavelengths can mix to produce new signals at other wavelengths, but with lower power. In the FWM process, light at two frequencies, ω_0 and ω_1, is injected into the amplifier. These injected signals are generally referred to as pump and probe beams. The pump and probe beams can be obtained from two single wavelength distributed feedback (DFB) lasers. The pump signal is of higher power than the probe signal. Consider the case when both the pump and the probe signals are CW. Propagation through the SOA results in the generation of two additional FWM signals with frequencies $2\omega_0$–ω_1 and $2\omega_1$–ω_0. The FWM signal at frequency $2\omega_0$–ω_1 has higher power if the pump signal strength (at frequency ω_0) is higher than that of the probe signal. If I_0 and I_1 are the intensities of the signals at frequencies ω_0 and ω_1, the intensities of the signals at frequencies $2\omega_0$–ω_1 and $2\omega_1$–ω_0 are proportional to $I_0^2 I_1$ and $I_0 I_1^2$,

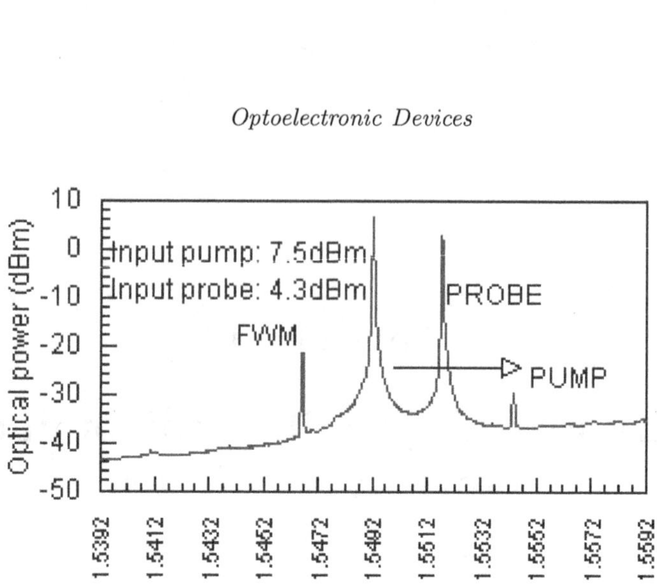

Figure A.13.1: Spectra of FWM. The wavelength separation λ_d between the pump and probe signal is 2.5 nm. FWM signals produced are of lower intensity than that of the pump and probe signals.

respectively. An example of the output spectrum from an SOA with a continuous wave (CW) co-propagating pump and probe signal is shown in Figure A.13.1.

FWM within an SOA is now analyzed. The susceptibility χ of the semiconductor is a tensor. χ also depends on the optical field intensity in the case of high input power. This could be described by expanding the susceptibility as a power series in the optical field E

$$\chi = \chi^{(1)} + \chi^{(2)}E + \chi^{(3)}EE + \cdots \tag{A.13.1}$$

where $\chi^{(1)}$ is the small signal or linear susceptibility and $\chi^{(2)}$ and $\chi^{(3)}$ are the higher order nonlinear susceptibilities. The third-order susceptibility $\chi^{(3)}$ gives rise to all the nonlinear effects in an SOA such as FWM and self-phase modulation. The induced nonlinear polarization due to $\chi^{(3)}$ is expressed as

$$P = \varepsilon_0\chi^{(3)}EE \cdot E \tag{A.13.2}$$

For the FWM case, the third-order susceptibility comes from the carrier density modulation due to the beating between the pump and the probe, so equation (A.13.2) is written as [1, 2]

$$P = \varepsilon_0 \frac{\partial \chi(n)}{\partial n} n_1 \cdot E \tag{A.13.3}$$

where n_1 is the modulated carrier density and $n \gg n_1$. It has also been deduced that the susceptibility χ(n) is related to the carrier density n by [2] (see also Eq. 7.15)

$$\chi(n) = -\frac{\bar{n}c}{\omega_0}(\alpha + i)g(n) \tag{A.13.4}$$

where g(n) = a $(n - n_0)$ is the gain and $\bar{n}$ is the effective mode index. The quantity α is the line width enhancement factor, which is defined as

$$\alpha = \frac{\delta n_{\mathrm{R}}}{\delta n_{\mathrm{I}}} \tag{A.13.5}$$

where n_{R} and n_{I} are the real and the imaginary parts of the refractive index. In FWM, the beat frequency between the pump and probe is

$$\omega_{\mathrm{d}} = \omega_0 - \omega_1 \tag{A.13.6}$$

It is relatively small compared to ω_0 and ω_1. The pump beam is generally more intense than the probe beam, i.e. $E_1 \ll E_0$. (E_0, E_1 correspond to the electrical field of the pump and probe). The carrier density n is described by the following rate equation in the semiconductor optical amplifier when a current I is applied,

$$\frac{dn}{dt} = \frac{I}{eV} - \frac{n}{\tau_c} - \frac{\Gamma g(n)}{\hbar\omega_0 A}|E|^2 \tag{A.13.7}$$

where τ_c is the carrier lifetime and A is the cross section of the waveguide mode. Since typically the length of SOA is less than 5 mm, nearly phase matching condition is satisfied for the pump and probe beams, then the total optical field is

$$E = E_0 \exp(-i\omega_0 t) + E_1 \exp(-i\omega_1 t) \tag{A.13.8}$$

hence,

$$|E|^2 = |E_0|^2 + |E_1|^2 + \{E_0^* E_1 \exp[i(\omega_0 - \omega_1)t] + c.c\} \tag{A.13.9}$$

where *c.c.* denotes complex conjugate. It is expected that the carrier density will be modulated by the beat frequency $\omega_{\rm d}$, then we have (ns is the steady state carrier density)

$$n(t) = n_s + n_1 \exp(-i\omega_d t) \tag{A.13.10}$$

Since $E_1 \ll E_0$ and $\bar{n}_s \gg n_1$, substituting (A.13.9) and (A.13.10) into (A.13.7), the following is obtained:

$$n_s = \frac{\frac{I}{eV}\tau_c + \frac{|E_0|^2}{P_s}n_0}{1 + \frac{|E_0|^2}{P_s}} \tag{A.13.11}$$

$$n_1 = \frac{-(n_s - n_0)\frac{E_0^* E_1}{P_s}}{1 + \frac{|E_0|^2}{P_s} - i\omega_d \tau_c} \tag{A.13.12}$$

where P_s is the saturation power defined as

$$P_s = \frac{Ahv}{\Gamma a \tau_c} \tag{A.13.13}$$

where A is the cross-sectional area of the waveguide mode, Γ is the confinement factor, and $h\nu$ is the photon energy of the signal. By substituting Eqs. (A.13.4) and (A.13.12) into Eq. (A.13.3), the induced nonlinear polarization due to the beating signal between the pump and the probe in the SOA is obtained. The optical fields in the SOA satisfy Maxwell's equation

$$\nabla^2 E - \frac{n^2}{c^2}\frac{\partial^2 E}{\partial t^2} = \frac{1}{\varepsilon_0 c^2}\frac{\partial^2 P}{\partial t^2} \tag{A.13.14}$$

For the FWM process, generally $I_{\rm FWM} \ll I_1 \ll I_0$, where $I_{\rm FWM}$, I_0, and I_1 are the FWM, the pump, and the probe power, respectively. Substituting equation (A.13.3) into (A.13.14) and following procedures similar to those used in Ref. [1], a set of coupled equations are obtained for the electric fields of the pump, the probe, and the FWM beams [1]

$$\frac{dE_0}{dz} = \frac{1}{2}g(1 - i\alpha)E_0$$
$$\frac{dE_1}{dz} = \frac{1}{2}g(1 - i\alpha)E_1$$
$$\frac{dE_{\text{FWM}}}{dz} = \frac{1}{2}g(1 - i\alpha)E_{\text{FWM}} - \frac{1}{2}g\sigma E_1^* E_0^2 \qquad \text{(A.13.15)}$$

with

$$\sigma = \frac{1 - i\alpha}{1 + \frac{|E_0|^2}{I_s} - i2\pi \cdot f_d\tau_c} \cdot \frac{1}{I_s} + \frac{1 - i\beta}{1 - i2\pi \cdot f_d\tau_n} \cdot \frac{1}{I_n}$$

where E_0, E_1 are the electric fields of the pump and probe signals, z is the distance traveled in the amplifier, and $f_d = f_0 - f_1$ is frequency separation between the pump and the probe. The quantity α is the line width enhancement factor and β is the ratio of the real and imaginary parts of the refractive index change induced by gain nonlinearity. The signals enter the amplifier at $z = 0$ and the amplification length is L. The quantities I_s, I_n, τ_c, and τ_n are the saturation powers and the relaxation time constants for carrier density modulation and nonlinear gain, respectively. The quantity g is the saturated material gain multiplied by the optical confinement factor, which is given by

$$g(z) = \frac{g_0}{1 + \frac{|E_0(z)|^2}{I_s}} \qquad \text{(A.13.16)}$$

where g_0 is the unsaturated gain. If $I_0(0) < I_s$, the E_0^2/I_s term in the denominator of σ can be neglected. Also, for $g \cong g_0$, the equations for E_0 and E_1 are easily solved and substituting the solution into the equation for E_{FWM}, the following results:

$$\frac{dE_{\text{FWM}}}{dz} = AE_{\text{FWM}} - Be^{Cz}$$
$$A = \frac{1}{2}g(1 - i\alpha) \qquad \text{(A.13.17)}$$

where

$$B = \frac{1}{2}g \cdot \left[\frac{1 - i\alpha}{1 - i2\pi \cdot f_d\tau_c} \cdot \frac{1}{I_s} + \frac{1 - i\beta}{1 - i2\pi \cdot f_d\tau_n} \cdot \frac{1}{I_n}\right] \cdot I_0(0) \cdot E_1(0)$$

$$C = \frac{1}{2}g(3 - i\alpha) \qquad \text{(A.13.18)}$$

$I_0(0)$ and $E_1(0)$ are, respectively, the pump intensity and the field of the probe at z =0. Using the boundary condition $E_{\mathrm{FWM}}(z = 0) = 0$, the solution of the above is

$$E_{\mathrm{FWM}}(z) = \frac{B}{A - C}(e^{Cz} - e^{Az}) \qquad \text{(A.13.19)}$$

The FWM power at $z = L$ is

$$I_{\mathrm{FWM}} = E_{\mathrm{FWM}}(L) \cdot E^{*}_{\mathrm{FWM}}(L)$$

$$I_{\mathrm{FWM}} = \frac{|B|^2}{g^2} \cdot f(G) \qquad \text{(A.13.20)}$$

where $G = \exp(gL)$ is the total gain of the SOA and,

$$f(G) = G^3 - G$$

In the above calculation, the absorption and reflection losses have been neglected. Although there are other simplifying assumptions, the above equation is a good approximation. The typical parameter values are $\tau_c = 200$ ps, $\tau_n = 650$ fs, $I_s = 20$ mW, $I_n = 52I_s$, $\alpha = 5$, $\beta = -2.2$ [1], $I_0(0) = 7.5$ dBm, $I_1(0) = 4.3$ dBm, and $G = 8.0$ dB. The FWM power is proportional to the $f(G)$. For $G > 5$ dB, $f(G)$ varies G^3. The FWM power is higher for SOA with high gain.

A.13.2 Cross-gain modulation

Cross-gain modulation is a result of gain saturation in SOA. It occurs when light at two different wavelengths, a pump and a probe, are injected into the semiconductor optical amplifier. When operated under gain saturation conditions, the available optical gain is distributed between the two wavelengths depending on their relative photon densities. The changes in the power level of the pump wavelength have an inverse effect on the gain available to the probe wavelength and result in data transfer.

The mechanism of cross-gain saturation or data transfer is shown in Figure A.13.2. When the pump ($\lambda_2 = 1550.68$ nm) is not present, the gain to the probe wavelength ($\lambda_1 = 1552.24$ nm) is high so that the output power of the probe is very high (dashed line in Figure A.13.2). When the pump (λ_2) and probe (λ_1) are injected into the SOA at the same time, the pump power is so high that it saturates the gain of the SOA. The available gain to the probe (λ_1) will be reduced. So the output power of the probe is much lower (solid line in Figure A.13.2). From Figure A.13.2 it follows, the compression efficiency is high, more than 50%. When the power of the pump is modulated with data, the gain of the probe is also modulated. Thus the output power of the probe is modulated. This results in transfer of data from pump to probe. Thus wavelength conversion is achieved.

The rate equations for the propagation of a single pulse has been derived previously in Chapter 7, Eqs. (7.27)–(7.29). They are

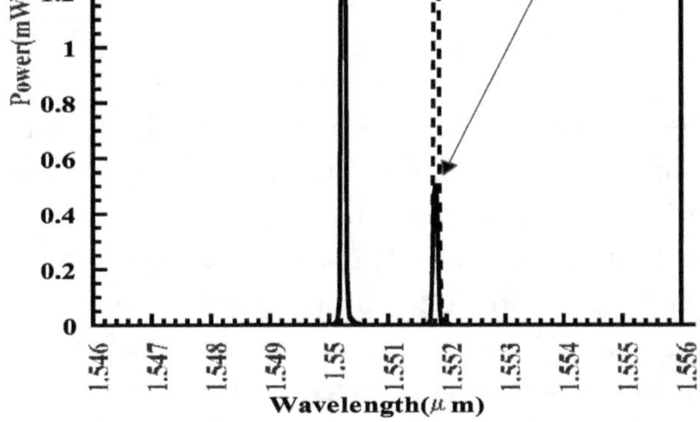

Figure A.13.2: Cross-gain saturation between two wavelengths in SOA.

rewritten here as follows:

$$\frac{\partial P}{\partial z} = (g - \alpha_{\text{int}})P \tag{A.13.21}$$

$$\frac{\partial \phi}{\partial z} = -\frac{1}{2}\alpha g \tag{A.13.22}$$

$$\frac{\partial g}{\partial \tau} = -\frac{g - g_0}{\tau_c} - \frac{gP}{E_s} \tag{A.13.23}$$

The quantity $E_s = \tau_c P_s$, where P_s is the saturation power of the amplifier, g_0 is the small signal mode gain, and α is the line width enhancement factor. The quantity g_0 is given by

$$g_0 = \Gamma a(I\tau_c/qV - n_0) \tag{A.13.24}$$

If there are two optical fields at wavelengths λ_1 and λ_2, injected into an SOA, the Eqs. (A.13.21)–(A.13.23) can be expanded to yield the following set of equations:

$$\frac{\partial P_i}{\partial z} = (g_i - \alpha_{\text{int}})P_i, \quad i = 1, 2 \tag{A.13.25}$$

$$\frac{\partial \phi_i}{\partial z} = -\frac{1}{2}\alpha g_i \tag{A.13.26}$$

$$\frac{\partial g_i}{\partial \tau} = -\frac{g_i - g_{0,i}}{\tau_c} - \frac{g_1 P_1}{E_{s,1}} - \frac{g_2 P_2}{E_{s,2}} \tag{A.13.27}$$

$$g_i = \Gamma_i a_i (n - n_{0,i}) \tag{A.13.28}$$

Wavelength conversion at high bit rate needs high bandwidth for the cross-gain modulation phenomenon. A small signal analysis of the above equations is now carried out [1].

Assume that there are only two wavelengths injected into the SOA, Eqs. (A.13.27) and (A.13.25) are rewritten using carrier density n instead of gain g noting $g = a\ (n - n_0)$, where a and n_0 are the gain coefficient and carrier density at transparency. The carrier density, $n(z, \tau)$, and optical power, $P(z, \tau)$, in the amplifier, evolve with time as

$$\frac{\partial n}{\partial \tau} = \frac{I}{eV} - \frac{n}{\tau_c} - \frac{(n - n_{0,1})P_1}{E_{s,1}} - \frac{(n - n_{0,2})P_2}{E_{s,2}} \tag{A.13.29}$$

$$\frac{\partial P_i}{\partial z} = [\Gamma a_i(n - n_{0,i}) - \alpha_{\text{int}}]P_i, \quad i = 1, 2 \tag{A.13.30}$$

where I is the injected current density, V is the volume of the gain region of SOA, $n_{0,i}$ is transparency carrier density, a_i is differential gain coefficient ($i = 1, 2$), P_2 is the pump power, P_1 is the probe power, and α_{int} is the internal loss. $E_{s,i}$ is the saturation field given by

$$E_{s,i} = \frac{\sigma_m \hbar \omega_i}{a_i \Gamma} \tag{A.13.31}$$

where σ_m is the active region area and Γ is the mode confinement factor. The quantity Γ is assumed to be the same for these two beams. Because of the small signal modulation, the optical powers may be written as $P_i = \bar{P}_i + \tilde{p}_i$ and the carrier density as $n = \bar{n} + \tilde{n}$ denoting the steady-state and perturbing signals. The following equations for $\bar{P}_i$, $\bar{n}$, $\tilde{p}_i$, and $\tilde{n}$ are then obtained:

$$0 = \frac{J}{ed} - \frac{\bar{n}}{\tau_c} - (\bar{n} - n_{0,1})\frac{\bar{P}_1}{E_{s,1}} - (\bar{n} - n_{0,2})\frac{\bar{P}_2}{E_{s,2}} \tag{A.13.32}$$

$$\frac{\partial \tilde{n}}{\partial t} = -\frac{\tilde{n}}{\tau_c} - \frac{[(\bar{n} - n_{0,1})\tilde{p}_1 + \tilde{n}P_1]}{E_{s,1}} - \frac{[(\bar{n} - n_{0,2})\tilde{p}_2 + \tilde{n}P_2]}{E_{s,2}} \tag{A.13.33}$$

$$\frac{\partial P_i}{\partial z} = [\Gamma a_i(\bar{n} - n_{0,i}) - \alpha_{\text{int}}]P_i \tag{A.13.34}$$

$$\frac{\partial \tilde{p}_i}{\partial z} = [\Gamma a_i(\bar{n} - n_{0,i}) - \alpha_{\text{int}}]\tilde{p}_i + \Gamma a_i \tilde{n} \bar{P}_i \tag{A.13.35}$$

For the case of sinusoidal modulation at angular frequency ω:

$$\tilde{n} = -\frac{(\bar{n} - n_{0,1})\frac{\tilde{p}_1}{E_{s,1}} + (\tilde{n} - n_{0,2})\frac{\tilde{p}_2}{E_{s,2}}}{1/\tau_c + 1/\tau_{s,1} + 1/\tau_{s,2} + j\omega} \tag{A.13.36}$$

At incident facet z = 0, $\tilde{p}_1(z = 0) = 0$. Assuming $g = g_1 = g_2 = \Gamma a(\bar{n} - n_0)$ and $E_{s1} = E_{s2} = E_s$. We also assume that the CW intensity of the probe at wavelength λ_1 is smaller than the CW portion of the pump at wavelength λ_2, then the conversion efficiency from pump (λ_2) to probe (λ_1) at position z is given by [1]

$$\begin{aligned}\eta &= \left|\frac{\tilde{p}_1(z,\omega)}{\tilde{p}_2(z,\omega)}\right| \\ &= \left|\exp\left(\frac{-gz}{\tau_{s,1}/\tau_e + 1 + j\omega\tau_{s,1}}\right) - 1\right| \end{aligned} \qquad \text{(A.13.37)}$$

where $\tau_{s,i}$ is the carrier lifetime due to stimulated emission, which is defined as

$$1/\tau_{s,i} = \bar{P}_i/E_s \qquad \text{(A.13.38)}$$

Thus the response of cross-gain modulation is determined by the carrier lifetime (τ).

References

1. N. K. Dutta and Q. Wang, *Semiconductor Optical Amplifier*, 2nd Edn., World Scientific, 2013.
2. G. P. Agrawal and N. A. Olsson, *IEEE J. Quant. Electron.*, **25** (1989) 11; **25** (1989) 2297.
3. D. A. O. Davies, *IEEE Photon. Technol. Lett.*, **7** (1995) 617.
4. A. Mecozzi, *IEEE Photon. Technol. Lett.*, **8** (1996) 1471.

Appendix 14

Energy Levels in Quantum Well

For the confined electrons in a double heterostructure

$$E_{\vec{k}} = \frac{\hbar^2 \vec{k}^2}{2m^*} \tag{A.14.1}$$

with $\vec{k}^2 = k_x^2 + k_y^2 + k_z^2$. The energy eigenvalues are solutions of

$$\left\{ -\frac{\hbar^2}{2m} \nabla^2 + U(r) \right\} \psi_{\vec{k}} = E_{\vec{k}} \psi_{\vec{k}} \tag{A.14.2}$$

where $U(r)$ is the periodic potential, one effect of which is to cause the electrons to have an effective mass at the zone edge. The Schrödinger equation for the free particle like electrons, with energies given by Eq. (A.14.1), is

$$-\frac{\hbar^2}{2m^*} \nabla^2 \psi_{\vec{k}} = E_{\vec{k}} \psi_{\vec{k}} \tag{A.14.3}$$

with eigenvalues $\psi_{\vec{k}} = e^{i\vec{k}\cdot\vec{r}}$, and m* is the effective mass of the electron.

In a quantum well, the electron motion normal to the well (z-direction) is not free particle like due to presence of a potential well, $V(z)$. The Schrödinger equation is given by,

$$\left\{ -\frac{\hbar^2}{2m^*} \nabla^2 + V(z) \right\} \psi_{\vec{k}} = E_{\vec{k}} \psi_{\vec{k}} \tag{A.14.4}$$

with $\psi_{\vec{k}} = e^{ik_x x + ik_y y} f(z)$. Substituting in Eq. (A.14.4), the following equation for $f(z)$ is obtained:

$$-\frac{\hbar^2}{2m^*}\frac{d^2 f}{dz^2} + V(z)f(z) = E_n f(z) \tag{A.14.5}$$

where

$$E_{\vec{k}} = \frac{\hbar^2}{2m^*}(k_x^2 + k_y^2) + E_n \tag{A.14.6}$$

A similar equation holds for holes. Consider a potential of the following form shown below, i.e.

$$V(z) \begin{array}{ll} = 0 & -\frac{L_z}{2} \leq z \leq \frac{L_z}{2} \\ = V_0 & |z| \geq \frac{L_z}{2} \end{array} \tag{A.14.7}$$

where L_z is the thickness of the well. For $V_0 = \infty$, i.e. for an infinite potential well, the equation for $f(z)$ is

$$-\frac{\hbar^2}{2m}\frac{d^2 f}{dz^2} - Ef = 0, \quad \text{for } |z| \leq L_z/2 \tag{A.14.8}$$

and $f = 0$ for $|z| \geq L_z/2$. The solutions of Eq. (A.14.8) are

$$\begin{aligned} f &= \cos\frac{\pi n z}{L_z} \quad n = 1, 3, 5, \ldots \\ &= \sin\frac{n\pi z}{L_z} \quad n = 2, 4, 6, \ldots \end{aligned} \tag{A.14.9}$$

Substituting Eq. (A.14.9) in (A.14.8), the energy levels for an infinite potential well is given by

$$E_n = \frac{\hbar^2 \pi^2 n^2}{2mL_z^2} \quad n = 1, 2, 3, \ldots \tag{A.14.10}$$

For the finite V_0 case, the Schrödinger equation for f is given by,

$$\begin{aligned} -\frac{\hbar^2}{2m}\frac{d^2 f}{dz^2} &= Ef \qquad -\frac{L_z}{2} \leq z \leq \frac{L_z}{2} \\ -\frac{\hbar^2}{2m}\frac{d^2 f}{dz^2} + V_0 f &= Ef \qquad |z| \geq \frac{L_z}{2} \end{aligned} \tag{A.14.11}$$

The solutions of the above for bound states are,

$$\begin{aligned} f &= A\cos\alpha z + B\sin\alpha z \quad -\frac{L_z}{2} \leq z \leq \frac{L_z}{2} \\ &= C\exp(\beta z) \quad z < -\frac{L_z}{2} \\ &= D\exp(-\beta z) \quad z > \frac{L_z}{2} \end{aligned} \tag{A.14.12}$$

where $\beta^2 = \frac{2m^*}{\hbar^2}(V_0 - E)$, $\alpha^2 = \frac{2m^*E}{\hbar^2}$. Consider the even solutions, i.e. $B = 0$. The quantities f and df/dz are continuous at $z = \pm L_z/2$. This gives four equations. The determinant of which is the eigenvalue equation for the energy E. It is as follows:

$$\alpha \tan\frac{\alpha L_z}{2} = \beta \tag{A.14.13}$$

The potential well for electrons and holes in a double heterostructure depends on the materials. For GaAs/AlGaAs double heterostructure, the conduction and valence band discontinuities (ΔE_c and ΔE_ν) are found experimentally to be

$$\begin{aligned} \Delta E_c/\Delta E_g &= 0.67 \\ \Delta E_\nu/\Delta E_g &= 0.33 \end{aligned} \tag{A.14.14}$$

where ΔE_g is the band gap difference between the two semiconductors. For InGaAsP/InP double heterostructure, the conduction and valence band discontinuities (ΔE_c and ΔE_ν) are found experimentally to be

$$\begin{aligned} \Delta E_c/\Delta E_g &= 0.39 \\ \Delta E_\nu/\Delta E_g &= 0.61 \end{aligned} \tag{A.14.15}$$

Thus, the splitting of the energy difference between the conduction and valence bands depends on the materials involved. The energy levels of a particle in a potential well of depth V are shown in Figure A.14.1. The plot is shown in units of $\hbar^2\pi^2/2mL_z^2$. The values for $V_0 = \infty$ and the corresponding n values are shown on the right axis.

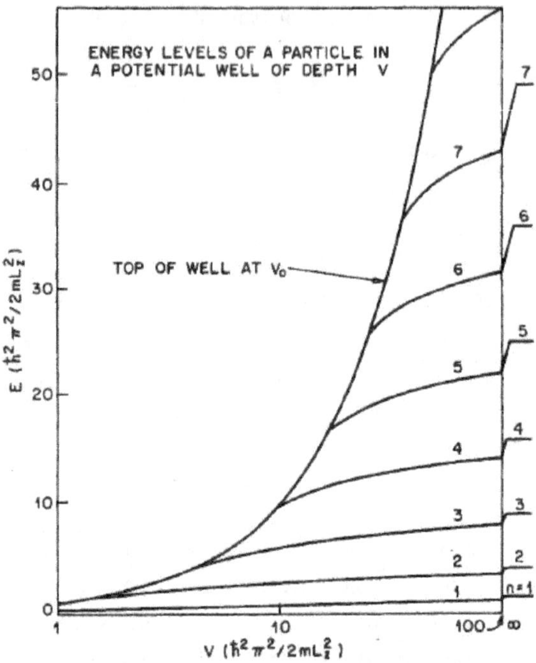

Figure A.14.1: The energy of the bound states in normalized units are shown. The energy values of the infinite potential well are on the right axis for different n values [1].

For a quantum wire, the particle motion is quantized along two directions (y, z). Thus, the solution of the Schrödinger equation, using separation of variables, is

$$\psi(x, y, z) = e^{ik_x x} f_1(y) f_2(z) \tag{A.14.16}$$

where the wire is assumed to be in the x-direction. The equations for f_1, f_2 are

$$\begin{aligned} -\frac{\hbar^2}{2m^*}\frac{d^2 f_1}{dy^2} + V(y) f_1 = E_1 f_1 \\ -\frac{\hbar^2}{2m^*}\frac{d^2 f_2}{dz^2} + V(y) f_2 = E_2 f_2 \end{aligned} \tag{A.14.17}$$

and the total energy

$$E = E_1 + E_2 + \frac{\hbar^2 k_x^2}{2m^*} \tag{A.14.18}$$

Equation (A.14.17) may be solved approximately the same way as that for a quantum well.

For a quantum dot, the carrier motion is restricted in all three dimensions. Thus, the energy levels are determined by a solution of Schrödinger equation in a three-dimensional potential well.

Reference

1. G. P. Agrawal and N. K. Dutta, Chapter 9, *Semiconductor Lasers*, van Nostrand Reinhold, NY, 1992.

Appendix 15

Problems

Chapter 2

2.1. (a) Using the results derived in Appendix 5, show that the total spontaneous emission rate in undoped direct gap semiconductors in the Boltzmann approximation is given by $R = Bn^2$, where B is a constant and n is the injected carrier density.
(b) Derive an expression for B.
(c) Calculate the value of B at $T = 300$ K, for a semiconductor which has a band gap of 1.42 eV with $m_c = 0.6\ m_0$ and $m_{hh} = 0.45\ m_0$. Neglect the contribution of light holes.

2.2. Calculate the carrier lifetime for an injected carrier density of 2×10^{18} cm^{-3}.

2.3. Show that the modulation bandwidth of an LED with an active region doping of 1×10^{17} cm^{-3} increases approximately as square root of injected current.

2.4. Assuming that the band gap difference between the light-emitting region and the cladding region must be greater than 0.2 eV, calculate the range of wavelengths for which an LED with InGaAsP active region and InP cladding region can be fabricated.

2.5. Determine the composition of AlGaN for an emission wavelength 420 nm that can be grown lattice matched over GaN using $b = -1.0$ eV in Eq. (2.8).

Chapter 3

3.1. (a) Using $\eta_i = 1$, calculate the differential quantum efficiency of a laser emitting (in mW/mA) at 1550 nm with an internal optical loss of 30 cm^{-1}, and mirror reflectivity of 35% each. The laser length is 250 μm. What is the efficiency for a laser length of 100 μm?

(b) Calculate the material gain at threshold of the laser if the confinement factor is 0.5.

(c) Using Eq. (3.7), calculate the threshold current density of the laser.

(d) Using an active layer thickness of 0.2 μm and width of 2 μm, calculate the threshold current of the laser.

3.2. Derive Eq. (3.12).

3.3. Derive Eq. (A.1.31) of Appendix 1 and show that for small D, confinement factor, $\Gamma = (1/2)\ D^2$.

3.4. (a) Calculate the full-width-at-half-maximum of the near field both parallel and perpendicular to junction plane for a InGaAsP/InP laser with 0.2 μm thick and 1 μm wide InGaAsP active region. The laser emits at 1550 nm. The refractive index of InP is 3.22 and that of InGaAsP is 3.51. Use results of Appendix 1.

(b) Calculate the numerical aperture of the waveguide.

(c) Calculate the full-width-at-half-maximum of the far field angular divergence.

3.5. Using the condition of transparency (Eq. (3.24)) and the results of Appendix 5, calculate the minimum carrier density needed for transparency (n_t) in an undoped active region using the Boltzmann approximation. Show that n_t varies as $T^{3/2}$, where T is the temperature.

3.6. Explain how an index grating is produced in a DFB laser. Calculate the grating period for first-order grating in DFB lasers for lasers operating at 850 and 1550 nm. A set up similar to Figure 9.13 is used to make a grating on a semiconductor. For $\theta = 45°$, in Eq. 9.15, calculate the wavelength of the laser needed (for grating fabrication) for the above cases.

3.7. Show that the small signal 3-dB bandwidth of a semiconductor laser is proportional to $P^{1/2}$, where P is the output power of one facet.

Chapter 4

4.1. (a) Calculate the reflectivity of a $Al_{0.1}Ga_{0.9}As/AlAs$ quarter wave stack as a function of number of pairs using Eq. (4.1). The refractive index of AlAs is 3.178 and that of $Al_{0.1}Ga_{0.9}As$ is 3.54.
(b) Calculate the reflectivity of a CaF_2/ZnS stack as a function of number of pairs. The index of CaF_2 is 1.7 and that of ZnS is 2.6.

4.2. The reflectivities of the mirrors of a VCSEL are 99.4% and 99.9%.
The effective single pass cavity length is 1 μm.

(a) Using Eq. (3.7), calculate the threshold current density of the laser.
(b) For a laser with 0.2 μm thick active region and 10 μm diameter, calculate the threshold current.

4.3. (a) Calculate the threshold gain of a QCL with an internal loss of 30 cm^{-1} and mirror reflectivity of 35% each. The laser length is 1 mm.
(b) Calculate the material gain at threshold of the laser if the confinement factor is 0.5.
(c) Assuming a current dependence of gain of 20 cm^{-1}/ (kA/ cm^2), calculate the threshold current density.
(d) Using an active layer width of 2 μm, calculate the threshold current of the laser.

4.4. (a) Calculate the threshold gain of a QD laser with an internal loss of 30 cm^{-1} and mirror reflectivity of 35% each. The laser length is 250 μm.
(b) Using Eq. (4.14) calculate the threshold current density of a QD laser.

(c) Using an active layer width of 2 μm, calculate the threshold current of the laser.

4.5. What are the ideas behind quantum well, quantum wire, and quantum dot lasers? Calculate the lowest energy in which an InGaAsP/InP quantum well laser with a well width of 8 nm can operate for an InGaAsP active region semiconductor well of bandgap 0.7 eV (~1550 nm). Calculate the lowest energy of operation for quantum wire of size 8 nm × 8 nm and quantum dot of size 8 nm × 8 nm × 8 nm for the above case.

4.6. The differential gain (rate of change of gain with carrier density) of a quantum dot laser is larger than that for a regular double heterostructure laser by a factor of 3. What is the expected increase in the small signal bandwidth?

Chapter 5

5.1. The optical loss of a modulator increases from −7 to −25 dB by the application of a voltage of 5V. Calculate the percentage change in light intensity going through the modulator when the voltage is changed from 0 to 5 V. Assuming the modulator is 250 μm long, calculate the change in loss coefficient in cm^{-1}, as the voltage is increased from 0 to 5V.

5.2. An integrated laser modulator has a laser which emits 5 mW and all of it is injected into the modulator. Calculate the output power in the ON state ($V = 0$ V) and the OFF state ($V = 5$ V) for the modulator of Problem 1.

5.3. The measured 3-dB bandwidth of a modulator is 30 GHz. Assuming, the bandwidth is primarily due to RC time constant, calculate the capacitance of the modulator using $R = 50$ ohm.

5.4. A $LiNbO_3$ modulator has an insertion loss of 2 dB and a V_π of 5 V. Plot the output power as a function of voltage when 3 mW of CW light is injected into the modulator.

5.5. Calculate V_π for a $LiNbO_3$ modulator which utilizes push-pull effect operating at 1550 nm using $L = 5$ cm, $\Gamma = 0.5$, and an electrode separation of 30 μm.

5.6. Why does the microwave index need to equal the optical index for high speed modulators? Using simple equations describe what will happen to the bandwidth of a modulator if the two indices are different.

Chapter 6

6.1. The absorbing region of a photodetector has an absorption coefficient of 10^4 cm^{-1}, has a thickness of 2 μm, and, an area of diameter 100 μm. The incident surface has an index of 3.2 and the surface is uncoated. The wavelength of incident light is 1550 nm. Calculate the photocurrent produced for 100 μW of incident light. Calculate the responsivity in W/A. Calculate the responsivity if the surface is antireflection coated with a dielectric film.

6.2. Calculate the width of the depletion layer and the capacitance of a photodiode of diameter 100 μm using a dielectric constant of 10, doping density of 10^{13} cm^{-3}, and a reverse bias of 3 V. What should be the area of the photodiode for a capacitance of 10 pF?

6.3. Derive an expression of the modulation response of the photocurrent J(ω) as a function of frequency ω by integrating Eq. (6.22) and Eq. (6.21). Plot $J(\omega)$ as a function of ω.

6.4. Calculate $H(\iota)$ as a function of ω for the parameter values noted in the text and plot it.

6.5. Calculate the r.m.s noise current of a photodiode for an equivalent resistance (R_{eq}) of 50 Ohms at 300 K, a dark current (I_D) of 10 nA, and a bandwidth of 0.5 GHz, when (i) no optical power is incident on the photodiode and (ii) when 1 μW of the power is incident on the photodiode at a wavelength of 1550 nm. Assume a responsivity of 0.9A/W.

6.6. Derive Eq. (6.38) using Eq. (6.33).

6.7. Calculate the r.m.s noise for cases (i) and (ii) of problem 6.5 for an APD with an excess noise factor of 10 and a gain of 10.

6.8. Plot the excess noise factor (F_e in Eq. (6.39)) as a function of M_e for various $k_p = 0.1$, 0.3, and 0.5.

Chapter 7

7.1. The Lorentzian gain spectrum of a semiconductor amplifier has full-width-at-half-maximum of 5 THz. Calculate the amplifier bandwidth in THz when the amplifier is operated with a peak gain of 30 dB, i.e. exp $(g_o L) = 1000$. Calculate the width of the amplifier spectrum in nm if the peak gain is centered at 1550 nm. Neglect gain saturation.

7.2. The small signal gain of a semiconductor amplifier is 30 dB. It has a saturation power of 30 mW and its length is 1 mm. Calculate the output power for incident powers of 0.1 mW and 10 mW.

7.3. Derive Eqs. (7.35) and (7.36) from earlier equations, e.g. (7.16) and (7.27).

7.4. Using Eq. (7.21) and $g(n) = a(n - n_0)$, where n is the injected carrier density and n_0 is the carrier density at transparency (i.e. $g = 0$), derive expressions for steady state amplifier gain (G) and saturation power, for amplifier of length L. Note $A^2 = P$, where P is the optical power. Express the power dependence in terms of ratio P/P_s (P_s is given by Eq. (7.30)).

7.5. Using Eq. (7.29) and the results of problem 7.4, calculate the phase change of light traveling through an amplifier of length L as a function of current and input power.

Chapter 9

9.1. Derive Eq. (9.6) from (9.3). Calculate the value of I_s in number of photons/cm^2/s for $\sigma = 10^{-20}$ cm^2 and $\tau = 2$ ms. Calculate P_0 (the threshold pump rate) in number of photons/cm^3/s for a laser with density of ions of 2×10^{19} cm^{-3} and $\alpha_t = 0.1$ cm^{-1}. Assuming a pump photon wavelength of 1 μm, calculate the threshold pump rate in W/cm^{-3} and the threshold pump power in W for a pump volume in the form of a cylinder of 2 mm diameter and 5 cm in length.

9.2. Consider a laser cavity 5 cm in length. The total internal optical loss due to scattering, etc. is 0.1 cm^{-1}. One mirror has a reflectivity of 99%. Calculate the reflectivity of the second mirror so that the total optical loss (including mirror losses) is 0.2 cm^{-1}.

9.3. Derive an equation similar to (A.4.17) for a four level laser. Show that the output is linearly proportional to pump power above threshold.

9.4. The threshold pump power for an Er-doped fiber laser is given by Eq. (A.4.26). Consider a 10-cm long fiber laser with mirror reflectivity of 98% and 99.5% and an internal optical loss of 0.02 cm^{-1} Calculate the threshold power in mW using Eq. (A.4.26) and the parameters stated there. How much total power (from both mirrors) is emitted at a pump power of five times the threshold pump power?

9.5. (a) Calculate the full-width-at-half-maximum (FWHM) of the waveguide mode of a typical single-mode fiber (core diameter = 10 μm) with a cladding index of 1.5 and with a core index higher (than that of the cladding) by 0.3%. (b) Calculate the FWHM of the waveguide mode for a typical core diameter of 4 μm of rare earth-doped fiber for the same index differences. (c) Calculate the mode confinement factor (fraction of power in the core) for the above cases in the Gaussian approximation. (d) Calculate the numerical aperture of the fiber waveguide of (a).

9.6. Derive Eq. (9.13) for the reflection coefficient of a Bragg grating.

9.7. Calculate K for $\delta n = 0.1$, $\eta = 0.1$, and $\lambda_B = 240$ nm. Plot the maximum reflectivity as a function of length L of the grating.

9.8. Calculate the threshold power at the input for stimulated Raman and Brillouin scattering using the results of Appendix 12 for fibers with cross section of 50 μm^2 and optical loss of 0.15 dB/km.

Chapter 10

10.1. Using the results of Appendix 1, (a) Calculate the maximum waveguide gain region thickness (d_m) for single transverse mode operation of a planar waveguide laser with cladding index of 1.5 (n_2) and core index of $n_1 = n_2(1 + \Delta)$ for $\Delta = 4\%$ and 8%. The wavelength is 1550 nm. (b) For a waveguide thickness of

$d_m/2$ μm, calculate the mode confinement factor. (c) Calculate the numerical aperture for $\Delta = 4\%$ and 8%.

10.2. Calculate the full-width-at-half-maximum of the near field (in μm) and far field (in deg.) of the planar waveguide of problem 1.

10.3. Calculate the optical gain (in cm^{-1}) needed for a planar waveguide laser with a confinement factor of 0.5, cavity length of 10 cm, round-trip internal loss of 20%, and optical reflectivity of 50% and 99% of the two mirrors. Using equations of section 9.1 above, calculate P_0 (the threshold pump rate) in number of photons/cm^3/s for a laser with density of ions of 2×10^{19} cm^{-3} for $\sigma = 10^{-20}$ cm^2 and $\tau = 2$ ms. Assuming a pump photon wavelength of 1 μm, calculate the threshold pump rate in W/cm^{-3}.

Chapter 11

11.1. Using $\sigma_t = 100$ nA for a bandwidth (B) of 0.5 GHz, and using σ_t varies as $B^{1/2}$, calculate the average received power needed for 10^{-9} and 10^{-12} BER for a transmission system operating at 10 Gb/s.

11.2. Carry out the calculation of average received power at 40 Gb/s for the case of problem 11.1.

11.3. An optical fiber transmission system is operating at 10 Gb/s. The fiber has a loss of 0.2 dB/km. It has eight splices over 80 km span with per-splice loss of 0.3 dB. The system margin for operation is 10 dB (i.e. assume an additional optical loss of 10 dB). Using the results of problem 1 calculate the transmitted average laser power (in mW) for 10^{-12} bit-error-rate (BER).

11.4. Derive an expression for the BER of a PAM-4 system assuming a decision threshold at the mid intensity for each intensity level and equal noise for each level.

Index

www.ingramcontent.com/pod-product-compliance
Lightning Source LLC
LaVergne TN
LVHW010536300926
849162LV00002B/155